ELEMENTS OF MATHEMATICS

NICOLAS BOURBAKI

ELEMENTS OF MATHEMATICS

# Integration II

Chapters 7–9

Translated by Sterling K. Berberian

Springer

Originally published as
Éléments de Mathématique, Intégration

Chapters 7 and 8, 1963
Chapter 9, 1969

Translated by
Sterling K. Berberian, Professor Emeritus
The University of Texas at Austin
USA

Mathematics Subject Classification (2000): 28-01, 28Bxx, 43-xx, 46Exx

Library of Congress Control Number: 2004109575

ISBN 978-3-642-05821-9    ISBN 978-3-662-07931-7 (eBook)
DOI 10.1007/978-3-662-07931-7



springeronline.com

Originally published by Springer-Verlag Berlin Heidelberg New York in 2004
Softcover reprint of the hardcover 1st edition 2004



Typesetting by the Translator
Cover design: design+production GmbH, Heidelberg
Printed on acid-free paper    41/3142/XT – 5 4 3 2 1 0

# To the reader

1. The Elements of Mathematics series takes up mathematics at their beginning, and gives complete proofs. In principle, it requires no particular knowledge of mathematics on the reader's part, but only a certain familiarity with mathematical reasoning and a certain capacity for abstract thought. Nevertheless, it is directed especially to those who have a good knowledge of at least the content of the first year or two of a university mathematics course.

2. The method of exposition we have chosen is axiomatic, and normally proceeds from the general to the particular. The demands of proof impose a rigorously fixed order on the subject matter. It follows that the utility of certain considerations will not be immediately apparent to the reader unless he already has a fairly extensive knowledge of mathematics.

3. The series is divided into Books, and each Book into chapters. The Books already published, either in whole or in part, in the French edition, are listed below. When an English translation is available, the corresponding English title is mentioned between parentheses. Throughout the volume a reference indicates the English edition, when available, and the French edition otherwise.

| | | | |
|---|---|---|---|
| Théorie des Ensembles (Theory of Sets) | designated by | E | (S) |
| Algèbre (Algebra) | — | A | (A) |
| Topologie Générale (General Topology) | — | TG | (GT) |
| Fonctions d'une Variable Réelle (Functions of a Real Variable) | — | FVR | (FRV) |
| Espaces Vectoriels Topologiques (Topological Vector Spaces) | — | EVT | (TVS) |
| Intégration (Integration) | — | INT | (INT) |
| Algèbre Commutative (Commutative Algebra) | — | AC | (CA) |
| Variétés Différentielles et Analytiques | — | VAR | |
| Groupes et Algèbres de Lie (Lie Groups and Lie Algebras) | — | LIE | (LIE) |
| Théories Spectrales | — | TS | |

In the *first six* Books (according to the above order), every statement in the text assumes as known only those results which have already been discussed in the same chapter, or in the *previous chapters ordered as follows*: S; A, Chapters I to III; GT, Chapters I to III; A, from Chapter IV on; GT, from Chapter IV on; FRV; TVS; INT.

From the seventh Book onward, the reader will usually find a precise indication of its logical relationship to the other Books (the first six Books being always assumed to be known).

4. However we have sometimes inserted examples in the text that refer to facts the reader may already know but which have not yet been discussed in the series. Such examples are placed between two asterisks: $^{*}\ldots_{*}$. Most readers will undoubtedly find that these examples help them to understand the text. In other cases, the passages between $^{*}\ldots_{*}$ refer to results that are discussed elsewhere in the series. We hope that the reader will be able to verify the absence of any vicious circle.

5. The logical framework of each chapter consists of the *definitions*, the *axioms*, and the *theorems* of the chapter. These are the parts that have mainly to be borne in mind for subsequent use. Less important results and those which can easily be deduced from the theorems are labelled as "propositions", "lemmas", "corollaries", "remarks", etc. Those which may be omitted on a first reading are printed in small type. A commentary on a particularly important theorem occasionally appears under the name of "scholium".

To avoid tedious repetitions it is sometimes convenient to introduce notations or abbreviations that are in force only within a certain chapter or a certain section of a chapter (for example, in a chapter that is concerned only with commutative rings, the word "ring" would always signify "commutative ring"). Such conventions are always explicitly mentioned, generally at the beginning of the chapter in which they occur.

6. Some passages in the text are designed to forewarn the reader against serious errors. These passages are signposted in the margin with the sign ("dangerous bend").

7. The Exercises are designed both to enable the reader to satisfy himself that he has digested the text and to bring to his attention results that have no place in the text but are nevertheless of interest. The most difficult exercises bear the sign ¶.

8. In general, we have adhered to the commonly accepted terminology, *except where there appeared to be good reasons for deviating from it.*

9. We have made a particular effort always to use rigorously correct language, without sacrificing simplicity. As far as possible we have drawn

attention in the text to *abuses of language*, without which any mathematical text runs the risk of pedantry, not to say unreadability.

10. Since in principle the text consists of the dogmatic exposition of a theory, it contains in general no references to the literature. Bibliographical references are gathered together in *Historical Notes*. The bibliography which follows each historical note contains in general only those books and original memoirs that have been of the greatest importance in the evolution of the theory under discussion. It makes no pretense of any sort to completeness.

As to the exercises, we have not thought it worthwhile in general to indicate their origins, since they have been drawn from many different sources (original papers, textbooks, collections of exercises).

11. References to a part of this series are given as follows:

a) If reference is made to theorems, axioms, or definitions presented *in the same section* (§), they are cited by their number.

b) If they occur *in another section of the same chapter*, this section is also cited in the reference.

c) If they occur *in another chapter in the same Book*, the chapter and section are cited.

d) If they occur *in another Book*, the Book is cited first, by the abbreviation of its title.

The *Summaries of Results* are cited by the letter R; thus S, R signifies the "*Summary of Results of the Theory of Sets*".

CHAPTER VII

# Haar measure

*In this chapter and in the next, when we speak of a function* (resp. *a measure*), *it will be understood to be either a real or a complex function* (resp. *measure*); *if* T *is a locally compact space, the notation* $\mathscr{K}(\mathrm{T})$ *will denote either the space* $\mathscr{K}_{\mathbf{R}}(\mathrm{T})$ *or the space* $\mathscr{K}_{\mathbf{C}}(\mathrm{T})$; *similarly for the notations* $\overline{\mathscr{K}(\mathrm{T})}$, $\mathscr{C}(\mathrm{T})$, $\mathrm{L}^p(\mathrm{T},\mu)$, $\mathscr{M}(\mathrm{T})$, *etc. It is of course understood that in a situation involving several functions, measures or vector spaces, the results obtained are valid when these functions, measures or vector spaces are all real or all complex. The space* $\overline{\mathscr{K}(\mathrm{T})}$ *will always be assumed to be equipped with the topology of uniform convergence, the space* $\mathscr{C}(\mathrm{T})$ *with the topology of compact convergence, and the space* $\mathscr{K}(\mathrm{T})$ *with the direct limit topology whose definition is reviewed at the beginning of Chapter* VI. *The notation* $\mathscr{K}_+(\mathrm{T})$ *will denote the set of functions* $\geqslant 0$ *of* $\mathscr{K}(\mathrm{T})$. *If* $\mathrm{A} \subset \mathrm{T}$, $\varphi_{\mathrm{A}}$ *will always denote the characteristic function of* A. *If* $t \in \mathrm{T}$, $\varepsilon_t$ *will denote the positive measure defined by the mass* $+1$ *at the point* $t$.

*All locally convex spaces will be assumed to be Hausdorff.*

*We will denote by* e *the neutral element of all of the groups considered, absent express mention to the contrary.*

## §1. CONSTRUCTION OF A HAAR MEASURE

### 1. Definitions and notations

Let G be a topological group operating continuously on the left (GT, III, §2, No. 4) in a locally compact space X; for $s \in \mathrm{G}$ and $x \in \mathrm{X}$, let $sx$ be the transform of $x$ by $s$. We denote by $\boldsymbol{\gamma}_{\mathrm{X}}(s)$, or $\boldsymbol{\gamma}(s)$, the homeomorphism of X onto X defined by

$$\boldsymbol{\gamma}(s)x = sx. \tag{1}$$

We have

$$\boldsymbol{\gamma}(st) = \boldsymbol{\gamma}(s)\boldsymbol{\gamma}(t). \tag{2}$$

If $f$ is a function defined on X, $\boldsymbol{\gamma}(s)f$ will be defined by transport of structure, that is, by the formula $\big(\boldsymbol{\gamma}(s)f\big)\big(\boldsymbol{\gamma}(s)x\big) = f(x)$; in other words,

$$(3) \qquad \big(\boldsymbol{\gamma}(s)f\big)(x) = f(s^{-1}x).$$

If $\mu$ is a measure defined on X, $\boldsymbol{\gamma}(s)\mu$ will also be defined by transport of structure, which leads to

$$(4) \qquad \langle f, \boldsymbol{\gamma}(s)\mu\rangle = \langle \boldsymbol{\gamma}(s^{-1})f, \mu\rangle \quad \text{for } f \in \mathscr{K}(\mathrm{X}).$$

In other words,

$$(5) \qquad \int_{\mathrm{X}} f(x)\, d\big(\boldsymbol{\gamma}(s)\mu\big)(x) = \int_{\mathrm{X}} f(sx)\, d\mu(x).$$

If A is a $\big(\boldsymbol{\gamma}(s)\mu\big)$-integrable set, then $s^{-1}\mathrm{A}$ is $\mu$-integrable and

$$(6) \qquad \big(\boldsymbol{\gamma}(s)\mu\big)(\mathrm{A}) = \mu(s^{-1}\mathrm{A}).$$

The measure $\boldsymbol{\gamma}(s)\mu$ may also be defined as the *image* of $\mu$ under $\boldsymbol{\gamma}(s)$.

Instead of writing $d\big(\boldsymbol{\gamma}(s)\mu\big)(x)$, it is sometimes useful to write $d\mu(s^{-1}x)$; the formula (5) then takes the following form:

$$\int_{\mathrm{X}} f(x)\, d\mu(s^{-1}x) = \int_{\mathrm{X}} f(sx)\, d\mu(x);$$

the expression on the right may then be deduced from that on the left by 'replacing $x$ by $sx$.'

DEFINITION 1. — *Let $\mu$ be a measure on* X.

a) *$\mu$ is said to be invariant under* G *if $\boldsymbol{\gamma}(s)\mu = \mu$ for every $s \in \mathrm{G}$.*

b) *$\mu$ is said to be relatively invariant under* G *if $\boldsymbol{\gamma}(s)\mu$ is proportional to $\mu$ for every $s \in \mathrm{G}$.*

c) *$\mu$ is said to be quasi-invariant under* G *if $\boldsymbol{\gamma}(s)\mu$ is equivalent to $\mu$ for every $s \in \mathrm{G}$.*

*Remarks.* — 1) Assume $\mu$ invariant. Then $|\mu|$, $\mathscr{R}(\mu)$, $\mathscr{I}(\mu)$ are invariant. If $\mu$ is real, then $\mu^+$ and $\mu^-$ are invariant.

2) Assume $\mu$ relatively invariant and nonzero. There exists, for every $s \in \mathrm{G}$, a unique complex number $\chi(s)$ such that

$$(7) \qquad \boldsymbol{\gamma}(s)\mu = \chi(s)^{-1}\mu,$$

and the function $\chi$ on G is a representation of G in $\mathbf{C}^*$, called the *multiplier* of $\mu$. The formula (5) then gives

$$\int_X f(sx)\,d\mu(x) = \chi(s)^{-1}\int_X f(x)\,d\mu(x)\,, \tag{8}$$

and formula (6) gives

$$\mu(sA) = \chi(s)\mu(A)\,. \tag{9}$$

With the conventions made above, (7) may also be written

$$d\mu(sx) = \chi(s)\,d\mu(x)\,. \tag{10}$$

3) Since $|\boldsymbol{\gamma}(s)\mu| = \boldsymbol{\gamma}(s)(|\mu|)$, to say that $\mu$ is quasi-invariant amounts to saying that $|\mu|$ is quasi-invariant.

If $\mu$ is quasi-invariant and $\mu'$ is another measure on X equivalent to $\mu$, then $\boldsymbol{\gamma}(s)\mu'$ is equivalent to $\boldsymbol{\gamma}(s)\mu$, hence to $\mu$, hence to $\mu'$, and so $\mu'$ is quasi-invariant. To say that $\mu$ is quasi-invariant under G therefore means that the *class* of $\mu$ is invariant under G.

For $\mu$ to be quasi-invariant, it is necessary and sufficient that the set of locally $\mu$-negligible subsets of X be invariant under G (Ch. V, §5, No. 5, Th. 2), or again that, for every $\mu$-negligible compact subset K of X and for every $s \in G$, $sK$ be $\mu$-negligible (*loc. cit.*, *Remark*).

If $\mu$ is quasi-invariant, then the support of $\mu$ is invariant under G. In particular if G is *transitive* in X (A, I, §5, No. 5, Def. 6), this support is either empty (if $\mu = 0$) or equal to X (if $\mu \neq 0$).

*Lemma* 1. — *Let* X, Y, Z *be three topological spaces, with* Y *locally compact. Let* $(x, y) \mapsto xy$ *be a continuous mapping of* $X \times Y$ *into* Z, *which defines a mapping* $x \mapsto u_x$ *of* X *into* $\mathscr{F}(Y; Z)$ *by the relation* $u_x(y) = xy$. *Let* $\mathbf{f}$ *be a continuous function on* Z *with values in* $\mathbf{R}$ *or in a Banach space,* S *the support of* $\mathbf{f}$, *and* $\mu$ *a measure on* Y. *Assume that for every* $x_0 \in X$, *there exists a neighborhood* V *of* $x_0$ *in* X *such that* $\bigcup_{x\in V} u_x^{-1}(S)$ *is relatively compact in* Y. *Then:*

a) *for every* $x \in X$, $\mathbf{f} \circ u_x$ *is continuous on* Y, *with compact support;*

b) *the mapping* $x \mapsto \int_Y \mathbf{f}(xy)\,d\mu(y)$, *which is defined by* a), *is continuous on* X.

The assertion a) is obvious. Let us prove b). Since continuity is a local property, we may reduce to the case that $\bigcup_{x\in X} u_x^{-1}(S)$ is contained in a compact subset Y′ of Y. Since the function $(x, y) \mapsto \mathbf{f}(xy)$ is continuous on $X \times Y$, $\mathbf{f} \circ u_x$ tends to $\mathbf{f} \circ u_{x_0}$ uniformly on Y′ as $x$ tends to $x_0$

(GT, X, §3, No. 4, Th. 3), therefore $\mu(\mathbf{f} \circ u_x)$ tends to $\mu(\mathbf{f} \circ u_{x_0})$. Whence the lemma.

Let us now return to the previous notations.

PROPOSITION 1. — *Assume that* G *is locally compact. Let* $\mu$ *be a nonzero relatively invariant measure on* X. *Then its multiplier* $\chi$ *is a continuous function on* G.

For, let $f \in \mathscr{K}(\mathrm{X})$, S the support of $f$, $s_0$ a point of G, and V a compact neighborhood of $s_0$ in G; then, the set

$$\bigcup_{s \in \mathrm{V}} \gamma(s)^{-1}(\mathrm{S}) = \mathrm{V}^{-1}\mathrm{S}$$

is compact in X; by Lemma 1 and formula (8), $\chi(s^{-1})\langle \mu, f\rangle$ depends continuously on $s$; if $f$ is chosen so that $\langle \mu, f\rangle \neq 0$, one sees that $\chi$ is continuous.

Now let G be a topological group operating continuously on the right in a locally compact space X; for $s \in \mathrm{G}$ and $x \in \mathrm{X}$, let $xs$ be the transform of $x$ by $s$. We denote by $\boldsymbol{\delta}_{\mathrm{X}}(s)$, or $\boldsymbol{\delta}(s)$, the homeomorphism of X defined by

$$\boldsymbol{\delta}(s)x = xs^{-1}. \tag{1'}$$

We have

$$\boldsymbol{\delta}(st) = \boldsymbol{\delta}(s)\boldsymbol{\delta}(t). \tag{2'}$$

By transport of structure, one defines the action of $\boldsymbol{\delta}(s)$ on functions and measures on X:

$$\big(\boldsymbol{\delta}(s)f\big)(x) = f(xs) \tag{3'}$$

$$\langle f, \boldsymbol{\delta}(s)\mu\rangle = \langle \boldsymbol{\delta}(s^{-1})f, \mu\rangle \tag{4'}$$

$$\int_{\mathrm{X}} f(x)\, d\big(\boldsymbol{\delta}(s)\mu\big)(x) = \int_{\mathrm{X}} f(xs^{-1})\, d\mu(x) \tag{5'}$$

$$\big(\boldsymbol{\delta}(s)\mu\big)(\mathrm{A}) = \mu(\mathrm{A}s). \tag{6'}$$

We agree to write $d\mu(xs)$ in place of $d\big(\boldsymbol{\delta}(s)\mu\big)(x)$, and (5′) then takes the form

$$\int_{\mathrm{X}} f(x)\, d\mu(xs) = \int_{\mathrm{X}} f(xs^{-1})\, d\mu(x).$$

One defines in an analogous manner the measures on X invariant, relatively invariant and quasi-invariant under G. If $\mu$ is relatively invariant, its multiplier $\chi$ is defined by the formulas

$$\boldsymbol{\delta}(s)\mu = \chi(s)\mu \tag{7'}$$

$$\int_X f(xs)\, d\mu(x) = \chi(s)^{-1} \int_X f(x)\, d\mu(x) \tag{8'}$$

$$\mu(\mathrm{A}s) = \chi(s)\mu(\mathrm{A}) \tag{9'}$$

$$d\mu(xs) = \chi(s)\, d\mu(x)\,. \tag{10'}$$

If one regards the group $\mathrm{G}^0$ opposite to G as operating in X by $(x, s) \mapsto xs$, then $\mu$ is relatively invariant under $\mathrm{G}^0$ with the same multiplier $\chi$.

Finally, let G be a locally compact group. It operates on itself by left and right translations, according to the formulas $\boldsymbol{\gamma}(s)x = sx$, $\boldsymbol{\delta}(s)x = xs^{-1}$. Then

$$\boldsymbol{\gamma}(s)\boldsymbol{\delta}(t) = \boldsymbol{\delta}(t)\boldsymbol{\gamma}(s)\,. \tag{11}$$

All of the foregoing is applicable here, thus we have on G the concepts of measures that are *left-invariant, right-invariant, relatively left-invariant, relatively right-invariant, left quasi-invariant, right quasi-invariant* (however, see Nos. 8 and 9).

The mapping $x \mapsto x^{-1}$ is a homeomorphism of G onto G. For every function $f$ on G, we define the function $\check{f}$ on G by

$$\check{f}(x) = f(x^{-1})\,. \tag{12}$$

For every measure $\mu$ on G, we define the measure $\check{\mu}$ by

$$\check{\mu}(f) = \mu(\check{f}) \quad \text{for } f \in \mathscr{K}(\mathrm{G})\,. \tag{13}$$

In other words,

$$\int_G f(x)\, d\check{\mu}(x) = \int_G f(x^{-1})\, d\mu(x)\,. \tag{14}$$

If A is a $\check{\mu}$-integrable set, then $\mathrm{A}^{-1}$ is $\mu$-integrable and

$$\check{\mu}(\mathrm{A}) = \mu(\mathrm{A}^{-1})\,. \tag{15}$$

We agree to write $d\mu(x^{-1})$ in place of $d\check{\mu}(x)$, and (14) then takes the form

$$\int_{G} f(x)\,d\mu(x^{-1}) = \int_{G} f(x^{-1})\,d\mu(x)\,.$$

## 2. The existence and uniqueness theorem

DEFINITION 2. — *Let* G *be a locally compact group. A nonzero positive measure on* G *that is left* (resp. *right*) *invariant is called a left* (resp. *right*) *Haar measure on* G.

THEOREM 1. — *On every locally compact group, there exists a left* (resp. *right*) *Haar measure, and, up to a constant factor, there exists only one.*

A) *Existence.* — Set $\mathcal{K}(G) = \mathcal{K}$, $\mathcal{K}_+(G) = \mathcal{K}_+$, $\mathcal{K}_+^* = \mathcal{K}_+ - \{0\}$. If C is a compact subset of G, we denote by $\mathcal{K}_+^*(C)$ the set of $f \in \mathcal{K}_+^*$ with support in C. For $f \in \mathcal{K}$ and $g \in \mathcal{K}_+^*$, there exist numbers $c_1, \ldots, c_n \geqslant 0$ and elements $s_1, \ldots, s_n$ of G such that $f \leqslant \sum_{i=1}^{n} c_i \gamma(s_i) g$: for, there exists a nonempty open set U in G such that $\inf_{s \in U} g(s) > 0$, and the support of $f$ can be covered by a finite number of left-translates of U. Let $(f : g)$ be the infimum of the numbers $\sum_{i=1}^{n} c_i$ for all systems $(c_1, \ldots, c_n, s_1, \ldots, s_n)$ of numbers $\geqslant 0$ and elements of G such that $f \leqslant \sum_{i=1}^{n} c_i \gamma(s_i) g$. Then:

(i) $(\gamma(s)f : g) = (f : g)$ for $f \in \mathcal{K}$, $g \in \mathcal{K}_+^*$, $s \in G$;

(ii) $(\lambda f : g) = \lambda (f : g)$ for $f \in \mathcal{K}$, $g \in \mathcal{K}_+^*$, $\lambda \geqslant 0$;

(iii) $((f + f') : g) \leqslant (f : g) + (f' : g)$ for $f \in \mathcal{K}$, $f' \in \mathcal{K}$, $g \in \mathcal{K}_+^*$;

(iv) $(f : g) \geqslant (\sup f)/(\sup g)$ for $f \in \mathcal{K}$, $g \in \mathcal{K}_+^*$;

(v) $(f : h) \leqslant (f : g)(g : h)$ for $f \in \mathcal{K}$, $g \in \mathcal{K}_+^*$, $h \in \mathcal{K}_+^*$;

(vi) $0 < \dfrac{1}{(f_0 : f)} \leqslant \dfrac{(f : g)}{(f_0 : g)} \leqslant (f : f_0)$ for $f$, $f_0$, $g$ in $\mathcal{K}_+^*$;

(vii) let $f$, $f'$, $h$ be in $\mathcal{K}_+$ with $h(s) \geqslant 1$ on the support of $f + f'$, and let $\varepsilon > 0$; there exists a compact neighborhood V of $e$ such that, for every $g \in \mathcal{K}_+^*(V)$,

$$(f : g) + (f' : g) \leqslant \big((f + f') : g\big) + \varepsilon (h : g)\,.$$

The properties (i), (ii), (iii) are obvious. Let $f \in \mathcal{K}$, $g \in \mathcal{K}_+^*$; if $f \leqslant \sum_{i=1}^{n} c_i \gamma(s_i) g$ with the $c_i \geqslant 0$, then $\sup f \leqslant \sum_{i=1}^{n} c_i g(s_i^{-1} s)$ for some

$s \in \mathrm{G}$, therefore $\sup f \leqslant \big(\sum_{i=1}^{n} c_i\big) \sup g$, whence (iv). Let us now prove (v); let $f \in \mathscr{K}$, $g, h$ in $\mathscr{K}_+^*$; if $f \leqslant \sum_{i=1}^{n} c_i \boldsymbol{\gamma}(s_i) g$ and $g \leqslant \sum_{j=1}^{p} d_j \boldsymbol{\gamma}(t_j) h$ ($c_i \geqslant 0$, $d_j \geqslant 0$, $s_i, t_j$ in G), then $f \leqslant \sum_{i,j} c_i d_j \boldsymbol{\gamma}(s_i t_j) h$, therefore $(f : h) \leqslant \sum_{i,j} c_i d_j = \big(\sum_i c_i\big)\big(\sum_j d_j\big)$; thus $(f : h) \leqslant (f : g)(g : h)$. Applying (v) to $f_0, f, g$ on the one hand and to $f, f_0, g$ on the other, one obtains (vi). Finally, let $f, f', h$ be in $\mathscr{K}_+$ with $h(s) \geqslant 1$ on the support of $f + f'$, and let $\varepsilon > 0$. Set $\mathrm{F} = f + f' + \frac{1}{2}\varepsilon h$; the functions $\varphi, \varphi'$, that coincide respectively with $f/\mathrm{F}$ and $f'/\mathrm{F}$ on the support of $f + f'$ and are zero outside it, belong to $\mathscr{K}_+$; for every $\eta > 0$, there exists a compact neighborhood V of $e$ such that $|\varphi(s) - \varphi(t)| \leqslant \eta$ and $|\varphi'(s) - \varphi'(t)| \leqslant \eta$ for $s^{-1}t \in \mathrm{V}$. Then let $g \in \mathscr{K}_+^*(\mathrm{V})$; for every $s \in \mathrm{G}$,

$$\varphi \cdot \boldsymbol{\gamma}(s) g \leqslant \big(\varphi(s) + \eta\big) \cdot \boldsymbol{\gamma}(s) g ;$$

for, this is obvious at the points where $\boldsymbol{\gamma}(s)g$ is zero, therefore outside of $s\mathrm{V}$; and in $s\mathrm{V}$, $\varphi \leqslant \varphi(s) + \eta$; similarly, $\varphi' \cdot \boldsymbol{\gamma}(s) g \leqslant \big(\varphi'(s) + \eta\big) \cdot \boldsymbol{\gamma}(s) g$. This stated, let $c_1, \ldots, c_n$ be numbers $\geqslant 0$ and $s_1, \ldots, s_n$ elements of G such that $\mathrm{F} \leqslant \sum_{i=1}^{n} c_i \boldsymbol{\gamma}(s_i) g$; then

$$f = \varphi \mathrm{F} \leqslant \sum_{i=1}^{n} c_i \varphi \cdot \boldsymbol{\gamma}(s_i) g \leqslant \sum_{i=1}^{n} c_i \big(\varphi(s_i) + \eta\big) \cdot \boldsymbol{\gamma}(s_i) g$$

and similarly for $f'$; consequently

$$(f : g) + (f' : g) \leqslant \sum_{i=1}^{n} c_i \big(\varphi(s_i) + \varphi'(s_i) + 2\eta\big) \leqslant (1 + 2\eta) \sum_{i=1}^{n} c_i$$

since $\varphi + \varphi' \leqslant 1$. Applying the definition of F, then (ii), (iii) and (v), one infers that

$$\begin{aligned}(f : g) + (f' : g) &\leqslant (1 + 2\eta)(\mathrm{F} : g) \leqslant \\ &(1 + 2\eta)\big[\big((f + f') : g\big) + \tfrac{1}{2}\varepsilon(h : g)\big] \leqslant \\ &\big((f + f') : g\big) + \tfrac{1}{2}\varepsilon(h : g) + 2\eta\big((f + f') : h\big)(h : g) + \varepsilon\eta(h : g)\end{aligned}$$

and, if $\eta$ has been chosen so that $\eta\big[2\big((f + f') : h\big) + \varepsilon\big] \leqslant \frac{1}{2}\varepsilon$, one obtains (vii).

As V runs over the set of compact neighborhoods of $e$, the $\mathscr{K}_+^*(\mathrm{V})$ form a base of a filter $\mathfrak{B}$ on $\mathscr{K}_+^*$. Let $\mathfrak{F}$ be an ultrafilter on $\mathscr{K}_+^*$ finer than $\mathfrak{B}$. On the other hand, let us fix $f_0 \in \mathscr{K}_+^*$ and let us set, for $f \in \mathscr{K}_+^*$ and $g \in \mathscr{K}_+^*$,

$$\mathrm{I}_g(f) = \frac{(f : g)}{(f_0 : g)}.$$

By (vi), $\lim_{g,\mathfrak{F}} \mathrm{I}_g(f) = \mathrm{I}(f)$ exists in the compact space $[1/(f_0 : f), (f : f_0)]$. By (iii), $\mathrm{I}(f + f') \leqslant \mathrm{I}(f) + \mathrm{I}(f')$. By (vii), $\mathrm{I}(f) + \mathrm{I}(f') \leqslant \mathrm{I}(f + f') + \varepsilon \mathrm{I}(h)$ for all $\varepsilon > 0$ provided $h$ is $\geqslant 1$ on the support of $f + f'$; it follows that $\mathrm{I}(f + f') = \mathrm{I}(f) + \mathrm{I}(f')$. By Ch. II, §2, No. 1, Prop. 2, I is extendible to a linear form on $\mathscr{K}$; this linear form is a nonzero positive measure on G, left-invariant by (i); this is the sought-for left Haar measure. Passing to the opposite group, one deduces from this the existence of a right Haar measure.

B) *Uniqueness.* — Let $\mu$ be a left Haar measure, $\nu$ a right Haar measure. Then $\check{\nu}$ is a left Haar measure. We are going to show that $\mu$ and $\check{\nu}$ are proportional. This will prove that any two left Haar measures are indeed proportional.

Let $f \in \mathscr{K}$ be such that $\mu(f) \neq 0$. By Lemma 1, the function $\mathrm{D}_f$ defined on G by the formula

$$\mathrm{D}_f(s) = \mu(f)^{-1} \int f(t^{-1}s)\, d\nu(t) \tag{16}$$

is continuous on G. Let $g \in \mathscr{K}$. The function $(s,t) \mapsto f(s)g(ts)$ is continuous with compact support in $\mathrm{G} \times \mathrm{G}$. By Ch. III, §4, No. 1, Th. 2,

$$\begin{aligned} \mu(f)\nu(g) &= \Big(\int f(s)\, d\mu(s)\Big)\Big(\int g(t)\, d\nu(t)\Big) \\ &= \int d\mu(s) \int f(s)g(ts)\, d\nu(t) = \int d\nu(t) \int f(s)g(ts)\, d\mu(s) \\ &= \int d\nu(t) \int f(t^{-1}s)g(s)\, d\mu(s) \\ &= \int g(s)\Big[\int f(t^{-1}s)\, d\nu(t)\Big] d\mu(s) = \mu\big(g \cdot \mu(f)\mathrm{D}_f\big), \end{aligned} \tag{17}$$

whence

$$\nu(g) = \mu(\mathrm{D}_f \cdot g). \tag{18}$$

This proves, first, that $\mathrm{D}_f$ does not depend on $f$. For, if $f' \in \mathscr{K}$ is such that $\mu(f') \neq 0$, then $\mathrm{D}_f \cdot \mu = \mathrm{D}_{f'} \cdot \mu$, therefore $\mathrm{D}_f = \mathrm{D}_{f'}$ locally almost everywhere for $\mu$, hence everywhere, since $D_f$ and $D_{f'}$ are continuous and the support of $\mu$ is G. We may therefore write $D_f = D$. The formula (16) gives

$$\mu(f)D(e) = \check{\nu}(f). \tag{19}$$

The formula (19) may be extended by linearity to the functions $f \in \mathcal{K}$ such that $\mu(f) = 0$. We have $D(e) \neq 0$ since $\check{\nu} \neq 0$. This indeed establishes the proportionality of $\mu$ and $\check{\nu}$.

COROLLARY. — *Every left-invariant* (resp. *right-invariant*) *measure on* G *is proportional to a left* (resp. *right*) *Haar measure.*

*Examples.* — 1) On the additive group $\mathbf{R}$, the Lebesgue measure $dx$ is a Haar measure (Ch. III, §1, No. 3, *Example*).

2) For every function $f \in \mathcal{K}(\mathbf{R}_+^*)$, we have (FRV, II, §1, formula (12))

$$\int_0^{+\infty} \frac{f(x)}{x}\, dx = \int_0^{+\infty} \frac{f(tx)}{tx}\, t\, dx = \int_0^{+\infty} \frac{f(tx)}{x}\, dx$$

for all $t > 0$; the measure $x^{-1}\, dx$ is thus a Haar measure on the multiplicative group $\mathbf{R}_+^*$.

3) Let us take for G the torus $\mathbf{T} = \mathbf{R}/\mathbf{Z}$. Let $\varphi$ be the canonical mapping of $\mathbf{R}$ onto $\mathbf{T}$. For $f \in \mathcal{K}(\mathbf{T})$, the function $f \circ \varphi$ is continuous and periodic with period 1 on $\mathbf{R}$, and the integral

$$I(f) = \int_a^{a+1} f(\varphi(x))\, dx$$

is independent of the choice of $a \in \mathbf{R}$; it is immediate that it is invariant under translation; it therefore defines a Haar measure on $\mathbf{T}$. By transport of structure, one deduces from this that $I(f) = \int_a^{a+1} f(e^{2\pi i t})\, dt$ is a Haar measure on the multiplicative group $\mathbf{U}$ of complex numbers of absolute value 1 (GT, VIII, §2, No. 1).

PROPOSITION 2. — *Let* G *be a locally compact group,* $\mu$ *a left or right Haar measure on* G. *For* G *to be discrete, it is necessary and sufficient that* $\mu(\{e\}) > 0$. *For* G *to be compact, it is necessary and sufficient that* $\mu^*(G) < +\infty$.

The conditions are obviously necessary. Let us show their sufficiency. Let V be a compact neighborhood of $e$. If $\mu(\{e\}) > 0$, then V is a finite set since $\mu(V) < +\infty$; since G is Hausdorff, it is therefore discrete. Suppose

$\mu^*(G) < +\infty$ and $\mu$ is, for example, left-invariant. Consider the set $\mathscr{E}$ of finite subsets $\{s_1, \ldots, s_n\}$ of G such that $s_i V \cap s_j V = \varnothing$ for $i \neq j$; then

$$n\mu(V) = \mu(s_1 V \cup \ldots \cup s_n V) \leqslant \mu^*(G),$$

therefore $n \leqslant \mu^*(G)/\mu(V)$. We may therefore choose in $\mathscr{E}$ a maximal element $\{s_1, \ldots, s_n\}$. Then, for every $s \in G$, there exists an $i$ such that $sV \cap s_i V \neq \varnothing$, hence such that $s \in s_i VV^{-1}$. Thus G is the union of the compact sets $s_i VV^{-1}$, hence is compact.

## 3. Modulus

Let $\mu$ be a left Haar measure on G. For every $s \in G$, $\boldsymbol{\delta}(s)\mu$ is also left-invariant (No. 1, formula (11)), therefore (Th. 1) there exists a unique number $\Delta_G(s) > 0$ such that $\boldsymbol{\delta}(s)\mu = \Delta_G(s)\mu$. By virtue of Th. 1, the number $\Delta_G(s)$ is independent of the choice of $\mu$.

DEFINITION 3. — *The function* $\Delta_G$ *on* G *is called the modulus* (or *modular function*) *of* G. *If* $\Delta_G = 1$, *the group* G *is said to be unimodular.*

One can also say that $\mu$ is relatively right-invariant with multiplier $\Delta_G$. Thus $\Delta_G$ is a *continuous representation of* G *in* $\mathbf{R}_+^*$ (No. 1, Prop. 1).

*Remark.* — If $\varphi$ is an isomorphism of G onto a locally compact group G′, then $\Delta_{G'} \circ \varphi = \Delta_G$. In particular:

1) Since $x \mapsto x^{-1}$ is an isomorphism of G onto the opposite group $G^0$, one has $\Delta_{G^0} = \Delta_G^{-1}$.

2) If $\varphi$ is an automorphism of G, then $\Delta_G \circ \varphi = \Delta_G$.

Let $s \in G$. Then:

$$\boldsymbol{\delta}(s)(\Delta_G^{-1} \cdot \mu) = \big(\boldsymbol{\delta}(s)\Delta_G^{-1}\big) \cdot \big(\boldsymbol{\delta}(s)\mu\big) = \big(\Delta_G(s)^{-1}\Delta_G^{-1}\big) \cdot \big(\Delta_G(s)\mu\big) = \Delta_G^{-1} \cdot \mu,$$

therefore $\Delta_G^{-1} \cdot \mu = \mu'$ is a right Haar measure. From this, one deduces that $\boldsymbol{\gamma}(s)\mu' = \big(\boldsymbol{\gamma}(s)\Delta_G^{-1}\big) \cdot \mu = \Delta_G(s)(\Delta_G^{-1} \cdot \mu) = \Delta_G(s)\mu'$, therefore, for every right Haar measure $\nu$, we have $\boldsymbol{\gamma}(s)\nu = \Delta_G(s)\nu$. Since $\check{\mu}$ is a right Haar measure, $\check{\mu} = a\Delta_G^{-1} \cdot \mu$ with a constant $a > 0$; it follows that

$$\mu = a(\Delta_G^{-1} \cdot \mu)^{\vee} = a\Delta_G \cdot \check{\mu} = a^2\mu,$$

therefore $a = 1$ and finally $\check{\mu} = \Delta_G^{-1} \cdot \mu$. One sees similarly that $\check{\nu} = \Delta_G \cdot \nu$. We therefore have the following results:

*Formulary.* — Let G be a locally compact group, $\Delta$ its modulus, $\mu$ a left Haar measure, and $\nu$ a right Haar measure.

1) One has

$$\boldsymbol{\gamma}(s)\mu = \mu \qquad \boldsymbol{\delta}(s)\mu = \Delta(s)\mu \qquad \check{\mu} = \Delta^{-1} \cdot \mu . \tag{20}$$

If $f$ is $\mu$-integrable on G, then the left and right translates of $f$ are $\mu$-integrable, and

$$\begin{aligned} \int f(sx)\, d\mu(x) &= \int f(x)\, d\mu(x) \\ \int f(xs)\, d\mu(x) &= \Delta(s)^{-1} \int f(x)\, d\mu(x) . \end{aligned} \tag{21}$$

Moreover, $\check{f}$ is integrable for $\Delta^{-1} \cdot \mu$ and

$$\int f(x^{-1})\Delta(x)^{-1}\, d\mu(x) = \int f(x)\, d\mu(x) . \tag{22}$$

If A is a $\mu$-integrable subset of G, then $s\mathrm{A}$ and $\mathrm{A}s$ are $\mu$-integrable and

$$\mu(s\mathrm{A}) = \mu(\mathrm{A}) \qquad \mu(\mathrm{A}s) = \Delta(s)\mu(\mathrm{A}) . \tag{23}$$

2) One has

$$\boldsymbol{\delta}(s)\nu = \nu \qquad \boldsymbol{\gamma}(s)\nu = \Delta(s)\nu \qquad \check{\nu} = \Delta \cdot \nu . \tag{24}$$

If $f$ is $\nu$-integrable on G, then the left and right translates of $f$ are $\nu$-integrable and

$$\begin{aligned} \int f(xs)\, d\nu(x) &= \int f(x)\, d\nu(x) \\ \int f(sx)\, d\nu(x) &= \Delta(s) \int f(x)\, d\nu(x) . \end{aligned} \tag{25}$$

Moreover, $\check{f}$ is integrable for $\Delta \cdot \nu$ and

$$\int f(x^{-1})\Delta(x)\, d\nu(x) = \int f(x)\, d\nu(x) . \tag{26}$$

If A is a $\nu$-integrable subset of G, then $s\mathrm{A}$ and $\mathrm{A}s$ are $\nu$-integrable and

$$\nu(\mathrm{A}s) = \nu(\mathrm{A}) \qquad \nu(s\mathrm{A}) = \Delta(s)^{-1}\nu(\mathrm{A}) . \tag{27}$$

3) $\nu$ is proportional to $\Delta^{-1} \cdot \mu$, and $\mu$ is proportional to $\Delta \cdot \nu$.

4) Suppose G is *unimodular*. Let $\mu$ be a Haar measure on G. Then:

$$\gamma(s)\mu = \boldsymbol{\delta}(s)\mu = \check{\mu} = \mu. \tag{28}$$

If $f$ is $\mu$-integrable on G, then the left and right translates of $f$ are $\mu$-integrable, as is $\check{f}$, and

$$\int f(sx)\,d\mu(x) = \int f(xs)\,d\mu(x) = \int f(x^{-1})\,d\mu(x) = \int f(x)\,d\mu(x). \tag{29}$$

If A is a $\mu$-integrable subset of G, then $s\mathrm{A}$, $\mathrm{A}s$ and $\mathrm{A}^{-1}$ are $\mu$-integrable and

$$\mu(s\mathrm{A}) = \mu(\mathrm{A}s) = \mu(\mathrm{A}^{-1}) = \mu(\mathrm{A}). \tag{30}$$

The analogous properties hold for the essential integral.

PROPOSITION 3. — *If there exists in* G *a compact neighborhood* V *of* $e$ *invariant under the inner automorphisms, then* G *is unimodular.*

For, let $\mu$ be a left Haar measure on G. For every $s \in \mathrm{G}$,

$$\mu(\mathrm{V}) = \mu(s^{-1}\mathrm{V}s) = \Delta_{\mathrm{G}}(s)\mu(\mathrm{V}),$$

whence $\Delta_{\mathrm{G}}(s) = 1$ since $0 < \mu(\mathrm{V}) < +\infty$.

From this, one deduces immediately:

COROLLARY. — *If* G *is discrete, or compact, or commutative, then* G *is unimodular.*

This is moreover trivial when G is *commutative*. Note also that if G is *discrete*, then the measure on G for which each point has mass 1 is obviously a left and right Haar measure on G, called the *normalized* Haar measure on G. If G is *compact*, there exists one and only one Haar measure on G such that $\mu(\mathrm{G}) = 1$; it is called the *normalized* Haar measure of G. The preceding two conventions are not in accord when G is both discrete and compact, that is, finite; when we are in this case we shall always explicitly specify what is meant by normalized Haar measure.

Subgroups and quotient groups of a unimodular group are not always unimodular (§2, Exer. 5). See, however, Prop. 10 of §2, No. 7.

*We shall see later that semi-simple or nilpotent connected Lie groups are unimodular.*

## 4. Modulus of an automorphism

Let G be a locally compact group, $\varphi$ an automorphism of G, and $\mu$ a left Haar measure on G. It is clear that $\varphi^{-1}(\mu)$ is also a left Haar measure on G. Therefore there exists (No. 2, Th. 1) one and only one number $a > 0$ such that $\varphi^{-1}(\mu) = a\mu$. By No. 2, Th. 1, this number is independent of the choice of $\mu$. Note that if one had started with a right Haar measure, for example $\Delta_G^{-1} \cdot \mu$ (No. 3), one would have arrived at the same scalar $a$: for, since $\varphi^{-1}$ leaves $\Delta_G$ invariant (No. 3, *Remark*), one has $\varphi^{-1}(\Delta_G^{-1} \cdot \mu) = \Delta_G^{-1} \cdot \varphi^{-1}(\mu) = a\Delta_G^{-1} \cdot \mu$.

DEFINITION 4. — *The number $a > 0$ such that $\varphi^{-1}(\mu) = a\mu$ is called the modulus of the automorphism $\varphi$ and is denoted* $\operatorname{mod}_G \varphi$ *or simply* $\operatorname{mod} \varphi$.

If $f$ is a $\mu$-integrable function on G, then

$$\int f(\varphi^{-1}(x))\, d\mu(x) = (\operatorname{mod}\varphi) \int f(x)\, d\mu(x). \tag{31}$$

If A is a $\mu$-integrable subset of G, then

$$\mu(\varphi(\mathrm{A})) = (\operatorname{mod}\varphi)\mu(\mathrm{A}). \tag{32}$$

In particular, for $s \in \mathrm{G}$, let $i_s$ be the inner automorphism $x \mapsto s^{-1}xs$. Then $i_s^{-1} = \boldsymbol{\delta}(s)\boldsymbol{\gamma}(s)$, therefore

$$i_s^{-1}(\mu) = \boldsymbol{\delta}(s)\mu = \Delta_G(s)\mu,$$

consequently

$$\operatorname{mod} i_s = \Delta_G(s). \tag{33}$$

If G is either discrete or compact, then its normalized Haar measure is transformed into itself by every automorphism $\varphi$ of G, as one sees immediately by transport of structure. Thus *an automorphism of a discrete or compact group has modulus* 1.

PROPOSITION 4. — *Let* G *be a locally compact group,* $\Gamma$ *a topological group, and* $\gamma \mapsto u_\gamma$ *a homomorphism of* $\Gamma$ *into the group* $\mathscr{G}$ *of automorphism of* G, *such that* $(\gamma, x) \mapsto u_\gamma(x)$ *is a continuous mapping of* $\Gamma \times \mathrm{G}$ *into* G. *Then, the mapping* $\gamma \mapsto \operatorname{mod}(u_\gamma)$ *is a continuous representation of* $\Gamma$ *in* $\mathbf{R}_+^*$.

This mapping is obviously a representation (algebraic) of $\Gamma$ in $\mathbf{R}_+^*$; it will suffice to prove its continuity. Let $f \in \mathscr{K}(\mathrm{G})$ and let S be its support.

Let $\gamma_0 \in \Gamma$ and let U be a relatively compact neighborhood of $u_{\gamma_0}^{-1}(S)$. The mapping $\gamma \mapsto u_\gamma$ is a continuous mapping of $\Gamma$ into $\mathscr{G}$ equipped with the topology of compact convergence (GT, X, §3, No. 4, Th. 3); therefore $u_\gamma^{-1}(S) \subset U$ for $\gamma$ sufficiently near $\gamma_0$. Lemma 1 of No. 1 then proves that $\int f(u_\gamma(x))\,d\mu(x)$ (where $\mu$ denotes a left Haar measure of G) depends continuously on $\gamma$; whence the proposition.

## 5. Haar measure of a product

PROPOSITION 5. — *Let* $(G_\iota)_{\iota\in I}$ *be a family of locally compact groups. For every* $\iota \in I$ *let* $\mu_\iota$ *be a left* (resp. *right*) *Haar measure on* $G_\iota$. *Assume that there exists a finite subset* J *of* I *such that, for every* $\iota \in I - J$, $G_\iota$ *is compact and* $\mu_\iota(G_\iota) = 1$. *Then the product measure* $\bigotimes_{\iota\in I} \mu_\iota$ *is a left* (resp. *right*) *Haar measure on* $G = \prod_{\iota\in I} G_\iota$. *If* $x = (x_\iota) \in G$ *then*

$$\Delta_G(x) = \prod_{\iota\in I} \Delta_{G_\iota}(x_\iota).$$

For every finite subset J of I, $\bigotimes_{\iota\in J} \mu_\iota$ is a left (resp. right) Haar measure on $\prod_{\iota\in J} G_\iota$, as follows immediately from the definitions. Therefore $\bigotimes_{\iota\in I} \mu_\iota$ is a left (resp. right) Haar measure on G (Ch. III, §4, No. 6, Prop. 9). On the other hand, if the $\mu_\iota$ are left Haar measures then

$$\boldsymbol{\delta}(x)\Big(\bigotimes_{\iota\in I} \mu_\iota\Big) = \bigotimes_{\iota\in I} \boldsymbol{\delta}(x_\iota)\mu_\iota = \bigotimes_{\iota\in I}\big(\Delta_{G_\iota}(x_\iota)\mu_\iota\big) = \Big(\prod_{\iota\in I} \Delta_{G_\iota}(x_\iota)\Big)\bigotimes_{\iota\in I} \mu_\iota,$$

whence $\Delta_G(x) = \prod_{\iota\in I} \Delta_{G_\iota}(x_\iota)$.

*Examples.* — 1) Lebesgue measure on $\mathbf{R}^n$ is a Haar measure of the additive group $\mathbf{R}^n$.

2) The mapping $(r,u) \mapsto ru$ is an isomorphism of $\mathbf{R}_+^* \times \mathbf{U}$ onto $\mathbf{C}^*$ (GT, VIII, §1, No. 3). If $\mathbf{C}^*$ is identified with $\mathbf{R}_+^* \times \mathbf{U}$ by means of this isomorphism, and if $du$ denotes a Haar measure on $\mathbf{U}$, then $r^{-1}\,dr\,du$ is a Haar measure on $\mathbf{C}^*$ by Example 2 of No. 2. On the other hand, the bijection $\theta \mapsto e^{2i\pi\theta}$ of $[0,1[$ onto $\mathbf{U}$ transforms the Lebesgue measure $d\theta$ on $[0,1[$ into a Haar measure on $\mathbf{U}$ by Example 3 of No. 2. It follows that if $f \in \mathscr{K}(\mathbf{C}^*)$, the integral

$$\int_0^{+\infty}\int_0^1 f(re^{2i\pi\theta})r^{-1}\,dr\,d\theta$$

defines a Haar measure on $\mathbf{C}^*$.

### 6. Haar measure of an inverse limit*

Let G be a locally compact group (hence complete). Let $(K_\alpha)_{\alpha\in A}$ be a decreasing directed family of compact normal subgroups of G, with intersection $\{e\}$ (so that the filter base formed by the $K_\alpha$ converges to $e$). Set $G_\alpha = G/K_\alpha$; let $\varphi_\alpha : G \to G_\alpha$ and $\varphi_{\beta\alpha} : G_\alpha \to G_\beta$ $(\alpha \geqslant \beta)$ be the canonical homomorphisms. Then, the inverse limit of the inverse system $(G_\alpha, \varphi_{\beta\alpha})$ may be identified with G, and the canonical mapping of this inverse limit into $G_\alpha$ is identified with $\varphi_\alpha$ (GT, III, §7, No. 3, Prop. 2). The mappings $\varphi_\alpha$ and $\varphi_{\beta\alpha}$ are proper (*loc. cit.*, §4, No. 1, Cor. 2 of Prop. 1). These assumptions remain fixed throughout this subsection.

*Lemma* 2. — a) *Let $f \in \mathscr{K}_+(G)$, S a compact subset of G containing* Supp $f$, U *an open neighborhood of* S *in* G, *and* $\varepsilon > 0$. *There exist an $\alpha \in A$ and a function $g \in \mathscr{K}_+(G)$, zero outside* U *and constant on the cosets of* $K_\alpha$, *such that* $|f - g| \leqslant \varepsilon$.

b) *Let $\mu$ and $\mu'$ be two measures on* G *such that* $\varphi_\alpha(\mu) = \varphi_\alpha(\mu')$ *for all* $\alpha \in A$. *Then* $\mu = \mu'$.

There exists an $\alpha_1 \in A$ such that $K_{\alpha_1}S \cap K_{\alpha_1}(G - U) = \varnothing$ (GT, II, §4, No. 3, Prop. 4). Augmenting S and diminishing U, we may therefore assume that S and U are unions of cosets of $K_{\alpha_1}$. Consider the continuous numerical functions $h$ on S having the following property: there exists an $\alpha \geqslant \alpha_1$ such that $h$ is constant on the cosets of $K_\alpha$. These functions form a subalgebra of $\mathscr{K}(S)$ (because $(K_\alpha)$ is a decreasing directed family) that contains the constants and separates the points of S: for, let $x, y$ be two distinct points of S; since the intersection of the $K_\alpha$ is $\{e\}$, there exists an $\alpha \geqslant \alpha_1$ such that $\varphi_\alpha(x) \neq \varphi_\alpha(y)$, then a numerical function $u$ continuous on $\varphi_\alpha(S)$ such that $u(\varphi_\alpha(x)) \neq u(\varphi_\alpha(y))$. By the Stone–Weierstrass theorem, there exist an $\alpha \geqslant \alpha_1$ and a continuous function $h \geqslant 0$ on S, constant on the cosets of $K_\alpha$, such that $|f - h| \leqslant \frac{\varepsilon}{2}$ on S. For every $t \in \mathbf{R}$, set $\delta(t) = \left(t - \frac{\varepsilon}{2}\right)^+$, and set $h' = \delta \circ h$. Then $h'$ is a function $\geqslant 0$, continuous on S, constant on the cosets of $K_\alpha$, and $|h - h'| \leqslant \frac{\varepsilon}{2}$ on S, therefore $|f - h'| \leqslant \varepsilon$ on S. On the other hand, $h'(x) = 0$ if $x$ belongs to the boundary of S in G, because then $h(x) \leqslant \frac{\varepsilon}{2}$. If $h'$ is extended by 0 on the complement of S, one obtains a function $g$ that meets the requirements, which proves a).

Now let $\mu$, $\mu'$ be two measures on G such that $\varphi_\alpha(\mu) = \varphi_\alpha(\mu')$ for all $\alpha \in A$. Let $v \in \mathscr{K}(G)$ be a function constant on the cosets of $K_\alpha$ for some $\alpha \in A$, so that we may write $v = w \circ \varphi_\alpha$ with $w \in \mathscr{K}(G_\alpha)$; then

* Cf. Ch. III, §4, No. 5.

$\mu(v) = \big(\varphi_\alpha(\mu)\big)(w) = \big(\varphi_\alpha(\mu')\big)(w) = \mu'(v)$; it follows that $\mu = \mu'$ by virtue of a).

PROPOSITION 6. — *For every* $\alpha \in \mathrm{A}$, *let* $\mu_\alpha$ *be a positive measure on* $\mathrm{G}_\alpha$. *Suppose that* $\varphi_{\beta\alpha}(\mu_\alpha) = \mu_\beta$ *for* $\alpha \geqslant \beta$. *Then, there exists one and only one positive measure* $\mu$ *on* G *such that* $\varphi_\alpha(\mu) = \mu_\alpha$ *for all* $\alpha \in \mathrm{A}$.

Uniqueness follows immediately from Lemma 2 b). Let us prove the existence of $\mu$. Let V be the vector space of functions belonging to $\mathscr{K}(\mathrm{G})$ and constant on the cosets of some $\mathrm{K}_\alpha$ ($\alpha$ may depend on the function). It follows from Lemma 2 a) that V satisfies the condition (P) of Ch. III, §1, No. 7, Prop. 9: for, let K be a compact set in G and choose $f \in \mathscr{K}_+(\mathrm{G})$ with $f(x) > 0$ for all $x \in \mathrm{K}$; let $a > 0$ be the smallest value of $f$ on K; by Lemma 2 a), there exists a function $g \in \mathrm{V} \cap \mathscr{K}_+(\mathrm{G})$ such that $|f-g| \leqslant a/2$, therefore $g(x) > 0$ for all $x \in \mathrm{K}$, and condition (P) is verified. Let $f \in \mathrm{V}$. There exists an $\alpha \in \mathrm{A}$ such that $f$ is constant on the cosets of $\mathrm{K}_\alpha$. By passage to the quotient, $f$ defines a function $f_\alpha \in \mathscr{K}(\mathrm{G}_\alpha)$. The number $\mu(f) = \mu_\alpha(f_\alpha)$ does not depend on the choice of $\alpha$: for, let $\beta$ be any index such that $f$ is constant on the cosets of $\mathrm{K}_\beta$; let $\gamma \in \mathrm{A}$ be such that $\gamma \geqslant \alpha$, $\gamma \geqslant \beta$; then $f$ defines functions $f_\beta \in \mathscr{K}(\mathrm{G}_\beta)$, $f_\gamma \in \mathscr{K}(\mathrm{G}_\gamma)$ such that $f = f_\beta \circ \varphi_\beta = f_\gamma \circ \varphi_\gamma$; then $f_\alpha \circ \varphi_{\alpha\gamma} = f_\gamma$, therefore $\mu_\gamma(f_\gamma) = \big(\varphi_{\alpha\gamma}(\mu_\gamma)\big)(f_\alpha) = \mu_\alpha(f_\alpha)$, and similarly $\mu_\gamma(f_\gamma) = \mu_\beta(f_\beta)$, whence our assertion. This established, it is clear that $\mu$ is a linear form on V and that $\mu(f) \geqslant 0$ for $f \geqslant 0$. By Prop. 9 of Ch. III, §1, No. 7, $\mu$ may be extended to a positive measure on G, which we again denote by $\mu$. One has $\varphi_\alpha(\mu) = \mu_\alpha$ for all $\alpha \in \mathrm{A}$ by the very construction of $\mu$.

DEFINITION 5. — *The measure* $\mu$ *is said to be the inverse limit* (or *projective limit*) *of the* $\mu_\alpha$.

PROPOSITION 7. — *We retain the notations of Prop. 6. If each* $\mu_\alpha$ *is a left* (resp. *right*) *Haar measure on* $\mathrm{G}_\alpha$, *then* $\mu$ *is a left* (resp. *right*) *Haar measure on* G.

Suppose, for example, that the $\mu_\alpha$ are left Haar measures. Let $s \in \mathrm{G}$. For every $x \in \mathrm{G}$,

$$\big(\varphi_\alpha \circ \boldsymbol{\gamma}(s)\big)(x) = \varphi_\alpha(sx) = \varphi_\alpha(s)\varphi_\alpha(x) = \big(\boldsymbol{\gamma}(\varphi_\alpha(s)) \circ \varphi_\alpha\big)(x);$$

therefore $\varphi_\alpha\big(\boldsymbol{\gamma}(s)\mu\big) = \boldsymbol{\gamma}\big(\varphi_\alpha(s)\big)\mu_\alpha = \mu_\alpha$. Therefore $\boldsymbol{\gamma}(s)\mu = \mu$ by Lemma 2 b), thus $\mu$ is a left Haar measure.

We suppose henceforth the $\mathrm{K}_\alpha$ to be not only compact, but *open* in G. The $\mathrm{G}_\alpha$ are then discrete and, for $\beta \geqslant \alpha$, $\mathrm{K}_\alpha/\mathrm{K}_\beta$ is a compact and discrete group, hence is finite. The group G is unimodular (Prop. 3).

PROPOSITION 8. — a) *Let* $\mu$ *and* $\mu'$ *be two positive measures on* G *such that, for every* $\alpha$ *and for every coset* C *of* $\mathrm{K}_\alpha$, $\mu(\mathrm{C}) = \mu'(\mathrm{C})$. *Then* $\mu = \mu'$.

b) *Fix an* $\alpha_0 \in \mathrm{A}$. *For every* $\alpha \geqslant \alpha_0$ *let* $n_\alpha$ *be the number of elements of the finite group* $\mathrm{K}_{\alpha_0}/\mathrm{K}_\alpha$. *There exists one and only one positive measure* $\mu$ *on* G *such that, for every* $\alpha \geqslant \alpha_0$, *each coset of* $\mathrm{K}_\alpha$ *has measure* $n_\alpha^{-1}$. *Moreover,* $\mu$ *is a Haar measure on* G, *such that* $\mu(\mathrm{K}_{\alpha_0}) = 1$.

Let $\mu$ and $\mu'$ be two positive measures on G satisfying the condition a). The points of the discrete group $\mathrm{G}_\alpha$ then have the same measure for $\varphi_\alpha(\mu)$ and $\varphi_\alpha(\mu')$, whence $\varphi_\alpha(\mu) = \varphi_\alpha(\mu')$, and this for all $\alpha$. Therefore $\mu = \mu'$ (Lemma 2 b)).

Let us prove b). For every $\alpha \geqslant \alpha_0$, let $\mu_\alpha$ be the Haar measure of the discrete group $\mathrm{G}_\alpha$ such that every point has measure $n_\alpha^{-1}$. Let $\alpha, \beta$ be such that $\alpha \geqslant \beta \geqslant \alpha_0$. Then $\mathrm{K}_\beta/\mathrm{K}_\alpha$ has $n_\alpha/n_\beta$ elements. Therefore $\varphi_{\beta\alpha}(\mu_\alpha)$ is the measure on $\mathrm{G}_\beta$ such that each point has measure $n_\alpha^{-1} \cdot \dfrac{n_\alpha}{n_\beta} = n_\beta^{-1}$; in other words, $\varphi_{\beta\alpha}(\mu_\alpha) = \mu_\beta$. It then suffices to apply Props. 6 and 7.

*Example.* — Let $\mathbf{Q}_p$ be the $p$-adic field, the completion of $\mathbf{Q}$ for the $p$-adic absolute value $|x|_p = p^{-v_p(x)}$ (GT, IX, §3, No. 2, *Example* 3). The elements of $\mathbf{Q}_p$ are called *p-adic numbers.* We denote again by $|x|_p$ the continuous extension of the $p$-adic absolute value to $\mathbf{Q}_p$. One has

$$|x + y|_p \leqslant \sup(|x|_p, |y|_p)$$

for $x, y$ in $\mathbf{Q}$ (*loc. cit.*), hence for $x, y$ in $\mathbf{Q}_p$; moreover, if $|y|_p < |x|_p$ then $|x + y|_p = |x|_p$, because $|x|_p = |(x+y) - y|_p \leqslant \sup(|x+y|_p, |y|_p)$. If $(x_n)$ is a sequence of points of $\mathbf{Q}_p$ tending to $x \in \mathbf{Q}_p^*$, then $|x - x_n|_p < |x|_p$ and $|x - x_n|_p < |x_n|_p$ for $n$ sufficiently large, therefore $|x|_p = |x_n|_p$. This proves that, for every $x \in \mathbf{Q}_p^*$, $|x|_p$ is a power of $p$.

Let $\mathbf{Z}_p$ be the closure of $\mathbf{Z}$ in $\mathbf{Q}_p$; this is a subring of $\mathbf{Q}_p$; its elements are called *p-adic integers.* One has $|x|_p \leqslant 1$ for every $x \in \mathbf{Z}_p$. Conversely, let $x$ be an element of $\mathbf{Q}_p$ such that $|x|_p \leqslant 1$, and let us show that $x \in \mathbf{Z}_p$; there exists a sequence $(x_n)$ of elements of $\mathbf{Q}$ tending to $x$, and $|x_n|_p \leqslant 1$ for $n$ sufficiently large by what we have seen above; it suffices to show that $x_n$ belongs to $\mathbf{Z}_p$ for $n$ sufficiently large; in other words, we are reduced to the case that $x \in \mathbf{Q}$; then $x = a/b$ with $b$ relatively prime to $p$; for every integer $n > 0$, there exist $b'_n \in \mathbf{Z}$ and $h_n \in \mathbf{Z}$ such that $bb'_n + h_n p^n = 1$, whence $x = \dfrac{abb'_n + ah_n p^n}{b} = ab'_n + \dfrac{ah_n p^n}{b}$ and $|x - ab'_n|_p \leqslant p^{-n}$, therefore $ab'_n$ tends to $x$.

From this, it follows that the closed ball with center 0 and radius $p^{-n}$, identical to the open ball with center 0 and radius $p^{-n+1}$, is $p^n\mathbf{Z}_p$. The topological space $\mathbf{Q}_p$ is therefore zero-dimensional, hence totally disconnected (GT, IX, §6, No. 4).

Let us show that the integers $0, 1, \ldots, p^n - 1$ constitute a system of representatives of $\mathbf{Z}_p$ modulo $p^n\mathbf{Z}_p$. First, $|k - k'|_p > p^{-n}$ for two such integers $k$ and $k'$, therefore the classes modulo $p^n\mathbf{Z}_p$ of these integers are distinct. On the other hand, let $x \in \mathbf{Z}_p$; there exists a $k \in \mathbf{Z}$ such that $|x - k|_p \leqslant p^{-n}$; adding a suitable multiple of $p^n$ to $k$, one can suppose that $k \in [0, p^n - 1]$, and $x$ is congruent to $k$ modulo $p^n\mathbf{Z}_p$. Whence our assertion. This shows that $\mathbf{Z}_p/p^n\mathbf{Z}_p$ is canonically isomorphic to $\mathbf{Z}/p^n\mathbf{Z}$. One sees, moreover, that $\mathbf{Z}_p$ is precompact, hence *compact* since it is complete. Since $\mathbf{Z}_p$ is an open subgroup of $\mathbf{Q}_p$, $\mathbf{Q}_p$ is *locally compact.* The topology of $\mathbf{Q}_p$ has a countable base (GT, IX, §2, No. 9, Cor. of Prop. 16). The additive group $\mathbf{Q}_p$ may be identified with the inverse limit of the discrete groups $\mathbf{Q}_p/p^n\mathbf{Z}_p$.

There exists one and only one Haar measure $\alpha$ on the additive group $\mathbf{Q}_p$ such that $\alpha(\mathbf{Z}_p) = 1$; it is called the *normalized Haar measure* on $\mathbf{Q}_p$. Since $\mathbf{Z}_p$ is the union of $p^n$ disjoint cosets of $p^n\mathbf{Z}_p$ ($n$ an integer $\geqslant 0$), one has $\alpha(p^n\mathbf{Z}_p) = p^{-n}$; similarly $\alpha(p^{-n}\mathbf{Z}_p) = p^n$, so that, finally, $\alpha(p^n\mathbf{Z}_p) = p^{-n}$ for every $n \in \mathbf{Z}$. By Prop. 8 b), *$\alpha$ is the only positive measure on $\mathbf{Q}_p$ such that every coset of $p^n\mathbf{Z}_p$ ($n$ an integer $\geqslant 0$) has measure $p^{-n}$.*

The restriction of $\alpha$ to $\mathbf{Z}_p$ is obviously a Haar measure on $\mathbf{Z}_p$.

## 7. Local definition of a Haar measure

PROPOSITION 9. — *Let G be a locally compact group, V an open subset of G, and $\mu$ a nonzero positive measure on V having the following property: if U is an open subset of V and if $s \in G$ is such that $sU \subset V$, then the image of the measure $\mu_U$ induced by $\mu$ on U, under the homeomorphism $x \mapsto sx$ of U onto $sU$, is $\mu_{sU}$. Then, there exists one and only one left Haar measure $\alpha$ on G that induces $\mu$ on V.*

For every $s \in G$, let $\mu_s$ be the image of $\mu$ under the homeomorphism $x \mapsto sx$ of V onto $sV$. The restriction of $\mu_s$ to $V \cap sV$ is the image of $\mu_{s^{-1}V\cap V}$ under the restriction of $x \mapsto sx$ to $s^{-1}V \cap V$; by hypothesis, this image is $\mu_{V\cap sV}$. By translation, one concludes from this that $\mu_s$ and $\mu_t$ have the same restriction to $sV \cap tV$ for any $s, t$. Therefore, by Prop. 1 of Ch. III, §2, No. 1, there exists a measure $\alpha$ on G that induces $\mu_s$ on $sV$ for every $s$. It is clear that $\alpha$ is the unique left Haar measure on G inducing $\mu$ on V.

COROLLARY. — *Let G, G′ be two locally compact groups,* V (resp. V′) *an open neighborhood of the neutral element of* G (resp. G′), *and $\varphi$ a local isomorphism of* G′ *with* G (GT, III, §1, No. 3, Def. 2) *defined on* V′, *such that $\varphi(V') = V$. Let $\alpha'$ be a left Haar measure on* G′, *and $\alpha'_{V'}$ its* *restriction to* $V'$. *Then* $\varphi(\alpha'_{V'})$ *is the restriction to* $V$ *of a unique left Haar measure* $\alpha$ *on* $G$.

Let $V_1$ be an open neighborhood of $e$ in $G$ such that $V_1V_1^{-1} \subset V$. Let $\mu$ be the restriction of $\varphi(\alpha'_{V'})$ to $V_1$. Let $U$ be an open subset of $V_1$ and let $s \in G$ be such that $sU \subset V_1$. Then $s \in V_1V_1^{-1} \subset V$, therefore $s = \varphi(s')$ for some $s' \in V'$. Let $x \in U$. Then $x = \varphi(x')$ for some $x' \in V'$, therefore $sx = \varphi(s')\varphi(x') = \varphi(s'x')$ since $sx \in sU \subset V$. Since the left translations in $G'$ preserve $\alpha'$, one sees that $V_1$ and $\mu$ satisfy the conditions of Prop. 9. Let $\alpha$ be the left Haar measure on $G$ inducing $\mu$ on $V_1$. For every $t \in V$, there exists an open neighborhood $W$ of $e$ in $V_1$ such that $tW \subset V$. Then the restriction of $\varphi(\alpha'_{V'})$ to $tW$ may be deduced by translation from the restriction of $\mu$ to $W$, hence is the restriction of $\alpha$ to $tW$. Therefore $\varphi(\alpha'_{V'})$ is the restriction of $\alpha$ to $V$.

One says that $\alpha$ *is deduced from* $\alpha'$ *by means of the local isomorphism* $\varphi$.

*Example.* — The Haar measure on $\mathbf{T}$ obtained in No. 2, *Example* 3 may be deduced from the Lebesgue measure on $\mathbf{R}$ by a local isomorphism of $\mathbf{R}$ with $\mathbf{T}$.

## 8. Relatively invariant measures

PROPOSITION 10. — *Let* $G$ *be a locally compact group,* $\mu$ *a relatively left-invariant measure on* $G$ *with multiplier* $\chi$. *If* $\chi_1$ *is a continuous representation of* $G$ *in* $\mathbf{C}^*$, *the measure* $\chi_1 \cdot \mu$ *is relatively left-invariant with multiplier* $\chi_1\chi$.

For,

$$\begin{aligned}\boldsymbol{\gamma}(s)(\chi_1 \cdot \mu) = (\boldsymbol{\gamma}(s)\chi_1) \cdot (\boldsymbol{\gamma}(s)\mu) &= (\chi_1(s^{-1})\chi_1) \cdot (\chi(s)^{-1}\mu) \\ &= (\chi_1\chi)(s)^{-1}(\chi_1 \cdot \mu).\end{aligned}$$

COROLLARY 1. — *Let* $\mu$ *be a left Haar measure on* $G$. *For a nonzero measure* $\nu$ *on* $G$ *to be relatively left-invariant, it is necessary and sufficient that it be of the form* $a\chi \cdot \mu$, *where* $a \in \mathbf{C}^*$ *and* $\chi$ *is a continuous representation of* $G$ *in* $\mathbf{C}^*$; *its multiplier is then* $\chi$.

The condition is sufficient (Prop. 10). On the other hand, if $\nu$ is a nonzero relatively left-invariant measure with multiplier $\chi$, then $\chi^{-1} \cdot \nu$ is left-invariant (Prop. 10) hence is of the form $a\mu$ with $a \in \mathbf{C}^*$ (No. 2, Cor. of Th. 1).

COROLLARY 2. — *Every relatively left-invariant measure is relatively right-invariant.*

For, with the notations of Cor. 1,

$$\begin{aligned}(34)\qquad \boldsymbol{\delta}(s)(\chi\cdot\mu) &= (\boldsymbol{\delta}(s)\chi)\cdot(\boldsymbol{\delta}(s)\mu) = (\chi(s)\chi)\cdot(\Delta_{\mathrm{G}}(s)\mu)\\ &= (\chi\Delta_{\mathrm{G}})(s)(\chi\cdot\mu).\end{aligned}$$

On account of Cor. 2, we shall henceforth speak of *relatively invariant measures on* G, without further specification. The relatively invariant measures admit as special cases the left Haar measures and the right Haar measures. Given a relatively invariant measure $\nu$ on G, it is convenient to distinguish between its *left multiplier* $\chi$ and its *right multiplier* $\chi'$, defined by $\boldsymbol{\gamma}(s)\nu = \chi(s)^{-1}\nu$, $\boldsymbol{\delta}(s)\nu = \chi'(s)\nu$. By (34), these multipliers satisfy the relation

$$(35)\qquad \chi' = \chi\Delta_{\mathrm{G}}.$$

Still denoting by $\mu$ a left Haar measure, we have

$$\check{\nu} = (\chi\cdot\mu)^{\vee} = \check{\chi}\cdot\check{\mu} = (\chi^{-1}\Delta_{\mathrm{G}}^{-1})\cdot\mu,$$

thus $\check{\nu}$ is relatively invariant with left multiplier $\chi^{-1}\Delta_{\mathrm{G}}^{-1}$ and right multiplier $\chi^{-1}$.

The concepts of negligible, locally negligible, measurable and locally integrable function are the same for every relatively invariant measure.

## 9. Quasi-invariant measures

PROPOSITION 11. — *Let* G *be a locally compact group,* $\mu$ *a left Haar measure on* G. *For a measure* $\nu \neq 0$ *on* G *to be left quasi-invariant, it is necessary and sufficient that* $\nu$ *be equivalent to* $\mu$.

Sufficiency is obvious. Let $\nu \neq 0$ be a left quasi-invariant measure, and let us show that $\nu$ is equivalent to $\mu$. We can restrict ourselves to the case that $\nu > 0$. Let A be a compact subset of G. We will show, as will establish the proposition, that the conditions $\mu(\mathrm{A}) = 0$, $\nu(\mathrm{A}) = 0$ are equivalent (Ch. V, §5, No. 5, Th. 2).

*a)* For every $f \in \mathscr{K}_+(\mathrm{G})$, the function $(x, y) \mapsto f(x)\varphi_{\mathrm{A}}(xy)$ on $\mathrm{G}\times\mathrm{G}$ is $(\nu\otimes\mu)$-integrable, because it is upper semi-continuous, bounded, and its support is contained in the compact set $\mathrm{K}\times\mathrm{K}^{-1}\mathrm{A}$ if one sets $\mathrm{K} = \operatorname{Supp} f$. Therefore, by the Lebesgue–Fubini theorem,

$$(36)\qquad \int d\nu(y)\int \varphi_{\mathrm{A}}(xy)f(x)\,d\mu(x) = \int f(x)\,d\mu(x)\int \varphi_{\mathrm{A}}(xy)\,d\nu(y).$$

*b*) Suppose $\nu(\mathrm{A}) = 0$. By hypothesis, $\nu(x\mathrm{A}) = 0$ for all $x \in \mathrm{G}$, therefore the right side of (36) is zero. Therefore there exists a $\nu$-negligible set $\mathrm{N}_f$ such that, for $y \notin \mathrm{N}_f$,

$$(37) \qquad 0 = \int \varphi_{\mathrm{A}}(xy) f(x)\, d\mu(x) = \Delta_{\mathrm{G}}(y)^{-1} \int \varphi_{\mathrm{A}}(x) f(xy^{-1})\, d\mu(x).$$

Let B be a compact subset of G such that $\nu(\mathrm{B}) \neq 0$, and take for $f$ a function in $\mathscr{K}_+(\mathrm{G})$ equal to 1 on $\mathrm{AB}^{-1}$. There then exists a $y \in \mathrm{B}$ such that (37) is verified. But since $\varphi_{\mathrm{A}}(x) f(xy^{-1}) = \varphi_{\mathrm{A}}(x)$ for $y \in \mathrm{B}$, this proves that $\mu(\mathrm{A}) = 0$.

*c*) Suppose $\mu(\mathrm{A}) = 0$. Then, for every $f \in \mathscr{K}_+(\mathrm{G})$, the left side of (36) is zero, hence also the right side. Consequently, there exists a locally $\mu$-negligible set M such that $\int \varphi_{\mathrm{A}}(xy)\, d\nu(y) = 0$ for $x \notin \mathrm{M}$. Since $\mu \neq 0$, it follows that $\nu(x\mathrm{A}) = 0$ for some $x \in \mathrm{G}$, whence $\nu(\mathrm{A}) = 0$.

Applying Prop. 11 to $\mathrm{G}^0$, one sees that the right quasi-invariant measures are identical with the left quasi-invariant measures. They are called simply *quasi-invariant* measures on G.

## 10. Locally compact fields

DEFINITION 6. — *Let* K *be a locally compact field*[1]. *For* $a \in \mathrm{K}^*$, *one calls modulus of* $a$, *and denotes by* $\mathrm{mod}_{\mathrm{K}}(a)$ *or simply* $\mathrm{mod}(a)$, *the modulus of the automorphism* $x \mapsto ax$ *of the additive group* $\mathrm{K}^+$ *underlying* K; *one sets* $\mathrm{mod}(0) = 0$.

*Examples.* — 1) Let $\mathrm{K} = \mathbf{R}$. If $s > 0$ then $s \cdot [0,1] = [0,s]$; if $s < 0$, $s \cdot [0,1] = [s,0]$. Thus $\mathrm{mod}_{\mathbf{R}}\, t = |t|$ for all $t \in \mathbf{R}$.

2) Let $\mathrm{K} = \mathbf{Q}_p$. If $s \in \mathbf{Q}_p^*$ is such that $|s|_p = p^{-n}$, then $s\mathbf{Z}_p$ is the set of $x \in \mathbf{Q}_p$ such that $|x|_p \leqslant p^{-n}$; therefore, if $\mu$ denotes the normalized Haar measure on $\mathbf{Q}_p$, then $\mu(s\mathbf{Z}_p) = p^{-n}$. Thus $\mathrm{mod}_{\mathbf{Q}_p} t = |t|_p$ for all $t \in \mathbf{Q}_p$.

PROPOSITION 12. — *The function* mod *is continuous on* K, *and* $\mathrm{mod}(ab) = \mathrm{mod}(a)\mathrm{mod}(b)$ *for all* $a, b$ *in* K.

The last assertion is obvious. Prop. 4 of No. 4 shows that the function mod is continuous at every point of $\mathrm{K}^*$. It remains only to show its continuity at 0. This is obvious for discrete K; we shall therefore assume K nondiscrete. Let $\alpha$ be a Haar measure on $\mathrm{K}^+$ and let C be a compact subset of K such that $\alpha(\mathrm{C}) > 0$; for $a \in \mathrm{K}^*$, we have $\alpha(a\mathrm{C}) = \mathrm{mod}(a)\alpha(\mathrm{C})$.

---

[1] *Corps* (A, I, §9, No. 1), also translated as "division ring" (GT, III, §6, No. 7).

Since K is not discrete, $\alpha(\{0\}) = 0$ (No. 2, Prop. 2); therefore, for every $\varepsilon > 0$ there exists an open neighborhood U of 0 such that $\alpha(\mathrm{U}) \leqslant \varepsilon$. Since the product in K is continuous, $a\mathrm{C} \subset \mathrm{U}$ for $a$ sufficiently near 0, and then $\mathrm{mod}(a) \leqslant \varepsilon/\alpha(\mathrm{C})$.

PROPOSITION 13. — *For every* $\mathrm{M} > 0$, *let* $\mathrm{V_M}$ *be the set of* $x \in \mathrm{K}$ *such that* $\mathrm{mod}(x) \leqslant \mathrm{M}$. *If* K *is nondiscrete, the* $\mathrm{V_M}$ *form a fundamental system of compact neighborhoods of* 0 *in* K.

The $\mathrm{V_M}$ are closed neighborhoods of 0 by Prop. 12. Let us show that they are compact. Let U be a compact neighborhood of 0. There exists an $r \neq 0$ in K such that $\mathrm{mod}(r) < 1$ and $r^n \in \mathrm{U}$ for all $n > 0$: for, let W be a neighborhood of 0 such that $\mathrm{WU} \subset \mathrm{U}$; by Prop. 12, there exists an $r \neq 0$ in K such that $\mathrm{mod}(r) < 1$ and $r \in \mathrm{U} \cap \mathrm{W}$; then $r^2 \in \mathrm{WU} \subset \mathrm{U}$, and $r^n \in \mathrm{U}$ for all $n > 0$ by induction on $n$. We are going to show that $\mathrm{V_M}$ is contained in a finite union of sets $r^{-q}\mathrm{U}$ ($q$ an integer $\geqslant 0$), which will prove that the $\mathrm{V_M}$ are indeed compact. If $x$ is a cluster point of the sequence $(r^n)$, then $\mathrm{mod}(x)$ is a cluster point of the sequence $\big(\mathrm{mod}(r)^n\big)$, therefore $\mathrm{mod}(x) = 0$, $x = 0$; since U is compact, it follows (GT, I, §9, No. 1, Cor. of Th. 1) that $\lim_{n\to\infty} r^n = 0$. Now let $a \in \mathrm{V_M}$. Since the sequence $(r^n a)_{n\geqslant 0}$ tends to 0, there exists a smallest integer $n \geqslant 0$ such that $r^n a \in \mathrm{U}$. If $n > 0$ then $r^{n-1}a \notin \mathrm{U}$, therefore $r^n a \in \mathrm{U} \cap \complement(r\mathrm{U})$; the closure X of $\mathrm{U} \cap \complement(r\mathrm{U})$ is compact since U is compact, and it does not contain 0 since $r\mathrm{U}$ is a neighborhood of 0; therefore, in X, $\mathrm{mod}(x)$ is bounded below by a number $m > 0$. Thus, if $n > 0$, we have $m \leqslant \mathrm{mod}(r^n a)$, whence $\mathrm{mod}(r^{-1})^n \leqslant \mathrm{M}/m$. Since $\mathrm{mod}(r^{-1}) > 1$, the integer $n$ can only take on a finite number of values, a number not depending on $a$, which completes the proof of our assertion.

This being so, since the intersection of the $\mathrm{V_M}$ reduces to $\{0\}$ the $\mathrm{V_M}$ form a fundamental system of neighborhoods of 0 (GT, I, §9, No. 2, Prop. 1).

COROLLARY. — *The topology of a nondiscrete locally compact field admits a countable base.*

For, K is the union of compact sets $\mathrm{V_1}, \mathrm{V_2}, \ldots$. On the other hand, K is metrizable by Prop. 1 of GT, IX, §3, No. 1. Therefore the topology of K admits a countable base (*loc. cit.*, §2, No. 9, Cor. of Prop. 16).

PROPOSITION 14. — *Let* $\alpha$ *be a Haar measure on* $\mathrm{K}^+$. *Then the measure* $\beta = (\mathrm{mod_K})^{-1} \cdot \alpha$ *on* $\mathrm{K}^*$ *is a left Haar measure on the multiplicative group* $\mathrm{K}^*$.

For, if $b \in \mathrm{K}^*$, the mapping $a \mapsto b^{-1}a$ of K into K transforms $\alpha$ into $(\mathrm{mod_K}\, b)\alpha$, hence $(\mathrm{mod_K})^{-1} \cdot \alpha$ into itself, whence the proposition.

COROLLARY. — *Let* $\mathbf{f}$ *be a function defined on* $\mathrm{K}^*$, *with values in* $\overline{\mathbf{R}}$ *or in a Banach space. For* $\mathbf{f}$ *to be* $\beta$*-integrable, it is necessary and sufficient* *that* $(\text{mod}_K)^{-1}\mathbf{f}$ *be* $\alpha$*-integrable, in which case*

$$\int_{K^*} \mathbf{f}(x)\, d\beta(x) = \int_{K^+} \big(\text{mod}_K(x)\big)^{-1}\mathbf{f}(x)\, d\alpha(x)\,.$$

This follows from Prop. 14, the Cor. of Prop. 13, and Ch. V, §5, No. 3, Th. 1.

PROPOSITION 15. — *Assume* K *to be commutative. Let* $u$ *be an automorphism of the vector space* $E = K^n$. *Then*

$$\text{mod}_E\, u = \text{mod}_K(\det u)\,.$$

It suffices to verify the formula when $u$ runs over a system of generators of $\mathbf{GL}(E)$. Now, $\mathbf{GL}(E)$ is generated by the following elements (A, II, §10, No. 13, Cor. 2 of Prop. 14):

($a$) The elements $u_1$ of the form

$$(x_1, \ldots, x_n) \mapsto (x_{\sigma(1)}, \ldots, x_{\sigma(n)})\,,$$

where $\sigma \in \mathfrak{S}_n$;

($b$) the elements $u_2$ of the form

$$(x_1, \ldots, x_n) \mapsto (ax_1, x_2, \ldots, x_n)$$

with $a \in K^*$;

($c$) the elements $u_3$ of the form

$$(x_1, \ldots, x_n) \mapsto (x_1 + \sum_{i=2}^{n} c_i x_i, x_2, \ldots, x_n)\,.$$

If $f \in \mathscr{K}(E)$ then, denoting by $\alpha$ a Haar measure on $K^+$,

$$\begin{aligned}
&\int \cdots \int_{K^n} f(x_1 + \sum_{i=2}^{n} c_i x_i, x_2, \ldots, x_n)\, d\alpha(x_1) d\alpha(x_2) \ldots d\alpha(x_n) \\
&\quad = \int \cdots \int_{K^{n-1}} d\alpha(x_2) \ldots d\alpha(x_n) \int_K f(x_1 + \sum_{i=2}^{n} c_i x_i, x_2, \ldots, x_n)\, d\alpha(x_1) \\
&\quad = \int \cdots \int_{K^{n-1}} d\alpha(x_2) \ldots d\alpha(x_n) \int_K f(x_1, x_2, \ldots, x_n)\, d\alpha(x_1) \\
&\quad = \int \cdots \int_{K^n} f(x_1, \ldots, x_n)\, d\alpha(x_1) \ldots d\alpha(x_n)\,,
\end{aligned}$$

and on the other hand $\text{mod}_K(\det u_3) = \text{mod}_K(1) = 1$, whence the result for $u_3$. It is established in an analogous manner for $u_1$ and $u_2$.

Let K be a commutative locally compact field, E a vector space of finite dimension $n$ over K. If $\varphi$ is an isomorphism of the vector space $K^n$ onto the vector space E, $\varphi$ transforms the topology of $K^n$ into a topology on E that makes E a locally compact vector space. This topology (called *canonical*) is independent of $\varphi$ since every automorphism of the vector space $K^n$ is bicontinuous. Absent express mention to the contrary, when we speak of E as a topological vector space, the topology will always be understood to be the one just defined. Every automorphism $u$ of the vector space E is bicontinuous, therefore $\operatorname{mod}_E u$ is defined. If, on the other hand, $u$ is a noninvertible endomorphism of E, one sets $\operatorname{mod}_E u = 0$. Then:

COROLLARY 1. — *Let* K *be a commutative locally compact field,* E *a finite-dimensional vector space over* K, *and* $u$ *an endomorphism of the vector space* E. *Then* $\operatorname{mod}_E(u) = \operatorname{mod}_K(\det u)$.

If $u$ is invertible, this follows from Prop. 15. If $u$ is not invertible then $\det u = 0$, therefore $\operatorname{mod}_K(\det u) = 0 = \operatorname{mod}_E u$.

COROLLARY 2. — *Let* E *be a real vector space of finite dimension* $n$, $(e_1, e_2, \ldots, e_n)$ *a basis of* E, P *the set of* $x = \sum_{i=1}^{n} \xi_i e_i \in E$ *such that* $0 \leqslant \xi_i \leqslant 1$ *for all* $i$, *and* $\mu$ *the unique Haar measure on the additive group* E *such that* $\mu(P) = 1$. *Let* $x_1, \ldots, x_n$ *be points of* E, S *the closed convex envelope in* E *of the set* $\{0, x_1, \ldots, x_n\}$. *Writing* $x_i = \sum_{j=1}^{n} \alpha_{ij} e_j$, *one has*

$$\mu(S) = \mu(\mathring{S}) = \frac{1}{n!} \,|\det(\alpha_{ij})|\,.$$

We shall identify E with $\mathbf{R}^n$ by means of the isomorphism that transforms $(e_i)$ into the canonical basis of $\mathbf{R}^n$. Then $\mu$ is identified with the Lebesgue measure $\mu_n$ on $\mathbf{R}^n$.

Suppose first that $x_i = e_i$ for all $i$. Then S is the set $S_n$ of $x = (\xi_i)$ in $\mathbf{R}^n$ such that

$$\xi_i \geqslant 0 \text{ for all } i \quad \text{ and } \quad \xi_1 + \cdots + \xi_n \leqslant 1\,.$$

Set $\mu_n(S_n) = a_n$. Let $\lambda \in \mathbf{R}$. Identifying $\mathbf{R}^n$ with $\mathbf{R}^{n-1} \times \mathbf{R}$, we may consider the section $C_\lambda$ of $S_n$ at $\lambda$. This section is empty if $\lambda < 0$ or $\lambda > 1$; if $0 \leqslant \lambda \leqslant 1$, $C_\lambda$ is the set of $(\xi_1, \ldots, \xi_{n-1}) \in \mathbf{R}^{n-1}$ such that

$$\xi_1 \geqslant 0, \ldots, \xi_{n-1} \geqslant 0\,, \quad \xi_1 + \cdots + \xi_{n-1} \leqslant 1 - \lambda\,,$$

hence may be deduced from $S_{n-1}$ by a homothety of ratio $1 - \lambda$, so that $\mu_{n-1}(C_\lambda) = (1-\lambda)^{n-1} a_{n-1}$. By the Lebesgue–Fubini theorem,

$$a_n = \int_0^1 (1-\lambda)^{n-1} a_{n-1}\, d\lambda = \frac{1}{n}\, a_{n-1}\,.$$

Since $a_1 = 1$, one sees that $a_n = \frac{1}{n!}$.

Let us return to the general case of the corollary. Let $u$ be the endomorphism of $\mathbf{R}^n$ such that $u(e_i) = x_i$ for all $i$. One has $u(\mathrm{S}_n) = \mathrm{S}$. If $u$ is invertible, Prop. 15 proves that

$$\mu_n(\mathrm{S}) = \frac{1}{n!}\,|\det u| = \frac{1}{n!}\,|\det(\alpha_{ij})|.$$

Since $\mathrm{S} - \mathring{\mathrm{S}}$ is contained in a finite number of hyperplanes, $\mu(\mathring{\mathrm{S}}) = \mu(\mathrm{S})$. Finally, if $u$ is not invertible, then S is contained in a hyperplane, so that $\mu(\mathrm{S}) = 0 = \det(\alpha_{ij})$.

## 11. Finite-dimensional algebras over a locally compact field

Let K be a commutative field, A a K-algebra of finite rank with unity element. For every $a \in \mathrm{A}$, let $L_a, R_a$ be the endomorphisms $x \mapsto ax$, $x \mapsto xa$ of the vector space A, and let $\mathrm{N}_{\mathrm{A/K}}(a) \in \mathrm{K}$, $\mathrm{N}_{\mathrm{A^0/K}}(a) \in \mathrm{K}$ be the norms of $a$ in the regular representations of A and the opposite algebra $\mathrm{A}^0$; recall that $\mathrm{N}_{\mathrm{A/K}}(a) = \det(L_a)$, $\mathrm{N}_{\mathrm{A^0/K}}(a) = \det(R_a)$ (A, III, §9, No. 3). The following conditions are equivalent: $a$ invertible, $L_a$ invertible in $\mathrm{Hom_K(A, A)}$, $R_a$ invertible in $\mathrm{Hom_K(A, A)}$, $\mathrm{N}_{\mathrm{A/K}}(a) \neq 0$, $\mathrm{N}_{\mathrm{A^0/K}}(a) \neq 0$. We denote by $\mathrm{A}^*$ the set of invertible elements of A.

Now assume the field K to be locally compact, hence the algebra A to be locally compact. Then $\mathrm{N}_{\mathrm{A/K}}$ and $\mathrm{N}_{\mathrm{A^0/K}}$ are continuous mappings of A into K, therefore $\mathrm{A}^*$ is *open* in A. By Cor. 1 of Prop. 15 of No. 10,

$$(38)\qquad \operatorname{mod}_{\mathrm{A}} L_a = \operatorname{mod}_{\mathrm{K}} \mathrm{N}_{\mathrm{A/K}}(a), \qquad \operatorname{mod}_{\mathrm{A}} R_a = \operatorname{mod}_{\mathrm{K}} \mathrm{N}_{\mathrm{A^0/K}}(a).$$

PROPOSITION 16. — *Let $\alpha$ be a Haar measure of the additive group of* A. *The measures*

$$\big(\operatorname{mod}_{\mathrm{K}} \mathrm{N}_{\mathrm{A/K}}(a)\big)^{-1} d\alpha(a), \qquad \big(\operatorname{mod}_{\mathrm{K}} \mathrm{N}_{\mathrm{A^0/K}}(a)\big)^{-1} d\alpha(a)$$

*on* $\mathrm{A}^*$ *are, respectively, left and right Haar measures of the multiplicative group* $\mathrm{A}^*$.

Let $\alpha'$ be the restriction of $\alpha$ to the open set $\mathrm{A}^*$. For $a \in \mathrm{A}^*$, one has $L_a(\alpha') = \big(\operatorname{mod}_{\mathrm{K}} \mathrm{N}_{\mathrm{A/K}}(a)\big)^{-1}\alpha'$, therefore $\big(\operatorname{mod}_{\mathrm{K}} \mathrm{N}_{\mathrm{A/K}}(a)\big)^{-1} d\alpha'(a)$ is a left Haar measure on $\mathrm{A}^*$ (No. 8, Cor. 1 of Prop. 10). Passing to the opposite algebra, one sees that $\big(\operatorname{mod}_{\mathrm{K}} \mathrm{N}_{\mathrm{A^0/K}}(a)\big)^{-1} d\alpha'(a)$ is a right Haar measure on $\mathrm{A}^*$.

PROPOSITION 17. — *Suppose* A *is a field (locally compact). For every* $a \in \mathrm{A}$, $\mathrm{mod}_{\mathrm{A}}(a) = \mathrm{mod}_{\mathrm{K}} \mathrm{N}_{\mathrm{A}/\mathrm{K}}(a)$.

This is a translation of the first formula of (38).

*Examples.* — 1) Take $\mathrm{K} = \mathbf{R}$, $\mathrm{A} = \mathbf{C}$. Taking into account *Alg.*, chap. VIII, §12, n° 2, prop. 4, we obtain $\mathrm{mod}_{\mathbf{C}}(z) = |z|^2$ for all $z \in \mathbf{C}$.[2]

2) Take $\mathrm{K} = \mathbf{R}$, and for A the *quaternion field* $\mathbf{H}$ (GT, VIII, §1, No. 4). Consider the following elements of $\mathbf{M}_2(\mathbf{C})$:

$$X_1 = \begin{pmatrix} 0 & i \\ i & 0 \end{pmatrix} \quad X_2 = \begin{pmatrix} 0 & -1 \\ 1 & 0 \end{pmatrix} \quad X_3 = \begin{pmatrix} i & 0 \\ 0 & -i \end{pmatrix}$$

which, together with $I_2$, form a basis of $\mathbf{M}_2(\mathbf{C})$ over $\mathbf{C}$. One verifies easily that

$$X_1^2 = X_2^2 = X_3^2 = -I_2, \qquad X_1X_2 = -X_2X_1 = X_3,$$
$$X_2X_3 = -X_3X_2 = X_1, \qquad X_3X_1 = -X_1X_3 = X_2.$$

The mapping $a + bi + cj + dk \mapsto aI_2 + bX_1 + cX_2 + dX_3$ may therefore be extended to a $\mathbf{C}$-isomorphism of the algebra $\mathbf{C} \otimes_{\mathbf{R}} \mathbf{H}$ onto the algebra $\mathbf{M}_2(\mathbf{C})$. Since $[\mathbf{H} : \mathbf{R}] = 4$, $\mathbf{C}$ is a neutralizing field of $\mathbf{H}$ (*Alg.*, chap. VIII, §10, n° 5), and the reduced norm of $q = a + bi + cj + dk \in \mathbf{H}$ is

$$\begin{aligned} \mathrm{Nrd}(q) &= \det(aI_2 + bX_1 + cX_2 + dX_3) \\ &= \det \begin{pmatrix} a + id & -c + ib \\ c + ib & a - id \end{pmatrix} = a^2 + b^2 + c^2 + d^2 = \|q\|^2 . \end{aligned}$$

By *Alg.*, chap. VIII, §12, n° 3, prop. 8, one has

$$\mathrm{N}_{\mathbf{H}/\mathbf{R}}(q) = \big(\mathrm{Nrd}_{\mathbf{H}/\mathbf{R}}(q)\big)^2 = \|q\|^4 .$$

This stated, Prop. 17 shows that

$$\mathrm{mod}_{\mathbf{H}}(q) = \|q\|^4 .$$

A deeper study of the structure of locally compact fields will be made in CA, VI, §9.

---

(2) This also follows from Cor. 1 of Prop. 15 and the fact that left-multiplication by $z = a + ib$ has matrix $\begin{pmatrix} a & -b \\ b & a \end{pmatrix}$ with respect to the canonical basis $1, i$ of $\mathbf{C}$ over $\mathbf{R}$.

## §2. QUOTIENT OF A SPACE BY A GROUP; HOMOGENEOUS SPACES

### 1. General results

Let X be a locally compact space in which a locally compact group H operates on the right, continuously and *properly*, by $(x, \xi) \mapsto x\xi$ $(x \in \mathrm{X}, \xi \in \mathrm{H})$. The equivalence relation in X defined by H is open (GT, III, §2, No. 4, Lemma 2) and X/H is Hausdorff (*loc. cit.*, §4, No. 2, Prop. 3) hence locally compact (GT, I, §10, No. 4, Prop. 10). We denote by $\pi$ the canonical mapping of X onto X/H. The saturation of a subset Y of X is $\mathrm{YH} = \overset{-1}{\pi}(\pi(\mathrm{Y}))$. If K is a compact subset of X, then $\pi(\mathrm{K})$ is compact and the saturation $\overset{-1}{\pi}(\pi(\mathrm{K}))$ of K is closed in X. Every compact subset of X/H is the image under $\pi$ of a compact subset of X (GT, I, §10, No. 4, Prop. 10). *We assume given once and for all a left Haar measure* $\beta$ *on* H.

Let $\chi$ be a continuous representation of H in $\mathbf{R}_+^*$. If a function $g$ on X satisfies $g(x\xi) = \chi(\xi)g(x)$ for all $x \in \mathrm{X}$ and $\xi \in \mathrm{H}$, its support S is invariant under H hence may be written $\overset{-1}{\pi}(\pi(\mathrm{S}))$. We shall denote by $\mathscr{K}^\chi(\mathrm{X})$ the Riesz space formed by the continuous real-valued functions $g$ on X that satisfy $g(x\xi) = \chi(\xi)g(x)$ $(x \in \mathrm{X}, \xi \in \mathrm{H})$ and whose support is the saturation of a compact subset of X; we denote by $\mathscr{K}_+^\chi(\mathrm{X})$ the set of elements $\geqslant 0$ of $\mathscr{K}^\chi(\mathrm{X})$. In particular, $\mathscr{K}^1(\mathrm{X})$ is none other than the set of continuous functions on X, constant on the orbits, whose support is the saturation of a compact subset.

PROPOSITION 1. — *Let* $f$ *be a continuous real-valued function on* X *whose support* S *has compact intersection with the saturation of every compact subset of* X.

a) *For every* $x \in \mathrm{X}$, *the function* $\xi \mapsto f(x\xi)$ *on* H *belongs to* $\mathscr{K}(\mathrm{H})$; *one sets*

$$f^\chi(x) = \int_{\mathrm{H}} f(x\xi)\chi(\xi)^{-1}\,d\beta(\xi). \tag{1}$$

b) *The function* $f^\chi$ *is continuous, is zero outside* SH, *and satisfies* $f^\chi(x\xi) = \chi(\xi)f^\chi(x)$.

c) *If* $g$ *is a continuous real-valued function on* X *and satisfies* $g(x\xi) = \chi(\xi)g(x)$, *then* $(fg)^\chi = f^1 g$ ($f^1$ being given by the formula (1) with $\chi$ replaced by the representation $\xi \mapsto 1$ of H in $\mathbf{R}_+^*$).

d) *If* $\eta \in \mathrm{H}$, *then* $(\boldsymbol{\delta}(\eta)f)^{\chi} = \chi(\eta)\Delta_{\mathrm{H}}(\eta)^{-1}f^{\chi}$.

Let $x_0 \in \mathrm{X}$ and let V be a compact neighborhood of $x_0$ in X. The set of $\xi \in \mathrm{H}$ such that $\mathrm{V}\xi$ intersects S is also the set of $\xi \in \mathrm{H}$ such that $\mathrm{V}\xi$ intersects $\mathrm{S} \cap \mathrm{VH}$, hence is compact in H since $\mathrm{S} \cap \mathrm{VH}$ is compact and H operates properly in X (GT, III, §4, No. 5, Th. 1); then Lemma 1 of §1, No. 1 proves a) and the continuity of $f^{\chi}$. The rest of b) is obvious. Finally, c) and d) result from the following calculations:

$$(fg)^{\chi}(x) = \int_{\mathrm{H}} f(x\xi)g(x\xi)\chi(\xi)^{-1}\,d\beta(\xi) = \int_{\mathrm{H}} f(x\xi)g(x)\chi(\xi)\chi(\xi)^{-1}\,d\beta(\xi)$$
$$= g(x)\int_{\mathrm{H}} f(x\xi)\,d\beta(\xi) = g(x)f^{1}(x)$$

$$(\boldsymbol{\delta}(\eta)f)^{\chi}(x) = \int_{\mathrm{H}} f(x\xi\eta)\chi(\xi)^{-1}\,d\beta(\xi)$$
$$= \Delta_{\mathrm{H}}(\eta)^{-1}\int_{\mathrm{H}} f(x\xi)\chi(\xi\eta^{-1})^{-1}\,d\beta(\xi)$$
$$= \chi(\eta)\Delta_{\mathrm{H}}(\eta)^{-1}\int_{\mathrm{H}} f(x\xi)\chi(\xi)^{-1}\,d\beta(\xi)\,.$$

PROPOSITION 2. — *The mapping* $f \mapsto f^{\chi}$ *of* $\mathscr{K}(\mathrm{X})$ *into* $\mathscr{K}^{\chi}(\mathrm{X})$ *is linear, and the image of* $\mathscr{K}(\mathrm{X})$ (resp. $\mathscr{K}_{+}(\mathrm{X})$) *is* $\mathscr{K}^{\chi}(\mathrm{X})$ (resp. $\mathscr{K}_{+}^{\chi}(\mathrm{X})$).

Linearity is immediate. It is clear that $f^{\chi} \geqslant 0$ for $f \geqslant 0$. It then suffices to apply the following lemma:

*Lemma* 1. — *Let* K *be a compact subset of* X, $u$ *a function of* $\mathscr{K}_{+}(\mathrm{X})$ *with* $u(x) > 0$ *for* $x \in \mathrm{K}$. *Let* $g \in \mathscr{K}^{\chi}(\mathrm{X})$ *be such that* $\mathrm{Supp}\, g \subset \mathrm{KH}$.

a) *One has* $\inf_{x \in \mathrm{KH}} u^{1}(x) > 0$.

b) *The function* $h$ *equal to* $g/u^{1}$ *on* KH, *and to* 0 *on* X − KH, *belongs to* $\mathscr{K}^{\chi}(\mathrm{X})$.

c) $g = (uh)^{\chi}$.

One has $u^{1}(x) > 0$ for $x \in \mathrm{K}$, therefore $\inf_{x \in \mathrm{KH}} u^{1}(x) = \inf_{x \in \mathrm{K}} u^{1}(x) > 0$. Assertion b) follows from this at once. Finally, $(uh)^{\chi} = u^{1}h$ by Prop. 1 c), and it is clear that $u^{1}h = g$.

Let I be a relatively bounded linear form (Ch. II, §2, No. 2) on $\mathscr{K}^{\chi}(\mathrm{X})$. Then $f \mapsto \mathrm{I}(f^{\chi})$ is a relatively bounded linear form on $\mathscr{K}(\mathrm{X})$, that is, a *measure* $\mu_{\mathrm{I}}$ on X. The mapping $\mathrm{I} \mapsto \mu_{\mathrm{I}}$ is injective by Prop. 2. The measures $\mu_{\mathrm{I}}$ on X so obtained may be characterized as follows:

PROPOSITION 3. — *Let* $\mu$ *be a measure on* X. *The following conditions are equivalent:*

a) *There exists a relatively bounded linear form* I *on* $\mathscr{K}^{\chi}(\mathrm{X})$ *such that* $\mathrm{I}(f^{\chi}) = \mu(f)$ *for all* $f \in \mathscr{K}(\mathrm{X})$.

b) $\boldsymbol{\delta}(\xi)\mu = \chi(\xi)^{-1}\Delta_{\mathrm{H}}(\xi)\mu$ *for all* $\xi \in \mathrm{H}$.

c) *For all* $f, g$ *in* $\mathscr{K}(\mathrm{X})$,

$$\mu(f \cdot g^1) = \mu(f^{\chi} \cdot g). \tag{2}$$

d) *If* $f \in \mathscr{K}(\mathrm{X})$ *is such that* $f^{\chi} = 0$, *then* $\mu(f) = 0$.

a) $\Rightarrow$ b): If $\mu(f) = \mathrm{I}(f^{\chi})$ then, taking into account Prop. 1 d),

$$\begin{aligned}\langle \boldsymbol{\delta}(\xi)\mu, f\rangle = \langle \mu, \boldsymbol{\delta}(\xi^{-1})f\rangle &= \mathrm{I}\Big(\big(\boldsymbol{\delta}(\xi^{-1})f\big)^{\chi}\Big)\\ &= \mathrm{I}\big(\chi(\xi)^{-1}\Delta_{\mathrm{H}}(\xi)f^{\chi}\big)\\ &= \chi(\xi)^{-1}\Delta_{\mathrm{H}}(\xi)\langle \mu, f\rangle,\end{aligned}$$

whence $\boldsymbol{\delta}(\xi)\mu = \chi(\xi)^{-1}\Delta_{\mathrm{H}}(\xi)\mu$.

b) $\Rightarrow$ c): Suppose hypothesis b) is satisfied. Note that the functions $(x, \xi) \mapsto f(x)g(x\xi)$ and $(x, \xi) \mapsto f(x\xi)g(x)$ on $\mathrm{X} \times \mathrm{H}$ are continuous with compact support (because H operates properly in X); this established, Th. 2 of Ch. III, §4, No. 1 permits us to write:

$$\begin{aligned}\int_{\mathrm{X}} f(x)\,d\mu(x) \int_{\mathrm{H}} g(x\xi)\,d\beta(\xi) &= \int_{\mathrm{H}} d\beta(\xi) \int_{\mathrm{X}} f(x)g(x\xi)\,d\mu(x)\\ &= \int_{\mathrm{H}} d\beta(\xi) \int_{\mathrm{X}} f(x\xi^{-1})g(x)\chi(\xi)\Delta_{\mathrm{H}}(\xi)^{-1}\,d\mu(x)\\ &= \int_{\mathrm{X}} g(x)\,d\mu(x) \int_{\mathrm{H}} f(x\xi^{-1})\chi(\xi)\Delta_{\mathrm{H}}(\xi)^{-1}\,d\beta(\xi)\\ &= \int_{\mathrm{X}} g(x)\,d\mu(x) \int_{\mathrm{H}} f(x\xi)\chi(\xi)^{-1}\,d\beta(\xi),\end{aligned}$$

which proves c).

c) $\Rightarrow$ d): If c) is verified and if $f^{\chi} = 0$, then $\mu(f \cdot g^1) = 0$ for all $g \in \mathscr{K}(\mathrm{X})$, thus $\mu(f) = 0$ on choosing $g \in \mathscr{K}(\mathrm{X})$ such that $g^1 = 1$ on $\operatorname{Supp} f$ (which is possible by Prop. 2 applied with $\chi = 1$).

d) $\Rightarrow$ a): If condition d) is satisfied, there exists a linear form I on $\mathscr{K}^{\chi}(\mathrm{X})$ such that $\mu(f) = \mathrm{I}(f^{\chi})$ for $f \in \mathscr{K}(\mathrm{X})$, and this form is relatively bounded by virtue of Prop. 2.

### 2. The case $\chi = 1$

If $f$ is a function on X/H, then $f \circ \pi$ is a function on X constant on the orbits, continuous if and only if $f$ is continuous. The mapping $f \mapsto f \circ \pi$ defines in particular a *bijection* of $\mathscr{K}(\mathrm{X/H})$ onto $\mathscr{K}^1(\mathrm{X})$.

We can then, in the case that $\chi = 1$, reformulate certain results of No. 1 in the following way:

Let $f$ be a continuous numerical function on X whose support has compact intersection with the saturation of every compact subset of X. The formula

$$f^\flat(\pi(x)) = \int_{\mathrm{H}} f(x\xi)\, d\beta(\xi) \tag{3}$$

defines a continuous function $f^\flat$ on X/H. If $g$ is a continuous function on X/H, then

$$(f \cdot g \circ \pi)^\flat = f^\flat \cdot g. \tag{4}$$

If $\eta \in \mathrm{H}$, then

$$(\boldsymbol{\delta}(\eta) f)^\flat = \Delta_{\mathrm{H}}(\eta)^{-1} f^\flat. \tag{5}$$

One must not forget that the definition of $f^\flat$ depends on the choice of $\beta$. If H is compact and $\beta$ is normalized, the function $f^\flat$ is sometimes called the *orbital mean* of $f$.

If $f \in \mathscr{K}(\mathrm{X})$, then $f^\flat \in \mathscr{K}(\mathrm{X/H})$. The mapping $f \mapsto f^\flat$ of $\mathscr{K}(\mathrm{X})$ into $\mathscr{K}(\mathrm{X/H})$ is linear, and the image of $\mathscr{K}(\mathrm{X})$ (resp. $\mathscr{K}_+(\mathrm{X})$) is $\mathscr{K}(\mathrm{X/H})$ (resp. $\mathscr{K}_+(\mathrm{X/H})$).

*Remark* 1. — We are going to show that the mapping $f \mapsto f^\flat$ is a *strict morphism* (GT, III, §2, No. 8) of $\mathscr{K}(\mathrm{X})$ onto $\mathscr{K}(\mathrm{X/H})$.

*a*) The mapping is continuous: it suffices to prove that, for every compact subset K of X, the restriction of $f \mapsto f^\flat$ to $\mathscr{K}(\mathrm{X, K})$ is a continuous mapping of $\mathscr{K}(\mathrm{X, K})$ into $\mathscr{K}(\mathrm{X/H}, \pi(\mathrm{K}))$ (TVS, II, §4, No. 4, Prop. 5); since H operates properly in X, the set P of $\xi \in \mathrm{H}$ such that $\mathrm{K}\xi$ intersects K is compact; one concludes from (3) that $\sup_{x\in\mathrm{K}} |f^\flat(\pi(x))| \leqslant \beta(\mathrm{P}) \sup_{x\in\mathrm{K}} |f(x)|$, and this proves our assertion.

*b*) Let K′ be a compact subset of X/H. Let us choose a compact subset K of X such that $\pi(\mathrm{K}) = \mathrm{K}'$, and let us show that the restriction of $f \mapsto f^\flat$ to $\mathscr{K}(\mathrm{X, K})$ is a strict morphism of $\mathscr{K}(\mathrm{X, K})$ onto $\mathscr{K}(\mathrm{X/H, K'})$. It suffices to construct a right inverse for this restriction (GT, III, §6, No. 2, Prop. 3). Now, by Lemma 1 of No. 1 (whose notations we adopt), one obtains such an inverse by composing the following mappings:

$\alpha$) the mapping $f' \mapsto f' \circ \pi$ of $\mathscr{K}(\mathrm{X/H, K'})$ into the set E of functions of $\mathscr{K}^1(\mathrm{X})$ whose support is contained in KH;

$\beta$) the mapping of E into E that, to every $g \in \mathrm{E}$, makes correspond the function equal to $g/u^1$ on KH, and to 0 on X − KH;

$\gamma$) the mapping of E into $\mathscr{K}(\mathrm{X})$ that, to every function $h \in \mathrm{E}$, makes correspond $uh$.

*c*) This established, if V is a convex neighborhood of 0 in $\mathscr{K}(\mathrm{X})$, then $\mathrm{V} \cap \mathscr{K}(\mathrm{X, K})$ is a convex neighborhood of 0 in $\mathscr{K}(\mathrm{X, K})$, therefore $\mathrm{V}^\flat \cap \mathscr{K}(\mathrm{X/H, K'})$

is a convex neighborhood of 0 in $\mathscr{K}(\mathrm{X/H}, \mathrm{K}')$ by *b*), therefore $\mathrm{V}^\flat$ is a neighborhood of 0 in $\mathscr{K}(\mathrm{X/H})$ (TVS, II, §4, No. 4). This completes the proof.

PROPOSITION 4. — a) *Let $\lambda$ be a measure on* X/H. *There exists one and only one measure $\lambda^\sharp$ on* X *such that*

$$\int_{\mathrm{X/H}} f^\flat \, d\lambda = \int_{\mathrm{X}} f \, d\lambda^\sharp \tag{6}$$

*for all $f \in \mathscr{K}(\mathrm{X})$. One has $\boldsymbol{\delta}(\xi)\lambda^\sharp = \Delta_{\mathrm{H}}(\xi)\lambda^\sharp$ for all $\xi \in \mathrm{H}$.*

b) *Conversely, let $\mu$ be a measure on* X *such that $\boldsymbol{\delta}(\xi)\mu = \Delta_{\mathrm{H}}(\xi)\mu$ for all $\xi \in \mathrm{H}$. There exists one and only one measure $\lambda$ on* X/H *such that $\mu = \lambda^\sharp$.*

This is a special case of No. 1, Prop. 3.

DEFINITION 1. — *With hypotheses and notations as in Prop. 4, $\lambda$ is called the quotient of $\mu$ by $\beta$ and is denoted $\dfrac{\mu}{\beta}$ or $\mu/\beta$.*

The mapping $\lambda \mapsto \lambda^\sharp$ of $\mathscr{M}(\mathrm{X/H})$ into $\mathscr{M}(\mathrm{X})$ is none other than the *transpose* of the mapping $f \mapsto f^\flat$ of $\mathscr{K}(\mathrm{X})$ into $\mathscr{K}(\mathrm{X/H})$. Let $\mathfrak{F}$ be a filter on $\mathscr{M}(\mathrm{X/H})$; to say that $\lim_{\lambda,\mathfrak{F}} \lambda^\sharp(f) = 0$ for all $f \in \mathscr{K}(\mathrm{X})$ is equivalent to saying that $\lim_{\lambda,\mathfrak{F}} \lambda(f') = 0$ for all $f' \in \mathscr{K}(\mathrm{X/H})$; therefore the mapping $\lambda \mapsto \lambda^\sharp$ is, for the vague topologies, an *isomorphism* of $\mathscr{M}(\mathrm{X/H})$ onto a linear subspace of $\mathscr{M}(\mathrm{X})$. This subspace is *vaguely closed*, since it is the set of $\mu \in \mathscr{M}(\mathrm{X})$ such that $\boldsymbol{\delta}(\xi)\mu = \Delta_{\mathrm{H}}(\xi)\mu$ for all $\xi \in \mathrm{H}$. It is clear that the conditions $\lambda \geqslant 0$ and $\lambda^\sharp \geqslant 0$ are equivalent.

The formula (6) may, by analogy with the usual notation for double integrals, be written

$$\int_{\mathrm{X}} f(x) \, d\lambda^\sharp(x) = \int_{\mathrm{X/H}} d\lambda(\dot{x}) \int_{\mathrm{H}} f(x\xi) \, d\beta(\xi) \qquad (\dot{x} = \pi(x)). \tag{7}$$

This involves an abuse of notation, the integral $\int_{\mathrm{H}} f(x\xi)\, d\beta(\xi)$ being regarded as a function of $\dot{x}$ and not of $x$; this manner of writing will be used frequently in what follows provided no confusion can arise.

*Remark* 2. — Let E be a locally convex vector space and let $\mathbf{m}$ be a vectorial measure on X/H with values in E. The mapping $f \mapsto \mathbf{m}(f^\flat)$ of $\mathscr{K}(\mathrm{X})$ into E is then a vectorial measure on X, with values in E, which we shall again denote by $\mathbf{m}^\sharp$. The mapping $\mathbf{m} \mapsto \mathbf{m}^\sharp$ is again an *isomorphism* of $\mathscr{L}(\mathscr{K}(\mathrm{X/H}); \mathrm{E})$ onto a linear subspace A of $\mathscr{L}(\mathscr{K}(\mathrm{X}); \mathrm{E})$ (when these spaces are equipped with the topology of pointwise convergence). Moreover, since the mapping $f \mapsto f^\flat$ is a surjective strict morphism, the subspace A consists precisely of the vectorial measures $\mathbf{n}$ on X that are zero on the kernel N of the mapping $f \mapsto f^\flat$. In order that $\mathbf{n} \in \mathrm{A}$, it is therefore necessary and sufficient that the scalar measures $z' \circ \mathbf{n}$ be zero on N for every $z' \in \mathrm{E}'$. One then deduces from Prop. 3 that $\mathbf{n} \in \mathrm{A}$ if and only if $\boldsymbol{\delta}(\xi)\mathbf{n} = \Delta_{\mathrm{H}}(\xi)\mathbf{n}$ for all $\xi \in \mathrm{H}$.

### 3. Another interpretation of $\lambda^\sharp$

For every $x \in X$, the mapping $\xi \mapsto x\xi$ of H into X is *proper* (GT, III, §4, No. 2, Prop. 4), therefore $\beta$ admits an image measure on X under this mapping, which image is concentrated on the orbit $x$H (Ch. V, §6, No. 2, Cor. 3 of Prop. 2); since $\beta$ is left-invariant, this image measure depends only on the class $u = \pi(x)$ of $x$ in X/H, and will be denoted $\beta_u$. By definition, for $f \in \mathscr{K}(X)$,

$$\int_X f(y)\, d\beta_u(y) = \int_H f(x\xi)\, d\beta(\xi) = f^\flat(u). \tag{8}$$

We thus see that

$$(\varepsilon_u)^\sharp = \beta_u. \tag{9}$$

*Lemma* 2. — *Let $f$ be a function on* X, *with values in a topological space.*

a) *If $f$ is a numerical function $\geqslant 0$ then, for $x \in X$,*

$$\int_X^* f(y)\, d\beta_{\dot{x}}(y) = \int_H^* f(x\xi)\, d\beta(\xi) \qquad (\dot{x} = \pi(x)).$$

b) *For $f$ to be $\beta_{\dot{x}}$-measurable, it is necessary and sufficient that the function $\xi \mapsto f(x\xi)$ on* H *be $\beta$-measurable.*

c) *Suppose that* $\mathbf{f}$ *is a function on* X, *with values in a Banach space or in $\overline{\mathbf{R}}$; then, for* $\mathbf{f}$ *to be $\beta_{\dot{x}}$-integrable* (resp. *essentially $\beta_{\dot{x}}$-integrable*), *it is necessary and sufficient that the function $\xi \mapsto \mathbf{f}(x\xi)$ on* H *be $\beta$-integrable* (resp. *essentially $\beta$-integrable*), *in which case* $\int_X \mathbf{f}(y)\, d\beta_{\dot{x}}(y) = \int_H \mathbf{f}(x\xi)\, d\beta(\xi)$.

This follows from Ch. V, §4, Prop. 2, Prop. 3 and Th. 2.

Since $f^\flat \in \mathscr{K}(X/H)$ for $f \in \mathscr{K}(X)$, formula (8) proves that the mapping $u \mapsto \beta_u$ of X/H into $\mathscr{M}(X)$ is vaguely continuous, that the family $(\beta_u)$ is $\lambda$-adequate[1] for any positive measure $\lambda$ on X/H, and that

$$\lambda^\sharp = \int_{X/H} \beta_u\, d\lambda(u), \tag{10}$$

which furnishes a new interpretation of $\lambda^\sharp$.

[1] In the sense of the first edition of Ch. V, hence *a fortiori* in the sense of the second (cf. the footnote to the Example of Ch. VI, §1, No. 1).

PROPOSITION 5. — *Let* $\lambda$ *be a positive measure on* $X/H$.

a) *Let* $f$ *be a* $\lambda^\sharp$*-measurable function on* $X$, *with values in a topological space, constant outside a countable union of* $\lambda^\sharp$*-integrable sets. Then, the set of* $\dot{x} \in X/H$ *such that the function* $\xi \mapsto f(x\xi)$ *is not* $\beta$*-measurable is locally* $\lambda$*-negligible.*

b) *Let* $f$ *be a* $\lambda^\sharp$*-measurable function* $\geqslant 0$ *on* $X$, *zero outside a countable union of* $\lambda^\sharp$*-integrable sets. Then, the function* $\dot{x} \mapsto \int^* f(x\xi)\, d\beta(\xi)$ *on* $X/H$ *is* $\lambda$*-measurable, and*

$$\int_X^* f(x)\, d\lambda^\sharp(x) = \int_{X/H}^* d\lambda(\dot{x}) \int_H^* f(x\xi)\, d\beta(\xi) \qquad (\dot{x} = \pi(x)).$$

c) *Let* $\mathbf{f}$ *be a* $\lambda^\sharp$*-integrable function on* $X$, *with values in a Banach space or in* $\overline{\mathbf{R}}$. *Then, the set of* $\dot{x} \in X/H$ *such that* $\xi \mapsto \mathbf{f}(x\xi)$ *is not* $\beta$*-integrable is* $\lambda$*-negligible; the function* $\mathbf{f}^\flat$ *on* $X/H$ *defined almost everywhere by the formula*

$$\mathbf{f}^\flat(\dot{x}) = \int_H \mathbf{f}(x\xi)\, d\beta(\xi) \qquad (\dot{x} = \pi(x)) \tag{11}$$

*is* $\lambda$*-integrable, and*

$$\int_{X/H} \mathbf{f}^\flat\, d\lambda = \int_X \mathbf{f}\, d\lambda^\sharp \tag{12}$$

*and*

$$\int_{X/H} |\mathbf{f}^\flat|\, d\lambda \leqslant \int_X |\mathbf{f}|\, d\lambda^\sharp. \tag{13}$$

d) *Let* $\mathbf{f}$ *be a* $\lambda^\sharp$*-measurable function on* $X$, *with values in a Banach space or in* $\overline{\mathbf{R}}$, *and zero outside a countable union of* $\lambda^\sharp$*-integrable sets. Then, for* $\mathbf{f}$ *to be* $\lambda^\sharp$*-integrable, it is necessary and sufficient that*

$$\int_{X/H}^* d\lambda(\dot{x}) \int_H^* |\mathbf{f}(x\xi)|\, d\beta(\xi) < +\infty \qquad (\dot{x} = \pi(x)).$$

Taking into account Lemma 2, the assertions a), b) and c) follow from Ch. V, §3, Prop. 4, Prop. 5 and Th. 1 (with the exception of (13), which follows from (12) because it is clear that $|\mathbf{f}^\flat| \leqslant |\mathbf{f}|^\flat$); d) follows from b).

PROPOSITION 6. — *Let* $\lambda$ *be a positive measure on* $X/H$.

a) *Let* $N$ *be a subset of* $X/H$. *For* $N$ *to be locally* $\lambda$*-negligible, it is necessary and sufficient that* $\overset{-1}{\pi}(N)$ *be locally* $\lambda^\sharp$*-negligible.*

b) *Let* $g$ *be a function on* $X/H$, *with values in a topological space. For* $g$ *to be* $\lambda$*-measurable, it is necessary and sufficient that* $g \circ \pi$ *be* $\lambda^\sharp$*-measurable.*

c) *Let* $\mathbf{h}$ *be a function on* $X/H$, *with values in a Banach space or in* $\overline{\mathbf{R}}$. *For* $\mathbf{h}$ *to be locally* $\lambda$*-integrable, it is necessary and sufficient that* $\mathbf{h} \circ \pi$ *be locally* $\lambda^\sharp$*-integrable, in which case* $(\mathbf{h} \cdot \lambda)^\sharp = (\mathbf{h} \circ \pi) \cdot \lambda^\sharp$.

Suppose $\mathbf{h} \circ \pi$ is locally $\lambda^\sharp$-integrable. For every $f \in \mathscr{K}(X)$, $f \cdot (\mathbf{h} \circ \pi)$ is $\lambda^\sharp$-integrable, therefore (Prop. 5) the function $\big(f \cdot (\mathbf{h} \circ \pi)\big)^\flat = f^\flat \cdot \mathbf{h}$ is $\lambda$-integrable and

$$\int_{X/H} f^\flat \cdot \mathbf{h}\, d\lambda = \int_X f \cdot (\mathbf{h} \circ \pi)\, d\lambda^\sharp .$$

Since $f \mapsto f^\flat$ is a surjective mapping of $\mathscr{K}(X)$ onto $\mathscr{K}(X/H)$, this shows that $\mathbf{h}$ is locally $\lambda$-integrable and that

$$(\mathbf{h} \cdot \lambda)^\sharp = (\mathbf{h} \circ \pi) \cdot \lambda^\sharp .$$

In particular, if $\overset{-1}{\pi}(N)$ is locally $\lambda^\sharp$-negligible, then $\varphi_N \circ \pi$ is locally $\lambda^\sharp$-negligible, therefore $(\varphi_N \cdot \lambda)^\sharp = (\varphi_N \circ \pi) \cdot \lambda^\sharp = 0$, consequently $\varphi_N \cdot \lambda = 0$ and $N$ is locally $\lambda$-negligible. Now suppose that $g \circ \pi$ is $\lambda^\sharp$-measurable. Let $K'$ be a compact subset of $X/H$. Let $f \in \mathscr{K}_+(X)$ be such that $f^\flat = 1$ on $K'$ (No. 1, Prop. 2), and let $K = \operatorname{Supp} f$; one has $\pi(K) \supset K'$. There exists a partition of $K$ formed of a $\lambda^\sharp$-negligible set $M$ and a sequence $(K_n)$ of compact sets such that $(g \circ \pi)\big| K_n$ is continuous for every $n$. Then $g\big|\pi(K_n)$ is continuous. Let $P$ be the set of points of $K'$ not belonging to $\pi(K_1) \cup \pi(K_2) \cup \ldots$; then $\overset{-1}{\pi}(P) \cap K$ is contained in $M$, hence is $\lambda^\sharp$-negligible; therefore $f \cdot \varphi_{\overset{-1}{\pi}(P)}$ is $\lambda^\sharp$-negligible; it follows (Prop. 5) that

$$0 = \int_X f \cdot \varphi_{\overset{-1}{\pi}(P)}\, d\lambda^\sharp = \int_{X/H} f^\flat \cdot \varphi_P\, d\lambda \geqslant \int_{X/H}^* \varphi_P\, d\lambda ,$$

therefore $P$ is $\lambda$-negligible, and $g$ is $\lambda$-measurable.

If $N$ is locally $\lambda$-negligible, then $\overset{-1}{\pi}(N)$ is locally $\lambda^\sharp$-negligible (Ch. V, §6, No. 6, Cor. 1 of Prop. 10). If $g$ is $\lambda$-measurable, then $g \circ \pi$ is $\lambda^\sharp$-measurable (*ibid.*). Finally, suppose $\mathbf{h}$ is locally $\lambda$-integrable. Then we already know that $\mathbf{h} \circ \pi$ is $\lambda^\sharp$-measurable. For every $f \in \mathscr{K}_+(X)$ we have, by Prop. 5,

$$\int_X^* f(x) |\mathbf{h}| \big(\pi(x)\big)\, d\lambda^\sharp(x) = \int_{X/H}^* |\mathbf{h}|(u) f^\flat(u)\, d\lambda(u) < +\infty ,$$

therefore $\mathbf{h} \circ \pi$ is locally $\lambda^\sharp$-integrable.

COROLLARY 1. — *Let* $\lambda, \lambda'$ *be two positive measures on* $X/H$. *For* $\lambda'$ *to have base* $\lambda$, *it is necessary and sufficient that* $\lambda'^{\sharp}$ *have base* $\lambda^{\sharp}$. *For* $\lambda$ *and* $\lambda'$ *to be equivalent, it is necessary and sufficient that* $\lambda^{\sharp}$ *and* $\lambda'^{\sharp}$ *be equivalent.*

The first assertion follows from Prop. 6, a) and c). The second follows from the first.

COROLLARY 2. — *Let* $\lambda$ *be a positive measure on* $X/H$, *and* $f$ *a* $\lambda^{\sharp}$*-measurable numerical function on* $X$. *Suppose that, for every* $\xi \in H$, $\boldsymbol{\delta}(\xi)f = f$ *locally* $\lambda^{\sharp}$*-almost everywhere. Then, there exists a* $\lambda$*-measurable function* $g$ *on* $X/H$ *such that* $f = g \circ \pi$ *locally* $\lambda^{\sharp}$*-almost everywhere.*

Replacing $f$ by $f/(1+|f|)$, one reduces to the case that $f$ is bounded, hence locally $\lambda^{\sharp}$-integrable. Let $\mu = f \cdot \lambda^{\sharp}$. The hypothesis on $f$ implies that $\boldsymbol{\delta}(\xi)\mu = f \cdot \boldsymbol{\delta}(\xi)\lambda^{\sharp} = \Delta_H(\xi)\mu$ for all $\xi \in H$. There then exists (Prop. 4) a measure $\lambda'$ on $X/H$ such that $\mu = \lambda'^{\sharp}$. By Cor. 1, there exists a locally $\lambda$-integrable function $g$ on $X/H$ such that $\lambda' = g \cdot \lambda$. By Prop. 6, $f \cdot \lambda^{\sharp} = \lambda'^{\sharp} = (g \circ \pi) \cdot \lambda^{\sharp}$, whence $f = g \circ \pi$ locally $\lambda^{\sharp}$-almost everywhere.

COROLLARY 3. — a) *Let* $(\lambda_\iota)_{\iota \in I}$ *be a family of real measures on* $X/H$. *For the family* $(\lambda_\iota)$ *to be bounded above in* $\mathscr{M}(X/H)$, *it is necessary and sufficient that the family* $(\lambda_\iota^{\sharp})$ *be bounded above in* $\mathscr{M}(X)$, *in which case*

$$\sup(\lambda_\iota^{\sharp}) = (\sup \lambda_\iota)^{\sharp}.$$

b) *Let* $\lambda$ *be a real measure on* $X/H$. *Then* $(\lambda^+)^{\sharp} = (\lambda^{\sharp})^+$ *and* $(\lambda^-)^{\sharp} = (\lambda^{\sharp})^-$.

c) *Let* $\lambda$ *be a complex measure on* $X/H$. *Then* $|\lambda|^{\sharp} = |\lambda^{\sharp}|$.

Assume the family $(\lambda_\iota)$ to be bounded above and let $\mu = \sup \lambda_\iota$. Since $\lambda \geqslant 0$ implies $\lambda^{\sharp} \geqslant 0$, we have $\mu^{\sharp} \geqslant \lambda_\iota^{\sharp}$ for all $\iota$, which shows that the family $(\lambda_\iota^{\sharp})$ is bounded above and that

$$(\sup \lambda_\iota)^{\sharp} \geqslant \sup(\lambda_\iota^{\sharp}).$$

Conversely, assume the family $(\lambda_\iota^{\sharp})$ to be bounded above and let $\nu = \sup(\lambda_\iota^{\sharp})$. Since $\boldsymbol{\delta}(\xi)\lambda_\iota^{\sharp} = \Delta_H(\xi)\lambda_\iota^{\sharp}$ for all $\xi \in H$, obviously $\boldsymbol{\delta}(\xi)\nu = \Delta_H(\xi)\nu$, therefore there exists a measure $\mu' \in \mathscr{M}(X/H)$ such that $\nu = \mu'^{\sharp}$. Since $\lambda^{\sharp} \geqslant 0$ implies $\lambda \geqslant 0$, we have $\mu' \geqslant \lambda_\iota$ for all $\iota$, which shows that the family $(\lambda_\iota)$ is bounded above and that $\nu = \mu'^{\sharp} \geqslant (\sup \lambda_\iota)^{\sharp}$, whence

$$\sup(\lambda_\iota^{\sharp}) \geqslant (\sup \lambda_\iota)^{\sharp},$$

which completes the proof of a). The assertion b) then follows at once since, for example, $\lambda^+$ is none other than $\sup(\lambda, 0)$. To prove c), it suffices to note that $|\lambda| = \sup \mathscr{R}(\alpha\lambda)$ over the complex numbers $\alpha$ of absolute value $1$, and on the other hand that $\mathscr{R}(\mu^\sharp) = (\mathscr{R}\mu)^\sharp$ for every $\mu \in \mathscr{M}(X/H)$.

*Remarks.* — 1) Prop. 6 a) may be expressed by saying that $\lambda$ is a *pseudo-image* measure of $\lambda^\sharp$ under $\pi$ (Ch. VI, §3, No. 2, Def. 1).

2) Suppose H is *compact* and $\beta$ is normalized. The saturation of every compact subset of X is compact. Therefore, if $f \in \mathscr{K}(X/H)$ then $f \circ \pi \in \mathscr{K}(X)$; and, for every positive measure $\lambda$ on X/H, Prop. 5 c) gives

$$\int_X (f \circ \pi)(x)\, d\lambda^\sharp(x) = \int_{X/H} f(u)\, d\lambda(u)\,.$$

In other words, $\lambda$ is the *image* of $\lambda^\sharp$ under $\pi$.

3) Cor. 3 c) of Prop. 6 shows at once that the results of this subsection remain valid in the case of complex measures (except for those that involve the upper integral).

4) Let $\mathbf{m}$ be a vectorial measure on X/H with values in E, and let $q$ be a lower semi-continuous semi-norm on E. For $\mathbf{m}$ to be $q$-majorizable (Ch. VI, §2, No. 3, Def. 3), it is necessary and sufficient that $\mathbf{m}^\sharp$ be so, in which case $q(\mathbf{m}^\sharp) = q(\mathbf{m})^\sharp$. This follows at once from the definitions and Cor. 3 a).

On the other hand, let $\mu$ be a positive measure on X/H. For $\mathbf{m}$ to be scalarly of base $\mu$, it is necessary and sufficient that $\mathbf{m}^\sharp$ be scalarly of base $\mu^\sharp$: this follows from Cor. 1.

Finally, if $\mathbf{m}$ has base $\mu$, with density $\mathbf{f}$ with respect to $\mu$ (Ch. VI, §2, No. 4, Def. 4), then $\mathbf{m}^\sharp$ has base $\mu^\sharp$, with density $\mathbf{f} \circ \pi$: this follows from Prop. 6 c).

## 4. The case that X/H is paracompact

If X/H is paracompact, we shall first see that the vector spaces $\mathscr{K}^\chi(X)$, for variable $\chi$, are all *isomorphic* to each other, and in particular isomorphic to $\mathscr{K}^1(X)$.

PROPOSITION 7. — *Assume* X/H *paracompact. Let* $\chi$ *be a continuous representation of* H *in* $\mathbf{R}_+^*$.

a) *There exists on* X *a continuous function* $r$, *with values* $> 0$, *such that* $r(x\xi) = \chi(\xi)r(x)$ *for all* $x \in X$ *and* $\xi \in H$.

b) *The mapping* $g \mapsto g/r$ *is an isomorphism of the vector space* $\mathscr{K}^\chi(X)$ *onto the vector space* $\mathscr{K}^1(X)$.

Let us apply Prop. 1 of No. 1 on taking $f$ to be a function $\geqslant 0$ that is not identically zero on any orbit (this is possible by Lemma 1 of Appendix 1); then $r = f^\chi$ satisfies the properties of a). The assertion b) is obvious.

PROPOSITION 8. — *Assume* X/H *paracompact. There exists a continuous function* $h \geqslant 0$ *on* X, *whose support has compact intersection with the saturation of every compact subset of* X, *and is such that* $h^\flat = 1$. *For such a function, one has* $g = \left(h \cdot (g \circ \pi)\right)^\flat$ *for every continuous function* $g$ *on* X/H.

Let us apply Prop. 1 of No. 1, with $\chi = 1$, on taking for $f$ a function $\geqslant 0$ that is not identically zero on any orbit. We have $f^1(x) > 0$ at every point $x$ of X. Set $h = f/f^1$. Then $h^1 = f^1/f^1 = 1$, therefore $h^\flat = 1$. It follows that if $g$ is a continuous function on X/H, then $\left(h \cdot (g \circ \pi)\right)^\flat = h^\flat \cdot g = g$.

*Remarks.* — 1) In particular, let X be a locally compact space on which a *discrete* group D operates continuously and properly on the right; suppose X/D is paracompact. Then, there exists a continuous function $h \geqslant 0$ on X whose support has compact intersection with the saturation of every compact subset of X, and is such that $\sum_{d \in \mathrm{D}} h(xd) = 1$ for every $x \in \mathrm{X}$ (all terms of the sum being zero except for a finite number of them).

2) Let us conserve the hypotheses and notations of Prop. 8. The mapping $g \mapsto h \cdot (g \circ \pi)$ is a continuous mapping of $\mathscr{K}(\mathrm{X/H})$ into $\mathscr{K}(\mathrm{X})$ that is a *right inverse* of the mapping $f \mapsto f^\flat$. Consequently, every bounded (resp. compact) subset of $\mathscr{K}(\mathrm{X/H})$ is the image of a bounded (resp. compact) subset of $\mathscr{K}(\mathrm{X})$. From this, one deduces immediately that the mapping $\lambda \mapsto \lambda^\sharp$ is again an isomorphism of $\mathscr{M}(\mathrm{X/H})$ onto a closed linear subspace of $\mathscr{M}(\mathrm{X})$ when these spaces are equipped with the topology of bounded (resp. compact) convergence.

PROPOSITION 9. — *We conserve the hypotheses and notations of Prop. 8. Let* $\lambda$ *be a positive measure on* X/H.

a) *The pair* $(\pi, h)$ *is* $\lambda^\sharp$*-adapted, and* $\int_{\mathrm{X}} h(x)\varepsilon_{\pi(x)}\, d\lambda^\sharp(x) = \lambda$.

b) *The mapping* $\pi$ *is proper for the measure* $h \cdot \lambda^\sharp$, *and* $\pi(h \cdot \lambda^\sharp) = \lambda$.

c) *Let* $\mathbf{k}$ *be a function on* X/H, *with values in a Banach space or in* $\overline{\mathbf{R}}$. *For* $\mathbf{k}$ *to be measurable* (resp. *locally integrable, essentially integrable, integrable*) *for* $\lambda$, *it is necessary and sufficient that* $h \cdot (\mathbf{k} \circ \pi)$ *be so for* $\lambda^\sharp$; *and, if* $\mathbf{k}$ *is essentially integrable for* $\lambda$, *then*

$$\int_{\mathrm{X/H}} \mathbf{k}\, d\lambda = \int_{\mathrm{X}} h \cdot (\mathbf{k} \circ \pi)\, d\lambda^\sharp. \tag{14}$$

Let $f \in \mathscr{K}(\mathrm{X/H})$. Then $h \cdot (f \circ \pi) \in \mathscr{K}(\mathrm{X})$ and

$$\int_{\mathrm{X}} h(x) f(\pi(x))\, d\lambda^\sharp(x) = \int_{\mathrm{X/H}} f(\dot{x})\, d\lambda(\dot{x}) \int_{\mathrm{H}} h(x\xi)\, d\beta(\xi) = \int_{\mathrm{X/H}} f(\dot{x})\, d\lambda(\dot{x}),$$

whence a). The assertion b) is proved similarly. The assertions of c) concerning measurability, essential integrability and formula (14) may then be obtained by applying the results of Ch. V (§4, Prop. 3, §5, Th. 1, §4, Th. 2). If $\mathbf{k}$ is $\lambda$-integrable, then $h \cdot (\mathbf{k} \circ \pi)$ is $\lambda^{\sharp}$-integrable (Ch. V, §3, No. 3, Th. 1). If $h \cdot (\mathbf{k} \circ \pi)$ is $\lambda^{\sharp}$-integrable, Prop. 5 proves that $\left(h \cdot (\mathbf{k} \circ \pi)\right)^{\flat} = h^{\flat} \cdot \mathbf{k} = \mathbf{k}$ is $\lambda$-integrable. If $\mathbf{k}$ is locally $\lambda$-integrable, then $h \cdot (\mathbf{k} \circ \pi)$ is locally $\lambda^{\sharp}$-integrable (Prop. 6). Finally, suppose $h \cdot (\mathbf{k} \circ \pi)$ locally $\lambda^{\sharp}$-integrable; for every $f \in \mathscr{K}(\mathrm{X}/\mathrm{H})$, $h \cdot (\mathbf{k} \circ \pi) \cdot (f \circ \pi)$ has compact support, and

$$|h \cdot (\mathbf{k} \circ \pi) \cdot (f \circ \pi)| \leqslant \mathrm{M} |h \cdot (\mathbf{k} \circ \pi)|,$$

where $\mathrm{M} = \sup |f|$; therefore $h \cdot \left((\mathbf{k}f) \circ \pi\right)$ is $\lambda^{\sharp}$-integrable, consequently $\mathbf{k}f$ is $\lambda$-integrable, by what has already been proved; this proves that $\mathbf{k}$ is indeed locally $\lambda$-integrable.

COROLLARY. — *The continuous linear mapping* $f \mapsto f^{\flat}$ *of* $\mathrm{L}^1(\mathrm{X}, \lambda^{\sharp})$ *into* $\mathrm{L}^1(\mathrm{X}/\mathrm{H}, \lambda)$ *defined by Prop.* 5 *is surjective.*

Suppose first that $\mathrm{X}/\mathrm{H}$ is paracompact and let $h$ be a function on $\mathrm{X}$ satisfying the conditions of Prop. 8. If $k$ is a $\lambda$-integrable numerical function on $\mathrm{X}/\mathrm{H}$, then $h \cdot (k \circ \pi)$ is $\lambda^{\sharp}$-integrable and $\left(h \cdot (k \circ \pi)\right)^{\flat} = k$ (Prop. 9).

In the general case, let $u \in \mathrm{L}^1(\mathrm{X}/\mathrm{H}, \lambda)$. There exists a function $f \in \mathscr{L}^1(\mathrm{X}/\mathrm{H}, \lambda)$ with class $u$ and zero outside a countable union of compact sets $\mathrm{K}_n$. Let us define recursively a sequence of relatively compact open sets $\mathrm{U}_n$ of $\mathrm{X}/\mathrm{H}$ such that $\mathrm{U}_{n+1} \supset \mathrm{K}_n \cup \overline{\mathrm{U}}_n$, and let $\mathrm{V}$ be the union of the $\mathrm{U}_n$. Then $\mathrm{V}$ is an open subset of $\mathrm{X}/\mathrm{H}$, a countable union of compact subsets $\overline{\mathrm{U}}_n$, hence is paracompact (GT, I, §9, No. 10, Th. 5). Set $\mathrm{Y} = \overset{-1}{\pi}(\mathrm{V})$ and let $\lambda_{\mathrm{V}}$ (resp. $\lambda^{\sharp}_{\mathrm{Y}}$) be the measure induced by $\lambda$ (resp. $\lambda^{\sharp}$) on $\mathrm{V}$ (resp. $\mathrm{Y}$). It is clear that $\mathrm{Y}/\mathrm{H}$ may be identified with $\mathrm{V}$ (GT, I, §3, Prop. 10) and that $\lambda^{\sharp}_{\mathrm{Y}}$ may be identified with $(\lambda_{\mathrm{V}})^{\sharp}$. Moreover, $f$ is zero outside $\mathrm{V}$ and belongs to $\mathscr{L}^1(\mathrm{V}, \lambda_{\mathrm{V}})$. Therefore, there exists $g \in \mathscr{L}^1(\mathrm{Y}, \lambda^{\sharp}_{\mathrm{Y}})$ such that $g^{\flat} = f$ almost everywhere in $\mathrm{V}$. Extending $g$ by $0$ on $\mathrm{X} - \mathrm{Y}$, one obtains a function $g_1 \in \mathscr{L}^1(\mathrm{X}, \lambda^{\sharp})$, and it is clear that the class of $g_1^{\flat}$ in $\mathrm{L}^1(\mathrm{X}/\mathrm{H}, \lambda)$ is none other than $u$.

*Remark* 3. — Suppose $\mathrm{X}/\mathrm{H}$ paracompact, and let us retain the notations of Proposition 9. The mapping $k \mapsto h \cdot (k \circ \pi)$ of $\mathrm{L}^1(\mathrm{X}/\mathrm{H}, \lambda)$ into $\mathrm{L}^1(\mathrm{X}, \lambda^{\sharp})$ is then *isometric* by (14) and is a *right inverse* of the mapping $f \mapsto f^{\flat}$ of $\mathrm{L}^1(\mathrm{X}, \lambda^{\sharp})$ onto $\mathrm{L}^1(\mathrm{X}/\mathrm{H}, \lambda)$.

## 5. Quasi-invariant measures on a homogeneous space

*Lemma* 3. — *Let* $\mathrm{G}$ *be a locally compact group,* $\mu$ *a left Haar measure on* $\mathrm{G}$, $\nu$ *and* $\nu'$ *two nonzero quasi-invariant measures on* $\mathrm{G}$. *If, for every* $s \in \mathrm{G}$, *the densities of* $\boldsymbol{\gamma}(s)\nu$ *with respect to* $\nu$ *and of* $\boldsymbol{\gamma}(s)\nu'$

*with respect to* $\nu'$ *are equal locally* $\mu$*-almost everywhere, then* $\nu$ *and* $\nu'$ *are proportional.*

Write $\nu = \rho\cdot\mu$, $\nu' = \rho'\cdot\mu$, where $\rho, \rho'$ are locally $\mu$-integrable functions on G and are everywhere nonzero (§1, No. 9, Prop. 11). For every $s \in \mathrm{G}$,

$$\boldsymbol{\gamma}(s)\nu = \big(\boldsymbol{\gamma}(s)\rho\big)\cdot\mu\,,\qquad \boldsymbol{\gamma}(s)\nu' = \big(\boldsymbol{\gamma}(s)\rho'\big)\cdot\mu\,,$$

and the hypothesis implies that $\rho^{-1}\cdot\boldsymbol{\gamma}(s)\rho = \rho'^{-1}\cdot\boldsymbol{\gamma}(s)\rho'$ locally $\mu$-almost everywhere. Set $\sigma = \rho'/\rho$, which is a $\mu$-measurable function on G. For every $s \in \mathrm{G}$, $\boldsymbol{\gamma}(s)\sigma = \sigma$ locally $\mu$-almost everywhere. Therefore $\sigma$ is equal to a constant locally $\mu$-almost everywhere, by Cor. 2 of Prop. 6 applied with $\mathrm{X} = \mathrm{H} = \mathrm{G}$.

Let G be a locally compact group, H a closed subgroup of G. Consider the homogeneous space G/H of left cosets with respect to H, on which G operates continuously on the left. We are going to show that there exists *one and only one class* of nonzero quasi-invariant measures on G/H.

Note that H operates on G continuously and properly by right translations; and the quotient space, which is none other than G/H is paracompact (GT, III, §4, No. 6, Prop. 13). We may therefore apply the results of Nos. 1 to 4, with $\mathrm{X} = \mathrm{G}$. We thus have mappings $f \mapsto f^{\flat}$ of $\mathscr{K}(\mathrm{G})$ onto $\mathscr{K}(\mathrm{G/H})$, and $\lambda \mapsto \lambda^{\sharp}$ of $\mathscr{M}(\mathrm{G/H})$ into $\mathscr{M}(\mathrm{G})$ (once a left Haar measure $\beta$ on H has been fixed). The fact that G operates on the left in G/H gives rise to a supplementary property:

$$\boldsymbol{\gamma}_{\mathrm{G/H}}(s)\cdot f^{\flat} = \big(\boldsymbol{\gamma}_{\mathrm{G}}(s)\cdot f\big)^{\flat} \qquad (s\in\mathrm{G},\ f\in\mathscr{K}(\mathrm{G})) \tag{15}$$

$$\big(\boldsymbol{\gamma}_{\mathrm{G/H}}(s)\cdot\lambda\big)^{\sharp} = \boldsymbol{\gamma}_{\mathrm{G}}(s)\cdot\lambda^{\sharp} \qquad (s\in\mathrm{G},\ \lambda\in\mathscr{M}(\mathrm{G/H})). \tag{16}$$

Indeed, for every $x \in \mathrm{G}$,

$$\begin{aligned}\big(\boldsymbol{\gamma}_{\mathrm{G/H}}(s)\cdot f^{\flat}\big)(\pi(x)) &= f^{\flat}(s^{-1}\pi(x)) = f^{\flat}(\pi(s^{-1}x))\\ &= \int_{\mathrm{H}} f(s^{-1}x\xi)\,d\beta(\xi) = \int_{\mathrm{H}} \big(\boldsymbol{\gamma}_{\mathrm{G}}(s)f\big)(x\xi)\,d\beta(\xi) = \big(\boldsymbol{\gamma}_{\mathrm{G}}(s)f\big)^{\flat}(\pi(x)),\end{aligned}$$

whence formula (15), which implies formula (16).

*Lemma* 4. — *Let* $\lambda$ *be a measure* $\neq 0$ *on* G/H, *and* $\mu$ *a left Haar measure on* G. *The following properties are equivalent*:

a) $\lambda$ *is quasi-invariant under* G;

b) *for a subset* A *of* G/H *to be locally* $\lambda$*-negligible, it is necessary and sufficient that* $\overset{-1}{\pi}(\mathrm{A})$ *be locally* $\mu$*-negligible*;

c) *the measure* $\lambda^{\sharp}$ *is equivalent to* $\mu$.

*Suppose this to be the case and let* $\lambda^\sharp = \rho \cdot \mu$, *where* $\rho$ *is a locally* $\mu$*-integrable function that is everywhere nonzero. Then, for every* $s \in \mathrm{G}$, *the density* $\theta_s$ *of* $\boldsymbol{\gamma}_{\mathrm{G/H}}(s)\lambda$ *with respect to* $\lambda$ *is such that*

$$\theta_s(\pi(x)) = \frac{\rho(s^{-1}x)}{\rho(x)} \tag{17}$$

*locally* $\mu$*-almost everywhere on* G.

c) $\Rightarrow$ b): This follows at once from Prop. 6 a).

b) $\Rightarrow$ a): If property b) holds, then the set of locally $\lambda$-negligible subsets of G/H is invariant under G, thus $\lambda$ is quasi-invariant under G.

a) $\Rightarrow$ c): Assume $\lambda$ is quasi-invariant under G; for every $s \in \mathrm{G}$, $\lambda$ and $\boldsymbol{\gamma}_{\mathrm{G/H}}(s)\lambda$ are equivalent, therefore $\lambda^\sharp$ and $\boldsymbol{\gamma}_{\mathrm{G}}(s) \cdot \lambda^\sharp = \big(\boldsymbol{\gamma}_{\mathrm{G/H}}(s) \cdot \lambda\big)^\sharp$ are equivalent (Cor. 1 of Prop. 6); since $\lambda^\sharp \neq 0$, $\lambda^\sharp$ is equivalent to $\mu$ (§1, No. 9, Prop. 11).

Moreover, for every $s \in \mathrm{G}$,

$$\begin{aligned}(\theta_s \circ \pi) \cdot \lambda^\sharp = (\theta_s \cdot \lambda)^\sharp = \big(\boldsymbol{\gamma}_{\mathrm{G/H}}(s)\lambda\big)^\sharp &= \boldsymbol{\gamma}_{\mathrm{G}}(s)\lambda^\sharp \\ &= \big(\boldsymbol{\gamma}_{\mathrm{G}}(s)\rho\big) \cdot \mu = \frac{\boldsymbol{\gamma}_{\mathrm{G}}(s)\rho}{\rho} \cdot \lambda^\sharp,\end{aligned}$$

whence (17).

The equivalence of a) and b) implies first the uniqueness result already announced, and even a more precise result:

THEOREM 1. — *Let* G *be a locally compact group,* H *a closed subgroup of* G.

a) *Any two nonzero quasi-invariant measures on* G/H *are equivalent; the subsets of* G/H *locally negligible for these measures are those whose inverse image in* G *is locally negligible for a Haar measure.*

b) *Let* $\lambda, \lambda'$ *be two nonzero quasi-invariant measures on* G/H. *If, for every* $s \in \mathrm{G}$, *the densities of* $\boldsymbol{\gamma}_{\mathrm{G/H}}(s)\lambda$ *with respect to* $\lambda$ *and of* $\boldsymbol{\gamma}_{\mathrm{G/H}}(s)\lambda'$ *with respect to* $\lambda'$ *are equal almost everywhere for* $\lambda$ *(or* $\lambda'$*), then* $\lambda$ *and* $\lambda'$ *are proportional.*

The assertion a) follows at once from Lemma 4. Let $\lambda$ and $\lambda'$ be two nonzero quasi-invariant measures satisfying the condition of b). Then, for every $s \in \mathrm{G}$, the densities of $\boldsymbol{\gamma}_{\mathrm{G}}(s)\lambda^\sharp$ with respect to $\lambda^\sharp$ and of $\boldsymbol{\gamma}_{\mathrm{G}}(s)\lambda'^\sharp$ with respect to $\lambda'^\sharp$ are equal locally $\mu$-almost everywhere, therefore (Lemma 3) $\lambda^\sharp$ and $\lambda'^\sharp$ are proportional, hence $\lambda$ and $\lambda'$ are proportional.

On the other hand, Lemma 4 reduces the search for nonzero quasi-invariant measures on G/H to that for the measures on G equivalent to

Haar measure and of the form $\lambda^\sharp$. On this subject we have the following lemma:

*Lemma 5. — Let $\mu$ be a left Haar measure on $\mathrm{G}$, and $\rho$ a locally $\mu$-integrable function. For $\rho \cdot \mu$ to be of the form $\lambda^\sharp$, it is necessary and sufficient that, for every $\xi \in \mathrm{H}$,*

$$\rho(x\xi) = \frac{\Delta_{\mathrm{H}}(\xi)}{\Delta_{\mathrm{G}}(\xi)}\,\rho(x) \tag{18}$$

*locally $\mu$-almost everywhere on $\mathrm{G}$.*

To say that $\rho \cdot \mu$ is of the form $\lambda^\sharp$ amounts to saying that, for every $\xi \in \mathrm{H}$, $\boldsymbol{\delta}(\xi)(\rho \cdot \mu) = \Delta_{\mathrm{H}}(\xi)\rho \cdot \mu$ (Prop. 4). Now,

$$\boldsymbol{\delta}(\xi)(\rho \cdot \mu) = \big(\boldsymbol{\delta}(\xi)\rho\big) \cdot \big(\boldsymbol{\delta}(\xi)\mu\big) = \Delta_{\mathrm{G}}(\xi)\big(\boldsymbol{\delta}(\xi)\rho\big) \cdot \mu\,,$$

whence the lemma.

We can now establish the announced existence result, and even a more precise result:

THEOREM 2. — *Let $\mathrm{G}$ be a locally compact group, $\mathrm{H}$ a closed subgroup of $\mathrm{G}$, $\mu$ a left Haar measure on $\mathrm{G}$, and $\beta$ a left Haar measure on $\mathrm{H}$.*

a) *There exist functions $\rho$ continuous and $> 0$ on $\mathrm{G}$, such that*

$$\rho(x\xi) = \frac{\Delta_{\mathrm{H}}(\xi)}{\Delta_{\mathrm{G}}(\xi)}\,\rho(x)$$

*for all $x \in \mathrm{G}$ and $\xi \in \mathrm{H}$.*

b) *Given such a function $\rho$, one can form the measure $\lambda = (\rho \cdot \mu)/\beta$ on $\mathrm{G}/\mathrm{H}$, and $\lambda$ is a nonzero positive measure quasi-invariant under $\mathrm{G}$.*

c) *For $s, x$ in $\mathrm{G}$, $\rho(sx)/\rho(x)$ depends only on $s$ and $\pi(x)$, hence defines a function $\chi$ continuous and $> 0$ on $\mathrm{G} \times (\mathrm{G}/\mathrm{H})$ such that*

$$\chi\big(s, \pi(x)\big) = \frac{\rho(sx)}{\rho(x)}\,. \tag{19}$$

*Then*

$$\gamma_{\mathrm{G}/\mathrm{H}}(s)\lambda = \chi(s^{-1}, .) \cdot \lambda \qquad \textit{for all } s \in \mathrm{G}\,. \tag{20}$$

a) follows from Prop. 7.

b) follows from Lemmas 5 and 4.

c) follows from (17).

*Remarks.* — 1) One deduces from Remark 1 of No. 3 that the nonzero quasi-invariant measures on G/H are none other than the pseudo-images under $\pi$ of a Haar measure on G.

*2) If G is a Lie group, we shall see later that the function $\rho$ of Th. 2 can be chosen to be infinitely differentiable.*

Under the conditions of Th. 2, certain results of Nos. 3 and 4 may be specialized as follows (on taking into account Ch. V, §4, Th. 2 and Prop. 2 for passing from properties relative to $\mu$ to properties relative to $\rho \cdot \mu$):

a) Let $f$ be a $\mu$-measurable function on G, with values in a topological space, constant outside a countable union of $\mu$-integrable sets; then, the set of $\dot{x} \in \mathrm{G/H}$ such that the function $\xi \mapsto f(x\xi)$ is not $\beta$-measurable is locally $\lambda$-negligible.

b) Let $f$ be a $\mu$-measurable function $\geqslant 0$ on G, zero outside a countable union of $\mu$-integrable sets. Then, the function

$$\dot{x} \mapsto \int_{\mathrm{H}}^{*} f(x\xi)\, d\beta(\xi)$$

on G/H is $\lambda$-measurable and

$$\int_{\mathrm{G}}^{*} f(x)\rho(x)\, d\mu(x) = \int_{\mathrm{G/H}}^{*} d\lambda(\dot{x}) \int_{\mathrm{H}}^{*} f(x\xi)\, d\beta(\xi) \qquad (\dot{x} = \pi(x)).$$

c) Let $\mathbf{f}$ be a $\rho \cdot \mu$-integrable function on G, with values in a Banach space or in $\overline{\mathbf{R}}$. Then, the set of $\dot{x} \in \mathrm{G/H}$ such that $\xi \mapsto \mathbf{f}(x\xi)$ is not $\beta$-integrable is $\lambda$-negligible; the function $\dot{x} \mapsto \int_{\mathrm{H}} \mathbf{f}(x\xi)\, d\beta(\xi)$ is $\lambda$-integrable, and

$$\int_{\mathrm{G}} \mathbf{f}(x)\rho(x)\, d\mu(x) = \int_{\mathrm{G/H}} d\lambda(\dot{x}) \int_{\mathrm{H}} \mathbf{f}(x\xi)\, d\beta(\xi).$$

d) There exists a continuous function $h \geqslant 0$ on G, whose support has compact intersection with the saturation KH of every compact subset K of G, and such that $\int_{\mathrm{H}} h(x\xi)\, d\beta(\xi) = 1$ for every $x \in \mathrm{G}$. For a function $\mathbf{g}$ on G/H to be measurable (resp. locally integrable, essentially integrable, integrable) for $\lambda$, it is necessary and sufficient that $h \cdot (\mathbf{g} \circ \pi)$ be so for $\rho \cdot \mu = \lambda^{\sharp}$; and, when $\mathbf{g}$ is essentially integrable for $\lambda$, one has

$$\int_{\mathrm{G/H}} \mathbf{g}(u)\, d\lambda(u) = \int_{\mathrm{G}} h(x)\mathbf{g}(\pi(x))\rho(x)\, d\mu(x).$$

## 6. Relatively invariant measures on a homogeneous space

Let G again be a locally compact group, H a closed subgroup, $\beta$ a left Haar measure on H.

*Lemma* 6. — *Let $\lambda$ be a measure on* G/H, $\chi$ *a continuous representation of* G *in* $\mathbf{C}^*$. *The following properties are equivalent:*

a) $\lambda$ *is relatively invariant on* G/H *with multiplier* $\chi$;

b) $\lambda^\sharp$ *is relatively invariant on* G *with left multiplier* $\chi$;

c) $\lambda^\sharp$ *is of the form* $a\chi \cdot \mu$ $(a \in \mathbf{C})$.

The condition a) means that, for every $s \in \mathrm{G}$,

$$\boldsymbol{\gamma}_{\mathrm{G/H}}(s)\lambda = \chi(s)^{-1}\lambda;$$

this is equivalent to $\left(\boldsymbol{\gamma}_{\mathrm{G/H}}(s)\lambda\right)^\sharp = \chi(s)^{-1}\lambda^\sharp$, that is, to

$$\boldsymbol{\gamma}_{\mathrm{G}}(s)\lambda^\sharp = \chi(s)^{-1}\lambda^\sharp .$$

Whence the equivalence of a) and b). The equivalence of b) and c) follows from §1, No. 8, Cor. 1 of Prop. 10.

THEOREM 3. — *Let* G *be a locally compact group,* H *a closed subgroup of* G, $\mu$ (resp. $\beta$) *a left Haar measure on* G (resp. H), $\chi$ *a continuous representation of* G *in* $\mathbf{C}^*$.

a) *In order that there exist on* G/H *a nonzero measure relatively invariant under* G *and with multiplier* $\chi$, *it is necessary and sufficient that* $\chi(\xi) = \Delta_{\mathrm{H}}(\xi)/\Delta_{\mathrm{G}}(\xi)$ *for all* $\xi \in \mathrm{H}$.

b) *This measure is then unique up to a constant factor; more precisely, it is proportional to* $(\chi \cdot \mu)/\beta$.

For there to exist on G/H a nonzero measure relatively invariant under G with multiplier $\chi$, it is necessary and sufficient (Lemma 6) that $\chi \cdot \mu$ be of the form $\lambda^\sharp$, hence (No. 2, Prop. 4) that $\boldsymbol{\delta}(\xi)(\chi \cdot \mu) = \Delta_{\mathrm{H}}(\xi)(\chi \cdot \mu)$ for all $\xi \in \mathrm{H}$. This condition may also be written $\chi(\xi)\chi \cdot \Delta_{\mathrm{G}}(\xi)\mu = \Delta_{\mathrm{H}}(\xi)\chi \cdot \mu$, that is,

$$\chi(\xi) = \Delta_{\mathrm{H}}(\xi)/\Delta_{\mathrm{G}}(\xi)$$

for all $\xi \in \mathrm{H}$. Whence a). The assertion b) follows at once from Lemma 6 and the fact that the mapping $\lambda \mapsto \lambda^\sharp$ is injective.

We shall see in §3 (No. 3, Example 4) some very simple examples where the representation $\xi \mapsto \Delta_{\mathrm{H}}(\xi)/\Delta_{\mathrm{G}}(\xi)$ cannot be extended to a continuous representation of G in $\mathbf{C}^*$. In this case, there therefore does not exist any nonzero complex measure on G/H relatively invariant under G.

COROLLARY 1. — *For there to exist on* G/H *a nonzero positive measure relatively invariant under* G, *it is necessary and sufficient that there* *exist a continuous representation of* G *in* $\mathbf{R}_+^*$ *extending the representation* $\xi \mapsto \Delta_{\mathrm{H}}(\xi)/\Delta_{\mathrm{G}}(\xi)$.

Note that this condition is fulfilled when H is *unimodular.*

COROLLARY 2. — *For there to exist on* G/H *a nonzero positive measure invariant under* G, *it is necessary and sufficient that* $\Delta_{\mathrm{G}}$ *coincide with* $\Delta_{\mathrm{H}}$ *on* H.

COROLLARY 3. — *Suppose that* H *is unimodular and that there exists on* G/H *a nonzero* bounded *positive measure* $\nu$ *relatively invariant under* G. *Then* $\nu$ *is invariant, and* G *is unimodular.*

Let $\chi$ be the multiplier of $\nu$. For every $s \in \mathrm{G}$, $\nu$ and $\boldsymbol{\gamma}(s)\nu$ have the same finite total mass (§1, No. 1, formula (6)); since $\boldsymbol{\gamma}(s)\nu = \chi(s)^{-1}\nu$, we have $\chi(s) = 1$. Thus $\nu$ is invariant. By Cor. 2, $\Delta_{\mathrm{G}}(s) = 1$ for all $s \in \mathrm{H}$. Let G′ be the set of $t \in \mathrm{G}$ such that $\Delta_{\mathrm{G}}(t) = 1$. This is a closed normal subgroup of G containing H. Let $\pi$ be the canonical mapping of G/H onto G/G′. Then $\pi(\nu)$ is a nonzero, bounded positive measure invariant under G. Therefore the left Haar measure of the group G/G′ is bounded, so that G/G′ is compact (§1, No. 2, Prop. 2). Consequently the image of G under $\Delta_{\mathrm{G}}$ is a compact subgroup of $\mathbf{R}_+^*$; this subgroup is reduced to $\{1\}$, thus $\Delta_{\mathrm{G}} = 1$ on all of G.

## 7. Haar measure on a quotient group

PROPOSITION 10. — *Let* G *be a locally compact group,* G′ *a closed normal subgroup,* G″ *the group* G/G′, $\pi$ *the canonical mapping of* G *onto* G/G′, *and* $\alpha, \alpha', \alpha''$ *left Haar measures on* G, G′, G″.

a) *Multiplying* $\alpha$ *by a constant factor if necessary, we can suppose that* $\alpha'' = \alpha/\alpha'$. *In particular, if* $f \in \mathscr{K}(\mathrm{G})$ *then*

$$\int_{\mathrm{G}} f(x)\,d\alpha(x) = \int_{\mathrm{G}''} d\alpha''(\dot{x}) \int_{\mathrm{G}'} f(x\xi)\,d\alpha'(\xi) \qquad (\dot{x} = \pi(x)).$$

b) *One has* $\Delta_{\mathrm{G}}(\xi) = \Delta_{\mathrm{G}'}(\xi)$ *for all* $\xi \in \mathrm{G}'$; *in particular, if* G *is unimodular then so is* G′.

c) *The kernel of the representation* $\Delta_{\mathrm{G}}$ *of* G *in* $\mathbf{R}_+^*$ *is the largest unimodular closed normal subgroup of* G.

Applying Th. 3 of No. 6 with $\chi = 1$ (and in the knowledge that here, there exists a measure on G/G′ invariant under G, namely $\alpha''$), we obtain a) and b); c) follows at once from b).

PROPOSITION 11. — *We maintain the notations of Prop.* 10. *Let* $u$ *be an automorphism of* G *such that* $u(\mathrm{G}') = \mathrm{G}'$. *Let* $u'$ *be the restriction* *of $u$ to $\mathrm{G}'$, and $u''$ the automorphism of $\mathrm{G}''$ deduced from $u$ by passage to the quotients. Then*

$$\mathrm{mod}_{\mathrm{G}}(u) = \mathrm{mod}_{\mathrm{G}'}(u')\,\mathrm{mod}_{\mathrm{G}''}(u'')\,.$$

For, if $\alpha'' = \alpha/\alpha'$ then $u''(\alpha'') = u(\alpha)/u'(\alpha')$, that is,

$$\begin{aligned}\mathrm{mod}_{\mathrm{G}''}(u'')^{-1}\alpha'' &= \mathrm{mod}_{\mathrm{G}}(u)^{-1}\alpha/\mathrm{mod}_{\mathrm{G}'}(u')^{-1}\alpha' \\ &= \frac{\mathrm{mod}_{\mathrm{G}'}(u')}{\mathrm{mod}_{\mathrm{G}}(u)}(\alpha/\alpha') = \frac{\mathrm{mod}_{\mathrm{G}'}(u')}{\mathrm{mod}_{\mathrm{G}}(u)}\alpha''\,,\end{aligned}$$

whence the proposition.

COROLLARY. — *For every $x \in \mathrm{G}$,*

$$\Delta_{\mathrm{G}}(x) = \Delta_{\mathrm{G}/\mathrm{G}'}(\dot{x})\,\mathrm{mod}(i_x)\,,$$

*where $\dot{x}$ denotes the canonical image of $x$ in $\mathrm{G}/\mathrm{G}'$, and $i_x$ the automorphism $s \mapsto x^{-1}sx$ of $\mathrm{G}'$.*

This follows from Prop. 11, and formula (33) of §1, No. 4.

## 8. A transitivity property

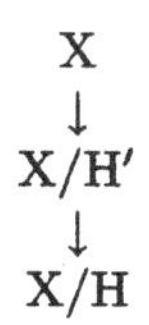

Let X be a locally compact space in which a locally compact group H acts on the right, continuously and *properly*, by $(x,\xi) \mapsto x\xi$ ($x \in \mathrm{X}$, $\xi \in \mathrm{H}$). Let $\mathrm{H}'$ be a closed subgroup of H; then $\mathrm{H}'$ operates on the right in X, continuously and *properly*. We shall denote by $\pi, \pi', p$ the canonical mappings of X onto X/H, of X onto $\mathrm{X}/\mathrm{H}'$, and of H onto $\mathrm{H}/\mathrm{H}'$.

Let $\beta, \beta'$ be left Haar measures on $\mathrm{H}, \mathrm{H}'$; suppose that $\Delta_{\mathrm{H}}$ and $\Delta_{\mathrm{H}'}$ *coincide on* $\mathrm{H}'$; one can then form the measure $\beta/\beta'$ on $\mathrm{H}/\mathrm{H}'$, left-invariant under H (No. 6, Th. 3). On the other hand, let $\mu$ be a positive measure on X such that

$$\boldsymbol{\delta}(\xi)\mu = \Delta_{\mathrm{H}}(\xi)\mu$$

for $\xi \in \mathrm{H}$; one can then form the measures $\mu/\beta$ on X/H and $\mu/\beta'$ on $\mathrm{X}/\mathrm{H}'$ (No. 2, Prop. 4). We are going to write $\mu/\beta'$ as the *integral*, with respect to $\mu/\beta$, of a family of measures on $\mathrm{X}/\mathrm{H}'$ indexed by the points of X/H. When $\mathrm{H}' = \{e\}$, we shall find ourselves again in the situation of No. 3.

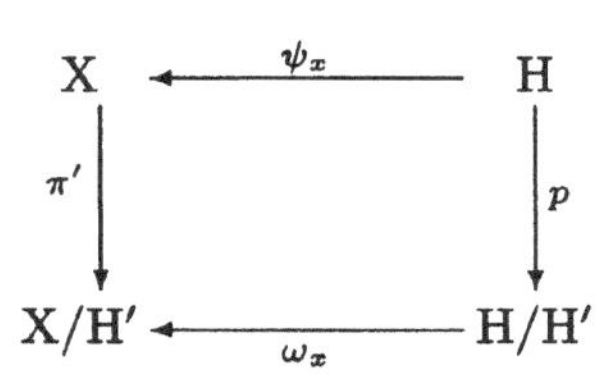

The mapping $(x,\xi) \mapsto \pi'(x\xi)$ of $\mathrm{X}\times\mathrm{H}$ into $\mathrm{X}/\mathrm{H}'$ is continuous; since $\pi'(x\xi) = \pi'(x\xi\xi')$ for all $\xi' \in \mathrm{H}'$ this mapping defines, by passage to the quotient, a *continuous* mapping of $\mathrm{X} \times (\mathrm{H}/\mathrm{H}')$ into $\mathrm{X}/\mathrm{H}'$; whence, for each fixed $x$ in X, a partial mapping $\omega_x$ of $\mathrm{H}/\mathrm{H}'$ into $\mathrm{X}/\mathrm{H}'$, deduced by passage to the quotient from the mapping $\psi_x : \xi \mapsto x\xi$ of H into X. Note that $\psi_{x\xi} = \psi_x \circ \gamma_{\mathrm{H}}(\xi)$, therefore that $\omega_{x\xi} = \omega_x \circ \gamma_{\mathrm{H/H'}}(\xi)$ for all $\xi \in \mathrm{H}$.

*Lemma 7. — Let* K *be a compact subset of* X/H′, *and* L *a compact subset of* X. *Then* $\bigcup_{x\in \mathrm{L}} \omega_x^{-1}(\mathrm{K})$ *is relatively compact in* H/H′.

Let $\mathrm{K}_1$ be a compact subset of X such that $\pi'(\mathrm{K}_1) = \mathrm{K}$. Let $\mathrm{K}_2$ be the set of $\xi \in \mathrm{H}$ such that $\mathrm{L}\xi$ intersects $\mathrm{K}_1$. Then $\mathrm{K}_2$ is compact (GT, III, §4, No. 5, Th. 1). Let $\xi \in \mathrm{H}$ be such that $p(\xi) \in \bigcup_{x\in \mathrm{L}} \omega_x^{-1}(\mathrm{K})$. Thus, there exists an $x \in \mathrm{L}$ such that $\omega_x\big(p(\xi)\big) \in \mathrm{K}$, in other words such that $\pi'(x\xi) \in \mathrm{K}$. Since $\pi'(\mathrm{K}_1) = \mathrm{K}$, there exists $\xi' \in \mathrm{H}'$ such that $x\xi\xi' \in \mathrm{K}_1$. Then $\xi\xi' \in \mathrm{K}_2$, therefore $p(\xi) = p(\xi\xi') \in p(\mathrm{K}_2)$. We have thus shown that $\bigcup_{x\in \mathrm{L}} \omega_x^{-1}(\mathrm{K}) \subset p(\mathrm{K}_2)$.

This lemma shows first of all that the mapping $\omega_x$ is *proper*. One can therefore form the measure $\omega_x(\beta/\beta')$ on X/H′, which is concentrated on $\omega_x(\mathrm{H/H'}) = \pi'\big(\psi_x(\mathrm{H})\big) = \pi'(x\mathrm{H})$. If $f \in \mathscr{K}(\mathrm{X/H'})$, Lemma 7 and §1, No. 1, Lemma 1 show that the function $x \mapsto \langle f, \omega_x(\beta/\beta')\rangle$ is continuous on X; moreover, $\langle f, \omega_x(\beta/\beta')\rangle$ is zero when $\mathrm{Supp}\, f$ does not intersect $\pi'(x\mathrm{H})$, in other words when $\pi(x)$ does not belong to the canonical image of $\mathrm{Supp}\, f$ in X/H.

Moreover, if $x \in \mathrm{H}$ then

$$\omega_{x\xi}(\beta/\beta') = \omega_x\big(\gamma_{\mathrm{H/H'}}(\xi)(\beta/\beta')\big) = \omega_x(\beta/\beta').$$

The mapping $x \mapsto \omega_x(\beta/\beta')$ of X into $\mathscr{M}(\mathrm{X/H'})$ therefore defines by passage to the quotient a mapping $u \mapsto (\beta/\beta')_u$ of X/H into $\mathscr{M}(\mathrm{X/H'})$. The foregoing shows that, for every $f \in \mathscr{K}(\mathrm{X/H'})$, the mapping $u \mapsto \langle f, (\beta/\beta')_u\rangle$ is continuous with compact support. Consequently the mapping $u \mapsto (\beta/\beta')_u$ *is a vaguely continuous and* $(\mu/\beta)$*-adequate family of measures on* X/H′, *with* X/H *as index set.*

Let $x \in \mathrm{X}$, and $u = \pi(x) \in \mathrm{X/H}$. Let $\mathbf{f}$ be a function on X/H′, with values in a Banach space or in $\overline{\mathbf{R}}$. By Ch. V, §4, Th. 2, for $\mathbf{f}$ to be $(\beta/\beta')_u$-integrable, it is necessary and sufficient that the function $p(\xi) \mapsto \mathbf{f}\big(\omega_x(p(\xi))\big) = \mathbf{f}\big(\pi'(x\xi)\big)$ on H/H′ be $(\beta/\beta')$-integrable, in which case

$$(21)\quad \int_{\mathrm{X/H'}} \mathbf{f}(u')\, d(\beta/\beta')_u(u') = \int_{\mathrm{H/H'}} \mathbf{f}\big(\pi'(x\xi)\big) d(\beta/\beta')(\dot{\xi}) \qquad (\dot{\xi} = p(\xi)).$$

One has analogous properties for measurability, the upper integral and the essential integral.

PROPOSITION 12. — *With the foregoing notations,*

$$\int_{X/H} (\beta/\beta')_u \, d(\mu/\beta)(u) = \mu/\beta' . \tag{22}$$

Let $f \in \mathscr{K}(X)$, and let $f^\flat \in \mathscr{K}(X/H')$, defined by

$$f^\flat(\pi'(x)) = \int_{H'} f(x\xi') \, d\beta'(\xi') .$$

It suffices (cf. No. 2) to prove that $f^\flat$ has the same integral with respect to the two members of (22). Now, $\langle \mu/\beta', f^\flat \rangle = \langle \mu, f \rangle$. On the other hand,

$$\langle \int_{X/H} (\beta/\beta')_u \, d(\mu/\beta)(u), f^\flat \rangle = \int_{X/H} \langle (\beta/\beta')_u, f^\flat \rangle \, d(\mu/\beta)(u) .$$

Now, let $x \in X$ and $u = \pi(x)$. We have

$$\begin{aligned}
\langle (\beta/\beta')_u, f^\flat \rangle = \langle \omega_x(\beta/\beta'), f^\flat \rangle &= \int_{H/H'} f^\flat(\omega_x(\dot\xi)) d(\beta/\beta')(\dot\xi) \\
&= \int_{H/H'} f^\flat(\pi'(x\xi)) d(\beta/\beta')(\dot\xi) \\
&= \int_{H/H'} d(\beta/\beta')(\dot\xi) \int_{H'} f(x\xi\xi') \, d\beta'(\xi') \\
&= \int_H f(x\xi) \, d\beta(\xi) .
\end{aligned}$$

Therefore

$$\langle \int_{X/H} (\beta/\beta')_u \, d(\mu/\beta)(u), f^\flat \rangle = \int_{X/H} d(\mu/\beta)(u) \int_H f(x\xi) \, d\beta(\xi) = \langle \mu, f \rangle ,$$

which proves the proposition.

COROLLARY 1. — a) *Let* $\mathbf{f}$ *be a function on* $X/H'$, *with values in a Banach space or in* $\overline{\mathbf{R}}$, *integrable for* $\mu/\beta'$. *There exists a* $(\mu/\beta)$*-negligible subset* $N$ *of* $X/H$ *having the following property: if* $x \in X$ *is such that* $\pi(x) \notin N$, *then the function* $\mathbf{f} \circ \omega_x$ *on* $H/H'$, *that is, the function* $\dot\xi \mapsto \mathbf{f}(\pi'(x\xi))$, *is integrable for* $\beta/\beta'$. *The integral* $\int_{H/H'} \mathbf{f}(\pi'(x\xi)) d(\beta/\beta')(\dot\xi)$ *depends only on* $\dot x = \pi(x)$, *and is a* $(\mu/\beta)$*-integrable function of* $\dot x$; *and*

$$\int_{X/H'} \mathbf{f} \, d(\mu/\beta') = \int_{X/H} d(\mu/\beta)(\dot x) \int_{H/H'} \mathbf{f}(\pi'(x\xi)) d(\beta/\beta')(\dot\xi) .$$

b) *Let* $f$ *be a function* $\geqslant 0$ *on* $X/H'$, *measurable for* $\mu/\beta'$ *and zero outside a countable union of* $(\mu/\beta')$*-integrable sets. Then*

$$\pi(x) \mapsto \int_{H/H'}^{*} f(\pi'(x\xi))\,d(\beta/\beta')(\dot{\xi})$$

*is* $(\mu/\beta)$*-measurable, and*

$$\int_{X/H'}^{*} f\,d(\mu/\beta') = \int_{X/H}^{*} d(\mu/\beta)(\dot{x}) \int_{H/H'}^{*} f(\pi'(x\xi))\,d(\beta/\beta')(\dot{\xi}).$$

c) *Let* $\mathbf{f}$ *be a function on* $X/H'$ *with values in a Banach space or in* $\overline{\mathbf{R}}$, *measurable for* $\mu/\beta'$ *and zero outside a countable union of* $(\mu/\beta')$*-integrable sets. Then, for* $\mathbf{f}$ *to be* $(\mu/\beta')$*-integrable, it is sufficient that*

$$\int_{X/H}^{*} d(\mu/\beta)(\dot{x}) \int_{H/H'}^{*} |\mathbf{f}(\pi'(x\xi))|\,d(\beta/\beta')(\dot{\xi}) < +\infty.$$

COROLLARY 2. — *Let* G *be a locally compact group,* A *and* B *closed subgroups of* G *such that* $A \supset B$. *Suppose that there exists, on the homogeneous space* G/B *of left cosets with respect to* B, *a nonzero positive measure* $\alpha$ *that is invariant under* G *and is bounded.*

a) *The canonical image of* $\alpha$ *in* G/A *is a nonzero positive measure, invariant under* G, *and bounded.*

b) $\Delta_G$ *coincides with* $\Delta_A$ *on* A *and with* $\Delta_B$ *on* B.

c) *There exists, on the homogeneous space* A/B *of left cosets of* A *with respect to* B, *a nonzero positive measure invariant under* A *and bounded.*

The assertion a) is immediate. The assertion b) follows from a) and No. 6, Cor. 2 of Th. 3. By b), $\Delta_A$ coincides with $\Delta_B$ on B, and one can therefore apply the results of the present subsection, on taking $X = G$, $H = A$, $H' = B$. The function 1 on G/B is $\alpha$-integrable. By a) of Cor. 1, the function 1 on A/B is integrable for $\beta/\beta'$, where $\beta$ and $\beta'$ denote left Haar measures on A and B; thus $\beta/\beta'$ is bounded.

### 9. Construction of the Haar measure of a group from the Haar measures of certain subgroups

Let G be a locally compact group, X and Y two closed subgroups of G *such that* $\Omega = XY$ *contains a neighborhood* U *of* $e$. Then $\Omega$ is open in G; for, for any $x_0 \in X$ and $y_0 \in Y$, $XY = (x_0X)(Yy_0) \supset x_0Uy_0$, and

$x_0 \mathrm{U} y_0$ is a neighborhood of $x_0 y_0$; thus $\Omega$ is a neighborhood of each of its points.

*When G is a Lie group with Lie algebra $\mathfrak{g}$, the condition imposed on X and Y is satisfied if the subalgebras corresponding to X and Y have sum $\mathfrak{g}$. *

The group $\mathrm{X} \times \mathrm{Y}$ operates continuously in G on the left, by the law $(x, y) \cdot s = xsy^{-1}$ ($x \in \mathrm{X}$, $y \in \mathrm{Y}$, $s \in \mathrm{G}$). Let $\mathrm{Z} = \mathrm{X} \cap \mathrm{Y}$. The stabilizer of $e$ in $\mathrm{X} \times \mathrm{Y}$ is the subgroup $\mathrm{Z}_0$ of $\mathrm{X} \times \mathrm{Y}$ formed by the pairs $(z, z)$, where $z \in \mathrm{Z}$, a subgroup canonically isomorphic to Z. Thus the set $\Omega$ may be identified with the homogeneous space $(\mathrm{X} \times \mathrm{Y})/\mathrm{Z}_0$ of left cosets; more precisely, the mapping $(x, y) \mapsto xy^{-1}$ of $\mathrm{X} \times \mathrm{Y}$ onto $\Omega$ defines, by passage to the quotient, a continuous bijection of $(\mathrm{X} \times \mathrm{Y})/\mathrm{Z}_0$ onto $\Omega$. We shall assume that this mapping *is a homeomorphism.* (This is notably the case if G is *countable at infinity*: cf. App. I.)

PROPOSITION 13. — *Suppose in addition that* Z *is compact. Let* $\mu_{\mathrm{G}}$, $\mu_{\mathrm{X}}$, $\mu_{\mathrm{Y}}$ *be left Haar measures on* G, X, Y, *and* $\Lambda$ *the restriction of* $\Delta_{\mathrm{G}}$ *to* Y. *Then the restriction* $\mu$ *of* $\mu_{\mathrm{G}}$ *to* $\Omega$ *is, up to a constant factor, the image of* $\mu_{\mathrm{X}} \otimes (\Lambda^{-1} \cdot \mu_{\mathrm{Y}})$ *under the mapping* $(x, y) \mapsto xy^{-1}$ *of* $\mathrm{X} \times \mathrm{Y}$ *onto* $\Omega$ *(a mapping that is proper).*

For $x \in \mathrm{X}$, $y \in \mathrm{Y}$,

$$\boldsymbol{\gamma}\big((x, y)\big)\mu = \boldsymbol{\delta}(y)\boldsymbol{\gamma}(x)\mu = \Delta_{\mathrm{G}}(y)\mu .$$

Identifying $\Omega$ with the homogeneous space $(\mathrm{X} \times \mathrm{Y})/\mathrm{Z}_0$ and choosing a suitable Haar measure on $\mathrm{Z}_0$, one sees that $\mu^\sharp$ is the product of the left Haar measure of $\mathrm{X} \times \mathrm{Y}$, namely $\mu_{\mathrm{X}} \otimes \mu_{\mathrm{Y}}$, by the function $(x, y) \mapsto \Delta_{\mathrm{G}}(y)^{-1}$ (No. 6, Lemma 6). On the other hand $\mu$ is, up to a constant factor, the image of $\mu^\sharp$ under the canonical mapping of $\mathrm{X} \times \mathrm{Y}$ onto $\Omega$ (No. 3, *Remark* 2).

COROLLARY. — *Let* $\mathbf{f}$ *be a function defined on* $\Omega$, *with values in a Banach space or in* $\overline{\mathbf{R}}$. *For* $\mathbf{f}$ *to be* $\mu$*-integrable, it is necessary and sufficient that the function* $(x, y) \mapsto \mathbf{f}(xy)\Delta_{\mathrm{G}}(y)\Delta_{\mathrm{Y}}(y)^{-1}$ *be* $(\mu_{\mathrm{X}} \otimes \mu_{\mathrm{Y}})$*-integrable, in which case*

$$\int_\Omega \mathbf{f}(\omega)\, d\mu(\omega) = a \iint_{\mathrm{X} \times \mathrm{Y}} \mathbf{f}(xy)\Delta_{\mathrm{G}}(y)\Delta_{\mathrm{Y}}(y)^{-1}\, d\mu_{\mathrm{X}}(x)\, d\mu_{\mathrm{Y}}(y), \tag{23}$$

*where* $a$ *is a constant* $> 0$ *independent of* $\mathbf{f}$.

By Prop. 13, and Ch. V, §4, No. 4, Th. 2, for $\mathbf{f}$ to be $\mu$-integrable, it is necessary and sufficient that the function $(x, y) \mapsto \mathbf{f}(xy^{-1})$ be integrable for $\mu_{\mathrm{X}} \otimes (\Lambda^{-1} \cdot \mu_{\mathrm{Y}})$, or again that the function $(x, y) \mapsto \mathbf{f}(xy^{-1})\Delta_{\mathrm{G}}(y)^{-1}$ be integrable for $\mu_{\mathrm{X}} \otimes \mu_{\mathrm{Y}}$, or again that the function $(x, y) \mapsto \mathbf{f}(xy)\Delta_{\mathrm{G}}(y)\Delta_{\mathrm{Y}}(y)^{-1}$ be integrable for $\mu_{\mathrm{X}} \otimes \mu_{\mathrm{Y}}$. Formula (23) results from an analogous argument.

PROPOSITION 14. — *Suppose that the conditions of Prop. 13 are fulfilled and that, in addition,* $\mathrm{Y}$ *is normal.*

a) *The restriction of* $\mu_{\mathrm{G}}$ *to* $\Omega$ *is, up to a constant factor, the image of* $\mu_{\mathrm{X}} \otimes \mu_{\mathrm{Y}}$ *under the mapping* $(x, y) \mapsto xy$ *of* $\mathrm{X} \times \mathrm{Y}$ *onto* $\Omega$.

b) *For* $x \in \mathrm{X}$ *and* $y \in \mathrm{Y}$,

$$\Delta_{\mathrm{G}}(xy) = \Delta_{\mathrm{X}}(x)\Delta_{\mathrm{Y}}(y) \bmod(i_x),$$

*where* $i_x$ *denotes the automorphism* $v \mapsto x^{-1}vx$ *of* $\mathrm{Y}$.

We have $\Delta_{\mathrm{G}} = \Delta_{\mathrm{Y}}$ on $\mathrm{Y}$ (Prop. 10 b)), therefore a) follows from (23). Let $x_0 \in \mathrm{X}$, $y_0 \in \mathrm{Y}$. Denote by $p$ the mapping $(x, y) \mapsto xy$ of $\mathrm{X} \times \mathrm{Y}$ onto $\Omega$. Since

$$xy(x_0y_0)^{-1} = xx_0^{-1}(x_0yy_0^{-1}x_0^{-1}) = xx_0^{-1}\, i_{x_0^{-1}}(yy_0^{-1}),$$

we have

$$\begin{aligned}\Delta_{\mathrm{G}}(x_0y_0)p(\mu_{\mathrm{X}} \otimes \mu_{\mathrm{Y}}) &= \boldsymbol{\delta}(x_0y_0)p(\mu_{\mathrm{X}} \otimes \mu_{\mathrm{Y}}) \\ &= p\big(\boldsymbol{\delta}(x_0)\mu_{\mathrm{X}} \otimes i_{x_0^{-1}}\boldsymbol{\delta}(y_0)\mu_{\mathrm{Y}}\big) \\ &= p\big(\Delta_{\mathrm{X}}(x_0)\mu_{\mathrm{X}} \otimes \Delta_{\mathrm{Y}}(y_0)(\bmod i_{x_0})\mu_{\mathrm{Y}}\big) \\ &= \Delta_{\mathrm{X}}(x_0)\Delta_{\mathrm{Y}}(y_0)(\bmod i_{x_0})p(\mu_{\mathrm{X}} \otimes \mu_{\mathrm{Y}}),\end{aligned}$$

whence b).

*Remark.* — Prop. 14 applies in particular when $\mathrm{G}$ is the topological semi-direct product of $\mathrm{X}$ by $\mathrm{Y}$ (GT, III, §2, No. 10). In this case, $\mathrm{Z} = \{e\}$ and $\Omega = \mathrm{G}$. Since $yx = x\,i_x(y)$ for $x \in \mathrm{X}$, $y \in \mathrm{Y}$, the measure $\mu_{\mathrm{G}}$ is also, up to a constant factor, the image of $(\bmod i_x)\mu_{\mathrm{X}} \otimes \mu_{\mathrm{Y}}$ under the mapping $(x, y) \mapsto yx$ of $\mathrm{X} \times \mathrm{Y}$ into $\mathrm{G}$.

## 10. Integration on a fundamental domain

Let $\mathrm{X}$ be a locally compact space, $\mathrm{H}$ a *discrete* group operating on the right continuously and *properly* in $\mathrm{X}$. Let $\pi$ be the canonical mapping of $\mathrm{X}$ onto $\mathrm{X}/\mathrm{H}$. For every $x \in \mathrm{X}$, we denote by $\mathrm{H}_x$ the stabilizer of $x$ in $\mathrm{H}$; this is a finite subgroup of $\mathrm{H}$ (GT, III, §4, No. 2, Prop. 4); its order will be denoted $n(x)$. For every $s \in \mathrm{H}$, $\mathrm{H}_{xs} = s^{-1}\mathrm{H}_x s$, therefore $n(xs) = n(x)$. There exists an open neighborhood $\mathrm{U}$ of $x$ such that $\mathrm{U} \cap \mathrm{U}s = \varnothing$ for $s \notin \mathrm{H}_x$ (*loc. cit.*, No. 4, proof of Prop. 8); for $y \in \mathrm{U}$, one has $\mathrm{H}_y \subset \mathrm{H}_x$; thus the function $n$ on $\mathrm{X}$ is upper semi-continuous. When $\mathrm{X}$ is *countable at infinity*, $\mathrm{H}$ is *countable*; for, let $(\mathrm{K}_1, \mathrm{K}_2, \ldots)$ be a covering of $\mathrm{X}$ by a sequence of compact subsets, and let $x_0 \in X$; the set of $s \in H$ such that $x_0 s \in K_i$ is finite (*loc. cit.*, No. 5, Th. 1), whence our assertion.

DEFINITION 2. — *Let* $F \subset X$. *One says that* $F$ *is a fundamental domain (for* $H$*) if the restriction of* $\pi$ *to* $F$ *is a bijection of* $F$ *onto* $X/H$ (in other words, $F$ is a *system of representatives* for the equivalence relation defined by $H$).

*Lemma* 8. — *Let* $F$ *be a fundamental domain. For every* $x \in X$,

$$\sum_{s \in H} \varphi_{Fs}(x) = n(x). \tag{24}$$

Since $\varphi_{Fs}(xt) = \varphi_{Fst^{-1}}(x)$ for all $s$ and $t$ in $H$, the two members of (24) remain invariant when $x$ is replaced by $xt$. We can therefore suppose that $x \in F$. We then have the equivalences

$$\varphi_{Fs}(x) = 1 \Leftrightarrow x \in Fs \Leftrightarrow xs^{-1} \in F \Leftrightarrow xs^{-1} = x \Leftrightarrow s \in H_x,$$

whence (24).

PROPOSITION 15. — *Assume that* $X$ *is countable at infinity. Let* $\mu$ *be a measure* $\geqslant 0$ *on* $X$. *Let* $F$ *be a fundamental domain such that* $Fs$ *is* $\mu$*-measurable for every* $s \in H$. *Let* $\mathbf{f}$ *be a* $\mu$*-integrable function on* $X$, *with values in a Banach space or in* $\overline{\mathbf{R}}$. *Then the family of the*

$$\int_{Fs} n(x)^{-1}\mathbf{f}(x)\,d\mu(x) \qquad (s \in H)$$

*is summable, and*

$$\int_X \mathbf{f}(x)\,d\mu(x) = \sum_{s \in H} \int_{Fs} n(x)^{-1}\mathbf{f}(x)\,d\mu(x).$$

If $A$ is a finite subset of $H$, then

$$\Big|\sum_{s \in A} n^{-1}\mathbf{f}\varphi_{Fs}\Big| \leqslant n^{-1}|\mathbf{f}| \sum_{s \in A} \varphi_{Fs} \leqslant |\mathbf{f}|$$

by Lemma 8. Lemma 8 also proves that $\sum_{s \in A} n^{-1}\mathbf{f}\varphi_{Fs}$ converges pointwise to $\mathbf{f}$ with respect to the increasing directed set of finite subsets of $H$. Prop. 15 then follows from Ch. IV, §4, No. 3, Th. 2.

THEOREM 4. — *Let* $X$ *be a locally compact space countable at infinity,* $H$ *a discrete group operating continuously and properly on the right in* $X$,

*$\pi$ the canonical mapping of* X *onto* X/H, *$\mu$ a positive measure on* X *invariant under* H, *$\beta$ the normalized Haar measure of* H, *and $\lambda = \mu/\beta$. Let* F *be a $\mu$-measurable fundamental domain.*

a) *The pair $(\pi, n^{-1}\varphi_{\mathrm{F}})$ is $\mu$-adapted, and*

$$\int_{\mathrm{X}} n(x)^{-1}\varphi_{\mathrm{F}}(x)\varepsilon_{\pi(x)}\,d\mu(x) = \lambda\,.$$

b) *The mapping $\pi$ is proper for $n^{-1}\varphi_{\mathrm{F}}\cdot\mu$, and $\pi(n^{-1}\varphi_{\mathrm{F}}\cdot\mu) = \lambda$.*

c) *Let $k$ be a function on* X/H. *For $k$ to be $\lambda$-measurable* (resp. *$\lambda$-integrable*), *it is necessary and sufficient that $n^{-1}\varphi_{\mathrm{F}}(k\circ\pi)$ be $\mu$-measurable* (resp. *$\mu$-integrable*); *and, if $k$ is $\lambda$-integrable then*

$$\int_{\mathrm{X/H}} k\,d\lambda = \int_{\mathrm{F}} n^{-1}(k\circ\pi)\,d\mu\,.$$

We have $\mu = \lambda^{\sharp}$. Let $f \in \mathscr{K}_+(\mathrm{X/H})$. Then $n^{-1}\varphi_{\mathrm{F}}(f\circ\pi)$ is $\mu$-measurable and $\geqslant 0$, and by Prop. 5 b) of No. 3 we have

$$\int_{\mathrm{X}}^{*} n(x)^{-1}\varphi_{\mathrm{F}}(x)f\big(\pi(x)\big)d\mu(x) = \int_{\mathrm{X/H}}^{*} f(\dot{x})\,d\lambda(\dot{x})\int_{\mathrm{H}}^{*} n(x\xi)^{-1}\varphi_{\mathrm{F}}(x\xi)\,d\beta(\xi)$$

and $\int_{\mathrm{H}}^{*} n(x\xi)^{-1}\varphi_{\mathrm{F}}(x\xi)\,d\beta(\xi) = n(x)^{-1}\sum_{\xi\in\mathrm{H}}\varphi_{\mathrm{F}}(x\xi) = 1$ by Lemma 8. Thus $n^{-1}\varphi_{\mathrm{F}}\cdot(f\circ\pi)$ is $\mu$-integrable and

$$\int_{\mathrm{X}} n(x)^{-1}\varphi_{\mathrm{F}}(x)f\big(\pi(x)\big)d\mu(x) = \int_{\mathrm{X/H}} f(\dot{x})\,d\lambda(\dot{x})\,.$$

This proves a). The assertion b) is proved similarly. The assertion c) may be deduced from a) and from Ch. V, §4, Prop. 3 and Th. 2.

COROLLARY. — *We maintain the hypotheses and notations of Th. 4. Let* F′ *be a second $\mu$-measurable fundamental domain. Let* **u** *be a function on* X, *with values in a Banach space or in $\overline{\mathbf{R}}$, invariant under* H. *Suppose that* **u** *is $\mu$-integrable on* F. *Then* **u** *is $\mu$-integrable on* F′ *and*

$$\int_{\mathrm{F}} \mathbf{u}(x)\,d\mu(x) = \int_{\mathrm{F}'} \mathbf{u}(x)\,d\mu(x)\,.$$

Since **u** and $n$ are invariant under H, there exists a function **v** on X/H such that $\mathbf{v}\circ\pi$ coincides with $n\mathbf{u}$ on F and on F′. Then

$n^{-1}\varphi_{\mathrm{F}}(\mathbf{v}\circ\pi) = \varphi_{\mathrm{F}}\mathbf{u}$, $n^{-1}\varphi_{\mathrm{F}'}(\mathbf{v}\circ\pi) = \varphi_{\mathrm{F}'}\mathbf{u}$. By hypothesis, $n^{-1}\varphi_{\mathrm{F}}(\mathbf{v}\circ\pi)$ is $\mu$-integrable. By Th. 4, $\mathbf{v}$ is $\lambda$-integrable, $\varphi_{\mathrm{F}'}\mu$ is $\mu$-integrable, and

$$\int_{\mathrm{F}} \mathbf{u}\,d\mu = \int_{\mathrm{X/H}} \mathbf{v}\,d\lambda = \int_{\mathrm{F}'} \mathbf{u}\,d\mu .$$

For the *existence* of $\mu$-measurable fundamental domains, see Exer. 12.

## §3. APPLICATIONS AND EXAMPLES

### 1. Compact groups of linear mappings

Let E be a finite-dimensional vector space over $\mathbf{R}$, $\mathbf{C}$ or $\mathbf{H}$. Then End(E) is a finite-dimensional algebra over $\mathbf{R}$, and the canonical topology on End(E) (§1, No. 10) is the topology of compact convergence. The group $\mathrm{Aut(E)} = \mathbf{GL}(\mathrm{E})$ is an open subset of End(E), hence is a locally compact group. Let $(e_1, e_2, \dots, e_n)$ be a basis of E and, for every endomorphism $u$ of E, let $M(u) = (\alpha_{ij}(u))$ be the matrix of $u$ with respect to this basis; to say that a subset S of End(E) is relatively compact in End(E) is equivalent to saying that the functions $\alpha_{ij}(u)$ are bounded in S.

PROPOSITION 1. — *Let* G *be a subgroup of* Aut(E). *The following three properties are equivalent:*

(i) G *is relatively compact in* End(E);

(ii) G *is relatively compact in* Aut(E);

(iii) G *leaves invariant a nondegenerate*[1] *positive hermitian form on* E.

(iii) $\Rightarrow$ (i): Suppose that G leaves invariant a nondegenerate positive hermitian form $\Psi$. Let $(e_1, \dots, e_n)$ be an orthonormal basis for $\Psi$ (*Alg.*, Ch. IX, §6, No. 1, Cor. 1 of Th. 1). For every $u \in \mathrm{G}$, let $(u_{ij})$ be its matrix with respect to $(e_i)$. For any $j$, we have $\sum_{i=1}^{n} |u_{ij}|^2 = 1$, thus $|u_{ij}| \leqslant 1$ for all $i$ and $j$, which proves (i).

(i) $\Rightarrow$(ii): This follows from GT, X, §3, No. 5, Cor. of Th. 4, taking into account the fact that the topology of End(E) is that of compact convergence.

(ii) $\Rightarrow$ (iii): Suppose that the closure $\overline{\mathrm{G}}$ of G in Aut(E) is compact. Let $\Phi$ be a nondegenerate positive hermitian form on E. If the field of scalars is $\mathbf{R}$ or $\mathbf{C}$, the giving of $\Phi$ makes E a finite-dimensional Hilbert space, and condition (iii) will result from the following lemma:

[1] *Non dégénérée*; in EVT, the term is replaced by *séparante*, subsequently translated as "separating" (TVS, V, §1, No. 1).

*Lemma 1. — Let* F *be a Hilbert space,* K *a compact group, and* $s \mapsto U(s)$ *a representation of* K *in the group of invertible elements of* $\mathscr{L}(\mathrm{F};\mathrm{F})$*, continuous for the topology of pointwise convergence. There exists a nondegenerate positive hermitian form* $\varphi$ *on* F *such that*

$$\varphi(U(s)x, U(s)y) = \varphi(x, y)$$

*for all* $s \in \mathrm{K}$, $x \in \mathrm{F}$, $y \in \mathrm{F}$, *and such that the topological vector space structure of* F *defined by* $\varphi$ (TVS, V, §1, No. 3) *is identical to the original structure of* F.

Let $\alpha$ be a Haar measure on K. For any $x, y$ in F, the mapping $s \mapsto (U(s)x|U(s)y)$ is continuous. Set

$$\varphi(x, y) = \int (U(s)x|U(s)y)\, d\alpha(s).$$

It is immediate that $\varphi(x, y)$ is a sesquilinear form on F. Since the set of endomorphisms $U(s)$ is compact in $\mathscr{L}_s(\mathrm{F};\mathrm{F})$, there exists a constant M such that $\|U(s)\| \leqslant \mathrm{M}$ for all $s \in \mathrm{K}$. For every $x \in \mathrm{F}$, we therefore have

$$\mathrm{M}^{-1}\|x\| \leqslant \|U(s)x\| \leqslant \mathrm{M}\|x\|,$$

whence the inequalities

$$\mathrm{M}^{-2}\alpha(\mathrm{K})\|x\|^2 \leqslant \varphi(x, x) \leqslant \mathrm{M}^2\alpha(\mathrm{K})\|x\|^2,$$

which shows that $\varphi$ is positive and nondegenerate, and that the norm $\varphi(x, x)^{1/2}$ is equivalent to the norm $\|x\|$. Finally, for all $t \in \mathrm{K}$,

$$\begin{aligned}\varphi(U(t)x, U(t)y) &= \int (U(st)x|U(st)y)\, d\alpha(s) \\ &= \int (U(s)x|U(s)y)\, d\alpha(s) = \varphi(x, y).\end{aligned}$$

When the field of scalars is $\mathbf{H}$, one argues exactly as before on replacing everywhere the function $s \mapsto (U(s)x|U(s)y)$ by the function $s \mapsto \Phi(sx, sy)$ defined on G, with values in $\mathbf{H}$. This completes the proof of the proposition.

*Remark.* — Let $\Phi$ be a nondegenerate positive hermitian form on E. The unitary group $\mathbf{U}(\Phi)$ is closed in Aut(E), hence is compact (Prop. 1). Prop. 1 also shows that every compact subgroup of Aut(E) is contained in a subgroup of the form $\mathbf{U}(\Phi)$. If now $\mathbf{U}(\Phi)$ is contained in a compact subgroup K of Aut(E), one sees that there exists a nondegenerate positive hermitian form $\Phi'$ on E such that $\mathbf{U}(\Phi) \subset \mathrm{K} \subset \mathbf{U}(\Phi')$, and it follows easily

(Exer. 1) that $\Phi$ and $\Phi'$ are proportional, whence $\mathbf{U}(\Phi) = \mathrm{K}$. Thus the maximal compact subgroups of $\mathrm{Aut(E)}$ are the subgroups of the form $\mathbf{U}(\Phi)$.

## 2. Triviality of fibered spaces and of group extensions

PROPOSITION 2. — *Let* X *be a locally compact space in which a locally compact group* H *acts on the right, continuously and properly, by* $(x, \xi) \mapsto x\xi$. *Assume that* X/H *is paracompact. Let* $g$ *be a continuous representation of* H *in* $\mathbf{R}^n$. *Then there exists a continuous mapping* $f$ *of* X *into* $\mathbf{R}^n$ *such that* $f(x\xi) = f(x) + g(\xi)$ *for all* $x \in \mathrm{X}$ *and* $\xi \in \mathrm{H}$.

One reduces immediately to the case that $n = 1$. Since the additive group $\mathbf{R}$ is isomorphic to the multiplicative group $\mathbf{R}_+^*$, the proposition is then an immediate consequence of Prop. 7 of §2, No. 4.

COROLLARY. — *Let* X *be a locally compact space in which a finite-dimensional real vector space* V *operates on the right, continuously and properly, by* $(x, v) \mapsto xv$. *Let* $\pi$ *be the canonical mapping of* X *onto* B = X/V. *Assume that* B *is paracompact.*

a) *There exists a continuous mapping* $f$ *of* X *into* V *such that* $f(xv) = f(x) + v$ *for all* $x \in \mathrm{X}$ *and* $v \in \mathrm{V}$.

b) *If* $f$ *is a mapping satisfying the conditions of* a), *then the mapping* $x \mapsto (\pi(x), f(x))$ *is a homeomorphism of* X *onto* B × V.

The assertion a) results from Prop. 2 in which $g$ is taken to be the identity mapping of V. Let $f$ be a mapping satisfying the conditions of a). The mapping $x \mapsto x\cdot(-f(x))$ of X into X is continuous, and is constant on each orbit, hence is of the form $\varphi \circ \pi$, where $\varphi$ is a continuous mapping of B into X; for every $b \in \mathrm{B}$, $\pi(\varphi(b)) = b$. The mappings $x \mapsto (\pi(x), f(x))$ of X into B × V and $(b, v) \mapsto \varphi(b) \cdot v$ of B × V into X are inverse to each other, because $\varphi(\pi(x)) \cdot f(x) = x \cdot (-f(x)) \cdot (f(x)) = x$, $\pi(\varphi(b) \cdot v) = \pi(\varphi(b)) = b$, and, if $b = \pi(y)$, then

$$f\big(\varphi(\pi(y)) \cdot v\big) = f\big(y \cdot (-f(y)) \cdot v\big) = f(y) - f(y) + v = v\,.$$

Since these mappings are continuous, they are homeomorphisms.

*Remark.* — Let E be a finite-dimensional real affine space, T a compact space, $\mu$ a measure on T of *total mass* 1, and $f$ a continuous mapping of T into E. If an origin $a$ in E is chosen, E becomes equipped with a vector space structure, and the integral $\int_{\mathrm{T}} f(t)\,d\mu(t)$ therefore has meaning; it represents the point $x$ of E such that

$$x - a = \int_{\mathrm{T}} (f(t) - a)\,d\mu(t)\,.$$

*This point is independent of the choice of* $a$. For, let $a' \in \mathrm{E}$ and $x' \in \mathrm{E}$ be such that $x' - a' = \int_{\mathrm{T}} (f(t) - a')\, d\mu(t)$. Then

$$x' - a = (x' - a') + (a' - a) = \int_{\mathrm{T}} (f(t) - a')\, d\mu(t) + \int_{\mathrm{T}} (a' - a)\, d\mu(t)$$
$$= \int_{\mathrm{T}} (f(t) - a)\, d\mu(t) = x - a\,,$$

whence $x' = x$. We may therefore employ the symbol $\int_{\mathrm{T}} f(t)\, d\mu(t)$ without specifying the choice of origin in E. If $u$ is an affine mapping of E into another finite-dimensional affine space E′, then

$$u\Big(\int_{\mathrm{T}} f(t) d\mu(t)\Big) = \int_{\mathrm{T}} u(f(t))\, d\mu(t)\,.$$

For, E and E′ may be identified with vector spaces in such a way that $u$ becomes a linear mapping, in which case the formula is known (Ch. III, §3, No. 2, Prop. 2 and No. 3, Prop. 7).

*Lemma* 2. — *Let* G *be a compact group,* $\mu$ *the normalized Haar measure of* G, E *a finite-dimensional real affine space,* A *the affine group of* E, *and* $\rho$ *a homomorphism of* G *into* A. *Assume that, for every* $x \in \mathrm{E}$, *the mapping* $s \mapsto \rho(s)x$ *of* G *into* E *is continuous. Then, for every* $x \in \mathrm{E}$, *the point*

$$x_0 = \int_{\mathrm{G}} \rho(s) x\, d\mu(s) \in \mathrm{E}$$

*is invariant under* G.

For, for every $t \in \mathrm{G}$,

$$\rho(t)x_0 = \int_{\mathrm{G}} \rho(t)\rho(s)x\, d\mu(s) = \int_{\mathrm{G}} \rho(ts)x\, d\mu(s) = \int_{\mathrm{G}} \rho(s)x\, d\mu(s) = x_0\,.$$

PROPOSITION 3. — *Let* G *be a locally compact group. Let* H *be a closed normal subgroup of* G, *isomorphic to* $\mathbf{R}^n$ *and such that* G/H *is compact.*

a) *There exists a closed subgroup* L *of* G *such that* G *is the topological semi-direct product of* L *and* H.

b) *If* M *is a compact subgroup of* G, *there exists an element* $x \in \mathrm{H}$ *such that* $x^{-1}\mathrm{M}x \subset \mathrm{L}$.

c) *Every compact subgroup of* G *is contained in a maximal compact subgroup.*

d) *The maximal compact subgroups of* G *are the subgroups that are the transforms of* L *by the inner automorphisms of* G.

Let $\pi$ be the canonical homomorphism of $\mathrm{G}$ onto $\mathrm{K} = \mathrm{G}/\mathrm{H}$. By passage to the quotient, the mapping $(s,h) \mapsto shs^{-1}$ of $\mathrm{G} \times \mathrm{H}$ into $\mathrm{H}$ defines a continuous mapping $(\sigma, h) \mapsto \sigma \cdot h$ of $\mathrm{K} \times \mathrm{H}$ into $\mathrm{H}$ such that $shs^{-1} = \pi(s) \cdot h$. We shall identify $\mathrm{H}$ with $\mathbf{R}^n$ (and will therefore employ, as the case may be, either the multiplicative or the additive notation for the group law in $\mathrm{H}$). By the Cor. of Prop. 2, there exists a continuous mapping $f$ of $\mathrm{G}$ into $\mathrm{H}$ such that $f(xh) = f(x) + h$ for $x \in \mathrm{G}$, $h \in \mathrm{H}$. For every $x \in \mathrm{G}$, let $p(x) = x \cdot (-f(x))$, which depends only on the coset of $x$ with respect to $\mathrm{H}$. Set

$$
\begin{aligned}
(1) \qquad \mathrm{F}(x,y) = p(xy)^{-1}p(x)p(y) &= f(xy)y^{-1}x^{-1}x(-f(x))y(-f(y)) \\
&= f(xy)[y^{-1}(-f(x))y](-f(y)) \\
&= f(xy) - \pi(y)^{-1}f(x) - f(y) .
\end{aligned}
$$

We see that *if* $\mathrm{F}(x,y) = 0$ *for all* $x, y$ *in* $\mathrm{G}$, then $p(\mathrm{G}) = \mathrm{L}$ is a subgroup of $\mathrm{G}$ that intersects each coset of $\mathrm{H}$ in one and only one point. Since $p$ is continuous, $\mathrm{G}$ is then the topological semi-direct product of $\mathrm{L}$ and $\mathrm{H}$ (GT, III, §2, No. 10).

Now, for any $h, h' \in \mathrm{H}$,

$$
\begin{aligned}
F(xh, yh') &= f(xhyh') - \pi(y)^{-1}f(xh) - f(yh') \\
&= f(xhy) + h' - \pi(y)^{-1}f(x) - \pi(y)^{-1}h - f(y) - h' \\
&= f\big(xy(\pi(y)^{-1}h)\big) - \pi(y)^{-1}f(x) - f(y) - \pi(y)^{-1}h \\
&= f(xy) - \pi(y)^{-1}f(x) - f(y) = F(x,y) .
\end{aligned}
$$

Therefore $\mathrm{F}$ defines, by passage to quotients, a continuous mapping $\varphi$ of $\mathrm{K} \times \mathrm{K}$ into $\mathrm{H}$.

On the other hand, for all $x, y, z$ in $\mathrm{G}$, we have

$$
\begin{aligned}
\mathrm{F}(z, xy) + \mathrm{F}(x,y) &= f(zxy) - \pi(xy)^{-1}f(z) - f(xy) + f(xy) \\
&\qquad - \pi(y)^{-1}f(x) - f(y) \\
&= \pi(y)^{-1}f(zx) - \pi(xy)^{-1}f(z) - \pi(y)^{-1}f(x) + f(zxy) \\
&\qquad - \pi(y)^{-1}f(zx) - f(y) \\
&= \pi(y)^{-1}\mathrm{F}(z,x) + \mathrm{F}(zx, y) ,
\end{aligned}
$$

therefore, for all $x', y', z'$ in $\mathrm{K}$,

$$
-\varphi(x', y') = \varphi(z', x'y') - y'^{-1}\varphi(z', x') - \varphi(z'x', y') .
$$

Let us integrate with respect to $z'$ by means of the normalized Haar measure $\alpha$ of $\mathrm{K}$. Setting $\psi(x') = \int \varphi(z', x')\, d\alpha(z')$, $\psi$ is a continuous function on K, and (on observing that the operations of K in $\mathbf{R}^n$ respect the vector space structure of $\mathbf{R}^n$ by GT, VII, §2, No. 1, Prop. 1), one obtains

$$-\varphi(x', y') = \psi(x'y') - y'^{-1}\psi(x') - \psi(y') .$$

In other words, setting $k = \psi \circ \pi$, which is a continuous function on G,

$$-\mathrm{F}(x, y) = k(xy) - \pi(y)^{-1}k(x) - k(y) . \tag{2}$$

Comparing (1) and (2), one sees that if $f$ is replaced by the continuous function $f + k$ (which leaves verified the property $f(xh) = f(x) + h$), F is replaced by 0 and, as we saw earlier, this completes the proof of a).

For every $g \in \mathrm{G}$, let $l_g$ (resp. $h_g$) be the unique element of L (resp. H) such that $g = h_g l_g$. If $h_1 \in \mathrm{H}$ and $g \in \mathrm{G}$, then

$$gh_1 = h_g l_g h_1 = h_g(l_g h_1 l_g^{-1})l_g ,$$

thus $h_{gh_1} = h_g + l_g h_1 l_g^{-1}$. For every $g \in \mathrm{G}$, let $\psi_g$ be the mapping of H into itself defined by

$$\psi_g(h_1) = h_g + l_g h_1 l_g^{-1} .$$

One sees that the mapping $(g, h_1) \mapsto \psi_g(h_1)$ of $\mathrm{G} \times \mathrm{H}$ into H is continuous and makes H a homogeneous space for G, in which the stabilizer of the origin is L. We observe moreover that when H is identified with $\mathbf{R}^n$, $\psi_g$ is an *affine* mapping of H into itself. This said, let M be a compact subgroup of G; by Lemma 2, there exists an $x \in \mathrm{H}$ such that $\psi_m(x) = x$ for *all* $m \in \mathrm{M}$. For $y \in \mathrm{H}$, $\psi_y$ is the translation with vector $y$; it follows that for every $m \in \mathrm{M}$, $\psi_{x^{-1}} \circ \psi_m \circ \psi_x$ transforms the origin of H into itself, therefore $x^{-1}mx \in \mathrm{L}$. This proves that $x^{-1}\mathrm{M}x \subset \mathrm{L}$, whence b).

Let L′ be a closed subgroup of G containing L. Then L′ is the topological semi-direct product of L and $\mathrm{L}' \cap \mathrm{H}$. If L′ is compact, then $\mathrm{L}' \cap \mathrm{H}$ is compact hence reduces to a point (GT, IV, §2, No. 2, Cor. 1 of Th. 2), therefore $\mathrm{L}' = \mathrm{L}$. This proves that L is a maximal compact subgroup of G; the same is therefore true of the subgroups that are the transforms of L by the inner automorphisms of G. The assertions c) and d) of Prop. 3 are then immediate consequences of b).

PROPOSITION 4. — *Let* G *be a locally compact group and* H *a closed normal subgroup of* G *such that* $\mathrm{K} = \mathrm{G}/\mathrm{H}$ *is compact. Then every continuous representation* $u$ *of* H *in* $\mathbf{R}$, *such that* $u(s\xi s^{-1}) = u(\xi)$ *for all* $\xi \in \mathrm{H}$ *and* $s \in \mathrm{G}$, *may be extended to a continuous representation of* G *in* $\mathbf{R}$.

Let $\mathrm{L} = \mathrm{G} \times \mathbf{R}$ and let M be the set of $(\xi, -u(\xi))$, where $\xi$ runs over H. It is clear that M is a closed normal subgroup of L. Let $\mathrm{L}' = \mathrm{L}/\mathrm{M}$ and let $\pi$ be the canonical mapping of L onto L′. The subgroup of L generated by M and $\mathbf{R}$ is $H\times\mathbf{R}$, hence is closed; therefore $\pi(\mathbf{R})$ is a closed subgroup N of $L'$. The restriction $\rho$ of $\pi$ to $\mathbf{R}$ is a bijective continuous representation of $\mathbf{R}$ onto N. Lemma 2 of Appendix 1 proves that $\rho$ is bicontinuous. Moreover, $L'/N$ is isomorphic to $L/(H\times\mathbf{R}) = G/H$, hence is compact. By Prop. 3, and taking into account the fact that N is in the center of $L'$, $L'$ is the product of N with another subgroup. Therefore there exists a continuous representation of $L'$ onto N that reduces on N to the identity mapping. Therefore there exists a continuous representation $v$ of L onto $\mathbf{R}$ that is trivial on M and reduces on $\mathbf{R}$ to the identity mapping. For $\xi \in H$, one has $v((\xi,0)) = v\big((\xi,-u(\xi))(e,u(\xi))\big) = u(\xi)$, which completes the proof.

*Lemma* 3. — *Let* G *be a topological group generated by a compact neighborhood of* $e$. *Let* H *be a closed subgroup of* G *such that the homogeneous space* G/H *is compact. Then* H *is generated by a compact neighborhood of* $e$ *in* H.

Let C be a compact set such that $G = CH$. Enlarging C if necessary, we can suppose that C generates G and that $G = \overset{\circ}{C}H$. Then $C^2$ is compact and is covered by the $\overset{\circ}{C}s$ ($s \in H$), which are open. Therefore there exist $s_1,\dots,s_n$ in H such that $C^2 \subset \overset{\circ}{C}s_1 \cup\cdots\cup \overset{\circ}{C}s_n$. Let $\Gamma$ be the subgroup of H generated by the $s_i$. Then $C^2 \subset C\Gamma$. By induction, it follows that $C^n \subset C\Gamma$ for every $n$, therefore $G = C\Gamma$. Every element of H may be put in the form $ab$ with $a \in C$, $b \in \Gamma$, whence $a \in H$, whence $a \in C\cap H$. Therefore H is generated by $C\cap H$ and the $s_i$, that is, by a compact set.

*Lemma* 4. — *Let* G *be a connected topological group,* D *a totally disconnected normal subgroup of* G. *Then* D *is contained in the center of* G.

For, let $d \in D$. The image of G under the continuous mapping $x \mapsto xdx^{-1}$ is a connected subset of D, hence reduces to $\{d\}$, which proves that $xd = dx$ for all $x \in G$.

PROPOSITION 5. — *Let* G *be a connected topological group admitting a discrete normal subgroup* D *such that* $K = G/D$ *is compact, and such that the commutator subgroup of* K *is dense in* K. *Then* D *is finite and* G *is compact.*

The group G is locally isomorphic to K (GT, III, §2, No. 6, Prop. 19), hence is locally compact; since it is connected, it is generated by a compact neighborhood of $e$. By Lemmas 3 and 4, D is a finitely generated abelian group, hence is isomorphic to a group $\mathbf{Z}^r \times D_1$ with $D_1$ finite (A, VII, §4, No. 7, Th. 3). Suppose that $r > 0$. Then there exists a representation $f$ of D *onto* $\mathbf{Z}$. By Prop. 4, $f$ may be extended to a continuous representation $g$ of G in $\mathbf{R}$. By passage to quotients, $g$ defines a continuous representation $g'$ of K in $\mathbf{R}/\mathbf{Z}$; since $\mathbf{R}/\mathbf{Z}$ is abelian, the kernel of $g'$ contains the commutator subgroup of K, therefore $g'$ is trivial; in other words, $g(\mathrm{G}) \subset \mathbf{Z}$. Since G is connected, it follows that $g(\mathrm{G}) = \{0\}$, which is absurd since $f(\mathrm{D}) = \mathbf{Z}$. Thus $r = 0$ and D is finite. Consequently G is compact (GT, III, §4, No. 1, Cor. 2 of Prop. 2).

### 3. Examples

*In this subsection* (with the exception of Examples 7 and 8), K *denotes a nondiscrete locally compact commutative field*; $dx$ *denotes a Haar measure on the additive group of* K.

Recall that $\operatorname{mod} x = |x|$ when $\mathrm{K} = \mathbf{R}$, $\operatorname{mod} x = |x|^2$ when $\mathrm{K} = \mathbf{C}$, $\operatorname{mod} x = |x|_p$ when $\mathrm{K} = \mathbf{Q}_p$.

*Example* 1. — *General linear group.*

Let A be the algebra $\mathbf{M}_n(\mathrm{K})$. The group $\mathrm{A}^*$ of invertible elements of A is none other than the general linear group $\mathbf{GL}(n, \mathrm{K})$. For every $X \in \mathrm{A}$, the reduced norm $\mathrm{Nrd}_{\mathrm{A}/\mathrm{K}}(X)$ is $\det X$; consequently $\mathrm{N}_{\mathrm{A}/\mathrm{K}}(X) = (\det X)^n$ (*Alg.*, Ch. VIII, §12, No. 3, Prop. 8; cf. A, III, §9, No. 3, Example 3). Since $X \mapsto {}^tX$ is an isomorphism of A onto the opposite algebra,

$$\mathrm{N}_{\mathrm{A}^0/\mathrm{K}}(X) = \mathrm{N}_{\mathrm{A}/\mathrm{K}}({}^tX) = \det({}^tX)^n = (\det X)^n .$$

Then, Prop. 16 of §1, No. 11 proves that the measure

$$\operatorname{mod}(\det X)^{-n} \cdot \bigotimes_{i,j} dx_{ij} \qquad (X = (x_{ij})) \tag{3}$$

is a left and right Haar measure on $\mathbf{GL}(n, \mathrm{K})$.

To determine the relatively invariant measures on $\mathbf{GL}(n, \mathrm{K})$, we shall rely on the following lemma:

*Lemma* 5. — *The continuous representations of* $\mathbf{GL}(n, \mathrm{K})$ *in* $\mathbf{C}^*$ *are the mappings of the form* $X \mapsto \chi(\det X)$, *where* $\chi$ *is a continuous representation of* $\mathrm{K}^*$ *in* $\mathbf{C}^*$.

Such a mapping is obviously a continuous representation of $\mathbf{GL}(n, \mathrm{K})$ in $\mathbf{C}^*$. Conversely, suppose that $\psi$ is a continuous representation of $\mathbf{GL}(n, \mathrm{K})$ in $\mathbf{C}^*$. For $x \in \mathrm{K}^*$, set

$$\widetilde{x} = \begin{pmatrix} x & & & & \\ & 1 & & 0 & \\ & & 1 & & \\ & 0 & & \ddots & \\ & & & & 1 \end{pmatrix}$$

and $\chi(x) = \psi(\widetilde{x})$. Then, for every matrix $X \in \mathbf{GL}(n,\mathrm{K})$, we have $(\det X^{-1})^{\sim} \cdot X \in \mathbf{SL}(n,\mathrm{K})$. Since $\mathbf{SL}(n,\mathrm{K})$ is the commutator subgroup of $\mathbf{GL}(n,\mathrm{K})$ (A, III, §8, No. 9, Cor. of Prop. 17), $\psi\big((\det X^{-1})^{\sim} \cdot X\big) = 1$, whence

$$\psi(X) = \psi\big((\det X)^{\sim}\big) = \chi(\det X)\,.$$

This established, Cor. 1 of Prop. 10 of §1, No. 8 proves that the relatively invariant measures on $\mathbf{GL}(n,\mathrm{K})$ are, up to a constant factor, the measures of the form

$$\chi(\det X) \cdot \bigotimes_{ij} dx_{ij} \qquad (X = (x_{ij}))\,, \tag{4}$$

where $\chi$ is a continuous representation of $\mathrm{K}^*$ in $\mathbf{C}^*$.

*Example 2. — Affine group.*

For every $X \in \mathbf{GL}(n,\mathrm{K})$ and every $x \in \mathrm{K}^n$, let $(X,x)$ be the affine linear mapping $\xi \mapsto X\xi + x$ in $\mathrm{K}^n$. The set of $(X,x)$ is the affine group G of $\mathrm{K}^n$ (A, II, §9, No. 4). The set T of translations is a closed normal subgroup of G, canonically isomorphic to $\mathrm{K}^n$; on the other hand, $\mathbf{GL}(n,\mathrm{K})$ is a closed subgroup of G, and G is the semi-direct product of $\mathbf{GL}(n,\mathrm{K})$ and $\mathrm{T} = \mathrm{K}^n$. One equips G with the (locally compact) topology for which G is the topological semi-direct product of $\mathbf{GL}(n,\mathrm{K})$ and T (GT, III, §2, No. 10). One has

$$(X,x) = (1,x) \cdot (X,0)\,.$$

On the other hand, if $X \in \mathbf{GL}(n,\mathrm{K})$ and $x \in \mathrm{T}$ then, for every $\xi \in \mathrm{K}^n$,

$$(X,0)(1,x)(X,0)^{-1}\xi = X(X^{-1}\xi + x) = \xi + Xx = (1,Xx)\xi\,,$$

therefore the automorphism $(1,x) \mapsto (X,0)(1,x)(X,0)^{-1}$ of T has modulus $\mathrm{mod}(\det X)$ (§1, No. 10, Prop. 15). In view of Example 1 and §2, No. 9, *Remark*, the measure

$$\mathrm{mod}(\det X)^{-n-1} \cdot \Big(\bigotimes_{ij} dx_{ij}\Big) \otimes \Big(\bigotimes_{i} dx_{i}\Big) \qquad (X = (x_{ij})\,,\ x = (x_i)) \tag{5}$$

is a left Haar measure on G. On the other hand, by Prop. 14 of §2, No. 9,

$$\Delta_{\mathrm{G}}\big((X,x)\big) = \Delta_{\mathbf{GL}(n,\mathrm{K})}(X)\,\Delta_{\mathrm{K}^n}(x)\,(\mathrm{mod}\ \det X)^{-1}\,,$$

or

$$\Delta_{\mathrm{G}}\big((X,x)\big) = \mathrm{mod}(\det X^{-1})\,. \tag{6}$$

Thus, a right Haar measure on G is given by

$$(\text{mod}\det X)^{-n}\cdot\Big(\bigotimes_{ij} dx_{ij}\Big)\otimes\Big(\bigotimes_{i} dx_i\Big). \tag{7}$$

*Example* 3. — *Strict triangular group.*

Let $[1,n]$ be the set of integers $m$ such that $1 \leqslant m \leqslant n$. Let J be a subset of $[1,n]\times[1,n]$ satisfying the following conditions:

1) if $(i,j)\in \mathrm{J}$ then $i<j$;

2) if $(i,j)\notin \mathrm{J}$ then, for every integer $k$ such that $i<k<j$, at least one of the two pairs $(i,k)$ and $(k,j)$ does not belong to J.

Let $\mathrm{T}_j$ be the set of matrices $Z=(z_{ij})_{1\leqslant i\leqslant n,\,1\leqslant j\leqslant n}$ with elements in K, such that $z_{ii}=1$, and $z_{ij}=0$ if $i\neq j$ and $(i,j)\notin \mathrm{J}$. This is a closed subset of $\mathbf{GL}(n,\mathrm{K})$. The mapping $Z\mapsto (z_{ij})_{(i,j)\in\mathrm{J}}$ is a homeomorphism of $\mathrm{T_J}$ onto $\mathrm{K}^s$ (where $s$ denotes the number of elements of J). If $Z'=(z'_{ij})\in \mathrm{T_J}$, then $Z'Z=(z''_{ij})$ with

$$z''_{ij}=z_{ij}+z'_{ij}+\sum_{i<h<j} z'_{ih}z_{hj}\qquad \text{for } i<j,$$

$$z''_{ij}=0 \text{ for } i>j,\qquad z''_{ii}=1,$$

whence $Z'Z\in\mathrm{T_J}$. If $\mathrm{T_J}$ is identified with $\mathrm{K}^s$, then the mapping $Z\mapsto Z'Z$ (for fixed $Z'$) is identified with an affine mapping, and its determinant is 1, as one sees by ordering the pairs $(i,j)\in\mathrm{J}$ lexicographically and applying the following lemma:

*Lemma* 6. — *Let* L *be a totally ordered finite set. For every* $\lambda\in\mathrm{L}$, *let* $\mathrm{V}_\lambda$ *be a free module of finite dimension over a commutative ring* $k$; *for* $\lambda,\mu$ *in* L *such that* $\lambda\leqslant\mu$, *let* $f_{\lambda\mu}\in\mathrm{Hom}_k(\mathrm{V}_\mu,\mathrm{V}_\lambda)$. *Then the linear mapping*

$$(v_\lambda)_{\lambda\in\mathrm{L}}\mapsto\Big(\sum_{\mu\geqslant\lambda} f_{\lambda\mu}(v_\mu)\Big)_{\lambda\in\mathrm{L}},$$

*from* $\prod_{\lambda\in\mathrm{L}}\mathrm{V}_\lambda$ *into* $\prod_{\lambda\in\mathrm{L}}\mathrm{V}_\lambda$, *has determinant* $\prod_{\lambda\in\mathrm{L}}\det f_{\lambda\lambda}$.

One reduces immediately to the case that L is an interval of integers, and the lemma then follows from A, III, §8, No. 6, formula (31).

If $Z\in\mathrm{T_J}$, one then sees that there exists $Z'\in\mathrm{T_J}$ such that $Z'Z=I_n$, whence $Z'=Z^{-1}$. Thus, $\mathrm{T_J}$ is a closed subgroup of $\mathbf{GL}(n,\mathrm{K})$. On the other hand, Prop. 15 of §1, No. 10 shows that the measure

$$\bigotimes_{(i,j)\in\mathrm{J}} dz_{ij}$$

is a left Haar measure on $T_J$. By calculating $ZZ'$ one sees in the same way that this measure is a right Haar measure on $T_J$.

There is an analogous result if, in the definition of $T_J$, the roles of rows and columns are interchanged.

When J is the set of pairs $(i,j)$ such that $i < j$, the group $T_J$ is called the *upper strict triangular group* of order $n$ over K, and is denoted $T_1(n, K)$. Its transpose is called the *lower strict triangular group.*

*Example* 4. — *Large triangular group.*

Let $n_1, \ldots, n_r$ be integers $\geqslant 1$. Set $p_k = n_1 + \ldots + n_{k-1}$ and $n = p_{r+1} = n_1 + \cdots + n_r$. Let $I_k$ be the set of integers $j$ such that $p_k < j \leqslant p_{k+1}$, and J the union of the $I_k \times I_l$ for $k < l$. Let G be the closed subgroup of $\mathbf{GL}(n, K)$ whose elements are the matrices $(Z_{kl})_{1\leqslant k\leqslant r,\, 1\leqslant l\leqslant r}$ such that:

1) each $Z_{kl}$ is a matrix $(z_{ij})_{i\in I_k,\, j\in I_l}$ of elements of K, with $n_k$ rows and $n_l$ columns;

2) $Z_{kl} = 0$ for $k > l$;

3) $Z_{kk} \in \mathbf{GL}(n_k, K)$ for $1 \leqslant k \leqslant r$.

The formula for block multiplication

$$(8)\qquad \begin{pmatrix} Z_{11} & 0 & \ldots & 0 \\ 0 & Z_{22} & \ldots & 0 \\ \vdots & \vdots & \ddots & \vdots \\ 0 & 0 & \ldots & Z_{rr} \end{pmatrix} \begin{pmatrix} 1 & Z_{12} & \ldots & Z_{1r} \\ 0 & 1 & \ldots & Z_{2r} \\ \vdots & \vdots & \ddots & \vdots \\ 0 & 0 & \ldots & 1 \end{pmatrix} = \begin{pmatrix} Z_{11} & Z_{11}Z_{12} & \ldots & Z_{11}Z_{1r} \\ 0 & Z_{22} & \ldots & Z_{22}Z_{2r} \\ \vdots & \vdots & \ddots & \vdots \\ 0 & 0 & \ldots & Z_{rr} \end{pmatrix}$$

shows that G is the topological semi-direct product of the subgroup D of elements $(Z_{kl}) \in G$ such that $Z_{kl} = 0$ for $k \neq l$ and the subgroup $T_J$ of Example 3. Moreover, D is isomorphic to the direct product of the groups $\mathbf{GL}(n_k, K)$ for $1 \leqslant k \leqslant r$.

Let J′ be the set of pairs $(j,i)$ for $(i,j) \in J$ and let H be the set of pairs $(i,j) \in [1,n] \times [1,n]$ not belonging to J′. Let $Z' = (z_{ij})_{1\leqslant i\leqslant n,\, 1\leqslant j\leqslant n}$ be an element of G. By Prop. 14 of §2, No. 9 and the above Examples 1 and 3, one obtains a left Haar measure on G by taking the image of the measure

$$\bigotimes_{k=1}^{r} \Big((\operatorname{mod} \det Z_{kk})^{-n_k} \cdot \bigotimes_{i,j\in I_k} dz_{ij}\Big) \otimes \Big(\bigotimes_{(i,j)\in J} dz_{ij}\Big)$$

under the mapping

$$((Z_{kk}),(Z_{kl})) \mapsto \begin{pmatrix} Z_{11} & Z_{11}Z_{12} & \dots & Z_{11}Z_{1r} \\ 0 & Z_{22} & \dots & Z_{22}Z_{2r} \\ \vdots & \vdots & \ddots & \vdots \\ 0 & 0 & \dots & Z_{rr} \end{pmatrix}.$$

Now, consider, for $k < l$, the vector space of matrices $Z_{kl} = (z_{ij})_{i\in \mathrm{I}_k, j\in \mathrm{I}_l}$. It is the direct sum of the $n_l$ subspaces $\mathrm{M}_j$ $(j \in \mathrm{I}_l)$ formed by the matrices such that $z_{ih} = 0$ for $h \neq j$. Each of these subspaces $\mathrm{M}_j$ is stable under the mapping $Z_{kl} \mapsto Z_{kk}Z_{kl}$, and the restriction of this mapping to $\mathrm{M}_j$ has matrix $Z_{kk}$. Consequently (§1, No. 10, Prop. 15) the image of the measure $\bigotimes_{i\in \mathrm{I}_k, j\in \mathrm{I}_l} dz_{ij}$ under the mapping $Z_{kl} \mapsto Z_{kk}Z_{kl}$ is

$$(\text{mod det } Z_{kk})^{-n_l} \cdot \bigotimes_{i\in \mathrm{I}_k, j\in \mathrm{I}_l} dz_{ij}\,.$$

A left Haar measure on G is therefore given by

$$\text{(9)} \qquad \prod_{k=1}^{r} (\text{mod det } Z_{kk})^{-q_k} \cdot \bigotimes_{(i,j)\in \mathrm{H}} dz_{ij}$$

with $q_k = \sum_{k\leqslant l\leqslant r} n_l = n - p_k$.

Let us calculate the *modulus* of G, again using Prop. 14 of §2. The groups D and $\mathrm{T_J}$ are unimodular; on the other hand:

$$\begin{pmatrix} Z_{11} & 0 & \dots & 0 \\ 0 & Z_{22} & \dots & 0 \\ \vdots & \vdots & \ddots & \vdots \\ 0 & 0 & \dots & Z_{rr} \end{pmatrix} \begin{pmatrix} 1 & Z_{12} & \dots & Z_{1r} \\ 0 & 1 & \dots & Z_{2r} \\ \vdots & \vdots & \ddots & \vdots \\ 0 & 0 & \dots & 1 \end{pmatrix} \begin{pmatrix} Z_{11} & 0 & \dots & 0 \\ 0 & Z_{22} & \dots & 0 \\ \vdots & \vdots & \ddots & \vdots \\ 0 & 0 & \dots & Z_{rr} \end{pmatrix}^{-1} = \begin{pmatrix} 1 & Z'_{12} & \dots & Z'_{1r} \\ 0 & 1 & \dots & Z'_{2r} \\ \vdots & \vdots & \ddots & \vdots \\ 0 & 0 & \dots & 1 \end{pmatrix},$$

where $Z'_{kl} = Z_{kk}Z_{kl}Z_{ll}^{-1}$. Taking into account Example 3, and Prop. 15 of §1, No. 10, and arguing as above, one sees that if $X = \text{diag}(Z_{11}, \dots, Z_{rr}) \in \mathrm{D}$ then the modulus of the automorphism $Z \mapsto X^{-1}ZX$ of $\mathrm{T_J}$ is

$$\prod_{k<l} (\text{mod det } Z_{kk})^{-n_l} (\text{mod det } Z_{ll})^{n_k}\,,$$

therefore

$$\Delta_{\mathrm{G}}(Z) = \prod_{k=1}^{r} (\text{mod det } Z_{kk})^{n+n_k-2q_k}. \tag{10}$$

The transposed group $\mathrm{G}'$ of $\mathrm{G}$ is studied in the same manner. For $\mathrm{G}'$, one finds as left Haar measure

$$\prod_{k=1}^{r} (\text{mod det } Z_{kk})^{-p_{k+1}} \cdot \bigotimes_{(j,i)\in \mathrm{H}} dz_{ij}\,,$$

and as modulus

$$\prod_{k=1}^{r} (\text{mod det } Z_{kk})^{n+n_k-2p_{k+1}}.$$

If in particular one takes $n_1 = \ldots = n_r = 1$, one finds as group $\mathrm{G}$ the group $T(n, \mathrm{K})^*$ of invertible elements of the subalgebra of $\mathbf{M}_n(\mathrm{K})$ formed by the matrices $X = (x_{ij})$ such that $x_{ij} = 0$ for $i > j$. This algebra, which we shall denote $\mathrm{T}(n, \mathrm{K})$, is called the *upper triangular algebra*, and the group $\mathrm{T}(n, \mathrm{K})^*$ is called the *upper large triangular group* of order $n$ over $\mathrm{K}$. The preceding formulas then take the following form: a left Haar measure on $\mathrm{T}(n, \mathrm{K})^*$ is

$$\prod_{i=1}^{n} (\text{mod } z_{ii})^{i-n-1} \cdot \bigotimes_{i\leqslant j} dz_{ij} \qquad (Z = (z_{ij})) \tag{9 bis}$$

and the modulus of $\mathrm{T}(n, \mathrm{K})^*$ is

$$\Delta_{\mathrm{T}(n,\mathrm{K})^*}(Z) = \prod_{i=1}^{n} (\text{mod } z_{ii})^{2i-n-1} \qquad (Z = (z_{ij})). \tag{10 bis}$$

For the transpose of $\mathrm{T}(n, \mathrm{K})^*$, or *lower large triangular group*, one finds as left Haar measure

$$\prod_{i=1}^{n} (\text{mod } z_{ii})^{-i} \cdot \bigotimes_{i\geqslant j} dz_{ij}\,,$$

and as modulus

$$\prod_{i=1}^{n} (\text{mod } z_{ii})^{n+1-2i}.$$

*Remark.* — The group $\mathrm{T}(n,\mathrm{K})^*$ is a closed subgroup of $\mathbf{GL}(n,\mathrm{K})$, and $\Delta_{\mathrm{T}(n,\mathrm{K})^*}\big((z_{ij})\big) = \prod_{i=1}^{n}(\operatorname{mod} z_{ii})^{2i-n-1}$. We saw in Example 1 that $\Delta_{\mathbf{GL}(n,\mathrm{K})} = 1$. If $n > 1$, the function

$$\Delta_{\mathrm{T}(n,\mathrm{K})^*}/\Delta_{\mathbf{GL}(n,\mathrm{K})}$$

on $\mathrm{T}(n,\mathrm{K})^*$ cannot be extended to a continuous representation of $\mathbf{GL}(n,\mathrm{K})$ in $\mathbf{C}^*$ (because such a representation would be equal to 1 on $\mathbf{SL}(n,\mathrm{K})$ by Lemma 5, whereas $\operatorname{mod}(z_{11})^{1-n} \neq 1$ for $z_{11}$ suitably chosen). It follows that the homogeneous space $\mathbf{GL}(n,\mathrm{K})/\mathrm{T}(n,\mathrm{K})^*$ *admits no relatively invariant measure* if $n > 1$ (§2, No. 6, Cor. 1 of Th. 3).

This homogeneous space may be identified, for $n = 2$, with the *projective line* over K. For, let $(e_1, e_2)$ be the canonical basis of $\mathrm{K}^2$. The group $\mathbf{GL}(2,\mathrm{K})$ operates transitively on the set of lines of $\mathrm{K}^2$ with 0 omitted, and the stabilizer of $\mathrm{K}e_1 - \{0\}$ is $\mathrm{T}(2,\mathrm{K})^*$.

*Example* 5. — *Special triangular group.*

Let us take up again the notations at the beginning of Example 4, and consider the subgroup $\mathrm{G}_1 = \mathrm{G} \cap \mathbf{SL}(n,\mathrm{K})$. This subgroup is the topological semi-direct product of the group $\mathrm{D}_1 = \mathrm{D} \cap \mathbf{SL}(n,\mathrm{K})$ with $\mathrm{T}_\mathrm{J}$. The group $\mathrm{D}_1$ has a normal subgroup A isomorphic to $\mathbf{SL}(n_r,\mathrm{K})$, namely the subgroup consisting of the elements $\operatorname{diag}(Z_{kk})$ with $Z_{kk} = 1$ for $k < r$. The homomorphism

$$\varphi : \operatorname{diag}(Z_{11}, \dots, Z_{rr}) \mapsto (Z_{11}, \dots, Z_{r-1,r-1})$$

of $\mathrm{D}_1$ into $\mathbf{GL}(n_1,\mathrm{K}) \times \cdots \times \mathbf{GL}(n_{r-1},\mathrm{K})$ is surjective and has kernel A. On the other hand, $\varphi$ is continuous. Taking into account Lemma 2 of Appendix I, $\mathrm{D}_1/\mathrm{A}$ may be identified with $\mathbf{GL}(n_1,\mathrm{K}) \times \cdots \times \mathbf{GL}(n_{r-1},\mathrm{K})$. We shall denote by $\mu$ the Haar measure of A (cf. Example 6) and by

$$\alpha = \bigotimes_{k=1}^{r-1}\Big((\operatorname{mod}\det Z_{kk})^{-n_k}\cdot \bigotimes_{i,j\in \mathrm{I}_k} dz_{ij}\Big) \otimes' d\mu(Z_{rr})$$

the Haar measure on $\mathrm{D}_1$ such that

$$\alpha/\mu = \bigotimes_{k=1}^{r-1}\Big((\operatorname{mod}\det Z_{kk})^{-n_k}\cdot \bigotimes_{i,j\in \mathrm{I}_k} dz_{ij}\Big)$$

(§2, No. 7, Prop. 10). One then shows as in Example 4 that a left Haar measure on $\mathrm{G}_1$ is given by

$$\operatorname{mod}\Big(\prod_{k=1}^{r-1}(\det Z_{kk})^{n_k-q_k}\Big)$$
$$\cdot\left[\bigotimes_{k=1}^{r-1}\big((\operatorname{mod}\det Z_{kk})^{-n_k}\cdot \bigotimes_{i,j\in\mathrm{I}_k} dz_{ij}\big)\otimes' d\mu(Z_{rr})\right]\otimes \bigotimes_{(i,j)\in \mathrm{J}} dz_{ij}\,.$$

Since $G_1$ is normal in $G$, the *modulus* of $G_1$ is the restriction of that of $G$ (§2, No. 7, Prop. 10 b)).

If $n_r = 1$, the subgroup $A$ reduces to the neutral element, and a left Haar measure on $G$ is

$$\operatorname{mod}\left(\prod_{k=1}^{r-1}(\det Z_{kk})^{-q_k}\right)\cdot\bigotimes_{k=1}^{r-1}\left(\bigotimes_{i,j\in I_k} dz_{ij}\right)\otimes\bigotimes_{(i,j)\in J} dz_{ij}\,.$$

If one takes $n_1 = n_2 = \ldots = n_r = 1$, the group $G_1$ obtained is called the *upper special triangular group* and its transpose $G_1'$ is called the *lower special triangular group.* A left Haar measure on $G_1$ is

$$\operatorname{mod}\left(\prod_{i=1}^{n-1} z_{ii}^{i-n-1}\right)\cdot\left(\bigotimes_{i=1}^{n-1} dz_{ii}\right)\otimes\left(\bigotimes_{1\leqslant i<j\leqslant n} dz_{ij}\right) \tag{11}$$

and the modulus of $G_1$ is

$$\operatorname{mod}\left(\prod_{i=1}^{n-1} z_{ii}^{2i-2n}\right). \tag{12}$$

For $G_1'$ one finds similarly the left Haar measure

$$\operatorname{mod}\left(\prod_{i=1}^{n-1} z_{ii}^{n-i-1}\right)\cdot\left(\bigotimes_{i=1}^{n-1} dz_{ii}\right)\otimes\left(\bigotimes_{1\leqslant j<i\leqslant n} dz_{ij}\right)$$

and modulus

$$\operatorname{mod}\left(\prod_{i=1}^{n-1} z_{ii}^{2n-2i}\right).$$

*Example 6. — Special linear group.*

The closed subgroups $T_1(n,K)$ and ${}^t(T(n,K)^*)$ of $\mathbf{GL}(n,K)$ have intersection $\{e\}$. Thus the mapping $(M,N)\mapsto M\cdot N$ is a continuous bijection $\varphi$ of $T_1(n,K)\times{}^t(T(n,K)^*)$ onto a subset $\Omega$ of $\mathbf{GL}(n,K)$.

*Lemma 7. —* a) *Let $U=(u_{ij})\in\mathbf{GL}(n,K)$. In order that $U\in\Omega$, it is necessary and sufficient that $\det(u_{ij})_{k\leqslant i,j\leqslant n}\neq 0$ for $k=2,3,\ldots,n$.*

b) *$\Omega$ is an open subset of $\mathbf{GL}(n,K)$.*

c) *The mapping $\varphi$ is a homeomorphism of $T_1(n,K)\times{}^t(T(n,K)^*)$ onto $\Omega$.*

In order that $U\in\Omega$, it is necessary and sufficient that there exist a $Z=(z_{ij})\in T_1(n,K)$ such that $ZU\in{}^t(T(n,K))$ (then necessarily

$ZU \in {}^t(\mathrm{T}(n,\mathrm{K})^*)$ since $U$ and $Z$ are invertible). By what we saw earlier, if $Z$ exists then it is unique. Thus, in order that $U \in \Omega$, it is necessary and sufficient that the linear system

$$\sum_{k=1}^{n} z_{ik}u_{kj} = 0 \qquad (1 \leqslant i < j \leqslant n)$$

(where $(z_{ij}) \in \mathrm{T}_1(n,\mathrm{K})$) admit a unique solution. Now, this system may be written

$$\sum_{k=i+1}^{n} z_{ik}u_{kj} = -u_{ij} \qquad (1 \leqslant i < j \leqslant n). \tag{13}$$

For fixed $i$, one has a system of $n-i$ equations in the unknowns $z_{i,i+1}$, $z_{i,i+2}$, $\ldots$, $z_{i,n}$; for these systems to admit unique solutions, it is necessary and sufficient that

$$\det(u_{kj})_{i+1\leqslant k\leqslant n,\ i+1\leqslant j\leqslant n} \neq 0$$

for $i = 1,2,\ldots,n-1$. This proves a). From this it follows that $\Omega$ is open in $\mathbf{GL}(n,\mathrm{K})$. On the other hand, on solving the system (13) by means of Cramer's formulas, the $z_{ij}$ are obtained as rational functions of the $u_{ij}$ with nonzero denominators in $\Omega$, therefore $Z$ depends continuously on $U$ in $\Omega$, which proves c).

Now let $\mathrm{G}'_1 \subset {}^t(\mathrm{T}(n,\mathrm{K})^*)$ be the lower special triangular group. The mapping $(M,N) \mapsto M \cdot N$ is a continuous bijection $\psi$ of $\mathrm{T}_1(n,\mathrm{K}) \times \mathrm{G}'_1$ onto a subset $\Omega'$ of $\mathbf{SL}(n,\mathrm{K})$.

*Lemma* 8. — a) *Let* $U = (u_{ij}) \in \mathbf{SL}(n,\mathrm{K})$. *In order that* $U \in \Omega'$, *it is necessary and sufficient that* $\det(u_{ij})_{k\leqslant i,j\leqslant n} \neq 0$ *for* $k = 2,3,\ldots,n$.

b) $\Omega'$ *is an open subset of* $\mathbf{SL}(n,\mathrm{K})$.

c) *The mapping* $\psi$ *is a homeomorphism of* $\mathrm{T}_1(n,\mathrm{K}) \times \mathrm{G}'_1$ *onto* $\Omega'$.

For, let $M \in \mathrm{T}_1(n,\mathrm{K})$ and $N \in {}^t(\mathrm{T}(n,\mathrm{K})^*)$. In order that $M \cdot N \in \mathbf{SL}(n,\mathrm{K})$, it is necessary and sufficient that $N \in \mathrm{G}'_1$. Therefore $\Omega' = \mathbf{SL}(n,\mathrm{K}) \cap \Omega$ and Lemma 8 follows at once from Lemma 7.

PROPOSITION 6. — a) *The group* $\mathbf{SL}(n,\mathrm{K})$ *is unimodular.*

b) *Let* $\mu_1$ *and* $\mu_2$ *be left Haar measures on the upper strict triangular group* $\mathrm{T}_1(n,\mathrm{K})$ *and the lower special triangular group* $\mathrm{G}'_1$, *respectively. The image of* $\mu_1 \otimes \mu_2$ *under the homeomorphism* $(M,N) \mapsto M \cdot N^{-1}$ *of* $\mathrm{T}_1(n,\mathrm{K}) \times \mathrm{G}'_1$ *onto* $\Omega'$ *is the restriction to* $\Omega'$ *of a Haar measure on* $\mathbf{SL}(n,\mathrm{K})$.

c) *The complement of* $\Omega'$ *in* $\mathbf{SL}(n,\mathrm{K})$ *is negligible for the Haar measure of* $\mathbf{SL}(n,\mathrm{K})$.

The group $\mathbf{GL}(n, \mathrm{K})$ is unimodular (Example 1), and $\mathbf{SL}(n, \mathrm{K})$ is a normal subgroup of $\mathbf{GL}(n, \mathrm{K})$, hence is unimodular (§2, No. 7, Prop. 10 b)). The assertion b) follows from a), Lemma 8, and Prop. 13 of §2, No. 9. Let us prove c). By Lemma 8 a), it suffices to prove the following: if $p\big((u_{ij})_{1 \leqslant i,j \leqslant n}\big)$ is a polynomial, not identically zero on $\mathbf{SL}(n, \mathrm{K})$, then the set $\mathrm{E}$ of $U \in \mathbf{SL}(n, \mathrm{K})$ such that $p(U) = 0$ is negligible for the Haar measure. Taking into account §1, No. 10, Cor. of Prop. 13, the topology of $\mathbf{SL}(n, \mathrm{K})$ has a countable base. It therefore suffices to prove that for every $U_0 \in \mathrm{E}$, there exists a neighborhood of $U_0$ in $\mathbf{SL}(n, \mathrm{K})$ whose intersection with $\mathrm{E}$ is negligible; or again that there exists a neighborhood $\mathrm{W}$ of $I$ in $\mathbf{SL}(n, \mathrm{K})$ such that $U_0^{-1}\mathrm{E} \cap \mathrm{W}$ is negligible. Let us take $\mathrm{W} = \Omega'$. In view of b), it all comes down to showing that the set of pairs $(M, N) \in \mathrm{T}_1(n, \mathrm{K}) \times \mathrm{G}_1'$ such that $p(U_0 MN) = 0$ is negligible for $\mu_1 \otimes \mu_2$. By the expressions for $\mu_1$ and $\mu_2$ (calculated in Examples 3 and 5), this will result from the following lemma:

*Lemma* 9. — *Let* $\psi$ *be a polynomial* $\neq 0$ *of* $\mathrm{K}[\mathrm{X}_1, \ldots, \mathrm{X}_r]$. *In the space* $\mathrm{K}^r$, *the set* $\mathrm{N}$ *defined by* $\psi(x_1, \ldots, x_r) = 0$ *is negligible for the Haar measure.*

Let us argue by induction on $r$. The lemma is evident for $r = 1$, since $\mathrm{N}$ is then a finite set. Changing if necessary the numbering of the variables, we can suppose that $\psi \notin \mathrm{K}[\mathrm{X}_1, \ldots, \mathrm{X}_{r-1}]$; write

$$\psi(\mathrm{X}_1, \ldots, \mathrm{X}_r) = \mathrm{X}_r^m \, \psi_0(\mathrm{X}_1, \ldots, \mathrm{X}_{r-1}) + \cdots + \psi_m(\mathrm{X}_1, \ldots, \mathrm{X}_{r-1})$$

with $m > 0$ and $\psi_0 \neq 0$. In the space $\mathrm{K}^{r-1}$, let $\mathrm{N}_0$ be the set defined by $\psi_0(x_1, \ldots, x_{r-1}) = 0$, which is negligible by the induction hypothesis. For every $(x_1, \ldots, x_{r-1}) \notin \mathrm{N}_0$, the set of $x_r \in \mathrm{K}$ such that $(x_1, \ldots, x_{r-1}, x_r) \in \mathrm{N}$ is finite, therefore negligible. Since $\mathrm{K}^r$ is countable at infinity (§1, No. 10, Cor. of Prop. 13), $\mathrm{N} \cap [(\mathrm{K}^{r-1} - \mathrm{N}_0) \times \mathrm{K}]$ is negligible in $\mathrm{K}^r$ (Ch. V, §8, No. 2, Prop. 4). Therefore $\mathrm{N}$ is negligible.

*Example* 7. — *Iwasawa decomposition of* $\mathbf{GL}(n, \mathrm{K})$.

In this example, $\mathrm{K}$ denotes one of the fields $\mathbf{R}$, $\mathbf{C}$, $\mathbf{H}$. If $\lambda \in \mathrm{K}$, $\overline{\lambda}$ is defined to be equal to $\lambda$ if $\mathrm{K} = \mathbf{R}$, and to the conjugate of $\lambda$ if $\mathrm{K} = \mathbf{C}$ or $\mathbf{H}$. Let $\mathrm{E}$ be a right vector space over $\mathrm{K}$ of dimension $n$, and let $\Phi$ be a nondegenerate positive hermitian form on $\mathrm{E}$.

*Lemma* 10. — *Let* $(f_1, f_2, \ldots, f_n)$ *be a basis of* $\mathrm{E}$.

a) *There exists one and only one orthonormal basis* $(e_1, e_2, \ldots, e_n)$ *of* $\mathrm{E}$ *such that* $f_i = e_1\alpha_{i1} + e_2\alpha_{i2} + \cdots + e_i\alpha_{ii}$ $(i = 1, 2, \ldots, n)$ *with* $\alpha_{ii} > 0$ *for all* $i$.

b) *For fixed* $\Phi$, *the* $e_i$ *and* $\alpha_{ij}$ *depend continuously on* $(f_1, \ldots, f_n) \in \mathrm{E}^n$.

Let $E_i = f_1K + f_2K + \cdots + f_iK$, which has dimension $i$. Let $g_i$ be a nonzero element of $E_i$ orthogonal to $E_{i-1}$ and such that $\Phi(g_i, g_i) = 1$. By induction on $i$, one sees that $(g_1, \ldots, g_i)$ is an orthonormal basis of $E_i$. In particular, $(g_1, \ldots, g_n)$ is an orthonormal basis of E. Let $\lambda_i = \Phi(f_i, g_i)$. Since $f_i \notin E_{i-1}$, one has $\lambda_i \neq 0$. Set $e_i = g_i\,|\lambda_i|\,\lambda_i^{-1}$. Then

$$\Phi(e_i, e_i) = |\lambda_i|^2\,\overline{\lambda_i}^{-1}\,\Phi(g_i, g_i)\lambda_i^{-1} = 1,$$

thus $(e_1, \ldots, e_i)$ is also an orthonormal basis of $E_i$; moreover, $\Phi(e_i, f_i) = |\lambda_i|\,\overline{\lambda_i}^{-1}\,\Phi(g_i, f_i) = |\lambda_i| > 0$, thus the $e_i$ have the properties of a). Let $(e'_1, \ldots, e'_n)$ be another orthonormal basis of E with the same properties. One sees by induction on $i$ that $(e'_1, \ldots, e'_i)$ must be a basis of $E_i$, therefore $e'_i = e_i\mu_i$ for some $\mu_i \in K$. Then

$$1 = \Phi(e'_i, e'_i) = \overline{\mu}_i\Phi(e_i, e_i)\mu_i = \overline{\mu}_i\mu_i,$$

and $0 < \Phi(e'_i, f_i) = \overline{\mu}_i\Phi(e_i, f_i)$, therefore $\mu_i > 0$ and $\mu_i^2 = 1$, thus $\mu_i = 1$, whence a). Suppose already proven that the $e_i$ and $\alpha_{ij}$ depend continuously on $(f_1, \ldots, f_n)$ for $i < i_0$, and let us prove that $e_{i_0}$ and the $\alpha_{i_0 j}$ depend continuously on $(f_1, \ldots, f_n)$. For $j < i_0$, $\overline{\alpha}_{i_0 j} = \Phi(f_{i_0}, e_j)$ depends continuously on $(f_1, \ldots, f_n)$ by the induction hypothesis. On the other hand,

$$\Phi(f_{i_0}, f_{i_0}) = |\alpha_{i_0 1}|^2 + |\alpha_{i_0 2}|^2 + \cdots + |\alpha_{i_0, i_0-1}|^2 + \alpha_{i_0 i_0}^2,$$

thus $\alpha_{i_0 i_0}$ depends continuously on $(f_1, \ldots, f_n)$. Therefore

$$e_{i_0} = (f_{i_0} - e_1\alpha_{i_0 1} - \cdots - e_{i_0-1}\alpha_{i_0, i_0-1})\alpha_{i_0 i_0}^{-1}$$

depends continuously on $(f_1, \ldots, f_n)$.

Henceforth let $E = K^n$ and let us take for $\Phi$ the form

$$\overline{x}_1 y_1 + \cdots + \overline{x}_n y_n.$$

Recall that $\mathbf{U}(n, K)$ denotes the corresponding unitary group. Even when K is noncommutative, we shall again denote by $T_1(n, K)$ the group of upper triangular matrices of $\mathbf{M}_n(K)$ whose diagonal entries are all 1.

PROPOSITION 7. — *Let $D_+^*$ be the group of diagonal matrices with diagonal elements $> 0$. The mapping $(U, D, T) \mapsto UDT$ is a homeomorphism of $U(n, K) \times D_+^* \times T_1(n, K)$ onto $\mathbf{GL}(n, K)$.*

Let $(\varepsilon_1, \ldots, \varepsilon_n)$ be the canonical basis of $\mathrm{K}^n$. Let $X \in \mathbf{GL}(n, \mathrm{K})$. Then the $X \cdot \varepsilon_i = f_i$ form a basis of E. To this basis $(f_i)$ one can associate a basis $(e_i)$ as in Lemma 10. Let $U$ be the matrix of the unitary automorphism of E that transforms $\varepsilon_i$ into $e_i$. Then

$$U^{-1} \cdot f_i = \varepsilon_1 \alpha_{i1} + \varepsilon_2 \alpha_{i2} + \cdots + \varepsilon_i \alpha_{ii}$$

with $\alpha_{ii} > 0$ for $i = 1, 2, \ldots, n$. Thus $X = UC$, where $C$ is the matrix

$$\begin{pmatrix} \alpha_{11} & \alpha_{21} & \ldots & \alpha_{n1} \\ 0 & \alpha_{22} & \ldots & \alpha_{n2} \\ \vdots & \vdots & \ddots & \vdots \\ 0 & 0 & \ldots & \alpha_{nn} \end{pmatrix}.$$

Moreover, $U$ and $C$ depend continuously on X by Lemma 10. On the other hand, formula (8) shows that $C$ may be put in the form $DT$ with $D \in \mathrm{D}_+^*$, $T \in \mathrm{T}_1(n, \mathrm{K})$, $D$ and $T$ depending continuously on $C$. The uniqueness of the decomposition $X = UDT$ follows from the uniqueness property of Lemma 10.

The homeomorphism of Prop. 7 is called the *Iwasawa decomposition* of $\mathbf{GL}(n, \mathrm{K})$.

The group $\mathrm{G} = \mathrm{D}_+^* \cdot \mathrm{T}_1(n, \mathrm{K})$ is the set of upper triangular matrices over K whose diagonal elements are $> 0$. Let us identity the element $(z_{ij})$ of this group with the element

$$\big((z_{ii})_{1 \leqslant i \leqslant n}, (z_{ij})_{1 \leqslant i < j \leqslant n}\big) \in (\mathbf{R}_+^*)^n \times \mathrm{K}^{n(n-1)/2}.$$

Arguing exactly as in Example 4, one finds as *right* Haar measure on this group the measure (when $\mathrm{K} = \mathbf{R}$)

$$\Big(\prod_{i=1}^{n} z_{ii}^{-i}\Big) \cdot \Big(\bigotimes_{i=1}^{n} dz_{ii}\Big) \otimes \Big(\bigotimes_{i<j} dz_{ij}\Big).$$

Then applying Prop. 13 of §2, No. 9, one sees that if $\mathbf{GL}(n, \mathrm{K})$ is identified with $\mathbf{U}(n, \mathrm{K}) \times \mathrm{G}$ by the mapping $(U, S) \mapsto US$, a Haar measure on $\mathbf{GL}(n, \mathrm{K})$ is given by (when $\mathrm{K} = \mathbf{R}$)

$$\Big(\prod_{i=1}^{n} z_{ii}^{-i}\Big) \cdot \alpha \otimes \Big(\bigotimes_{i=1}^{n} dz_{ii}\Big) \otimes \Big(\bigotimes_{i<n} dz_{ij}\Big), \tag{14}$$

where $\alpha$ denotes a Haar measure on $\mathbf{U}(n, \mathrm{K})$.

*Example 8.* — *Spaces of hermitian forms.*

In this example, K always denotes one of the fields $\mathbf{R}, \mathbf{C}, \mathbf{H}$. We write $\delta = \dim_{\mathbf{R}} \mathrm{K}$ (thus $\delta = 1$, 2 or 4). A hermitian form $\Phi$ on the right vector space $\mathrm{K}^n$ may be written

$$\Phi(x, y) = \Phi(x_1, \ldots, x_n, y_1, \ldots, y_n) = \sum_{i,j=1}^{n} \overline{x}_i h_{ij} y_j$$

with $h_{ij} = \overline{h}_{ji}$ for all $i$ and $j$. We denote by $\mathfrak{H}$ the vector space over $\mathbf{R}$ formed by the hermitian matrices of $\mathbf{M}_n(\mathrm{K})$. The mapping $(h_{ij}) \mapsto \Phi$ is an isomorphism of $\mathfrak{H}$ onto the vector space of hermitian forms on $\mathrm{K}^n$, by means of which we shall identify these two spaces. Let $\mathfrak{H}_+^* \subset \mathfrak{H}$ be the set of nondegenerate positive hermitian forms on $\mathrm{K}^n$. The set $\mathfrak{H}_+^*$ is *convex* in $\mathfrak{H}$; for, if $\Phi_1, \Phi_2$ are in $\mathfrak{H}_+^*$ and if $\lambda, \mu$ are two numbers $> 0$ such that $\lambda + \mu = 1$, it is clear that $\lambda\Phi_1 + \mu\Phi_2$ is a positive hermitian form; on the other hand, if $(\lambda\Phi_1 + \mu\Phi_2)(x, x) = 0$, then $\Phi_1(x, x) = \Phi_2(x, x) = 0$, therefore $x = 0$, so that $\lambda\Phi_1 + \mu\Phi_2$ is nondegenerate. Let us now show that $\mathfrak{H}_+^*$ is an *open* subset of $\mathfrak{H}$. Let S be the set of $x = (x_1, \ldots, x_n) \in \mathrm{K}^n$ such that $x_1\overline{x}_1 + \cdots + x_n\overline{x}_n = 1$; this is a compact subset of $\mathrm{K}^n$; if $\Phi \in \mathfrak{H}_+^*$, the function $x \mapsto \Phi(x, x)$ is continuous and $> 0$ on S, hence its infimum is $> 0$; if $\Phi' \in \mathfrak{H}$ is sufficiently near $\Phi$, it follows that $\Phi'(x, x) > 0$ for all $x \in \mathrm{S}$, so that $\Phi'$ is positive and nondegenerate.

The general linear group $\mathbf{GL}(n, \mathrm{K})$ operates continuously on the right in $\mathfrak{H}$ by $(X, \Phi) \mapsto \Phi \circ X$, that is, by $(X, H) \mapsto {}^t\overline{X} \cdot H \cdot X$, where $H$ denotes the hermitian matrix corresponding to $\Phi$. It is clear that $\mathfrak{H}_+^*$ is stable under $\mathbf{GL}(n, \mathrm{K})$. More precisely, by *Alg.*, Ch. IX, §6, No. 1, Cor. 1 of Th. 1,[1] $\mathfrak{H}_+^*$ is the orbit under $\mathbf{GL}(n, \mathrm{K})$ of the form $\sum_{i=1}^{n} \overline{x}_i y_i$ corresponding to the identity matrix $I_n$. The stabilizer of this form is $\mathbf{U}(n, \mathrm{K})$. By Lemma 2 of App. I, $\mathfrak{H}_+^*$ may be identified, as a topological homogeneous space, with $\mathbf{GL}(n, \mathrm{K})/\mathbf{U}(n, \mathrm{K})$.

For every $X \in \mathbf{GL}(n, \mathrm{K})$, let $\widetilde{X}$ be the automorphism $H \mapsto {}^t\overline{X} \cdot H \cdot X$ of the *real* vector space $\mathfrak{H}$. If $\mu$ denotes the Haar measure of the additive group $\mathfrak{H}$, one has $\widetilde{X}^{-1}(\mu) = |\det \widetilde{X}| \cdot \mu$ (§1, No. 10, Cor. 1 of Prop. 15). Let us show that

$$|\det \widetilde{X}| = |\mathrm{N}(X)|^{\lambda}, \tag{15}$$

where N denotes the norm in $\mathbf{M}_n(\mathrm{K})$ *regarded as an* $\mathbf{R}$-*algebra*, and where $\lambda = 1 - \dfrac{\delta - 2}{\delta n}$. It suffices to verify (15) for $X$ running over a system of

[1] Cf. TVS, V, §2, No. 4, Cor. 1 of Th. 2.

generators of $\mathbf{GL}(n, \mathrm{K})$, hence (A, II, §10, No. 13, Cor. 2 of Prop. 14) for $X$ of the following types:

a) $X$ is the matrix of a mapping of the form

$$(x_1, \ldots, x_n) \mapsto (x_{\sigma(1)}, \ldots, x_{\sigma(n)}),$$

where $\sigma \in \mathfrak{S}_n$. In this case, a power of $X$ is equal to $1$, therefore $|\det \widetilde{X}| = |\mathrm{N}(X)| = 1$.

b) $X$ is the matrix of a mapping of the form

$$(x_1, \ldots, x_n) \mapsto (ax_1, x_2, \ldots, x_n).$$

Then, if $(h_{ij}) \in \mathfrak{H}$, one has $\widetilde{X}((h_{ij})) = (h'_{ij})$ with $h'_{11} = \bar{a}h_{11}a = |a|^2 h_{11}$, $h'_{1i} = \bar{a}h_{1i}$ for $i > 1$, $h'_{ij} = h_{ij}$ for $i > 1$, $j > 1$; therefore

$$|\det \widetilde{X}| = |a|^2 |a|^{\delta(n-1)} = |a|^{2+\delta(n-1)}.$$

On the other hand, if $Y = (y_{ij}) \in \mathbf{M}_n(\mathrm{K})$, then $XY = (y'_{ij})$ with $y'_{1j} = ay_{1j}$ and $y'_{ij} = y_{ij}$ for $i > 1$; therefore $|\mathrm{N}(X)| = |a|^{\delta n}$. The formula (15) is again verified.

c) $X$ is the matrix of a mapping of the form

$$(x_1, \ldots, x_n) \mapsto (x_1 + bx_2, x_2, \ldots, x_n).$$

Then $\widetilde{X}((h_{ij})) = (h'_{ij})$ with $h'_{11} = h_{11}$, $h'_{12} = h_{12} + h_{11}b$, $h'_{1i} = h_{1i}$ for $i > 2$, $h'_{22} = h_{22} + \bar{b}h_{12} + \overline{h}_{12}b + \bar{b}h_{11}b$, $h'_{2i} = h_{2i} + \bar{b}h_{1i}$ for $i > 2$, $h'_{ij} = h_{ij}$ for $i > 2$, $j > 2$. Taking into account Lemma 6, one sees that $|\det \widetilde{X}| = 1$. One verifies similarly that $|\mathrm{N}(X)| = 1$, which completes the proof of formula (15).

This established, the measure $|\mathrm{N}(H)|^{-\lambda/2}\, d\mu(H)$ on $\mathfrak{H}$ is invariant under $\mathbf{GL}(n, \mathrm{K})$, since

$$\begin{aligned}\widetilde{X}^{-1}(|\mathrm{N}(H)|^{-\lambda/2}\, d\mu(H)) &= \big|\mathrm{N}(\widetilde{X}(H))\big|^{-\lambda/2} \cdot |\det \widetilde{X}|\, d\mu(H)\\ &= |\mathrm{N}(H)|^{-\lambda/2}\, |\mathrm{N}(X)|^{-\lambda}\, |\det \widetilde{X}|\, d\mu(H) = |\mathrm{N}(H)|^{-\lambda/2}\, d\mu(H).\end{aligned}$$

If $H \in \mathfrak{H}_+^*$, then $H = \widetilde{X}(\mathrm{I}_n) = {}^t\overline{X}X$ for some $X \in \mathbf{GL}(n, \mathrm{K})$, therefore $\mathrm{N}(H) = \overline{\mathrm{N}(X)}\,\mathrm{N}(X) > 0$. Consequently, on $\mathfrak{H}_+^*$, the *unique* (up to a constant factor) measure invariant under $\mathbf{GL}(n, \mathrm{K})$ (cf. §2, No. 6, Th. 3) is the measure

$$d\gamma(H) = \mathrm{N}(H)^{-\lambda/2}\, d\mu(H).$$

In particular,

$$d\gamma(H) = (\det H)^{-(n+1)/2}\, d\mu(H) \quad \text{when } \mathrm{K} = \mathbf{R},$$
$$d\gamma(H) = (\det H)^{-n}\, d\mu(H) \quad \text{when } \mathrm{K} = \mathbf{C}.$$

# APPENDIX I

**Lemma* 1. — *Let* X *be a locally compact space,* R *an open equivalence relation in* X, *such that the quotient space* X/R *is paracompact; let* $\pi$ *be the canonical mapping of* X *onto* X/R. *There exists a continuous real-valued function* $F \geqslant 0$ *on* X *such that:*

a) F *is not identically zero on any equivalence class with respect to* R;

b) *for every compact subset* K *of* X/R, *the intersection of* $\overset{-1}{\pi}(K)$ *with* Supp F *is compact.*

To each point $u \in X/R$ let us associate a function $f_u \in \mathscr{K}_+(X)$ such that $f_u$ is not identically zero on $\overset{-1}{\pi}(u)$; let $\Omega_u$ be the open set of points where $f_u > 0$; thus $u \in \pi(\Omega_u)$. Since $\pi$ is an open mapping, the $\pi(\Omega_u)$ form an open covering of X/R. There exists a locally finite open covering $(U_\iota)_{\iota \in I}$, finer than the covering by the $\pi(\Omega_u)$, then (GT, IX, §4, No. 3, Prop. 3) a partition of unity $(g_\iota)_{\iota \in I}$ on X/R subordinate to the covering $(U_\iota)$. For every $\iota \in I$, choose a $u_\iota$ such that $U_\iota \subset \pi(\Omega_{u_\iota})$. The function $F_\iota = (g_\iota \circ \pi) \cdot f_{u_\iota}$ belongs to $\mathscr{K}(X)$ and has support contained in $\overset{-1}{\pi}(U_\iota)$. The supports of the $F_\iota$ therefore form a locally finite family, so that one can define a continuous function $F \geqslant 0$ on X by setting $F = \sum_{\iota \in I} F_\iota$. For every $u \in X/R$, there exists an $\iota$ such that $g_\iota(u) > 0$ and therefore $u \in U_\iota$; next, there exists an $x \in \Omega_{u_\iota}$ such that $\pi(x) = u$; then $f_{u_\iota}(x) > 0$ and $g_\iota(\pi(x)) > 0$, therefore $F_\iota(x) > 0$ and *a fortiori* $F(x) > 0$; this proves that F has the property a). Finally, let K be a compact subset of X/R. There exists a finite subset J of I such that, for $\iota \in I - J$, one

*Same as TG, IX, §4, n° 4, prop. 5 (absent from GT).

has $U_\iota \cap K = \varnothing$, therefore $\overset{-1}{\pi}(K) \cap \operatorname{Supp} F_\iota = \varnothing$. Then the set

$$\overset{-1}{\pi}(K) \cap \operatorname{Supp} F = \overset{-1}{\pi}(K) \cap \Big(\bigcup_{\iota \in I} \operatorname{Supp} F_\iota\Big) = \overset{-1}{\pi}(K) \cap \Big(\bigcup_{\iota \in J} \operatorname{Supp} F_\iota\Big)$$

is compact.

*Lemma 2. — Let* $G$ *be a locally compact group countable at infinity,* $M$ *a Baire space. Suppose that* $G$ *operates on the left continuously and transitively in* $M$. *For every* $x \in M$, *let* $H_x$ *be the stabilizer of* $x$ *in* $G$, *so that the mapping* $s \mapsto sx$ *of* $G$ *onto* $M$ *defines, by passage to the quotient, a continuous bijection* $\varphi_x$ *of* $G/H_x$ *onto* $M$. *Then* $\varphi_x$ *is a homeomorphism of* $G/H_x$ *onto* $M$ (in other words (GT, III, §2, No. 5) $M$ is a topological homogeneous space).

Let $x_0 \in M$. It suffices to prove (*loc. cit.*, Prop. 15) that the mapping $s \mapsto sx_0$ transforms every neighborhood $V$ of $e$ in $G$ into a neighborhood of $x_0$ in $M$. Let $W$ be a symmetric compact neighborhood of $e$ such that $W^2 \subset V$. By hypothesis, $G$ is the union of a sequence of compact sets, hence of a sequence $(s_n W)$ of translates of $W$. Then $M$ is the union of the sequence of compact sets $(s_n W x_0)$. Since $M$ is a Baire space, there exists an index $n$ such that $s_n W x_0$ has an interior point $s_n w x_0$ (where $w \in W$). Consequently $x_0$ is an interior point of

$$w^{-1} s_n^{-1}(s_n W x_0) = w^{-1} W x_0 \subset V x_0 ,$$

so that $Vx_0$ is a neighborhood of $x_0$ in $M$.

# APPENDIX II

**Lemma* 1. — *Let* X, B *be two locally compact spaces,* $\pi$ *a mapping of* X *into* B, $\nu$ *a positive measure on* B. *Let* $b \mapsto \lambda_b$ $(b \in \mathrm{B})$ *be a* $\nu$*-adequate family of positive measures on* X *such that, for every* $b \in \mathrm{B}$, *the measure* $\lambda_b$ *is concentrated on* $\overset{-1}{\pi}(b)$. *Set* $\mu = \int \lambda_b \, d\nu(b)$, *and assume that the mapping* $\pi$ *is* $\mu$*-measurable.*

a) *If* $\mathrm{N} \subset \mathrm{B}$ *is locally* $\nu$*-negligible, then* $\overset{-1}{\pi}(\mathrm{N})$ *is locally* $\mu$*-negligible.*

b) *If* $f$ *is a* $\nu$*-measurable function on* B (*with values in a topological space*), *then* $f \circ \pi$ *is* $\mu$*-measurable on* X.

Let K be a compact subset of X. We are to prove that $\overset{-1}{\pi}(\mathrm{N}) \cap \mathrm{K}$ is $\mu$-negligible and that the restriction of $f \circ \pi$ to K is $\mu$-measurable. Now, K is the union of a $\mu$-negligible set and a sequence of compact subsets $\mathrm{K}_n$ such that $\pi|\mathrm{K}_n$ is continuous. It suffices to show that $\overset{-1}{\pi}(\mathrm{N}) \cap \mathrm{K}_n$ is $\mu$-negligible and that the restriction of $f \circ \pi$ to $\mathrm{K}_n$ is $\mu$-measurable. We may therefore assume henceforth that $\pi|\mathrm{K}$ is continuous. Then $\pi(\mathrm{K}) = \mathrm{K}'$ is compact. Since $\overset{-1}{\pi}(\mathrm{N}) \cap \mathrm{K} = \overset{-1}{\pi}(\mathrm{N} \cap \mathrm{K}') \cap \mathrm{K}$ and $\mathrm{N} \cap \mathrm{K}'$ is $\nu$-negligible, we may assume henceforth that N is $\nu$-negligible. Then N is contained in a $\nu$-negligible set N′ that is a countable intersection of open sets (Ch. IV, §4, No. 6, Cor. 2 of Th. 4). Since $\pi$ is $\mu$-measurable, $\overset{-1}{\pi}(\mathrm{N}')$ is $\mu$-measurable (Ch. IV, §5, No. 5, Prop. 7), therefore $\overset{-1}{\pi}(\mathrm{N}') \cap \mathrm{K}$ is $\mu$-integrable, and (Ch. V, §3, No. 4, Th. 1)

$$\mu(\overset{-1}{\pi}(\mathrm{N}') \cap \mathrm{K}) = \int_{\mathrm{B}} \lambda_b(\overset{-1}{\pi}(\mathrm{N}') \cap \mathrm{K}) d\nu(b).$$

---

*Cf. Ch. V, §6, No. 6, Cor. 1 of Prop. 10; the second edition of Ch. V was not available when Ch. VII was published.

Now, if $b \notin \mathrm{N}'$, $\overset{-1}{\pi}(\mathrm{N}') \cap \mathrm{K}$ is $\lambda_b$-negligible since $\lambda_b$ is concentrated on $\overset{-1}{\pi}(b)$ by hypothesis. Therefore $\mu\big(\overset{-1}{\pi}(\mathrm{N}') \cap \mathrm{K}\big) = 0$. A fortiori, $\overset{-1}{\pi}(\mathrm{N}) \cap \mathrm{K}$ is $\mu$-negligible and a) has indeed been proved. On the other hand, there exists a partition of $\mathrm{K}'$ formed by a $\nu$-negligible set M and a sequence $(\mathrm{K}'_n)$ of compact sets such that $f|\mathrm{K}'_n$ is continuous. Then the restriction of $f \circ \pi$ to each set $\overset{-1}{\pi}(\mathrm{K}) \cap \mathrm{K}'_n$ is continuous, and $\overset{-1}{\pi}(\mathrm{M}) \cap \mathrm{K}$ is $\mu$-negligible by a), therefore the restriction of $f \circ \pi$ to K is indeed $\mu$-measurable.

# Exercises

§1

1) Let G be a compact group. Show that every continuous representation $\varphi$ of G in $\mathbf{R}_+^*$ is such that $\varphi(G) = \{1\}$. From this, deduce that a relatively invariant positive measure on G is invariant.

2) Let G be a topological group, such that the commutator subgroup of G is dense in G. Show that every continuous representation $\varphi$ of G in a Hausdorff abelian group satisfies $\varphi(G) = \{e\}$. From this, deduce that if G is locally compact, then every relatively invariant complex measure on G is invariant. In particular, G is unimodular.

3) Let G be a locally compact group, $\mu$ a left Haar measure on G, and $\nu = \Delta_G^{-1/2} \cdot \mu$. Show that

$$\gamma(s)\nu = \Delta_G(s)^{1/2}\nu, \qquad \delta(s)\nu = \Delta_G(s)^{1/2}\nu, \qquad \check{\nu} = \nu.$$

4) Let $G, G'$ be two locally compact groups, V (resp. $V'$) an open neighborhood of the neutral element of G (resp. $G'$), $\varphi$ a local isomorphism of $G'$ with G, defined on $V'$, such that $\varphi(V') = V$. Show that $\Delta_G \circ \varphi$ is the restriction of $\Delta_{G'}$ to $V'$.

5) For every $a$ belonging to the multiplicative group $\mathbf{Q}^*$, let $\varphi(a)$ be the automorphism of the additive group $\mathbf{R}$ defined by $\varphi(a)x = ax$. Equip $\mathbf{Q}^*$ with the discrete topology. Let G be the topological semi-direct product of $\mathbf{Q}^*$ and $\mathbf{R}$ defined by $\varphi$ (GT, III, §2, No. 10). Show that G is locally isomorphic to $\mathbf{R}$, but is not unimodular.

6) Let G be a locally compact group having an open and compact subgroup H. Show that for every automorphism $\varphi$ of G, $\operatorname{mod}\varphi \in \mathbf{Q}_+^*$. (Observe that $\varphi(H) \cap H$ has finite index in H and in $\varphi(H)$.) Show that the equality $\Delta_G(s) = 1$ defines an open subgroup of G containing H.

7) Let G be a locally compact group, $\beta$ a left Haar measure on G, A a subset of G, and B a relatively compact $\beta$-integrable subset of G such that $\beta(B) > 0$. Show

that if the compact subsets of $A \cdot B$ have bounded measures for $\beta$, then A is relatively compact. (Imitate the argument of Prop. 2.)

8) Let G be a locally compact group, A a dense subset of G, $\alpha$ a left Haar measure on G, and H an $\alpha$-measurable subset of G having the following property: for every $s \in A$, $sH \cap (\complement H)$ and $H \cap (\complement sH)$ are locally $\alpha$-negligible. Show that either H is locally $\alpha$-negligible or its complement is locally $\alpha$-negligible. (Show that $\varphi_H \cdot \alpha$ is left invariant.)

9) Adopt the notations of the proof of Th. 1 of No. 2. Let $\beta$ be the unique left Haar measure on G such that $\beta(f_0) = 1$. Show that, for every $f \in \mathscr{K}(G)$, $I_g(f)$ tends to $\beta(f)$ with respect to $\mathfrak{B}$. (Let $a$ be a cluster point of $g \mapsto I_g(f)$ with respect to $\mathfrak{B}$. There exists an ultrafilter $\mathfrak{U}$ finer than $\mathfrak{B}$ such that $I_g(f)$ tends to $a$ with respect to $\mathfrak{U}$. On the other hand, $I_g(f)$ tends to $\beta(f)$ with respect to $\mathfrak{U}$.)

10) Let G be a locally compact group, $\mu$ a left Haar measure on G. Show that every $\mu$-integrable set A is contained in a countable union of compact sets. (Reduce to the case that A is open, and observe that there exists an open subgroup of G that is a countable union of compact subsets.)

¶ 11) Let G be a nondiscrete locally compact group countable at infinity, $\beta$ a left Haar measure on G. Show that every compact neighborhood V of $e$ contains a normal subgroup H of G, $\beta$-negligible and such that G/H is a Polish locally compact group. (One can proceed as follows: let L be the open subgroup of G generated by V; let $b_1, b_2, \ldots$ be representatives of the left cosets with respect to L. Form a decreasing sequence $(V_n)$ of symmetric neighborhoods of $e$ such that: 1) $V_n^2 \subset V_{n-1}$; 2) $xV_nx^{-1} \subset V_{n-1}$ for every $x \in V$; 3) $b_iV_nb_i^{-1} \subset V_{n-1}$ for $1 \leqslant i \leqslant n$; 4) $\beta(V_n) \leqslant 1/n$. Set $H = V_1 \cap V_2 \cap \cdots$.)

Let $(f_n)$ be a sequence of numerical functions uniformly continuous for the left uniform structure of G. Show that H can be constructed in such a way that, in addition, the $f_n$ are constant on the cosets of H. (In the preceding construction, take $V_n$ to be such that $|f_i(x) - f_i(y)| \leqslant 1/n$ for $x^{-1}y \in V_n$ and $1 \leqslant i \leqslant n$.)

Let $(g_n)$ be a sequence of $\beta$-integrable numerical functions on G. Show that H can be constructed in such a way that, in addition, each $g_n$ is equal almost everywhere to a function that is constant on the cosets of H.

12) Let K be a nondiscrete locally compact field, E a left topological vector space over K.

*a*) If E is 1-dimensional, then E is isomorphic to $K_s$ (imitate the proof of Prop. 2 of TVS, I, §2, No. 2, replacing the absolute value by the function $\operatorname{mod}_K$).

*b*) If E is $n$-dimensional, then E is isomorphic to $K_s^n$ (imitate the proof of Th. 2 of TVS, I, §2, No. 3).

*c*) Assume that E is locally compact. Let F be a linear subspace of E of finite dimension $n$. For every $a \in K$, let $\operatorname{mod}_E(a)$ be the modulus of $x \mapsto ax$ in E. Since F is closed in E by *b*), one can form $\operatorname{mod}_{E/F}(a)$. Show that $\operatorname{mod}_E(a) = \operatorname{mod}_K(a)^n \operatorname{mod}_{E/F}(a)$. Show that if $\operatorname{mod}_K(a) < 1$ and $F \neq E$, then $\operatorname{mod}_{E/F}(a) < 1$. (One has $a^n \to 0$ as $n \to +\infty$; from this, deduce that $\operatorname{mod}_{E/F}(a^n) \to 0$ by arguing as in Prop. 12 of No. 10.) Deduce from this that $n \leqslant \log \operatorname{mod}_E(1/a)/\log \operatorname{mod}_K(1/a)$, hence that E is finite-dimensional.

(We shall see in CA, VI, that the topology of a locally compact field can be defined by an absolute value. The final result of *c*) will then be a special case of Th. 3 of TVS, I, §2, No. 4.)

¶ 13) Let G be a locally compact group operating continuously on the left in a locally compact Polish space T, $\beta$ a left Haar measure on G, and $\nu$ a positive measure on T quasi-invariant under G.

*a*) There exists a function $(s, x) \mapsto \chi(s, x) > 0$ on $G \times T$, locally $(\beta \otimes \nu)$-integrable, such that for every $s \in G$, the function $x \mapsto \chi(s^{-1}, x)$ is locally $\nu$-integrable and satisfies $\gamma(s)\nu = \chi(s^{-1}, \cdot) \cdot \nu$. (Show that $(s,x) \mapsto (s,sx)$ transforms $\beta \otimes \nu$ into an equivalent measure, by making use of criterion 3) of Th. 2 of Ch. V, §5, No. 5; let $\chi(s^{-1},x)d\beta(s)\,d\nu(x)$ be this equivalent measure. For $\varphi \in \mathscr{K}(\mathrm{G})$ and $\psi \in \mathscr{K}(\mathrm{T})$,

$$\int \varphi(s)\,d\beta(s) \int \psi(sx)\,d\nu(x) = \int \varphi(s)\,d\beta(s) \int \psi(x)\chi(s^{-1},x)\,d\nu(x),$$

whence $\int \psi(sx)\,d\nu(x) = \int \psi(x)\chi(s^{-1},x)\,d\nu(x)$ except on a locally $\beta$-negligible set $\mathrm{N}(\psi)$. Next make use of Lemma 1 of Ch. VI, §3, No. 1; then modify $\chi$ on a $(\beta \otimes \nu)$-negligible set.)

*b*) Show that for any $s, t$ in G, one has

$$\chi(st,x) = \chi(s,tx)\chi(t,x)$$

except on a $\nu$-negligible set of values of $x$ (make use of the relation $\gamma(st)\nu = \gamma(s)(\gamma(t)\nu)$).

*c*) Show that the function $\chi$ of *a*) is determined up to a locally $(\beta \otimes \nu)$-negligible subset of $\mathrm{G} \times \mathrm{T}$.

¶ 14) Let T be a locally compact space, $\mathscr{U}$ a uniform structure on T for which there exists an entourage $\mathrm{V}_0$ such that $\mathrm{V}_0(t)$ is compact for all $t \in \mathrm{T}$ (GT, II, §4, Exer. 9). On the other hand, let $\Gamma$ be a uniformly equicontinuous group of homeomorphisms of T (for the uniform structure $\mathscr{U}$), such that there exists an $a \in \mathrm{T}$ whose orbit under $\Gamma$ is dense. Show that there exists on T a positive measure $\neq 0$ invariant under $\Gamma$, and that this measure is unique up to a constant factor. One may proceed as follows:

1° As for the existence of the invariant measure, follow the method of No. 2, Th. 1; one proves first that if K is a compact subset of T and U is an open neighborhood of $a$, there exists a finite number of elements $\sigma_i \in \Gamma$ such that $\mathrm{K} \subset \bigcup_i \sigma_i(\mathrm{U})$.

2° As for uniqueness, observe first that the entourages of $\mathscr{U}$ that are invariant under every homeomorphism $\sigma \times \sigma$ of $\mathrm{T} \times \mathrm{T}$, for $\sigma \in \Gamma$, form a fundamental system of entourages $\mathfrak{S}$. For every relatively compact set $\mathrm{A} \subset \mathrm{T}$ and every relatively compact set $\mathrm{B} \subset \mathrm{T}$ with nonempty interior, let $(\mathrm{A} : \mathrm{B})$ be the smallest number of elements of a covering of A by sets of the form $\sigma\mathrm{B}$, where $\sigma \in \Gamma$; if $\mathrm{C} \subset \mathrm{T}$ is a third relatively compact set with nonempty interior, then $(\mathrm{A}:\mathrm{C}) \leqslant (\mathrm{A}:\mathrm{B})(\mathrm{B}:\mathrm{C})$. Let K be a compact subset of T, $\mathrm{L} \supset \mathrm{K}$ a relatively compact open set, $\mathrm{V} \in \mathfrak{S}$ a symmetric open entourage contained in $\mathrm{V}_0$ such that $\mathrm{V}(\mathrm{K}) \subset \mathrm{L}$, and $\mathrm{W} \in \mathfrak{S}$ a symmetric closed entourage contained in V. For every symmetric entourage $\mathrm{U} \in \mathfrak{S}$ such that $\mathrm{UW} \subset \mathrm{V}$ and $\mathrm{WU} \subset \mathrm{V}$, show that, for every positive measure $\nu$ on T invariant under $\Gamma$,

$$(\mathrm{W}(a):\mathrm{U}(a))\nu(\mathrm{K}) \leqslant (\mathrm{L}:\mathrm{U}(a))\nu(\mathrm{V}(a)) \tag{1}$$

$$(\mathrm{K}:\mathrm{U}(a))\nu(\mathrm{W}(a)) \leqslant (\mathrm{V}(a):\mathrm{U}(a))\nu(\mathrm{L}). \tag{2}$$

For this, one proves that if $(\sigma_i(\mathrm{U}(a)))$ is a covering of L formed by $(\mathrm{L}:\mathrm{U}(a))$ sets, every $x \in \mathrm{K}$ belongs to at least $(\mathrm{W}(a):\mathrm{U}(a))$ sets of the form $\sigma_i(\mathrm{V}(a))$, and that every $y \in \mathrm{L}$ belongs to at most $(\mathrm{V}(a):\mathrm{U}(a))$ sets of the form $\sigma_i(\mathrm{W}(a))$ (note that if $y \in \sigma_i(\mathrm{W}(a))$, then $\sigma_i(\mathrm{U}(a))$ is contained in $\mathrm{V}(x)$).

Let $\mathfrak{F}$ be an ultrafilter finer than the section filter of $\mathfrak{S}$; let $\mathrm{A}_0$ be a nonempty, relatively compact open set in T, and set $\lambda_{\mathrm{U}}(\mathrm{A}) = (\mathrm{A}:\mathrm{U}(a))/(\mathrm{A}_0:\mathrm{U}(a))$ for every symmetric entourage $\mathrm{U} \in \mathfrak{S}$ and every relatively compact subset A of T. Let $\lambda(\mathrm{A}) = \lim_{\mathfrak{F}} \lambda_{\mathrm{U}}(\mathrm{A})$; for every compact subset K of T, set $\lambda'(\mathrm{K}) = \inf \lambda(\mathrm{B})$, where B runs over the set of relatively compact open neighborhoods of K. Deduce from (1) and (2) that

$$\lambda'(\mathrm{W}(a))\nu(\mathrm{K}) \leqslant \lambda(\mathrm{L})\nu(\mathrm{W}(a))$$
$$\lambda'(\mathrm{K})\nu(\mathrm{W}(a)) \leqslant \lambda'(\mathrm{W}(a))\nu(\mathrm{L}).$$

Conclude from this that if $K_1, K_2$ are two compact subsets of T, and $L_1 \supset K_1$, $L_2 \supset K_2$ are two relatively compact open sets, then

$$\lambda'(K_1)\nu(K_2) \leqslant \lambda(L_2)\nu(L_1),$$

and finally that $\lambda'(K_1)\nu(K_2) = \lambda'(K_2)\nu(K_1)$.

¶ 15) Let G be a locally compact group, H a closed subgroup of G, so that G operates continuously on G/H. Assume that H satisfies the following condition: for every neighborhood V of $e$ in G, there exists a neighborhood U of $e$ such that $HU \subset VH$ (note that this condition is satisfied for every subgroup H of G if $e$ admits a fundamental system of neighborhoods invariant under every inner automorphism of G). Show that there then exists a nonzero positive measure on G/H invariant under G. (Use the same method as in the proof of Th. 1 of No. 2, on observing, on the one hand, that as U runs over the set of neighborhoods of $e$ in G, the images of the sets HUH under the canonical mapping $\pi : G \to G/H$ form a fundamental system of neighborhoods of $\pi(H)$; on the other hand, for every neighborhood V of $e$ in G and every compact subset K of G, there exist a neighborhood U of $e$ in G and a compact subset L of G such that every set of the form $s$HUH that intersects KH is of the form $s'$HUH with $s' \in L$.)

¶ 16) Let G be a compact abelian group, E a dense subset of G that is stable under the law of composition of G. Assume that, for *every* bounded numerical function $f$ on E, there is defined a number $M(f)$ having the following property: for any two bounded functions $f, g$ on E, for any elements $a_i, b_k$ of E ($1 \leqslant i \leqslant m$, $1 \leqslant k \leqslant n$), and for any real numbers $\alpha_i, \beta_k$ ($1 \leqslant i \leqslant m$, $1 \leqslant k \leqslant n$) such that

$$\sum_i \alpha_i f(xa_i) \leqslant \sum_k \beta_k g(xb_k)$$

for all $x \in E$, one then has the inequality

$$\Big(\sum_i \alpha_i\Big) M(f) \leqslant \Big(\sum_k \beta_k\Big) M(g).$$

Assume in addition that $M(1) = 1$ (cf. TVS, IV, Appendix).

*a*) Let $\mu$ be the normalized Haar measure on G. Show that for every continuous numerical function $f$ on G, $\int f\,d\mu = M(f|E)$. (By considering a suitable partition of unity, approximate the constant function equal to $\int f\,d\mu$ on E by means of functions of the form $x \mapsto \sum_i \alpha_i f(xa_i)$, where $a_i \in E$.)

*b*) Deduce from *a*) that if one sets $\nu(B) = M(\varphi_B)$ for every subset B of E, then $\nu(U \cap E) \geqslant \mu(U)$ for every open subset U of G, and $\nu(F \cap E) \leqslant \mu(F)$ for every closed subset F of G. For every $\mu$-quadrable subset P of G (Ch. IV, §5, Exer. 17 d)), one has $\nu(P \cap E) = \mu(P)$.

¶ 17) Let T be a locally compact space, S a subset of T equipped with a law of composition $(x, y) \mapsto xy$ that makes it a monoid[1] (not necessarily having *a priori* a neutral element), and which is continuous on S×S when S is equipped with the topology

[1] *Monoïde*, as defined in the early editions of *Alg.*, Ch. I, is not required to have a neutral element, as it is in the bound edition of *Algèbre*; what is meant here is an associative magma (A, I, §1, No. 3, Def. 5).

induced by that of T. Assume in addition that every element of S is cancellable[2] (A, I, §2, No. 2). Let $\mu$ be a bounded positive measure on T, concentrated on S (so that S is $\mu$-measurable) and of total mass 1. Assume that $\mu$ is *left invariant* in the following sense: for every numerical function $f$ defined on S, continuous and bounded (hence $\mu$-measurable by Ch. IV, §5, No. 5, Cor. of Prop. 8), one has $\int f(sx)\,d\mu(x) = \int f(x)d\mu(x)$ for all $s \in \mathrm{S}$. It comes to the same to say that $\mu(s\mathrm{K}) = \mu(\mathrm{K})$ for every compact subset K of S.

*a*) Consider the product measure $\mu \otimes \mu$ on $\mathrm{T} \times \mathrm{T}$; show that there exist compact subsets K of S such that the images of $\mathrm{K} \times \mathrm{K}$ under the two continuous mappings $(x, y) \mapsto (x, xy)$ and $(x, y) \mapsto (xy, x)$ have measure arbitrarily close to 1 (use the Lebesgue–Fubini theorem). Conclude from this that these images have a common point, and deduce therefrom that S has a neutral element (cf. A, I, §2, Exer. 9).

*b*) Show that S is a compact group. (First prove that for every $x \in \mathrm{S}$, the inner measure $\mu_*(x\mathrm{S})$ is equal to 1, hence that $x\mathrm{S}$ is measurable and that the measure is concentrated on $x\mathrm{S}$; $x\mathrm{S}$ is then a monoid to which the result of *a*) can be applied, which shows that S is a group. Argue as in Prop. 2 to prove that S is compact. Finally, make use of Exer. 21 of GT, III, §4.[3])

18) Let X be a locally compact space, G a compact group operating continuously on X, E the orbit of a point of X, and $\mathscr{F}$ a vector space of continuous numerical functions on X such that, for every function $f \in \mathscr{F}$ and every $s \in \mathrm{G}$, one has $\gamma(s)f \in \mathscr{F}$; assume in addition that $\mathscr{F}$ contains the constant functions on X. Let $x_0 \in \mathrm{X}$ be a point invariant under G and such that $|f(x_0)| \leqslant \sup_{y \in \mathrm{E}} |f(y)|$ for *every* function $f \in \mathscr{F}$. Show that one then has $f(x_0) = \int_{\mathrm{G}} f(s \cdot z)\,ds$ for every $z \in \mathrm{E}$ and every $f \in \mathscr{F}$. (Show that there exists a positive measure $\nu$ on E, of total mass 1, such that $f(x_0) = \int_{\mathrm{E}} f(z)\,d\nu(z)$ for every $f \in \mathscr{F}$, and apply the Lebesgue–Fubini theorem.) *The case that $\mathrm{X} = \mathbf{R}^n$, G is the orthogonal group, and $x_0 = 0$; apply the formula (7) of §2.*

19) Let X be a compact space, A a normed algebra over **R** with unity element, and G a compact group; assume that G operates continuously on A and on X and that, for every $s \in \mathrm{G}$, $a \mapsto s \cdot a$ is an automorphism of the algebra A such that $\|s \cdot a\| = \|a\|$ for all $a \in \mathrm{A}$. A mapping $f$ of X into A is said to be *covariant* under G if

$$f(s \cdot x) = s \cdot f(x)$$

for all $s \in \mathrm{G}$ and $x \in \mathrm{X}$. Let B be a subring of $\mathrm{A}^{\mathrm{X}}$ consisting of covariant continuous functions and containing all of the covariant continuous mappings of X into **R** (**R** being identified with a subalgebra of A; one observes that for such a mapping $g$, $g(s \cdot x) = g(x)$ for all $s \in \mathrm{G}$ and $x \in \mathrm{X}$). Let $f$ be a covariant continuous mapping of X into A, and assume that for every $y \in \mathrm{X}$, there exists a mapping $g_y \in \mathrm{B}$ such that $f(y) = g_y(y)$. Then, for every $\varepsilon > 0$, there exists a mapping $g \in \mathrm{B}$ such that $\|f(x) - g(x)\| \leqslant \varepsilon$ for all $x \in \mathrm{X}$. (Make use of a suitable continuous partition of unity $(\varphi_i)$ on X and introduce the functions $h_i$ such that $h_i(x) = \int_{\mathrm{G}} \varphi_i(s \cdot x)\,ds$.)

¶ 20) Let G be a locally compact group, $\mu$ a left Haar measure on G, A and B two subsets of G.

*a*) Assume that one of the following two conditions holds:

$\alpha$) A is $\mu$-integrable;

$\beta$) $\mu^*(\mathrm{A}) < +\infty$ and B is $\mu$-measurable.

---

[2] *Régulier*, renamed *simplifiable* in the bound edition of *Algèbre*.

[3] In the cited exercise, *monoïde* of the French original is translated as "semigroup"; but *semi-groupe* refers to another concept (A, I, §2, Exer. 11).

Show that, in each of these two cases, the function $f(s) = \mu^*(s\mathrm{A} \cap \mathrm{B})$ is uniformly continuous on G for the right uniform structure of G. (For any two subsets M, N of G, one sets

$$d(\mathrm{M},\mathrm{N}) = \mu^*((\mathrm{M} \cap \complement\mathrm{N}) \cup (\mathrm{N} \cap \complement\mathrm{M})).$$

First consider the case that A is compact; making use of Ch. IV, §4, No. 6, Th. 4, show that for every $\varepsilon > 0$, there exists a neighborhood U of $e$ in G such that, for every $s \in \mathrm{G}$ and $t \in \mathrm{U}$,

$$d(s\mathrm{A} \cap \mathrm{B},\ st\mathrm{A} \cap \mathrm{B}) \leqslant \varepsilon.$$

Then apply Exer. 13 of Ch. IV, §5. If B is $\mu$-measurable, $\mu^*(\mathrm{A}) < +\infty$ and $(\mathrm{A}_n)$ is a decreasing sequence of integrable subsets of G containing A such that $\inf(\mu(\mathrm{A}_n)) = \mu^*(\mathrm{A})$, show that

$$\mu^*(s\mathrm{A} \cap \mathrm{B}) = \inf(\mu^*(s\mathrm{A}_n \cap \mathrm{B}))$$

(Ch. V, 1st edn., §2, No. 2, Lemma 1);[4] note on the other hand that $\mu(\mathrm{A}_n - \mathrm{A}_{n+1})$ tends to 0 with $1/n$.)

*b*) If $\mathrm{A}^{-1}$ is $\mu$-integrable and $\mu^*(\mathrm{B}) < +\infty$, then the function $f$ is uniformly continuous for both the right and left uniform structures of G; moreover, one then has $\int_{\mathrm{G}} f(s)\,d\mu(s) = \mu(\mathrm{A}^{-1})\mu^*(\mathrm{B})$. (Reduce to the case that B is integrable; observe then that $\mu(s\mathrm{A} \cap \mathrm{B}) = \mu(\mathrm{A} \cap s^{-1}\mathrm{B})$, that $\varphi_{s\mathrm{A}\cap\mathrm{B}} = \varphi_{s\mathrm{A}}\varphi_{\mathrm{B}}$ and that $\varphi_{s\mathrm{A}}(t) = \varphi_{t\mathrm{A}^{-1}}(s)$.)

*c*) Deduce from *a*) that in the two cases considered, the interiors of AB and BA are not empty if A and B are not negligible. (Cf. Ch. VIII, §4, No. 6, Prop. 17.)

*d*) In the group $\mathrm{G} = \mathbf{SL}_2(\mathbf{R})$, give an example of a compact set A and a $\mu$-measurable set B such that $f(s) = \mu(s\mathrm{A} \cap \mathrm{B})$ is not uniformly continuous for the left uniform structure. (Observe that there exist a sequence $(t_n)$ of elements of G tending to $e$ and a sequence $(s_n)$ of elements of G such that the sequence $(s_n^{-1}t_ns_n)$ tends to the point at infinity.)

*e*) Give an example of two nonmeasurable sets A, B in a locally compact group G, of finite outer measure and such that the function $s \mapsto \mu^*(s\mathrm{A} \cap \mathrm{B})$ is not continuous (cf. Ch. IV, §4, Exer. 8).

21) *a*) Let G be a locally compact group, $\mu$ a left Haar measure on G. Show that if S is a *stable* subset of G such that $\mu_*(\mathrm{S}) > 0$, then the interior of S is nonempty (cf. Exer. 20). In particular, if a subgroup H of G has nonzero inner measure, then H is an open subgroup of G.

*b*) If G is compact, then every stable subset S of G such that $\mu_*(\mathrm{S}) > 0$ is a compact open subgroup of G (observe that the interior of S is stable, and make use of Exer. 17 *b*)).

*c*) Assume that G is compact and abelian, written additively, and that there exists in G an element $a$ of infinite order. Show that there exists a stable subset S of G such that $a \in \mathrm{S}$, $-a \notin \mathrm{S}$ and $\mathrm{G} = \mathrm{S} \cup (-\mathrm{S})$ (use Zorn's lemma); deduce from *b*) that S is not measurable and that $\mu_*(\mathrm{S}) = 0$.

22) Let G be a locally compact group, $\mu$ a left Haar measure on G, and A an integrable subset of G such that $\mu(\mathrm{A}) > 0$. Show that the set H(A) of $s \in \mathrm{G}$ such that $\mu(\mathrm{A}) = \mu(\mathrm{A} \cap s\mathrm{A})$ is a compact group. (Observe that H(A) is closed in G, with the help of Exer. 20. To see that H(A) is compact, consider a compact subset B of A such that $\mu(\mathrm{B}) > \mu(\mathrm{A})/2$ and prove that $\mathrm{H}(\mathrm{A}) \subset \mathrm{BB}^{-1}$.)

¶ 23) Let G be a locally compact abelian group, written additively, $\mu$ a Haar measure on G, and A, B two integrable subsets of G.

---

[4] Suppressed from the second edition, the lemma asserts that if $f, g$ are numerical functions $\geqslant 0$ with $g$ measurable, then $\int^* fg\,d\mu = \inf \int^* \varphi g\,d\mu$ as $\varphi$ runs over the set of measurable functions $\geqslant f$.

*a*) For every $s \in G$, let

$$A' = \sigma_s(A, B) = A \cup (B + s), \qquad B' = \tau_s(A, B) = (A - s) \cap B.$$

Show that $\mu(A') + \mu(B') = \mu(A) + \mu(B)$ and $A' + B' \subset A + B$.

*b*) Assume that 0 belongs to $A \cap B$. A pair $(A', B')$ of integrable subsets of G is said to be *derived* from $(A, B)$ if there exist a sequence $(s_k)_{1 \leqslant k \leqslant n}$ of elements of G and two sequences $(A_k)_{0 \leqslant k \leqslant n}$, $(B_k)_{0 \leqslant k \leqslant n}$ of subsets of G such that $A_0 = A$, $B_0 = B$, $A_k = \sigma_{s_k}(A_{k-1}, B_{k-1})$, $B_k = \tau_{s_k}(A_{k-1}, B_{k-1})$ for $1 \leqslant k \leqslant n$, $s_k \in A_{k-1}$ for $1 \leqslant k \leqslant n$, and $A' = A_n$, $B' = B_n$. Show that there exists a sequence of pairs $(E_n, F_n)$ such that $E_0 = A$, $F_0 = B$, $(E_{n+1}, F_{n+1})$ is derived from $(E_n, F_n)$, and $\mu((E_n - s) \cap F_n) \geqslant \mu(F_{n+1}) - 2^{-n}$ for every $n$ and every $s \in E_n$. Set $E_\infty = \bigcup_n E_n$, $F_\infty = \bigcap_n F_n$. Show that for every $s \in E_\infty$,

$$\mu((E_\infty - s) \cap F_\infty) = \mu(F_\infty).$$

*c*) Assume that $\mu(F_\infty) > 0$. Show that the function

$$f(s) = \mu((E_\infty - s) \cap F_\infty)$$

can only take on the values 0 and $\mu(F_\infty)$, and that the set C of $s \in G$ such that $f(s) = \mu(F_\infty)$ is *open* and *closed*, is such that $\mu(C) = \mu(E_\infty)$, and is the closure of $E_\infty$ (use Exer. 20 *a*) and *b*)). On the other hand let D be the set of $s \in F_\infty$ such that the intersection of $F_\infty$ with every neighborhood of $s$ has measure $> 0$. Show that $\mu(D) = \mu(F_\infty)$ and that

$$E_\infty + D \subset C;$$

from this, deduce that D is contained in the subgroup $H(C)$ defined in Exer. 22, and that $H(C)$ is compact and open in G. Finally, show that $C + H(C) = C$, that $\mu(C) \geqslant \mu(A) + \mu(B) - \mu(H(C))$, and that $C \subset A + B$ (consider the measure of $E_\infty \cap (c - F_\infty)$ for every $c \in C$).

*d*) Deduce from *c*) that, for two integrable subsets A, B of G: *either* $\mu_*(A + B) \geqslant \mu(A) + \mu(B)$; *or else* there exists a compact open subgroup H of G such that $A + B$ contains a coset of H, in which case $\mu_*(A + B) \geqslant \mu(A) + \mu(B) - \mu(H)$. The case that G is connected.

¶ 24 *a*) In **R**, let A (resp. B) be the set of numbers $x$ whose proper dyadic expansion $x = x_0 + \sum_{i=1}^{\infty} x_i 2^{-i}$ ($x_0$ an integer, $x_i = 0$ or $x_i = 1$ for $i \geqslant 1$) is such that $x_i = 0$ for $i$ even and $> 0$ (resp. $i$ odd). Show that, for Lebesgue measure, A and B have measure zero but $A + B = \mathbf{R}$.

*b*) Deduce from *a*) that there exists a basis H of **R** (over **Q**) contained in $A \cup B$, hence of measure zero. The set $P_1$ of numbers $rh$, where $r \in \mathbf{Q}$ and $h \in H$, is also of measure zero.

*c*) Denote by $P_n$ the set of real numbers at most $n$ of whose coordinates relative to the basis H are nonzero. Show that if $P_n$ is negligible and $P_{n+1}$ is measurable, then $P_{n+1}$ is negligible. (Let $h_0 \in H$; show first that the set S of $x \in P_{n+1}$ whose coordinate relative to $h_0$ is $\neq 0$, is negligible. Using Exer. 20, show that if $P_{n+1}$ were not negligible, there would exist two distinct points $x', x''$ of $P_{n+1} \cap \complement S$ such that $(x' - x'')/h_0$ is rational, and deduce from this a contradiction.)

*d*) Deduce from *a*) and *b*) that there exist in **R** two negligible sets C, D such that $C + D$ is not measurable.

¶ 25) *a*) Let $f$ be a positive numerical function defined on $\mathbf{R}$, integrable (for the Lebesgue measure $\mu$ on $\mathbf{R}$), bounded, and with compact support. Let $\gamma = \sup_{t\in\mathbf{R}} f(t)$. For every $w \in \mathbf{R}$, denote by $\mathrm{U}_f(w)$ the set of $t \in \mathbf{R}$ such that $f(t) \geqslant w$; set $\nu_f(w) = \mu^*(\mathrm{U}_f(w))$. Show that for every $\alpha > 1$,

$$\int_{-\infty}^{+\infty} f^\alpha(t)\,dt = \int_0^\gamma \nu_f(w)\alpha w^{\alpha-1}\,dw.$$

*b*) Let $g$ be a second numerical function satisfying the same conditions as $f$, and set $\delta = \sup_{t\in\mathbf{R}} g(t)$. Let $h$ be the function defined on $\mathbf{R}^2$ by $h(u,v) = f(u) + g(v)$ if $f(u)g(v) \neq 0$, and $h(u,v) = 0$ otherwise; finally, set $k(t) = \sup_{u+v=t} h(u,v)$, so that $k$ is positive, integrable, bounded, and has compact support. Show that for every $\alpha > 1$,

$$\int_{-\infty}^{+\infty} k^\alpha(t)\,dt \geqslant (\gamma+\delta)^\alpha\left(\frac{1}{\gamma^\alpha}\int_{-\infty}^{+\infty} f^\alpha(t)\,dt + \frac{1}{\delta^\alpha}\int_{-\infty}^{+\infty} g^\alpha(t)\,dt\right).$$

(Observe that for $0 < w < 1$, one has $\mathrm{U}_k(\gamma w + \delta w) \supset \mathrm{U}_f(\gamma w) + \mathrm{U}_g(\delta w)$, and use *a*) and Exer. 23 *d*).)

*c*) Let $\mu_n$ be Lebesgue measure on $\mathbf{R}^n$, A and B two integrable subsets of $\mathbf{R}^n$. Show that (*Brunn–Minkowski inequality*)

$$((\mu_n)_*(\mathrm{A}+\mathrm{B}))^{1/n} \geqslant (\mu_n(\mathrm{A}))^{1/n} + (\mu_n(\mathrm{B}))^{1/n}.$$

(Reduce to the case that A and B are compact. Then argue by induction on $n$ using Exer. 23 *d*), the Lebesgue–Fubini theorem, the inequality proved in *b*), as well as Hölder's inequality.)

¶ 26) Let G be a locally compact group, $\mu$ a left Haar measure on G, $k$ an integer $\geqslant 1$, and A an integrable set. Show that for every $\varepsilon > 0$, there exists a neighborhood U of $e$ in G having the following property: for every finite subset S of $k$ elements in U, the set of $s \in \mathrm{G}$ such that $s\mathrm{S} \subset \mathrm{A}$ has measure $\geqslant (1-\varepsilon)\mu(\mathrm{A})$. (Reduce to the case that A is compact and take U to be such that

$$\mu(\mathrm{AU}) \leqslant \left(1 - \frac{\varepsilon}{k-1}\right)\mu(\mathrm{A}).)$$

For every finite subset H of G, set $p(\mathrm{H}) = \mathrm{Card}(\mathrm{H})$ and $q(\mathrm{H}) = 1$ if $\mathrm{H} \neq \varnothing$, $q(\mathrm{H}) = 0$ if $\mathrm{H} = \varnothing$; by evaluating the integrals

$$\int_{\mathrm{G}} p(\mathrm{A} \cap s\mathrm{S})\,ds \quad \text{and} \quad \int_{\mathrm{G}} q(\mathrm{A} \cap s\mathrm{S})\,ds,$$

show that if, for $h \leqslant k$, $\mathrm{M}_h$ denotes the set of $s \in \mathrm{G}$ for which $\mathrm{Card}(\mathrm{A} \cap s\mathrm{S}) = h$, then

$$\sum_{h=1}^{k} h\mu(\mathrm{M}_h) = k \cdot \mu(\mathrm{A}) \quad \text{and} \quad \sum_{h=1}^{k} \mu(\mathrm{M}_h) \leqslant \mu(\mathrm{AU}).$$

From this, conclude that $\sum_{h=1}^{k-1} h\mu(\mathrm{M}_h) \leqslant k\varepsilon \cdot \mu(\mathrm{A})$.

27) Let G be a group operating on a set X. A subset P (resp. C) of X is said to be a G-*filling* (resp. G-*covering*) if, for every $s \neq e$ in G, one has $s\mathrm{P} \cap \mathrm{P} = \varnothing$ (resp. if $\mathrm{X} = \bigcup_{s\in\mathrm{G}} s\mathrm{C}$). One calls G-*paving* a subset P that is both a G-filling and a G-covering.

*a)* Assume that X is locally compact, that G is countable, operates continuously in X, and that there exists a nonzero positive measure $\mu$ on X invariant under G. Let P and C be a G-filling and a G-covering that are $\mu$-integrable. Show that $\mu(\mathrm{C}) \geqslant \mu(\mathrm{P})$. (Observe that $\mu(\mathrm{C}) \geqslant \sum_{s\in\mathrm{G}} \mu(\mathrm{C} \cap s\mathrm{P})$.)

*b)* Assume in addition that there exists on X a uniform structure compatible with the topology of X and admitting a fundamental system $\mathfrak{S}$ of open entourages invariant under G. On the other hand, one denotes by $\Delta(\mathrm{G})$ the infimum of the numbers $\mu(\mathrm{C})$ over all the integrable G-coverings C of X. Let V be an entourage belonging to $\mathfrak{S}$, and let $a \in \mathrm{X}$ be such that $\mu(\mathrm{V}(a)) > \Delta(\mathrm{G})$; show that there exists an $s \in \mathrm{G}$ such that $s \neq e$ and $(a, sa) \in \overset{-1}{\mathrm{V}} \circ \mathrm{V}$.

*c)* Assume that X is a locally compact group, $\mu$ a left Haar measure, and G a countable subgroup of X, operating by left translation. With the same definition of $\Delta(\mathrm{G})$ as in *b*), show that if A is an integrable subset of X such that $\mu(\mathrm{A}) > \Delta(\mathrm{G})$, then there exists $s \in \mathrm{G} \cap \mathrm{AA}^{-1}$ such that $s \neq e$. Special case that $\mathrm{X} = \mathbf{R}^n$ and G is a discrete subgroup of rank $n$ in $\mathbf{R}^n$: $\Delta(\mathrm{G})$ is then equal to the absolute value of the determinant of a basis of G over $\mathbf{Z}$ with respect to the canonical basis of $\mathbf{R}^n$; if A is a symmetric closed convex set with nonempty interior[5] in $\mathbf{R}^n$ such that $\mu(\mathrm{A}) \geqslant 2^n\Delta(\mathrm{G})$, show that there exists a point of $\mathrm{A} \cap \mathrm{G}$ distinct from 0 (*Minkowski's theorem*).

*d)* With hypotheses as in *a*), let $f$ be a function $\geqslant 0$ and $\mu$-integrable on X. Show that there exist two points $a, b$ of X such that $\mu(\mathrm{C}) \sum_{s\in\mathrm{G}} f(sa) \geqslant \int_{\mathrm{X}} f(x)\, d\mu(x)$ and $\mu(\mathrm{P}) \sum_{s\in\mathrm{G}} f(sb) \leqslant \int_{\mathrm{X}} f(x)\, d\mu(x)$. (Observe that if $g$ is an integrable function $\geqslant 0$, E a $\mu$-integrable set in X, then there exists a $c \in \mathrm{E}$ such that $\int_{\mathrm{E}} g(x)\, d\mu(x) \leqslant g(c)\mu(\mathrm{E})$, and a $c' \in \mathrm{E}$ such that $\int_{\mathrm{E}} g(x)\, d\mu(x) \geqslant g(c')\mu(\mathrm{E})$.)

¶ 28) With hypotheses as in Exer. 27 *a*), assume in addition that there exists a $\mu$-integrable G-paving F.

*a)* For every $\mu$-integrable numerical function $f \geqslant 0$ defined on X, set $\widetilde{f}(x) = \sum_{s\in\mathrm{G}} f(sx)$, so that $\int_{\mathrm{F}} \widetilde{f}(x)\, d\mu(x) = \int_{\mathrm{X}} f(x)\, d\mu(x)$ (§2, No. 10, Prop. 15). Let $(f_i)_{1\leqslant i\leqslant n}$ be a family of numerical functions $\geqslant 0$, $\mu$-integrable on X; for $i \neq j$ set

$$m_{ij} = \int_{\mathrm{F}} \widetilde{f_i}(x)\, \widetilde{f_j}(x)\, d\mu(x),$$

and let $c_i = \sup_{x\in\mathrm{X}} \widetilde{f_i}(x)$. Show that there exists at least one pair of distinct indices $(i, j)$ such that

$$m_{ij} \geqslant \frac{1}{n(n-1)\mu(\mathrm{F})}\left(\left(\sum_{i=1}^{n} \int_{\mathrm{X}} f_i(x)\, d\mu(x)\right)^2 - \mu(\mathrm{F}) \sum_{i=1}^{n} c_i \int_{\mathrm{X}} f_i(x)\, dx\right).$$

[5] Cf. TVS II, §5, No. 2, Prop. 3; a closed convex set with nonempty interior was called a convex body (*corps convexe*) in the first edition of Ch. II of *Esp. vect. top.* (§3, No. 2, Def. 4).

(Bound the sum $\sum_{i\neq j} m_{ij}$ from below, using the Cauchy–Schwarz inequality.) From this, deduce that if $(A_i)_{1\leqslant i\leqslant n}$ is a finite family of $\mu$-integrable G-fillings, and f

$$\sum_{i=1}^{n} \mu(A_i) > \mu(F),$$

then there exist a pair $(i,j)$ of distinct indices and an $s \in G$ such that $\mu(A_i s \cap A_j) > 0$.

*b*) If $G_0$ is a subgroup of G of finite index $(G : G_0) = h$, and if $(s_1,\ldots,s_h)$ is a system of representatives of the right cosets of $G_0$ in G, then $F_0 = \bigcup_{1\leqslant i\leqslant h} s_i F$ is a $G_0$-paving.

*c*) In $X = \mathbf{R}^n$, let A be a symmetric closed convex set with nonempty interior; let $u_i : (x_j) \mapsto \sum_{j=1}^{n} c_{ij} x_j$ be $m$ linear forms on X, with integer coefficients $c_{ij}$ $(m < n)$. Show that for every integer $p \geqslant 1$ and every number $r \geqslant 0$ such that $\mu(A) r^n \geqslant 2^n p^m$, there exists a point $x \in rA \cap \mathbf{Z}^n$ distinct from 0 and such that $u_i(x) \equiv 0 \pmod p$ for $1 \leqslant i \leqslant m$ (apply Minkowski's theorem (Exer. 27 *c*)) to the subgroup $G_C$ of $\mathbf{Z}^n$ formed by the $z \in \mathbf{Z}^n$ such that $u_i(z) \equiv 0 \pmod p$ for $1 \leqslant i \leqslant m$, and make use of *b*)). Special case that $n = 2$, $m = 1$ and A is defined by $|x_1| \leqslant 1$, $|x_2| \leqslant 1$ (*Thue's theorem*).

¶ 29) *a*) Let $p$ be a prime number; there exist two integers $a, b$ such that $a^2+b^2+1 \equiv 0 \pmod p$ (*Alg.*, Ch. V, 1st edn., §11, No. 5, Cor. of Th. 3). Show that there exist integers $x_1, x_2, x_3, x_4$ not all zero such that $ax_1 + bx_2 \equiv x_3 \pmod p$, $bx_1 - ax_2 \equiv x_4 \pmod p$ and

$$y = x_1^2 + x_2^2 + x_3^2 + x_4^2 \leqslant \sqrt{2}\,p$$

(same method as in Exer. 28 *c*)). Show that $y$ is divisible by $p$, and deduce therefrom that $y = p$.

*b*) Deduce from *a*) that every integer $n \geqslant 0$ is the sum of four squares (*Lagrange's theorem*; make use of *Alg.*, Ch. IV, 1st edn., §2, Exer. 11).

30) Let G be a locally compact group, $\mu$ a left Haar measure on G, and $\nu$ a bounded measure on G. Assume that the mapping $s \mapsto \gamma(s)\nu$ of G into the Banach space $\mathscr{M}^1(G)$ is continuous. Show that the measure $\nu$ has base $\mu$. (Let A be a $\mu$-negligible compact subset of G. Arguing as for Prop. 11, show that $\nu(xA) = 0$ for $x$ running over a dense subset of G. From this, deduce that $\nu(A) = 0$.)

Conversely, if $\nu$ has base $\mu$, then the mapping $s \mapsto \gamma(s)\nu$ of G into $\mathscr{M}^1(G)$ is continuous: cf. Ch. VIII, §2, No. 5, Prop. 8.

## §2

1) Let G be a locally compact group, H a closed subgroup of G. For $\xi \in H$, set $\chi(\xi) = \Delta_H(\xi)\Delta_G(\xi)^{-1}$. Regard H as operating on the right in G by translations. Show that there exists on $\mathscr{K}^{\chi}(G)$ a nonzero positive linear form I and, up to a constant factor, only one, that is invariant under left translation. (Make use of Prop. 3 with $X = G$, while taking $\mu$ to be a left Haar measure of G.)

2) Let X and X′ be two locally compact spaces in which a locally compact group H operates on the right, continuously and properly. Let $\theta$ be a proper continuous mapping of X into X′, compatible with the identity mapping of H (GT, III, §2, No. 4), and let $\theta' : X/H \to X'/H$ be the mapping deduced from $\theta$ by passage to quotients. Let $f'$ be

a continuous function on $X'$ whose support has compact intersection with the saturation of each compact set. Then $f = f' \circ \theta$ has the same properties in $X$, and

$$f^\flat = f'^\flat = \theta'$$

(the mappings $f \mapsto f^\flat$, $f' \mapsto f'^\flat$ being relative to the choice of a same Haar measure on $H$).

3) Let $B$ be a locally compact space, $H$ a locally compact group. Set $X = B \times H$, the group $H$ operating in $X$ by

$$(b, \xi)\xi' = (b, \xi\xi').$$

Let $\lambda$ be a measure on $B = X/H$, and $\beta$ a left Haar measure on $H$. Then $\lambda^\sharp = \lambda \otimes \beta$.

4) Let $X$ be a locally compact space in which a locally compact group $H$ acts on the right continuously and properly. Let $\beta$ be a left Haar measure on $H$, $\mu$ a measure $\geqslant 0$ on $X$. Let $h$ be a continuous function $\geqslant 0$ on $X$ such that $h^\flat = 1$. If $f$ is a $\mu$-negligible function $\geqslant 0$ on $X$, the function $(x, \xi) \mapsto f(x)h(x\xi)$ is $(\mu \otimes \beta)$-negligible on $X \times H$. (There exist decreasing open sets $\Omega_1, \Omega_2, \ldots$ such that $\mu(\Omega_n)$ tends to $0$ and such that $f$ is zero outside $\Omega_1 \cap \Omega_2 \cap \cdots$. Show that $\int^* \varphi_{\Omega_n}(x)h(x\xi)\, d\mu(x) d\beta(\xi) \leqslant \mu(\Omega_n)$, on observing that the function $(x, \xi) \mapsto \varphi_{\Omega_n}(x)h(x\xi)$ is lower semi-continuous.)

5) Let $G$ be a locally compact group. For every $s \in G$, let $\psi_s$ be the automorphism of the additive group $\mathbf{R}$ defined by $\psi_s(x) = \Delta_G(s)x$. Let $\Gamma$ be the topological semi-direct product of $G$ and $\mathbf{R}$ defined by $s \mapsto \psi_s$ (GT, III, §2, No. 10). Show that $\Gamma$ is unimodular.

6) Let $G$ be a locally compact group, $G_1, G_2, G_3$ closed subgroups such that $G_3 \subset G_1 \cap G_2$. Assume that $G, G_1, G_2, G_3$ are unimodular, and that $G_1/G_3$ has finite total measure (for every measure invariant under $G_1$). Let $\lambda, \mu, \nu$ be invariant measures on $G/G_1$, $G/G_2$, $G_2/G_3$, and $\varphi$ the canonical mapping of $G$ onto $G/G_1$. Let $f \in \mathscr{K}(G/G_1)$. Show that

$$a \int_{G/G_1} f(u)\, d\lambda(u) = \int_{G/G_2} d\mu(\dot{x}) \int_{G_2/G_3} f(\varphi(x\xi))\, d\nu(\dot{\xi})$$

($\dot{x} = xG_2$, $\dot{\xi} = \xi G_3$), where $a$ is a constant independent of $f$.

¶ 7) *a*) Let $E$ be a locally compact space, $\Gamma$ a locally compact group operating on the left continuously in $E$. Assume that for every $x \in E$, the mapping $s \mapsto s \cdot x$ of $\Gamma$ into $E$ is proper, and that there exists a nonzero bounded positive measure $\mu$ on $E$ invariant under $\Gamma$. Then $\Gamma$ is compact. (Let $(s_n)$ be a sequence of points of $\Gamma$. Let $K$ be a compact subset of $E$ that is not $\mu$-negligible. Show, using Exer. 15 of Ch. IV, §4, that there exists an $x_0 \in E$ such that $s_n x_0 \in K$ for infinitely many values of $n$. From this, deduce that $(s_n)$ has a cluster point in $\Gamma$. Then apply Exer. 6 of GT, II, §4.)

*b*) Let $G$ be a locally compact group, $H$ and $K$ two closed subgroups of $G$. Assume that $G$ and $H$ are unimodular and that $G/H$ has finite measure for the measure invariant under $G$. Show that for $H$ and $K$ to satisfy the equivalent conditions of Exer. 11 *c*) of GT, III, §4, it is necessary and sufficient that $K$ be compact.

8) Let $\mathfrak{G}$ be a locally compact group, $G$ and $H$ two closed subgroups of $\mathfrak{G}$. Assume that every $s \in \mathfrak{G}$ can be written in one and only one way in the form $s = \xi x = y\eta$

($\xi, \eta \in \mathrm{G}$, $x, y \in \mathrm{H}$), and that $\xi, x, y, \eta$ depend continuously on $s$. Every $\xi \in \mathrm{G}$ defines a homeomorphism $\widehat{\xi}$ of H onto H by $\xi x \in \widehat{\xi}(x)\mathrm{G}$ $(x \in \mathrm{H})$. Every $x \in \mathrm{H}$ defines a homeomorphism $\widehat{x}$ of G onto G by $\xi x \in \mathrm{H}\widehat{x}(\xi)$ $(\xi \in \mathrm{G})$. Let $\mu, \alpha, \beta$ be left Haar measures on $\mathfrak{G}$, G, H. Show that

$$d\widehat{x}^{-1}(\alpha)(\xi) = \frac{\Delta_{\mathfrak{G}}(\widehat{\xi}(x))\Delta_{\mathrm{G}}(\widehat{x}(\xi))}{\Delta_{\mathrm{H}}(\widehat{\xi}(x))\Delta_{\mathrm{G}}(\xi)}\, d\alpha(\xi),$$

$$d\widehat{\xi}^{-1}(\beta)(x) = \frac{\Delta_{\mathrm{G}}(\widehat{x}(\xi))}{\Delta_{\mathfrak{G}}(\widehat{x}(\xi))}\, d\beta(x).$$

(Let $f \in \mathscr{K}(\mathrm{G})$. Express the fact that $\int f(u)\, d\mu(u)$, calculated with the help of (23), either for $\mathrm{X} = \mathrm{G}$, $\mathrm{Y} = \mathrm{H}$, or for $\mathrm{X} = \mathrm{H}$, $\mathrm{Y} = \mathrm{G}$, is invariant when $f$ is subjected to left translation by an element of G or an element of H.)

9) Let X be a locally compact space, H a *compact* group operating continuously (hence properly) on the right in X. Denote by $\beta$ the normalized Haar measure on H, by $\pi : \mathrm{X} \to \mathrm{X/H}$ the canonical mapping. Let $\lambda$ be a positive measure on X/H, $\lambda^\sharp$ the corresponding measure on X.

*a*) Show that if N is a $\lambda$-negligible subset of X/H, then $\overset{-1}{\pi}(\mathrm{N})$ is $\lambda^\sharp$-negligible. (Calculate the measure of a set $\overset{-1}{\pi}(\mathrm{U})$, where U is open in X/H.)

*b*) Let $p$ be a finite number $\geqslant 1$, and let $f$ be a function in $\mathscr{L}^p_{\mathrm{F}}(\mathrm{X}, \lambda^\sharp)$ (F a Banach space or $\mathrm{F} = \overline{\mathbf{R}}$). Show that the set of $x \in \mathrm{X}$ such that $\xi \mapsto f(x\xi)$ is not $\beta$-integrable on H is of the form $\overset{-1}{\pi}(\mathrm{N})$, where N is $\lambda$-negligible. Moreover, if $f^\flat$ is the function on X/H, defined almost everywhere (for $\lambda$), such that $f^\flat(\pi(x)) = \int_{\mathrm{H}} f(x\xi)\, d\beta(\xi)$, then $f^\flat$ belongs to $\mathscr{L}^p_{\mathrm{F}}(\mathrm{X/H}, \lambda)$, and $\mathrm{N}_p(f^\flat) \leqslant \mathrm{N}_p(f)$.

*c*) Conversely, if $g \in \mathscr{L}^p_{\mathrm{F}}(\mathrm{X/H}, \lambda)$, then the function $g \circ \pi$ belongs to $\mathscr{L}^p_{\mathrm{F}}(\mathrm{X}, \lambda^\sharp)$. If $p, q$ are conjugate exponents, F′ is the dual of F, $f \in \mathscr{L}^p_{\mathrm{F}}(\mathrm{X}, \lambda^\sharp)$ and $g \in \mathscr{L}^q_{\mathrm{F}'}(\mathrm{X/H}, \lambda)$, then

$$\int_{\mathrm{X}} \langle f(x), g(\pi(x))\rangle d\lambda^\sharp(x) = \int_{\mathrm{X/H}} \langle f^\flat(z), g(z)\rangle\, d\lambda(z).$$

10) With the notations of §1, No. 6, Prop. 6 let, for every $\alpha \in \mathrm{A}$, $\mu_\alpha$ be a left Haar measure on $\mathrm{G}_\alpha$, so that the inverse limit of the $\mu_\alpha$ is a left Haar measure $\mu$ on G. For every $\alpha \in \mathrm{A}$, denote by $\lambda_\alpha$ the normalized Haar measure on $\mathrm{K}_\alpha$.

*a*) Let $p$ be a finite number $\geqslant 1$ and let $f$ be a function in $\mathscr{L}^p_{\mathrm{F}}(\mathrm{G}, \mu)$ (F a Banach space or $\mathrm{F} = \overline{\mathbf{R}}$). For every $\alpha \in \mathrm{A}$, the function

$$f_\alpha(s) = \int_{\mathrm{K}_\alpha} f(s\xi)\, d\lambda_\alpha(\xi),$$

defined almost everywhere for $\mu$, belongs to $\mathscr{L}^p_{\mathrm{F}}(\mathrm{G}, \mu)$ (Exer. 9). Show that, with respect to the directed set A, $f_\alpha$ *tends in mean of order* $p$ to $f$ (make use of Lemma 2 of §1, No. 6).

*b*) Assume that $\mathrm{A} = \mathbf{N}$ and $p = 1$. Show that $f_n$ then tends to $f$ almost everywhere (for $\mu$) in G (use a method analogous to that of Ch. V, §8, Exer. 16).

¶ 11) Let G be a locally compact group, H a closed subgroup of G, $\lambda$ a left Haar measure on G; assume that there exists on G/H a measure $\mu$ invariant under G such that $\lambda = \mu^\sharp$ and $\mu(\mathrm{G/H}) < +\infty$. Let $\nu$ be a nonzero positive measure on G/H such that $\nu(\mathrm{G/H}) < +\infty$. Finally, let $h$ be a function on G having the properties of No. 4, Prop. 8.

*a*) Let A be a Borel subset of G/H. Show that

$$\int_{\mathrm{G}} h(s)\nu(s^{-1}\mathrm{A})\,d\lambda(s) = \nu(\mathrm{G/H})\mu(\mathrm{A})$$

(make use of Prop. 9 *a*) of No. 4).

*b*) Let $\mathrm{A}_i$ $(1 \leqslant i \leqslant n)$ be Borel subsets of G/H and, for every $i$, let $b_i = \sup_{s\in\mathrm{G}} \nu(s^{-1}\mathrm{A}_i)$. Show that there exist two distinct indices $i, j$ such that

$$\int_{\mathrm{G}} h(s)\nu(s^{-1}\mathrm{A}_i)\nu(s^{-1}\mathrm{A}_j)\,d\lambda(s) \geqslant \frac{(\nu(\mathrm{G/H}))^2}{\mu(\mathrm{G/H})}\left(\left(\sum_{k=1}^{n}\mu(\mathrm{A}_k)\right)^2 - \frac{\mu(\mathrm{G/H})}{\nu(\mathrm{G/H})}\sum_{k=1}^{n} b_k\mu(\mathrm{A}_k)\right)$$

(argue as in §1, Exer. 28 *a*)).

¶ 12) *a*) Let X be a topological space, $f : \mathrm{X} \to \mathbf{N}$ an upper semi-continuous function. Let $\mathrm{X}_0$ be the set of points of X where $f$ is locally constant, and $\mathrm{Y}_0 = \mathrm{X} - \mathrm{X}_0$. Define recursively two sequences $(\mathrm{X}_i)_{i\geqslant 0}$, $(\mathrm{Y}_i)_{i\geqslant 0}$ of subsets of X, in the following way: $\mathrm{X}_i$ is the set of points of $\mathrm{Y}_{i-1}$ where $f|\mathrm{Y}_{i-1}$ is locally constant, and $\mathrm{Y}_i = \mathrm{Y}_{i-1} - \mathrm{X}_i$. Show that $\mathrm{X}_i$ is a dense open subset of $\mathrm{Y}_{i-1}$, and that $\bigcap_i \mathrm{Y}_i = \varnothing$. (Show that $f(x) > i$ at every point of $\mathrm{Y}_i$.)

*b*) Let X be a set, H a group operating on the right in X, $\pi$ the canonical mapping of X onto X/H, and $\mathfrak{A}$ (resp. $\mathfrak{B}$) the set of $\mathrm{A} \subset \mathrm{X}$ such that $\pi|\mathrm{A} : \mathrm{A} \to \mathrm{X/H}$ is injective (resp. surjective). Let $(\mathrm{V}_1, \mathrm{V}_2, \ldots)$ be a countable covering of X by elements of $\mathfrak{A}$. Let

$$\mathrm{V}'_i = \mathrm{V}_i \cap \complement(\mathrm{V}_1\mathrm{H} \cup \mathrm{V}_2\mathrm{H} \cup \cdots \cup \mathrm{V}_{i-1}\mathrm{H}).$$

Show that $\mathrm{F} = \bigcup_{i\geqslant 0} \mathrm{V}'_i \in \mathfrak{A} \cap \mathfrak{B}$.

*c*) Let X be a locally compact space, H a countable discrete group operating on the right in X continuously and properly, $\pi$ the canonical mapping of X onto X/H, and $\mu$ a measure $\geqslant 0$ on X. Show that if the sets $\mathrm{V}_i$ of *b*) are quadrable (Ch. IV, §5, Exer. 17 *d*)), then the set F of *b*) is a quadrable fundamental domain.

*d*) Let us maintain the hypotheses of *c*), and use the notations $\mathrm{H}_x$, $n(x)$ of No. 10. An $x \in \mathrm{X}$ is said to be *general* if there exists a neighborhood V of $x$ in X such that $\mathrm{H}_y = \mathrm{H}_x$ for all $y \in \mathrm{V}$. Show that the general points of X are those where the function $n$ is locally constant. Show that a general point admits an open neighborhood that belongs to $\mathfrak{A}$ and is quadrable (make use of Ch. IV, §5, Exer. 17 *d*)).

*e*) Let us maintain the hypotheses of *c*), and assume moreover that X is countable at infinity. Show that if $\mathrm{X} - \mathrm{X}_0$ is $\mu$-negligible, then there exists a fundamental domain F that is Borel and quadrable. (Apply the construction of *a*) to the function $n$. Then apply the results of *c*) and *d*) in $\mathrm{X}_0$. Argue similarly in each $\mathrm{X}_i$.)

*f*) Let $\mathrm{U} \in \mathfrak{B}$ be an open set. Show that one can impose on the set F of *e*) the following supplementary properties: 1) $\mathrm{F} \subset \mathrm{U}$; 2) for every compact subset K of X, the set of $s \in \mathrm{H}$ such that $\mathrm{F}s$ intersects K is finite.

13) Let G be a compact group, $\mu$ a Haar measure on G, $u$ an endomorphism of G such that $u(\mathrm{G})$ is an *open* subgroup of G and the kernel $\overset{-1}{u}(e)$ (denoted $\mathrm{G}_u$) a *finite* subgroup of G.

*a*) Show that there exist a real number $h(u) > 0$ and an open neighborhood U of $e$ in G such that, for every open set $V \subset U$, $u(V)$ is open in G and $\mu(u(V)) = h(u)\mu(V)$ (cf. §1, No. 7, Cor. of Prop. 9).

*b*) Show that $h(u) = \mathrm{Card}(G/u(G))/\mathrm{Card}(G_u)$ (calculate $\mu(u(G))$ in two ways, using *a*) and Prop. 10 of No. 7).

## §3

1) Let E be a finite-dimensional vector space over **R**, **C** or **H**, $\Phi$ and $\Phi'$ two nondegenerate positive hermitian forms on E. Assume that $\mathbf{U}(\Phi) \subset \mathbf{U}(\Phi')$. Show that $\Phi$ and $\Phi'$ are proportional. (Make use of the fact that there exists an orthonormal basis $(e_1, \ldots, e_n)$ for $\Phi$ that is orthogonal for $\Phi'$. The mapping $u \in \mathscr{L}(E, E)$ such that $u(e_i) = e_j$, $u(e_j) = e_i$, $u(e_k) = e_k$ for $k \neq i, j$ belongs to $\mathbf{U}(\Phi)$.)

2) Adopt the notations of Lemma 7. Show that $\Omega$ has negligible complement in $\mathbf{GL}(n, K)$ for the Haar measure. (Argue as for Prop. 6.)

¶ 3) Let X be a locally compact space in which a locally compact group H operates on the right, continuously and properly, by $(x, \xi) \mapsto x\xi$ ($x \in X$, $\xi \in H$). Let $\pi$ be the canonical mapping $X \to X/H$. Let $\rho$ be a continuous representation of H in $\mathbf{GL}(n, \mathbf{C})$. Show that for every $b \in X/H$, there exist a neighborhood U of $b$ in X/H and a continuous mapping $r$ of $\overset{-1}{\pi}(U)$ into $\mathbf{GL}(n, \mathbf{C})$ such that $r(x\xi) = r(x)\rho(\xi)$ for all $x \in \overset{-1}{\pi}(U)$ and $\xi \in H$. (One may assume X/H to be compact. Let $f$ be a continuous function $\geqslant 0$ on X, not identically zero on any orbit, and with compact support. Set

$$r(x) = \int_H f(x\xi)\rho(\xi)^{-1}\, d\beta(\xi) \in \mathbf{M}_n(\mathbf{C}),$$

where $\beta$ denotes a left Haar measure on H. Show that, for $f$ and U suitably chosen, $r(x) \in \mathbf{GL}(n, \mathbf{C})$ for every $x \in \overset{-1}{\pi}(U)$.)

4) Let K be a nondiscrete commutative locally compact field. Let A be an algebra of finite rank over K.

*a*) Let $T(n, A)$ be the algebra of matrices $(x_{ij}) \in \mathbf{M}_n(A)$ such that $x_{ij} = 0$ for $i > j$. Show that if $M = (m_{ij}) \in T(n, A)$, then

$$N_{T(n,A)/K}(M) = \prod_{i=1}^{n} N_{A/K}(m_{ii}^{i-n+1}).$$

(Use Lemma 6.)

*b*) Let $T(n, A)^*$ be the group of invertible elements of $T(n, A)$. Let $M = (m_{ij}) \in T(n, A)$. Show that $M \in T(n, A)^*$ if and only if $m_{ii}$ is invertible in A for all $i$.

*c*) Show that a left Haar measure on $T(n, A)^*$ is given by

$$\left(\prod_{i=1}^{n} \operatorname{mod} N_{A/K}(m_{ii})^{i-n-1}\right) \cdot \bigotimes_{i \leqslant j} d\alpha(m_{ij}),$$

where $\alpha$ denotes a Haar measure of the additive group of A.

*d*) Show that if $\mathrm{N}_{\mathrm{A/K}} = \mathrm{N}_{\mathrm{A^0/K}}$, then

$$\Delta_{\mathrm{T}(n,\mathrm{A})^*}((m_{ij})) = \prod_{i=1}^{n} \operatorname{mod} \mathrm{N}_{\mathrm{A/K}}(m_{ii})^{2i-n-1}.$$

5) Equip $\mathbf{R}^n$ with the quadratic form $x_1^2 + x_2^2 + \cdots + x_n^2$. Let $\mathrm{G}_n$ be the group of displacements of $\mathbf{R}^n$ (*Alg.*, Ch. IX, §6, No. 6, Def. 3) of determinant 1.

*a*) Show that $\mathrm{G}_n$ is unimodular.

*b*) Every element of $\mathrm{G}_2$ may be uniquely written in the form

$$(x, y) \mapsto (u + x \cos\omega - y \sin\omega,\ v + x \sin\omega + y \cos\omega),$$

where $u, v$ are in $\mathbf{R}$, and $\omega$ is an element of the group $\Theta$ of angles of half-lines, isomorphic to $\mathbf{U}$. Show that $du \otimes dv \otimes d\omega$ (where $d\omega$ is a Haar measure on $\Theta$) is a Haar measure on $\mathrm{G}_2$.

6) Let P be the set of complex numbers $z = x + iy$ whose imaginary part is $> 0$. Show that $\mathbf{SL}(2, \mathbf{R})$ operates transitively and continuously on the left in P by

$$\begin{pmatrix} a & b \\ c & d \end{pmatrix} \cdot z \mapsto \frac{az + b}{cz + d}$$

and that P thus becomes a topological homogeneous space for $\mathbf{SL}(2, \mathbf{R})$. Show that $y^{-2}\, dx\, dy$ is an invariant measure on P.

¶ 7) Let G be the unimodular group $\mathbf{SL}(n, \mathbf{R})$, $\Gamma$ the subgroup of G formed by the matrices of determinant 1 with integer entries. Show, by induction on $n$, that every invariant measure $\mu$ on the homogeneous space $\mathrm{G}/\Gamma$ is bounded and that, for $f \in \mathscr{K}(\mathbf{R}^n)$, one has, with $\mu$ suitably chosen,

$$(1) \qquad \int_{\mathbf{R}^n} f(x)\, dx = \int_{\mathrm{G}/\Gamma} \Big( \sum_{z \in \mathbf{Z}^n,\, z \neq 0} f(X \cdot z) \Big) d\mu(\dot{X})$$

($\dot{X} = X\Gamma$). (Let G′ be the subgroup of G leaving invariant the basis vector $(1, 0, 0, \ldots, 0)$, so that $\mathrm{G}/\mathrm{G}'$ may be identified with $\mathbf{R}^n$. Let $\mathrm{G}'' = \mathrm{G}' \cap \Gamma$. Let H be the subgroup of matrices $\begin{pmatrix} 1 & w \\ 0 & Y \end{pmatrix}$ of G′ for which $Y \in \mathbf{SL}(n-1, \mathbf{R})$ has integer entries. Then $\mathrm{H}/\mathrm{G}''$ is compact and, by the induction hypothesis, $\mathrm{G}'/\mathrm{H}$ has a finite invariant measure, therefore by Prop. 12, §2, No. 8, $\mathrm{G}'/\mathrm{G}''$ has a finite invariant measure. Applying Exer. 6 of §2, and Exer. 20 *b*) of A, VII, §4, show that, for $f \in \mathscr{K}(\mathbf{R}^n)$,

$$a \int_{\mathbf{R}^n} f(x)\, dx = \int_{\mathrm{G}/\Gamma} \Big( \sum f(X \cdot m) \Big) d\mu(\dot{X}),$$

where the summation is over the set of vectors $m = (m_1, \ldots, m_n)$ whose coordinates are setwise coprime integers (A, VII, §1, No. 2, Th. 1). Applying this to the functions $f(2x)$, $f(3x), \ldots$ and summing, prove (1). Then, replacing $f$ by the function $x \mapsto \varepsilon^n f(\varepsilon x)$ with $f \geqslant 0$, and letting $\varepsilon$ tend to 0, show that every compact subset of $\mathrm{G}/\Gamma$ has measure $\leqslant 1$.)

8) *a*) Show that there exists on $\mathbf{S}_{n-1}$ a measure $\omega_{n-1}$ invariant under $\mathbf{O}(n,\mathbf{R})$ and, up to a constant factor, only one (regard $\mathbf{S}_{n-1}$ as a homogeneous space for $\mathbf{O}(n,\mathbf{R})$).

*b*) The mapping $(t,z) \mapsto tz$ permits identifying the topological space $\mathbf{R}^n - \{0\}$ with the topological space $\mathbf{R}_+^* \times \mathbf{S}_{n-1}$. Show that the measure induced on $\mathbf{R}^n - \{0\}$ by the Lebesgue measure on $\mathbf{R}^n$ may then be identified, up to a constant factor, with $t^{n-1}\,dt \otimes d\omega_{n-1}(z)$. (Make use of the fact that the Lebesgue measure on $\mathbf{R}^n$ is invariant under $\mathbf{O}(n,\mathbf{R})$ and the fact that the Lebesgue measure of the closed ball with center 0 and radius $r$ is proportional to $r^n$.)

*c*) Let $\mathrm{L}_h$ be the intersection of $\mathbf{S}_{n-1}$ and the hyperplane $x_n = h$. The non-empty $\mathrm{L}_h$ are the intransitivity classes of $\mathbf{O}(n-1,\mathbf{R})$ in $\mathbf{S}_{n-1}$. Show that $\omega_{n-1}$ may be put in the form $\int \lambda_h\, d\nu(h)$, where $\lambda_h$ is a measure invariant under $\mathbf{O}(n-1,\mathbf{R})$ and carried by $\mathrm{L}_h$, and $\nu$ is a measure on $[-1,1]$. Identifying $\mathrm{L}_h$ with $\mathbf{S}_{n-2}$ by $z \mapsto he_n + \sqrt{1-h^2}\,z$, one can assume that $\lambda_h = \omega_{n-2}$. Let $\mathrm{K}_\theta$ be the subset of $\mathbf{S}_{n-1}$ defined by $\sin\theta \leqslant x_n \leqslant 1$. By calculating the Lebesgue measure of the set of $tz$ $(0 \leqslant t \leqslant 1,$ $z \in \mathrm{K}_\theta)$, show that $\omega_{n-1}(\mathrm{K}_\theta)$ is proportional to $\int_\theta^{\pi/2} \cos^{n-2}\varphi\, d\varphi$. From this, deduce that if one sets $h = \sin\theta$ $(-\pi/2 \leqslant \theta \leqslant \pi/2)$, then the measure $\nu$ may be identified with $\cos^{n-2}\theta\, d\theta$.

*d*) Conclude, by induction on $n$, that

$$d\omega_{n-1} = \cos^{n-2}\theta_1 \cos^{n-3}\theta_2 \cdots \cos\theta_{n-2}\, d\theta_1\, d\theta_2 \cdots d\theta_{n-1},$$

where

$$\begin{aligned} x_1 &= \sin\theta_1 \\ x_2 &= \cos\theta_1 \sin\theta_2 \\ &\cdots \\ x_{n-1} &= \cos\theta_1 \cos\theta_2 \cdots \cos\theta_{n-2} \sin\theta_{n-1} \\ x_n &= \cos\theta_1 \cos\theta_2 \cdots \cos\theta_{n-2} \cos\theta_{n-1} \end{aligned}$$

with $-\pi/2 \leqslant \theta_i \leqslant \pi/2$ for $1 \leqslant i \leqslant n-2$, $0 \leqslant \theta_{n-1} < 2\pi$.

9) Let $(e_1, e_2, e_3)$ be the canonical basis of $\mathbf{R}^3$. Every matrix $\sigma \in \mathbf{SO}(3,\mathbf{R})$ such that $\sigma(e_3) \neq e_3$ and $\sigma(e_3) \neq -e_3$ may be written in a unique way as

$$\sigma(\varphi,\psi,\theta) = \begin{pmatrix} \cos\varphi\cos\psi - \sin\varphi\sin\psi\cos\theta & -\cos\varphi\sin\psi - \sin\varphi\cos\psi\cos\theta & \sin\varphi\sin\theta \\ \sin\varphi\cos\psi + \cos\varphi\sin\psi\cos\theta & -\sin\varphi\sin\psi + \cos\varphi\cos\psi\cos\theta & -\cos\varphi\sin\theta \\ \sin\psi\sin\theta & \cos\psi\sin\theta & \cos\theta \end{pmatrix},$$

where $0 < \theta < \pi$, $0 \leqslant \varphi < 2\pi$, $0 \leqslant \psi < 2\pi$ (*Euler's angles*). Show that $\sin\theta\, d\theta\, d\varphi\, d\psi$ is a Haar measure on $\mathbf{SO}(3,\mathbf{R})$. (Identify the homogeneous space $\mathbf{SO}(3,\mathbf{R})/\mathbf{SO}(2,\mathbf{R})$ with $\mathbf{S}_2$, use Exer. 8, and Th. 3 of §2.)

10) Let D be the group of displacements of $\mathbf{R}^3$ with determinant 1, the semi-direct product of $\mathbf{SO}(3,\mathbf{R})$ and the group T of translations. We employ the notation $\sigma(\varphi,\psi,\theta)$ of Exer. 9, and denote by $t(\xi,\eta,\zeta)$ the translation with vector $(\xi,\eta,\zeta)$.

*a*) Let H be the closed subgroup of D that leaves $\mathbf{R}e_3$ stable and preserves orientation on $\mathbf{R}e_3$. Show that H is the product of the group of translations $t(0,0,\zeta)$ and the group of rotations $\sigma(\omega,0,0)$. From this, deduce that the homogeneous space $\mathrm{E} = \mathrm{D}/\mathrm{H}$, which may be identified with the space of oriented affine lines of $\mathbf{R}^3$, possesses a measure invariant under D, and, up to a constant factor, only one. (Make use of Exer. 5 *a*).)

*b*) Let $\mathrm{E}_1$ be the open subspace of E formed by the oriented lines not orthogonal to $e_3$; such an oriented line can be determined by the coordinates $\xi', \eta'$ of its point of intersection with $\mathbf{R}e_1 + \mathbf{R}e_2$, and the coordinates $(\sin\varphi \sin\theta, -\cos\varphi\sin\theta, \cos\theta)$ of a direction vector. Show that a measure on $E_1$ invariant under D is given by

$$\sin\theta\,|\cos\theta|\,d\xi'\,d\eta'\,d\varphi\,d\theta\,.$$

11) Let G be the compact group $\mathbf{O}(n, R)$, $\lambda$ the normalized Haar measure on G. For $0 < k < n$, let H be the closed subgroup of G leaving invariant (globally) the subspace $\mathbf{R}^k$ of $\mathbf{R}^n$; the homogeneous space G/H, equipped with its quotient topology, may be identified canonically with the Grassmannian $E = G_{n-1, k-1}(\mathbf{R})$ (GT, VI, §3, No. 6) of the $k$-dimensional linear subspaces of $\mathbf{R}^n$. There is a measure $\mu$ on E invariant under G, determined up to a constant. For every subspace $P \in E$, let $\sigma_P$ be the image of the Lebesgue measure of $\mathbf{R}^k$ under an $s \in G$ such that $s\cdot\mathbf{R}^k = P$ (independent of the element $s \in G$ satisfying this relation). Show that $\mu$ can be chosen in such a way that the following property is satisfied: for every continuous function $f$ on $\mathbf{R}^n$ with compact support, and for every $P \in E$, one sets $F(P) = \int_P f(x)\,d\sigma(x)$; then

$$\int_E F(P)\,d\mu(P) = \int_{\mathbf{R}^n} \|x\|^{k-n} f(x)\,dx\,.$$

(If $P_0 = \mathbf{R}^k$, note that one can write $\int_E F(P)\,d\mu(P) = c\int_G F(s\cdot P_0)\,d\lambda(s)$ for a suitable constant $c$, and on the other hand,

$$\int_G f(s\cdot x)\,d\lambda(s) = c'\int_{\mathbf{S}_{n-1}} f(\|x\|z)\,d\omega_{n-1}(z)$$

with the notations of Exercise 8; finally, make use of Exer. 8 *b*).)

¶ 12) Let K be a commutative field, E a vector space over K, F a linear subspace of E, $p$ the canonical homomorphism of E onto E/F, and A the set of homomorphisms $f : E/F \to E$ such that $p\circ f$ is the identity homomorphism of E/F.

*a*) If $f \in A$ and $h \in \mathrm{Hom}(E/F, F)$, then $f + h \in A$. Show that if $f, f' \in A$, there exists one and only one $h \in \mathrm{Hom}(E/F, F)$ such that $f + h = f'$. One can therefore regard A as an affine space of which $\mathrm{Hom}(E/F, F)$ is the space of translations.

*b*) Let B be the set of linear subspaces of E that are supplementary to F in E. Show that $f \mapsto f(E/F)$ is a bijection of A onto B. One can therefore regard B as an affine space of which $\mathrm{Hom}(E/F, F)$ is the space of translations.

*c*) Show that if $u$ is an automorphism of E such that $u(F) = F$, then the mapping $f \mapsto u\circ f$ is an affine bijection of A onto A. (Choose an origin in A, and observe that the mapping $h \mapsto u\circ h$ is an automorphism of the vector space $\mathrm{Hom}(E/F, F)$.) From this, deduce that the mapping $F' \mapsto u(F')$ is an affine bijection of B onto B.

*d*) Assume that $K = \mathbf{R}$ and $\dim E < +\infty$. Let G be a compact group, and $\rho$ a continuous linear representation of G in E such that $\rho(s)(F) = F$ for all $s \in G$. Show that there exists an $F' \in B$ such that $\rho(s)(F') = F'$ for all $s \in G$. (Make use of *c*) and Lemma 2 of No. 2.) Obtain this result anew using Prop. 1 of No. 1.

CHAPTER VIII

# Convolution and representations

## §1. CONVOLUTION

### 1. Definition and examples

Recall (Ch. V, §6, Nos. 1 and 4; Ch. VI, §2, No. 10) that, if X and Y are locally compact spaces, $\mu$ a measure on X, and $\varphi$ a mapping of X into Y, $\varphi$ is said to be *$\mu$-proper* if: *a*) $\varphi$ is $\mu$-measurable; *b*) for every compact subset K of Y, $\overset{-1}{\varphi}(\mathrm{K})$ is essentially $\mu$-integrable. Then the image measure $\nu = \varphi(\mu)$ on Y exists and has the following property: for a function $\mathbf{f}$ on Y, with values in a Banach space or in $\overline{\mathbf{R}}$, to be essentially integrable for $\nu$, it is necessary and sufficient that $\mathbf{f} \circ \varphi$ be so for $\mu$, in which case,

$$\int_{\mathrm{Y}} \mathbf{f}(y)\, d\nu(y) = \int_{\mathrm{X}} \mathbf{f}\big(\varphi(x)\big) d\mu(x)\,.$$

DEFINITION 1. — *Let* $\mathrm{X}_1, \ldots, \mathrm{X}_n$ *be locally compact spaces,* $\mu_i$ *a measure on* $\mathrm{X}_i$ $(1 \leqslant i \leqslant n)$; *let* X *be the product of the* $\mathrm{X}_i$, $\mu$ *that of the* $\mu_i$. *Let* $\varphi$ *be a mapping of* X *into a locally compact space* Y. *One says that the sequence* $(\mu_i)$ *is $\varphi$-convolvable, or that* $\mu_1, \ldots, \mu_n$ *are $\varphi$-convolvable, if* $\varphi$ *is $\mu$-proper; in this case, the image* $\nu = \varphi(\mu)$ *of* $\mu$ *under* $\varphi$ *is called the convolution product of the* $\mu_i$ *for* $\varphi$, *and is denoted* $*_{\varphi}(\mu_i)_{1 \leqslant i \leqslant n}$, *or* $\mathop{*}\limits_{i=1}^{n} \mu_i$, *or* $\mu_1 * \mu_2 * \cdots * \mu_n$.

The last two notations are of course used only when there can be no doubt as to $\varphi$.

Let $\mathbf{f}$ be a function on Y, with values in a Banach space or in $\overline{\mathbf{R}}$. In order that $\mathbf{f}$ be essentially integrable for $\mu_1 * \cdots * \mu_n$, it is necessary and sufficient that the function

$$(x_1, \ldots, x_n) \mapsto \mathbf{f}\big(\varphi(x_1, \ldots, x_n)\big)$$

be essentially integrable for $\mu_1 \otimes \mu_2 \otimes \cdots \otimes \mu_n$, in which case

$$(1) \qquad \int \mathbf{f}\, d(\mu_1 * \cdots * \mu_n) = \int \mathbf{f}\big(\varphi(x_1, \ldots, x_n)\big) d\mu_1(x_1) \ldots d\mu_n(x_n),$$

a formula that may be regarded as *defining* $\mu_1 * \cdots * \mu_n$ when one takes $\mathbf{f} \in \mathscr{K}(\mathrm{Y})$.

The definitions imply at once that the $\mu_i$ are convolvable if and only if the $|\mu_i|$ are. When this is the case,

$$|\varphi(\mu_1 \otimes \cdots \otimes \mu_n)| \leqslant \varphi(|\mu_1 \otimes \cdots \otimes \mu_n|) = \varphi(|\mu_1| \otimes \cdots \otimes |\mu_n|)$$

(Ch. VI, §2, No. 10), that is,

$$(2) \qquad \big| \mathop{\ast}_i \mu_i \big| \leqslant \mathop{\ast}_i |\mu_i| .$$

If the $\mu_i$ are convolvable and positive, and if $\nu_i$ is a measure on $\mathrm{X}_i$ such that $0 \leqslant \nu_i \leqslant \mu_i$, then the $\nu_i$ are convolvable and

$$\mathop{\ast}_i \nu_i \leqslant \mathop{\ast}_i \mu_i .$$

Suppose $\mu_1, \mu_2, \ldots, \mu_n$ are convolvable, and that $\mu'_1, \mu_2, \ldots, \mu_n$ are convolvable ($\mu'_1$ being a measure on $\mathrm{X}_1$). By Ch. V, §6, No. 3, Cor. 1 of Prop. 6, $\mu_1 + \mu'_1, \mu_2, \ldots, \mu_n$ are convolvable and

$$(\mu_1 + \mu'_1) * \mu_2 * \cdots * \mu_n = \mu_1 * \mu_2 * \cdots * \mu_n + \mu'_1 * \mu_2 * \cdots * \mu_n .$$

*Examples.* — 1) For any $\varphi$, the measures $\varepsilon_{x_i}$, where $x_i \in \mathrm{X}_i$ for $1 \leqslant i \leqslant n$, are always convolvable and have convolution product $\varepsilon_y$, with $y = \varphi(x_1, x_2, \ldots, x_n)$. Consequently, if each of the $\mu_i$ has finite support, then the $\mu_i$ are convolvable and $\mu_1 * \cdots * \mu_n$ has finite support. In particular, let M be a monoid[1] equipped with a locally compact topology; if one takes $\varphi$ to be the law of composition in M then the measures on M with finite support form, for convolution, an algebra that is none other than the *algebra of the monoid* M (over $\mathbf{R}$ or over $\mathbf{C}$, according as one considers real or complex measures) (A, III, §2, No. 6).

2) Let M be a monoid equipped with the discrete topology; assume that for each $m \in \mathrm{M}$, there are only finitely many pairs $(m', m'') \in \mathrm{M} \times \mathrm{M}$ such that $m'm'' = m$; this amounts to saying that the law of composition in M is a proper mapping of $\mathrm{M} \times \mathrm{M}$ into M; the measures on M then form an algebra for convolution, an algebra that is none other than the *total*

[1] *Monoïde*, in the sense of Exer. 17 of Ch. VII, §1.

*algebra of the monoid* M (A, III, §2, No. 10); we note the following two special cases:

*a*) $M = \mathbf{N}$, the law of composition being addition. To every measure $\mu$ on $\mathbf{N}$, let us associate the formal series

$$S(\mu) = \sum_{n=0}^{\infty} \mu(\{n\})t^n$$

in an indeterminate $t$. Then $S(\mu * \mu') = S(\mu)S(\mu')$. An analogous remark holds for formal series in any number of indeterminates.

$_*b$) $M = \mathbf{N}^*$, the law of composition being multiplication. To every measure $\mu$ on $\mathbf{N}^*$, let us associate the formal Dirichlet series

$$D(\mu) = \sum_{n=1}^{\infty} \mu(\{n\})n^{-s}.$$

Then $D(\mu * \mu') = D(\mu)D(\mu')$.$_*$

3) Let $X, Y, Z$ be locally compact spaces, $\varphi$ a continuous mapping of $X \times Y$ into $Z$. If $x \in X$ and $\mu$ is a measure on $Y$, to say that $\varepsilon_x$ and $\mu$ are $\varphi$-convolvable comes to saying that the mapping $\varphi(x, \cdot)$ of $Y$ into $Z$ is $\mu$-proper. One then has $\varepsilon_x * \mu = \varphi(x, \cdot)(\mu)$.

## 2. Associativity

The following lemma completes Prop. 11 of Ch. V, §8, No. 5:[2]

*Lemma* 1. — *For* $1 \leqslant i \leqslant n$, *let* $X_i, Y_i$ *be two locally compact spaces,* $\mu_i$ *a measure on* $X_i$, *and* $\varphi_i$ *a continuous mapping of* $X_i$ *into* $Y_i$. *Let* $X = \prod_i X_i$, $Y = \prod_i Y_i$, $\mu = \bigotimes_i \mu_i$, *and* $\varphi$ *the mapping of* $X$ *into* $Y$ *that is the product of the* $\varphi_i$. *If* $\varphi$ *is* $\mu$*-proper and* $\mu_i \neq 0$ *for each* $i$, *then the* $\varphi_i$ *are* $\mu_i$*-proper and* $\varphi(\mu) = \bigotimes_i \varphi_i(\mu_i)$.

We can suppose that the $\mu_i$ are positive and $n = 2$. Let $f_1 \in \mathscr{K}_+(Y_1)$. Since $\mu_2 \neq 0$, there exists an $f_2 \in \mathscr{K}_+(Y_2)$ such that $f_2 \circ \varphi_2$ is not $\mu_2$-negligible. The function $(x_1, x_2) \mapsto f_1(\varphi_1(x_1))f_2(\varphi_2(x_2))$ is essentially $\mu$-integrable and continuous, hence $\mu$-integrable. Therefore there exists an $x_2 \in X_2$ such that $f_2(\varphi_2(x_2)) \neq 0$ and such that the function $x_1 \mapsto f_1(\varphi_1(x_1))f_2(\varphi_2(x_2))$ is $\mu_1$-integrable. Therefore $f_1 \circ \varphi_1$ is $\mu_1$-integrable,

[2]The lemma follows by induction on part *b*) of the cited Prop. 11, which is the case $n = 2$; the corresponding result in the first edition of Ch. V (§8, No. 3, Prop. 7) did not include the result of part *b*).

which proves that $\varphi_1$ is $\mu_1$-proper. One argues similarly for $\varphi_2$. Then $\varphi(\mu) = \bigotimes_i \varphi_i(\mu_i)$ by Prop. 11 of Ch. V, §8, No. 5.

The following lemma completes Prop. 4 of Ch. V, §6, No. 3.[3]

*Lemma 2. — Let* $\mathrm{T}, \mathrm{T}', \mathrm{T}''$ *be three locally compact spaces,* $\mu$ *a measure on* $\mathrm{T}$, $\pi$ *a* $\mu$*-measurable mapping of* $\mathrm{T}$ *into* $\mathrm{T}'$, $\pi'$ *a continuous mapping of* $\mathrm{T}'$ *into* $\mathrm{T}''$, *and* $\pi'' = \pi' \circ \pi$. *If* $\pi''$ *is* $\mu$*-proper, then* $\pi$ *is* $\mu$*-proper,* $\pi'$ *is* $\pi(\mu)$*-proper, and* $\pi''(\mu) = \pi'(\pi(\mu))$.

Let $\mathrm{K}'$ be a compact subset of $\mathrm{T}'$. Then $\mathrm{K}'' = \pi'(\mathrm{K}')$ is compact, therefore $\overset{-1}{\pi''}(\mathrm{K}'')$ is essentially $\mu$-integrable, therefore $\overset{-1}{\pi}(\mathrm{K}') \subset \overset{-1}{\pi''}(\mathrm{K}'')$ is essentially $\mu$-integrable, thus $\pi$ is $\mu$-proper. Then $\pi'$ is $\pi(\mu)$-proper and $\pi''(\mu) = \pi'(\pi(\mu))$ by Ch. V, §6, No. 3, Prop. 4.

PROPOSITION 1. — *Let* $\mathrm{X}_{ij}$ $(1 \leqslant i \leqslant m,\ 1 \leqslant j \leqslant n_i)$, $\mathrm{Y}_i$ $(1 \leqslant i \leqslant m)$, *and* $\mathrm{Z}$ *be locally compact spaces; for each* $i$, *let* $\varphi_i$ *be a mapping of* $\mathrm{X}_i = \prod_j \mathrm{X}_{ij}$ *into* $\mathrm{Y}_i$; *let* $\varphi$ *be the product of the* $\varphi_i$, *mapping* $\mathrm{X} = \prod_i \mathrm{X}_i$ *into* $\mathrm{Y} = \prod_i \mathrm{Y}_i$; *let* $\psi$ *be a mapping of* $\mathrm{Y}$ *into* $\mathrm{Z}$.

(i) *Let* $\mu_{ij}$ *be measures given, respectively, on the* $\mathrm{X}_{ij}$, *such that, for each* $i$, *the* $\mu_{ij}$ $(1 \leqslant j \leqslant n_i)$ *are* $\varphi_i$*-convolvable, and such that the measures* $\ast_j |\mu_{ij}|$ *are* $\psi$*-convolvable; then the* $\mu_{ij}$, *for* $1 \leqslant i \leqslant m$, $1 \leqslant j \leqslant n_i$, *are* $(\psi \circ \varphi)$*-convolvable and*

$$\mathop{\ast}_{i,j} \mu_{ij} = \mathop{\ast}_i \Big( \mathop{\ast}_j \mu_{ij} \Big). \tag{3}$$

(ii) *Assume* $\psi$ *and the* $\varphi_i$ *continuous, and let* $\mu_{ij}$ *be measures* $\neq 0$ *given, respectively, on the* $\mathrm{X}_{ij}$ *and* $(\psi \circ \varphi)$*-convolvable; then, for each* $i$, *the* $\mu_{ij}$ $(1 \leqslant j \leqslant n_i)$ *are* $\varphi_i$*-convolvable, the measures* $\ast_j |\mu_{ij}|$ *are* $\psi$*-convolvable, and the formula* (3) *holds.*

It suffices to consider the case that all of the measures in question are $\geqslant 0$.

Let us place ourselves under the hypotheses of (i). The mapping $\varphi$ is proper for $\bigotimes_{i,j} \mu_{ij}$, and

$$\varphi\Big(\bigotimes_{i,j} \mu_{ij}\Big) = \bigotimes_i \varphi_i\Big(\bigotimes_j \mu_{ij}\Big) = \bigotimes_i \Big(\mathop{\ast}_j \mu_{ij}\Big)$$

[3] The assertion of this lemma is in fact part *b*) of the cited Prop. 4, a part that was not included in the first edition of Ch. V.

(Ch. V, §8, No. 5, Prop. 11). The mapping $\psi \circ \varphi$ is proper for $\bigotimes_{i,j} \mu_{ij}$, and

$$(\psi \circ \varphi)\Big(\bigotimes_{i,j} \mu_{ij}\Big) = \psi\Big(\bigotimes_{i} \big(\mathop{*}_{j} \mu_{ij}\big)\Big) = \mathop{*}_{i} \Big(\mathop{*}_{j} \mu_{ij}\Big)$$

(Ch. V, §6, Prop. 4). Therefore the $\mu_{ij}$ $(1 \leqslant i \leqslant m,\ 1 \leqslant j \leqslant n_i)$ are $(\psi \circ \varphi)$-convolvable and formula (3) holds.

Let us place ourselves under the hypotheses of (ii). First of all, Lemma 2 proves that $\varphi$ is proper for $\bigotimes_{i,j} \mu_{ij}$. Lemma 1 then proves that for every $i$, $\varphi_i$ is proper for $\bigotimes_{j} \mu_{ij}$, and that

$$\varphi\Big(\bigotimes_{i,j} \mu_{ij}\Big) = \bigotimes_{i} \Big(\mathop{*}_{j} \mu_{ij}\Big).$$

By Lemma 2, $\psi$ is proper for $\bigotimes_{i} \Big(\mathop{*}_{j} \mu_{ij}\Big)$. Whence the proposition.

COROLLARY. — *Let* $\mathrm{X}_i, \mathrm{X}'_i$ $(1 \leqslant i \leqslant n)$, $\mathrm{Y}, \mathrm{Y}'$ *be locally compact spaces; let* $\varphi, \varphi'$ *be continuous mappings of* $\mathrm{X} = \prod_i \mathrm{X}_i$ *into* $\mathrm{Y}$ *and of* $\mathrm{X}' = \prod_i \mathrm{X}'_i$ *into* $\mathrm{Y}'$, *respectively; let* $f_i$ *be continuous mappings of* $\mathrm{X}_i$ *into* $\mathrm{X}'_i$ $(1 \leqslant i \leqslant n)$ *and* $g$ *a continuous mapping of* $\mathrm{Y}$ *into* $\mathrm{Y}'$, *such that* $\varphi' \circ f = g \circ \varphi$, $f$ *being the mapping of* $\mathrm{X}$ *into* $\mathrm{X}'$ *that is the product of the* $f_i$. *Let* $\mu_i$ *be measures given respectively on the* $\mathrm{X}_i$, *all* $\neq 0$. *Then the following two assertions are equivalent:*

(i) $f_i$ *is* $\mu_i$*-proper for all* $i$, *and the measures* $f_i(|\mu_i|)$ *are* $\varphi'$*-convolvable;*

(ii) *the* $\mu_i$ *are* $\varphi$*-convolvable, and* $g$ *is proper for* $*_{\varphi}(|\mu_i|)$.

*Moreover, when these assertions are verified,*

$$*_{\varphi'}(f_i(\mu_i)) = g(*_{\varphi}(\mu_i)) = *_{g\circ\varphi}(\mu_i). \tag{4}$$

For, let $h = \varphi' \circ f = g \circ \varphi$. By Prop. 1, the conditions (i) and (ii) are each equivalent to the following condition:

(iii) the $\mu_i$ are $h$-convolvable.

When this is so,

$$*_{\varphi'}(f_i(\mu_i)) = *_{h}(\mu_i) = g(*_{\varphi}(\mu_i)).$$

## 3. The case of bounded measures

PROPOSITION 2. — *Let* $X_1, \ldots, X_n, Y$ *be locally compact spaces,* $\mu_i$ *a bounded measure on* $X_i$ $(1 \leqslant i \leqslant n)$, $\mu$ *the product of the* $\mu_i$, $\varphi$ *a* $\mu$*-measurable mapping of* $\prod_i X_i$ *into* $Y$. *Then the* $\mu_i$ *are* $\varphi$*-convolvable and*

$$\Big\| \mathop{*}_{i=1}^{n} \mu_i \Big\| \leqslant \prod_{i=1}^{n} \|\mu_i\|.$$

*If the* $\mu_i$ *are moreover positive, then* $\| \mathop{*}_{i=1}^{n} \mu_i \| = \prod_{i=1}^{n} \|\mu_i\|$.

For, $\mu'_i = |\mu_i|$ is bounded and $\|\mu'_i\| = \|\mu_i\|$ (Ch. III, §1, No. 8, Cor. 1 of Prop. 10). One has $|\mu_1 \otimes \cdots \otimes \mu_n| = \mu'_1 \otimes \cdots \otimes \mu'_n$ (Ch. III, §4, Nos. 2, 4), therefore $\mu_1 \otimes \cdots \otimes \mu_n$ is bounded and

$$\|\mu_1 \otimes \cdots \otimes \mu_n\| = \|\mu_1\| \cdots \|\mu_n\|$$

(*ibid.*, Prop. 4). Therefore $\varphi$ is $\mu$-proper (Ch. V, §6, No. 1, *Remark* 1), that is, the $\mu_i$ are $\varphi$-convolvable. One has $\| \mathop{*}_{1}^{n} \mu'_i \| = \|\mu'_1 \otimes \cdots \otimes \mu'_n\|$ (Ch. V, §6, No. 2, Th. 1), consequently $\| \mathop{*}_{i=1}^{n} \mu'_i \| = \|\mu'_1\| \cdots \|\mu'_n\|$. Finally, $| \mathop{*}_i \mu_i | \leqslant \mathop{*}_i \mu'_i$ (No. 1, formula (2)), therefore

$$\Big\| \mathop{*}_i \mu_i \Big\| \leqslant \Big\| \mathop{*}_i \mu'_i \Big\| = \prod_{i=1}^{n} \|\mu_i\|.$$

PROPOSITION 3. — *Let* $X_1, \ldots, X_n, Y$ *be locally compact spaces,* $\varphi$ *a continuous mapping of* $\prod_{i=1}^{n} X_i$ *into* $Y$. *Then the mapping*

$$(\mu, \ldots, \mu_n) \mapsto *_{\varphi}(\mu_i)$$

*of* $\prod_{i=1}^{n} \mathscr{M}^1(X_i)$ *into* $\mathscr{M}^1(Y)$ *is a continuous multilinear mapping.*

This follows from Prop. 2 and what has been said in No. 1.

## 4. Properties concerning supports

PROPOSITION 4. — *Let* $X_1, \ldots, X_n, Y$ *be locally compact spaces,* $\mu_i$ *a measure on* $X_i$ $(1 \leqslant i \leqslant n)$, $S_i$ *its support, and* $\varphi$ *a continuous mapping of* $\prod_i X_i$ *into* Y *such that the restriction of* $\varphi$ *to* $\prod_i S_i$ *is proper. Then the* $\mu_i$ *are* $\varphi$*-convolvable.*

For, let K be a compact subset of Y. The support of $\mu = \mu_1 \otimes \cdots \otimes \mu_n$ is $S = \prod_i S_i$ (Ch. III, §4, No. 2, Prop. 2). Therefore $\overset{-1}{\varphi}(K) \cap \left(\prod_i X_i - S\right)$ is $\mu$-negligible. On the other hand, $\overset{-1}{\varphi}(K) \cap S$ is compact. Therefore $\overset{-1}{\varphi}(K)$ is $\mu$-integrable.

PROPOSITION 5. — *Let* $X_1, \ldots, X_n, Y$ *be locally compact spaces,* $\mu_i$ *a measure on* $X_i$ $(1 \leqslant i \leqslant n)$, $\mu$ *the product of the* $\mu_i$, $\varphi$ *a* $\mu$*-proper mapping of* $\prod_i X_i$ *into* Y, *and* $S_i$ *the support of* $\mu_i$.

a) *The support of* $\mathop{*}_i \mu_i$ *is contained in the closure of* $\varphi\left(\prod_i S_i\right)$.

b) *If* $\varphi$ *is continuous and the* $\mu_i$ *are positive, then the support of* $\mathop{*}_i \mu_i$ *is the closure of* $\varphi\left(\prod_i S_i\right)$.

Let $S = \prod_i S_i$ be the support of $\mu$. The support of $\mathop{*}_i \mu_i$ is contained in $\overline{\varphi(S)}$ by Ch. V, §6, No. 2, Cor. 3 of Prop. 2. If $\varphi$ is continuous and the $\mu_i$ are positive, then the support of $\mathop{*}_i \mu_i$ is $\overline{\varphi(S)}$ (*loc. cit.*, Cor. 4 of Prop. 2).

COROLLARY. — *If* $\varphi$ *is continuous and the* $\mu_i$ *have compact support, then the* $\mu_i$ *are convolvable and* $\mathop{*}_i \mu_i$ *has compact support.*

## 5. Vectorial expression of the convolution product

PROPOSITION 6. — *Let* X, Y, Z *be locally compact spaces,* $\varphi$ *a continuous mapping of* $X \times Y$ *into* Z, *and* $\lambda, \mu$ *measures on* X, Y. *For* $\lambda$ *and* $\mu$ *to be* $\varphi$*-convolvable, it is necessary and sufficient that the mapping* $(x, y) \mapsto \varepsilon_{\varphi(x,y)} = \varepsilon_x * \varepsilon_y$ *of* $X \times Y$ *into* $\mathscr{M}(Z)$ *be scalarly* $(\lambda \otimes \mu)$*-integrable for the topology* $\sigma(\mathscr{M}(Z), \mathscr{K}(Z))$, *in which case*

$$\lambda * \mu = \int_{X \times Y} (\varepsilon_x * \varepsilon_y)\, d\lambda(x)\, d\mu(y).$$

To say that $\lambda$ and $\mu$ are $\varphi$-convolvable signifies that, for every $f \in \mathscr{K}(Z)$, $f \circ \varphi$ is $(\lambda \otimes \mu)$-integrable, that is, for every $f \in \mathscr{K}(Z)$ the function $(x, y) \mapsto \langle f, \varepsilon_{\varphi(x,y)} \rangle$ is $(\lambda \otimes \mu)$-integrable, that is, again, that the mapping $(x, y) \mapsto \varepsilon_{\varphi(x,y)}$ of $X \times Y$ into $\mathscr{M}(Z)$ is scalarly $(\lambda \otimes \mu)$-integrable for $\sigma(\mathscr{M}(Z), \mathscr{K}(Z))$. If this is the case, then

$$\langle \lambda * \mu, f \rangle = \int f(\varphi(x,y))\, d\lambda(x)\, d\mu(y) = \int_{X \times Y} \langle \varepsilon_{\varphi(x,y)}, f \rangle\, d\lambda(x)\, d\mu(y),$$

whence $\lambda * \mu = \int_{X \times Y} \varepsilon_{\varphi(x,y)}\, d\lambda(x)\, d\mu(y)$.

PROPOSITION 7. — *Let* $X, Y, Z$ *be locally compact spaces,* $\varphi$ *a continuous mapping of* $X \times Y$ *into* $Z$, *and* $\lambda, \mu$ *measures on* $X, Y$. *Assume that for every* $x \in X$, $\varepsilon_x$ *and* $\mu$ *are* $\varphi$*-convolvable. For* $\lambda$ *and* $\mu$ *to be* $\varphi$*-convolvable, it is necessary and sufficient that the mapping* $x \mapsto \varepsilon_x * |\mu|$ *of* $X$ *into* $\mathscr{M}(Z)$ *be scalarly* $\lambda$*-integrable for the topology* $\sigma(\mathscr{M}(Z), \mathscr{K}(Z))$, *in which case* $\lambda * \mu = \int_X (\varepsilon_x * \mu)\, d\lambda(x)$.

Suppose that $\lambda$ and $\mu$ are $\varphi$-convolvable. For every $f \in \mathscr{K}(Z)$, $f \circ \varphi$ is $(|\lambda| \otimes |\mu|)$-integrable, therefore the function $x \mapsto \int_Y f(\varphi(x,y))\, d|\mu|(y) = \langle f, \varepsilon_x * |\mu| \rangle$ (which by hypothesis is defined for all $x \in X$) is $\lambda$-integrable; thus $x \mapsto \varepsilon_x * |\mu|$ is scalarly $\lambda$-integrable for $\sigma(\mathscr{M}(Z), \mathscr{K}(Z))$, and

$$\langle f, \lambda * \mu \rangle = \int_X d\lambda(x) \int_Y f(\varphi(x,y))\, d\mu(y) = \int_X \langle f, \varepsilon_x * \mu \rangle\, d\lambda(x),$$

whence $\lambda * \mu = \int_X (\varepsilon_x * \mu)\, d\lambda(x)$. Conversely, suppose that the mapping $x \mapsto \varepsilon_x * |\mu|$ of $X$ into $\mathscr{M}(Z)$ is scalarly $\lambda$-integrable for $\sigma(\mathscr{M}(Z), \mathscr{K}(Z))$. Let $f \in \mathscr{K}_+(Z)$. Then the function $(x,y) \mapsto f(\varphi(x,y))$ is continuous and (Ch. V, §8, No. 3, Prop. 5)

$$\begin{aligned}\iint^* f(\varphi(x,y))\, d|\lambda|(x)\, d|\mu|(y) &= \int^* d|\lambda|(x) \int^* f(\varphi(x,y))\, d|\mu|(y) \\ &= \int^* \langle f, \varepsilon_x * |\mu| \rangle\, d|\lambda|(x) < +\infty.\end{aligned}$$

Therefore $f \circ \varphi$ is $(\lambda \otimes \mu)$-integrable, so that $\lambda$ and $\mu$ are $\varphi$-convolvable.

## §2. LINEAR REPRESENTATIONS OF GROUPS

### 1. Continuous linear representations

Let $G$ be a topological group, $E$ a locally convex space, $U$ a linear representation of $G$ in $E$.

DEFINITION 1. — (i) *$U$ is said to be separately continuous if, for every $s \in \mathrm{G}$, $U(s)$ is a continuous endomorphism of* E, *and if, for every $x \in \mathrm{E}$, the mapping $s \mapsto U(s)x$ of* G *into* E *is continuous.*

(ii) *$U$ is said to be continuous if $(s,x) \mapsto U(s)x$ is a continuous mapping of $\mathrm{G} \times \mathrm{E}$ into* E.

(iii) *$U$ is said to be equicontinuous if it is continuous and if the set of endomorphisms $U(s)$, where $s$ runs over* G, *is equicontinuous.*

*Remarks.* — 1) To say that $U$ is separately continuous means that $s \mapsto U(s)$ is a continuous mapping of G into the space $\mathscr{L}(\mathrm{E};\mathrm{E})$ of continuous endomorphisms of E, equipped with the topology of pointwise convergence.

2) To say that $U$ is continuous is equivalent to the following set of three conditions:

*a*) for every $s \in \mathrm{G}$, $U(s)$ is continuous; *b*) there exists a neighborhood V of $e$ such that $U(\mathrm{V})$ is equicontinuous; *c*) there exists a total set D in E such that, for every $x \in \mathrm{D}$, the mapping $s \mapsto U(s)x$ is continuous.

These conditions are obviously necessary. Conversely, suppose that the conditions *a*), *b*), *c*) are satisfied. On $U(\mathrm{V})$, the topology of pointwise convergence is identical to the topology of pointwise convergence in D (TVS, III, §3, No. 4, Prop. 5). Therefore the mapping $(s,x) \mapsto U(s)x$ of $\mathrm{V} \times \mathrm{E}$ into E is continuous (GT, X, §2, No. 1, Cor. 3 of Prop. 1). Since $U(s_0 s)x = U(s_0)\big(U(s)x\big)$ for all $s_0 \in \mathrm{G}$, $s \in \mathrm{G}$, $x \in \mathrm{E}$, one sees that $U$ is continuous.

When G is locally compact, the conditions *a*) and *b*) are equivalent to the condition:

*a'*) for every compact subset K of G, $U(\mathrm{K})$ is equicontinuous.

3) Suppose that $U$ is a continuous linear representation of G in E. For every $s \in \mathrm{G}$, let $\widehat{U}(s)$ be the continuous extension of $U(s)$ to the completion $\widehat{\mathrm{E}}$ of E. Then $\widehat{U}$ is a linear representation of G in $\widehat{\mathrm{E}}$, satisfying conditions *a*) and *c*) of *Remark* 2, and also condition *b*) by GT, X, §2, No. 2, Prop. 4. Therefore $\widehat{U}$ is a *continuous* linear representation of G in $\widehat{\mathrm{E}}$.

4) When E is a normed space, $U$ is said to be *isometric* if $\|U(s)\| = 1$ for every $s \in \mathrm{G}$. For this, it suffices that $\|U(s)\| \leqslant 1$ for all $s \in \mathrm{G}$, because one then has

$$1 = \|1\| \leqslant \|U(s)\| \cdot \|U(s^{-1})\|,$$

whence $\|U(s)\| = \|U(s^{-1})\| = 1$ for all $s \in \mathrm{G}$.

PROPOSITION 1. — *If* G *is a locally compact group and* E *is barreled, then every separately continuous linear representation $U$ of* G *in* E *is continuous.*

For every compact subset K of G, $U(\mathrm{K})$ is compact for the topology of pointwise convergence (*Remark* 1), therefore is equicontinuous (TVS, III, §4, No. 2, Th. 1); one then applies *Remark* 2.

*Lemma* 1. — *Let* G *be a locally compact group,* $\rho$ *a lower semi-continuous finite numerical function* $\geqslant 0$ *on* G *such that* $\rho(st) \leqslant \rho(s)\rho(t)$ *for all* $s, t \in G$. *Then* $\rho$ *is bounded above on every compact subset of* G.

There exists a nonempty open subset U of G such that $\rho$ is bounded above on U (GT, IX, §5, No. 4, Th. 2). Let K be a compact subset of G. Then K is covered by a finite number of sets $s_1 U, \ldots, s_n U$. For every $x \in U$, one has $\rho(s_i x) \leqslant \rho(s_i)\rho(x)$, therefore $\rho$ is bounded above on the $s_i U$, hence on K.

*Lemma* 2. — *Let* G *be a topological group,* $U$ *a linear representation of* G *in a normed space* E, *and* A *a dense subset of* E. *Assume that for every* $s \in G$, $U(s)$ *is continuous, and that, for every* $x \in A$, $s \mapsto U(s)x$ *is a continuous mapping of* G *into* E. *Then the function* $s \mapsto g(s) = \|U(s)\|$ *on* G *is lower semi-continuous and satisfies* $g(st) \leqslant g(s)g(t)$.

Let B be the unit ball of E. Then $g(s) = \sup_{x \in B \cap A} \|U(s)x\|$, and each function $s \mapsto \|U(s)x\|$ is continuous on G, therefore $g$ is lower semi-continuous. On the other hand,

$$g(st) = \|U(s)U(t)\| \leqslant \|U(s)\| \cdot \|U(t)\| = g(s)g(t).$$

PROPOSITION 2. — *Let* G *be a locally compact group,* $U$ *a linear representation of* G *in a normed space* E. *Let* A *be a dense subset of* E. *Assume that for every* $s \in G$, $U(s)$ *is continuous and that, for every* $x \in A$, $s \mapsto U(s)x$ *is a continuous mapping of* G *into* E. *Then* $U$ *is continuous.*

For, $\|U(s)\|$ is bounded on every compact subset of G by Lemmas 1 and 2, and one then applies *Remark* 2.

## 2. Contragredient representation

Let $U$ be a separately continuous linear representation of G in E. Let E′ be the dual of E. The mapping $s \mapsto {}^tU(s)$ is a linear representation in E′ of the group $G^0$ opposite G; we shall say that this representation is the *transpose* of $U$. The mapping $s \mapsto {}^tU(s^{-1}) = {}^tU(s)^{-1}$ is a linear representation of G in E′, called the *contragredient* of $U$.

*Lemma* 3. — *Let* X *be a locally compact space,* Y *and* Z *topological spaces,* $\varphi$ *a continuous mapping of* $X \times Y$ *into* Z, *and* $\varphi_x$ *the mapping* $y \mapsto \varphi(x, y)$ *of* Y *into* Z. *The spaces* $\mathscr{C}(Y), \mathscr{C}(Z)$ *being equipped with the topology of compact convergence, the mapping* $(x, f) \mapsto f \circ \varphi_x$ *of* $X \times \mathscr{C}(Z)$ *into* $\mathscr{C}(Y)$ *is continuous.*

It clearly suffices to consider the case that X is compact. Let $(x_0, f_0) \in X \times \mathscr{C}(Z)$, K a compact subset of Y, and $\varepsilon > 0$. Let $K' = \varphi(X \times K)$. Since

$f_0 \circ \varphi$ is uniformly continuous in $X \times K$, there exists a neighborhood W of $x_0$ such that $|f_0(\varphi(x,y)) - f_0(\varphi(x_0,y))| \leqslant \varepsilon$ for $x \in W$ and $y \in K$. On the other hand, if one takes $f \in \mathscr{C}(Z)$ to be such that $|f(z) - f_0(z)| \leqslant \varepsilon$ for all $z \in K'$, one will have $|f(\varphi(x,y)) - f_0(\varphi(x,y))| \leqslant \varepsilon$ for $x \in X$, $y \in K$, and therefore $|f(\varphi(x,y)) - f_0(\varphi(x_0,y))| \leqslant 2\varepsilon$ for $x \in W$, $y \in K$. Whence the lemma.

Let us now return to the earlier notations.

PROPOSITION 3. — (i) *If $U$ is separately continuous, then ${}^tU$ is separately continuous when* E′ *is equipped with the weak topology* $\sigma(E', E)$.

(ii) *If* G *is locally compact and $U$ is continuous, then ${}^tU$ is continuous when* E′ *is equipped with the topology of compact convergence.*

The assertion (i) is immediate. The assertion (ii) follows from Lemma 3 where one has taken $X = G$, $Y = Z = E$, $\varphi(s,x) = U(s)x$.

## 3. Example: linear representations in spaces of continuous functions

Let G be a discrete group operating on the left on a set X. A complex function $\chi$ on $G \times X$ is called a *multiplier* if

$$\chi(e,x) = 1 \quad \text{for all } x \in X; \tag{1}$$

$$\chi(st,x) = \chi(s,tx)\chi(t,x) \quad \text{for all } s,t \text{ in } G,\ x \in X. \tag{2}$$

It follows that

$$\chi(t^{-1},tx)\chi(t,x) = 1 \quad \text{for all } t \in G,\ x \in X, \tag{3}$$

and in particular $\chi(t,x) \neq 0$ for all $t \in G$, $x \in X$.

For every complex function $f$ defined on X and every $s \in G$, let $\boldsymbol{\gamma}_\chi(s)f$ be the complex function on X defined by

$$(\boldsymbol{\gamma}_\chi(s)f)(x) = \chi(s^{-1},x)f(s^{-1}x). \tag{4}$$

Then $\boldsymbol{\gamma}_\chi(e)f = f$ and

$$\begin{aligned}(\boldsymbol{\gamma}_\chi(s)\boldsymbol{\gamma}_\chi(s')f)(x) &= \chi(s^{-1},x)(\boldsymbol{\gamma}_\chi(s')f)(s^{-1}x)\\ &= \chi(s^{-1},x)\chi(s'^{-1},s^{-1}x)f(s'^{-1}s^{-1}x)\\ &= \chi((ss')^{-1},x)f((ss')^{-1}x) = (\boldsymbol{\gamma}_\chi(ss')f)(x),\end{aligned}$$

thus $\boldsymbol{\gamma}_\chi$ *is a linear representation* of G. For $\chi = 1$, one recovers the endomorphisms $\boldsymbol{\gamma}(s)$ (Ch. VII, §1, No. 1, formula (3)).

Suppose now that G and X are locally compact, G operating continuously on X, and $\chi$ *continuous* on $G \times X$. Then $\mathscr{C}(X)$ and $\mathscr{K}(X)$ are stable for the $\boldsymbol{\gamma}_\chi(s)$, whence linear representations of G in $\mathscr{C}(X)$ and $\mathscr{K}(X)$ which we shall again denote $\boldsymbol{\gamma}_\chi$.

PROPOSITION 4. — *The linear representations* $\boldsymbol{\gamma}_\chi$ *of* G *in* $\mathscr{C}(X)$ *and* $\mathscr{K}(X)$ *are continuous.*

The mapping $(s, f) \mapsto (s, \gamma(s)f)$ of $G \times \mathscr{C}(X)$ into $G \times \mathscr{C}(X)$ is continuous (No. 2, Lemma 3). On the other hand, the mapping $(s, f) \mapsto \chi(s, \cdot)f$ of $G \times \mathscr{C}(X)$ into $\mathscr{C}(X)$ is continuous; for, if $s$ tends to $s_0$ in G, then $\chi(s, \cdot)$ tends to $\chi(s_0, \cdot)$ uniformly on every compact subset of X; if, moreover, $f$ tends to $f_0$ in $\mathscr{C}(X)$, then $\chi(s, \cdot)f$ tends to $\chi(s_0, \cdot)f_0$ uniformly on every compact subset of X, whence our assertion. Thus the representation $\boldsymbol{\gamma}_\chi$ of G in $\mathscr{C}(X)$ is continuous.

Let us show that the representation $\boldsymbol{\gamma}_\chi$ of G in $\mathscr{K}(X)$ is continuous. Since $\mathscr{K}(X)$ is the direct limit of Banach spaces, it is barreled (TVS, III, §4, No. 1, Cor. 3 of Prop. 3), thus it suffices to prove that $\boldsymbol{\gamma}_\chi$ is separately continuous (No. 1, Prop. 1). Now, let H be a compact subset of X and let $s_0 \in G$. Let V be a compact neighborhood of $s_0$ in G, and let $L = VH$, which is compact in X. For every $f \in \mathscr{K}(X, H)$, the support of $\boldsymbol{\gamma}_\chi(s_0)f$ is contained in L, and

$$\sup_{x \in X} \left| \left(\boldsymbol{\gamma}_\chi(s_0)f\right)(x) \right| \leqslant \sup_{x \in L} |\chi(s_0^{-1}, x)| \cdot \sup_{x \in X} |f(x)|,$$

therefore $f \mapsto \boldsymbol{\gamma}_\chi(s_0)f$ is a continuous linear mapping of $\mathscr{K}(X, H)$ into $\mathscr{K}(X, L)$; it follows that $f \mapsto \boldsymbol{\gamma}_\chi(s_0)f$ is a continuous linear mapping of $\mathscr{K}(X)$ into itself (TVS, II, §4, No. 4, Prop. 5). On the other hand, the topology of $\mathscr{K}(X, L)$ is induced by that of $\mathscr{C}(X)$. By what has already been proved, the mapping $s \mapsto \boldsymbol{\gamma}_\chi(s)f$ of V into $\mathscr{K}(X, L)$ is continuous. This completes the proof that $\boldsymbol{\gamma}_\chi$ is separately continuous.

PROPOSITION 5. — *Suppose that each function* $\chi(s, \cdot)$ *is bounded. Then* $\boldsymbol{\gamma}_\chi$ *leaves* $\overline{\mathscr{K}(X)}$ *stable, and the linear representation* $\boldsymbol{\gamma}_\chi$ *of* G *in* $\overline{\mathscr{K}(X)}$ *is continuous.*

It is clear that $\boldsymbol{\gamma}_\chi$ leaves $\overline{\mathscr{K}(X)}$ stable and that each of the $\boldsymbol{\gamma}_\chi(s)$ is *continuous* in $\overline{\mathscr{K}(X)}$. On the other hand, for every $f \in \mathscr{K}(X)$, $s \mapsto \boldsymbol{\gamma}_\chi(s)f$ is a continuous mapping of G into $\mathscr{K}(X)$ and *a fortiori* into $\overline{\mathscr{K}(X)}$. Therefore the representation $\boldsymbol{\gamma}_\chi$ in $\overline{\mathscr{K}(X)}$ is continuous (No. 1, Prop. 2).

## 4. Example: linear representations in spaces of measures

Again let G be a locally compact group, operating continuously on the left in a locally compact space X, and let $\chi$ be a *continuous* multiplier on $G \times X$. The linear representation $\boldsymbol{\gamma}_\chi$ of G in $\mathscr{K}(X)$ admits a contragredient representation in $\mathscr{M}(X)$, which we shall again denote by $\boldsymbol{\gamma}_\chi$, and which is defined by the following formula (where $\mu \in \mathscr{M}(X)$, $f \in \mathscr{K}(X)$):

$$\langle \boldsymbol{\gamma}_\chi(s)\mu, f\rangle = \langle \mu, \boldsymbol{\gamma}_\chi(s^{-1})f\rangle = \langle \chi(s,\cdot)\cdot\mu, \boldsymbol{\gamma}(s^{-1})f\rangle = \langle \boldsymbol{\gamma}(s)\big(\chi(s,\cdot)\cdot\mu\big), f\rangle\,,$$

whence

$$\boldsymbol{\gamma}_\chi(s)\mu = \boldsymbol{\gamma}(s)\big(\chi(s,\cdot)\cdot\mu\big) = \big(\boldsymbol{\gamma}(s)\chi(s,\cdot)\big)\cdot\big(\boldsymbol{\gamma}(s)\mu\big).$$

We note that

$$\big(\boldsymbol{\gamma}(s)\chi(s,\cdot)\big)(x) = \chi(s, s^{-1}x)\,.$$

The linear representation $\boldsymbol{\gamma}_\chi$ of G in $\mathscr{C}(X)$ admits a contragredient representation in the space $\mathscr{C}'(X)$ of measures on X with compact support, a representation which we again denote by $\boldsymbol{\gamma}_\chi$; the endomorphisms $\boldsymbol{\gamma}_\chi(s)$ of $\mathscr{C}'(X)$ are the restrictions of the endomorphisms $\boldsymbol{\gamma}_\chi(s)$ of $\mathscr{M}(X)$.

PROPOSITION 6. — *If one equips $\mathscr{M}(X)$ (resp. $\mathscr{C}'(X)$) with the topology of uniform convergence in the compact subsets of $\mathscr{K}(X)$ (resp. $\mathscr{C}(X)$), then the linear representation $\boldsymbol{\gamma}_\chi$ of G in $\mathscr{M}(X)$ (resp. $\mathscr{C}'(X)$) is continuous.*

PROPOSITION 7. — *Suppose that each function $\chi(s,\cdot)$ is bounded. Then $\boldsymbol{\gamma}_\chi$ leaves stable $\mathscr{M}^1(X)$ and, if $\mathscr{M}^1(X)$ is equipped with the topology of uniform convergence in the compact subsets of $\overline{\mathscr{K}(X)}$, then the linear representation $\boldsymbol{\gamma}_\chi$ of G in $\mathscr{M}^1(X)$ is continuous.*

These propositions result from Props. 3, 4, 5.

## 5. Example: linear representations in the spaces $L^p$

Again let G be a locally compact group, operating continuously on the left in a locally compact space X. Let $\beta$ be a positive measure on X with support X. Let us assume that there exists a *continuous* function $\chi > 0$ on $G \times X$ such that, for every $s \in G$,

$$\boldsymbol{\gamma}(s)\beta = \chi(s^{-1},\cdot)\cdot\beta$$

(which implies in particular that $\beta$ is quasi-invariant under G). *Then, $\chi$ is a multiplier.* For, given $s, t$ in G, one has

$$\begin{aligned}\boldsymbol{\gamma}(s)\boldsymbol{\gamma}(t)\beta = \boldsymbol{\gamma}(s)\big(\chi(t^{-1},\cdot)\cdot\beta\big) &= \big(\boldsymbol{\gamma}(s)\chi(t^{-1},\cdot)\big)\cdot\big(\boldsymbol{\gamma}(s)\beta\big)\\ &= \big(\boldsymbol{\gamma}(s)\chi(t^{-1},\cdot)\big)\cdot\chi(s^{-1},\cdot)\cdot\beta\,,\\ \boldsymbol{\gamma}(st)\beta &= \chi(t^{-1}s^{-1},\cdot)\cdot\beta\,,\end{aligned}$$

therefore

$$\chi(t^{-1}, s^{-1}x)\chi(s^{-1}, x) = \chi(t^{-1}s^{-1}, x)$$

locally $\beta$-almost everywhere, consequently everywhere, since $\chi$ is continuous and $\beta$ has support $\mathrm{X}$.

Let $p \in [1, +\infty[$. For every $f \in \mathscr{L}^p_{\mathrm{C}}(\mathrm{X}, \beta)$ and every $s \in \mathrm{G}$, let $\boldsymbol{\gamma}_{\chi,p}(s)f$ be the function on X defined by

$$\big(\boldsymbol{\gamma}_{\chi,p}(s)f\big)(x) = \chi(s^{-1}, x)^{1/p} f(s^{-1}x).$$

One has

$$\begin{aligned}\int^* |\chi(s^{-1}, x)^{1/p} f(s^{-1}x)|^p \, d\beta(x) &= \int^* |f(s^{-1}x)|^p \chi(s^{-1}, x)\, d\beta(x) \\ &= \int |f(x)|^p \, d\beta(x),\end{aligned}$$

therefore $\boldsymbol{\gamma}_{\chi,p}(s)f \in \mathscr{L}^p_{\mathrm{C}}(\mathrm{X}, \beta)$. One sees that $\boldsymbol{\gamma}_{\chi,p}(s)$ is an *isometric* endomorphism of $\mathscr{L}^p_{\mathrm{C}}(\mathrm{X}, \beta)$ and defines, by passage to the quotient, an isometric endomorphism of $\mathrm{L}^p_{\mathrm{C}}(\mathrm{X}, \beta)$, also denoted $\boldsymbol{\gamma}_{\chi,p}(s)$. On the other hand, $\chi^{1/p}$ is obviously a multiplier, therefore $\boldsymbol{\gamma}_{\chi,p}$ is a linear representation of G in $\mathrm{L}^p_{\mathrm{C}}(\mathrm{X}, \beta)$ by what we have seen in No. 3.

PROPOSITION 8. — *The linear representation* $\boldsymbol{\gamma}_{\chi,p}$ *of* G *in* $\mathrm{L}^p_{\mathrm{C}}(\mathrm{X}, \beta)$ *is continuous and isometric.*

Let $f \in \mathscr{K}(\mathrm{X})$. When $s$ tends to $s_0$ in G, $\boldsymbol{\gamma}_{\chi,p}(s)f$ tends to $\boldsymbol{\gamma}_{\chi,p}(s_0)f$ in $\mathscr{K}(\mathrm{X})$, hence in $\mathrm{L}^p_{\mathrm{C}}(\mathrm{X}, \beta)$. Since the $\boldsymbol{\gamma}_{\chi,p}(s)$ are isometric, Prop. 8 is obtained by applying *Remark* 2 of No. 1.

For the case that $\chi$ is not assumed continuous, cf. §4, Exer. 13.

PROPOSITION 9. — *Suppose that each function* $\chi(s, \cdot)$ *is bounded. Then* $\boldsymbol{\gamma}_{\chi}$ *leaves* $\mathrm{L}^p_{\mathrm{C}}(\mathrm{X}, \beta)$ *stable, and the linear representation* $\boldsymbol{\gamma}_{\chi}$ *of* G *in* $\mathrm{L}^p_{\mathrm{C}}(\mathrm{X}, \beta)$ *is continuous.*

Let $f \in \mathscr{L}^p_{\mathrm{C}}(\mathrm{X}, \beta)$. Then

$$\begin{aligned}&\int^* |\chi(s^{-1}, x) f(s^{-1}x)|^p \, d\beta(x) \\ &\qquad \leqslant \sup_{x \in \mathrm{X}} \chi(s^{-1}, x)^{p-1} \int^* |f(s^{-1}x)|^p \chi(s^{-1}, x)\, d\beta(x) \\ &\qquad = \sup_{x \in \mathrm{X}} \chi(s^{-1}, x)^{p-1} \int |f(x)|^p \, d\beta(x),\end{aligned}$$

therefore $\boldsymbol{\gamma}_{\chi}(s)f \in \mathscr{L}^p_{\mathrm{C}}(\mathrm{X}, \beta)$, and

$$\|\boldsymbol{\gamma}_{\chi}(s)\| \leqslant \sup_{x \in \mathrm{X}} \chi(s^{-1}, x)^{1/q}, \tag{5}$$

where $q$ denotes the exponent conjugate to $p$. If $f \in \mathscr{K}(\mathrm{X})$, then $\gamma_\chi(s)f$ tends to $\gamma_\chi(s_0)f$ in $\mathscr{K}(\mathrm{X})$, hence in $\mathscr{L}^p_{\mathrm{C}}(\mathrm{X},\beta)$, as $s$ tends to $s_0$. Therefore the representation $\gamma_\chi$ of G in $\mathrm{L}^p_{\mathrm{C}}(\mathrm{X},\beta)$ is continuous (No. 1, Prop. 2).

Properties analogous to those of Nos. 3, 4, 5 hold if G operates on the right in X.

In particular, if one regards G as operating on itself by left or right translations, and if one takes $\chi = 1$, one obtains the *left* and *right regular representations* of G in $\mathscr{C}(\mathrm{G})$, $\mathscr{K}(\mathrm{G})$, $\overline{\mathscr{K}(\mathrm{G})}$, $\mathscr{C}'(\mathrm{G})$, $\mathscr{M}(\mathrm{G})$, $\mathscr{M}^1(\mathrm{G})$. If one takes $\beta$ to be a left (resp. right) Haar measure on G, and if one takes $\chi = 1$, one obtains the *left* (resp. *right*) *regular representation* of G in $\mathrm{L}^p_{\mathrm{C}}(\mathrm{G},\beta)$.

## 6. Extension of a linear representation of G to the measures on G

Let G be a locally compact group, E a locally convex space, $U$ a linear representation of G in E. Assume $U$ to be continuous and E quasi-complete. Then, for every measure $\mu \in \mathscr{C}'(\mathrm{G})$, one has

$$\int_{\mathrm{G}} U(s)\,d\mu(s) \in \mathscr{L}(\mathrm{E};\mathrm{E})$$

(Ch. VI, §1, No. 7). We shall write $U(\mu) = \int_{\mathrm{G}} U(s)\,d\mu(s)$. We equip $\mathscr{C}'(\mathrm{G})$ with the topology of compact convergence in $\mathscr{C}(\mathrm{G})$. The mapping $(\mu, x) \mapsto U(\mu)x$ of $\mathscr{C}'(\mathrm{G}) \times \mathrm{E}$ into E is *hypocontinuous* relative to the equicontinuous subsets of $\mathscr{C}'(\mathrm{G})$ and the compact subsets of E; in particular, the mapping $\mu \mapsto U(\mu)$ of $\mathscr{C}'(\mathrm{G})$ into $\mathscr{L}(\mathrm{E};\mathrm{E})$ (equipped with the topology of compact convergence) is continuous (*loc. cit.*, Prop. 16).

In order to be able to apply these results later on, we note that if X is a locally compact space then $\mathscr{C}(\mathrm{X})$, equipped with the topology of compact convergence, is complete (GT, X, §1, No. 6, Cor. 3 of Th. 2). On the other hand, $\mathscr{K}(\mathrm{X})$ is barreled, therefore its dual $\mathscr{M}(\mathrm{X})$, equipped with the topology of compact convergence in $\mathscr{K}(\mathrm{X})$, is quasi-complete (TVS, III, §4, No. 2, Cor. 4 of Th. 1). Of course, $\overline{\mathscr{K}(\mathrm{X})}$ is complete for the topology deduced from its norm, therefore its dual $\mathscr{M}^1(\mathrm{X})$ is quasi-complete for the topology of compact convergence in $\overline{\mathscr{K}(\mathrm{X})}$ (*loc. cit.*).

Let us now assume that $U$ is a continuous linear representation of the locally compact group G on a *Banach space* E. Set $g(s) = \|U(s)\|$ for all $s \in \mathrm{G}$. Then, if $\mu$ is a measure on G such that $g$ is $\mu$-integrable, one has $\int_{\mathrm{G}} U(s)\,d\mu(s) \in \mathscr{L}(\mathrm{E};\mathrm{E})$ and $\left\| \int_{\mathrm{G}} U(s)\,d\mu(s)\right\| \leqslant \int g(s)\,d|\mu|(s)$ (Ch. VI, §1, No. 7, *Remark* 1). We again write $U(\mu) = \int_{\mathrm{G}} U(s)\,d\mu(s)$.

## 7. Relations between the endomorphisms $U(\mu)$ and the endomorphisms $U(s)$

*Lemma* 4. — *Let* T *be a locally compact space,* $a$ *a point of* T, M *a subset of* $\mathscr{M}(\mathrm{T})$, *and* $\mathfrak{F}$ *a filter on* M. *Assume that:*

(i) *for every compact subset* K *of* T, *the numbers* $|\mu|(\mathrm{K})$, *for* $\mu \in \mathrm{M}$, *are bounded above;*

(ii) $\lim_{\mu,\mathfrak{F}} |\mu|(\mathrm{K}) = 0$ *for every compact subset* K *of* $\mathrm{T} - \{a\}$.

(iii) *there exists a compact neighborhood* V *of* $a$ *in* T *such that* $\lim_{\mu,\mathfrak{F}} \mu(\mathrm{V}) = 1$.

*Then the filter* $\mathfrak{F}$ *converges to* $\varepsilon_a$ *in* $\mathscr{M}(\mathrm{T})$ *equipped with the topology of compact convergence in* $\mathscr{K}(\mathrm{T})$.

By the hypothesis (i), M is an equicontinuous subset of $\mathscr{M}(\mathrm{T})$ since it is vaguely bounded and $\mathscr{K}(\mathrm{T})$ is barreled (TVS, III, §4, No. 2, Th. 1). It therefore suffices (GT, X, §2, No. 4, Th. 1) to prove that if $f \in \mathscr{K}(\mathrm{T})$, then $\lim_{\mu,\mathfrak{F}} \mu(f) = f(a)$. Let K be the union of V and the support of $f$; if K′ is the closure of K − V, one has

$$|\mu(\mathrm{K}) - \mu(\mathrm{V})| = |\mu(\mathrm{K} - \mathrm{V})| \leqslant |\mu|(\mathrm{K}');$$

since K′ is compact and does not contain $a$, one concludes from this that $\lim_{\mu,\mathfrak{F}} \mu(\mathrm{K}) = 1$. Let $\varepsilon > 0$, and let W be an open neighborhood of $a$ in K such that $|f(t) - f(a)| \leqslant \varepsilon$ for $t \in \mathrm{W}$; one can write

$$\mu(f) - f(a) = f(a)\big(\mu(\mathrm{K}) - 1\big) + \int_{\mathrm{K}} \big(f(t) - f(a)\big)d\mu(t);$$

the integral over K may be written as the sum of the analogous integrals over W and K − W; if $\mathrm{C} = \sup|f|$, one therefore has

$$|\mu(f) - f(a)| \leqslant \mathrm{C}|\mu(\mathrm{K}) - 1| + \varepsilon \cdot |\mu|(\mathrm{K}) + 2\mathrm{C} \cdot |\mu|(\mathrm{K} - \mathrm{W}).$$

Since the first and third terms on the right side tend to 0 with respect to $\mathfrak{F}$, one sees that indeed $\lim_{\mu,\mathfrak{F}} \mu(f) = f(a)$.

COROLLARY 1. — *With hypotheses as in Lemma* 4, *suppose in addition that there exists a compact subset* $\mathrm{K}_0$ *of* T *containing the supports of all of the measures* $\mu \in \mathrm{M}$. *Then* $\mathfrak{F}$ *also converges to* $\varepsilon_a$ *in* $\mathscr{C}'(\mathrm{T})$ *equipped with the topology of compact convergence in* $\mathscr{C}(\mathrm{T})$.

For, the restriction mapping of $\mathscr{C}(\mathrm{T})$ into $\mathscr{C}(\mathrm{K}_0)$ is continuous; therefore, if H is a compact subset of $\mathscr{C}(\mathrm{T})$, then the restrictions to $\mathrm{K}_0$ of the functions in H form a compact subset of $\mathscr{C}(\mathrm{K}_0)$. It then suffices to apply Lemma 4 on replacing T by $\mathrm{K}_0$.

COROLLARY 2. — *With hypotheses as in Cor.* 1, *let* $f$ *be a continuous mapping of* T *into a quasi-complete locally convex space* E. *Then*

$$\lim_{\mu,\mathfrak{F}} \int f(t)\, d\mu(t) = f(a).$$

This follows from Cor. 1, and Prop. 14 of Ch. VI, §1, No. 6.

COROLLARY 3. — *Let* G *be a locally compact group,* E *a quasi-complete locally convex space, and* $U$ *a continuous linear representation of* G *in* E. *Let* $\beta$ *be a positive measure on* G, $a$ *an element of* G, *and* $\mathfrak{B}$ *a base for the filter of neighborhoods of* $a$, *formed of compact neighborhoods. For every* $\mathrm{V} \in \mathfrak{B}$, *let* $f_{\mathrm{V}}$ *be a continuous function* $\geqslant 0$ *on* G, *with support contained in* V, *and such that* $\int f_{\mathrm{V}}\, d\beta = 1$. *Then, for every* $x \in \mathrm{E}$,

$$U(a)x = \lim_{\mathrm{V}} U(f_{\mathrm{V}} \cdot \beta)x,$$

*the limit being taken with respect to the section filter of* $\mathfrak{B}$.

The mapping $s \mapsto U(s)x$ of G into E is continuous. By Cor. 2, $U(a)x = \lim_{\mathrm{V}} \int \big(U(s)x\big) \cdot f_{\mathrm{V}}(s)\, d\beta(s)$ with respect to the section filter of $\mathfrak{B}$, that is, $U(a)x = \lim_{\mathrm{V}} U(f_{\mathrm{V}} \cdot \beta)x$.

PROPOSITION 10. — *Let* G *be a locally compact group,* E *a quasi-complete locally convex space,* $U$ *a continuous linear representation of* G *in* E, *and* $\beta$ *a positive measure on* G *with support* G.

(i) *The vectors* $U(f \cdot \beta)x$, *where* $f$ *runs over* $\mathscr{K}(\mathrm{G})$ *and* $x$ *runs over* E, *are dense in* E.

(ii) *Let* F *be a closed linear subspace of* E. *If* F *is stable for* $U$, *then* $U(\mu)(\mathrm{F}) \subset \mathrm{F}$ *for every* $\mu \in \mathscr{C}'(\mathrm{G})$. *Conversely, if* $U(f \cdot \beta) \subset \mathrm{F}$ *for every* $f \in \mathscr{K}(\mathrm{G})$, *then* F *is stable for* $U$.

The first part of (ii) is immediate, since the restrictions of the $U(s)$ to F $(s \in \mathrm{G})$ define a continuous linear representation of G in the quasi-complete locally convex space F. The second part of (ii), and (i), follow from Cor. 3 of Lemma 4.

# §3. CONVOLUTION OF MEASURES ON GROUPS

## 1. Algebras of measures

Let G be a locally compact group. It will be understood, once and for all, that the measures $\mu_1, \ldots, \mu_n$ on G are said to be convolvable if they are so for the mapping

$$(x_1, x_2, \ldots, x_n) \mapsto x_1 x_2 \cdots x_n ;$$

and it is by means of this mapping that the convolution product $\mathop{*}_i \mu_i$ will always be taken. If $s \in \mathrm{G}$, $t \in \mathrm{G}$, then

$$\varepsilon_s * \varepsilon_t = \varepsilon_{st} . \tag{1}$$

If $s \in \mathrm{G}$ and $\mu \in \mathscr{M}(\mathrm{G})$, then

$$\varepsilon_s * \mu = \boldsymbol{\gamma}(s)\mu \tag{2}$$
$$\mu * \varepsilon_s = \boldsymbol{\delta}(s^{-1})\mu \tag{3}$$

by §1, No. 1, Example 3. If G is abelian, to say that $\mu_1$ and $\mu_2$ are convolvable is equivalent to saying that $\mu_2$ and $\mu_1$ are convolvable, and one then has $\mu_1 * \mu_2 = \mu_2 * \mu_1$ . When G is not abelian, it can happen that $\mu_1$ and $\mu_2$ are convolvable, without $\mu_2$ and $\mu_1$ being so (Exer. 12).

PROPOSITION 1. — *Let* G *be a locally compact group,* $\lambda, \mu, \nu$ *measures* $\neq 0$ *on* G.

(i) *If* $\lambda, \mu, \nu$ *are convolvable, then so are* $\lambda$ *and* $\mu$, $|\lambda| * |\mu|$ *and* $\nu$, $\mu$ *and* $\nu$, $\lambda$ *and* $|\mu| * |\nu|$, *and one has*

$$\lambda * \mu * \nu = (\lambda * \mu) * \nu = \lambda * (\mu * \nu) .$$

(ii) *If* $\lambda$ *and* $\mu$ *are convolvable, as well as* $|\lambda| * |\mu|$ *and* $\nu$, *then* $\lambda, \mu, \nu$ *are convolvable. Similarly if* $\mu$ *and* $\nu$ *are convolvable, as well as* $\lambda$ *and* $|\mu| * |\nu|$.

This follows from Prop. 1 of §1, No. 2.

There can exist measures $\lambda, \mu, \nu$ on G such that the convolution products $\lambda * \mu$, $(\lambda * \mu) * \nu$, $\mu * \nu$, $\lambda * (\mu * \nu)$ are all defined, and yet $(\lambda * \mu) * \nu \neq \lambda * (\mu * \nu)$ (cf. Exer. 4).

Let $\rho$ be a lower semi-continuous finite function $> 0$ on G such that $\rho(st) \leqslant \rho(s)\rho(t)$ for all $s, t$ in G. We denote by $\mathscr{M}^{\rho}(\mathrm{G})$ the vector space of measures $\lambda$ on G such that $\rho$ is $\lambda$-integrable, and by $\|\lambda\|_{\rho}$ (or simply $\|\lambda\|$) the norm $\int_{\mathrm{G}} \rho(s)\, d|\lambda|(s)$ on this space. When $\rho = 1$, one recovers the set $\mathscr{M}^1(\mathrm{G})$ of bounded measures on G.

PROPOSITION 2. — (i) *Any two elements of $\mathscr{M}^{\rho}(\mathrm{G})$ are convolvable.*

(ii) *For convolution, and for the norm $\|\lambda\|$, $\mathscr{M}^{\rho}(\mathrm{G})$ is a complete normed algebra, admitting $\varepsilon_e$ as unity element.*

(iii) *$\mathscr{C}'(\mathrm{G})$ is a subalgebra of $\mathscr{M}^{\rho}(\mathrm{G})$.*

Let $\lambda, \mu$ be in $\mathscr{M}^{\rho}(\mathrm{G})$, and let us show that $\lambda$ and $\mu$ are convolvable. Let $f \in \mathscr{K}_+(\mathrm{G})$. Since $\rho$ is $> 0$ and lower semi-continuous, there exists a constant $k > 0$ such that $f \leqslant k\rho$. Then

$$\begin{aligned}\int^* f(st)\, d|\lambda|(s)\, d|\mu|(t) &\leqslant k \int^* \rho(st)\, d|\lambda|(s)\, d|\mu|(t) \\ &\leqslant k \int^* \rho(s)\rho(t)\, d|\lambda|(s)\, d|\mu|(t) \\ &= k\Big(\int^* \rho(s)\, d|\lambda|(s)\Big)\Big(\int^* \rho(t)\, d|\mu|(t)\Big)\end{aligned}$$

(Ch. V, §8, No. 3, Cor. 1 of Prop. 8). Therefore $(s,t) \mapsto f(st)$ is $(\lambda \otimes \mu)$-integrable, so that $\lambda$ and $\mu$ are convolvable. On the other hand, using Ch. V (§1, Prop. 4, §6, Prop. 2, §8, Cor. 1 of Prop. 8) and the fact that $(s,t) \mapsto \rho(s)\rho(t)$ is lower semi-continuous in $\mathrm{G} \times \mathrm{G}$, one has

$$\begin{aligned}\int_{\mathrm{G}}^* \rho(s)\, d|\lambda * \mu|(s) &= \int_{\mathrm{G}}^{\bullet} \rho(s)\, d|\lambda * \mu|(s) \\ &\leqslant \int_{\mathrm{G}\times\mathrm{G}}^{\bullet} \rho(st)\, d|\lambda|(s)\, d|\mu|(t) \leqslant \int_{\mathrm{G}\times\mathrm{G}}^{\bullet} \rho(s)\rho(t)\, d|\lambda|(s)\, d|\mu|(t) \\ &= \int_{\mathrm{G}\times\mathrm{G}}^* \rho(s)\rho(t)\, d|\lambda|(s)\, d|\mu|(t) = \|\lambda\| \cdot \|\mu\|.\end{aligned}$$

One sees that $\lambda * \mu \in \mathscr{M}^{\rho}(\mathrm{G})$ and that $\|\lambda * \mu\| \leqslant \|\lambda\| \cdot \|\mu\|$. In view of Prop. 1, $\mathscr{M}^{\rho}(\mathrm{G})$ is an algebra. The mapping $\lambda \mapsto \rho \cdot \lambda$ is an isometric linear mapping $\theta$ of $\mathscr{M}^{\rho}(\mathrm{G})$ into $\mathscr{M}^1(\mathrm{G})$; if $\mu \in \mathscr{M}^1(\mathrm{G})$ then $1/\rho$, which is locally bounded and upper semi-continuous, is locally $\mu$-integrable, and $\rho$ is $(1/\rho)\cdot\mu$-integrable, thus $(1/\rho)\cdot\mu \in \mathscr{M}^{\rho}(\mathrm{G})$; this proves that $\theta$ is surjective; therefore $\mathscr{M}^{\rho}(\mathrm{G})$ is a complete normed algebra. Finally, it is clear that $\varepsilon_e$ is a unity element for $\mathscr{M}^{\rho}(\mathrm{G})$ and that $\mathscr{C}'(\mathrm{G})$ is a subalgebra of $\mathscr{M}^{\rho}(\mathrm{G})$ (§1, No. 4, Cor. of Prop. 5).

If $\rho = 1$, Prop. 2, (i) and (ii) also follow from §1, Prop. 2.

PROPOSITION 3. — *Let $\mu_1, \dots, \mu_n$ be measures on* G. *If all of the $\mu_i$, except at most one, have compact support, then the $\mu_i$ are convolvable.*

For, let $S_i$ be the support of $\mu_i$, and suppose that $S_i$ is compact for $i \neq i_0$. Let K be a compact subset of G. The set of $(x_1, \dots, x_n) \in \prod_i S_i$ such that $x_1 x_2 \cdots x_n \in K$ is compact, because the conditions $x_i \in S_i$ for all $i$, $x_1 x_2 \cdots x_n \in K$ imply

$$x_{i_0} \in S_{i_0-1}^{-1} \cdots S_1^{-1} K S_n^{-1} \cdots S_{i_0+1}^{-1} .$$

Therefore the $\mu_i$ are convolvable (§1, No. 4, Prop. 4).

PROPOSITION 4. — *The mapping $(\lambda, \mu) \mapsto \lambda * \mu$* (resp. $(\lambda, \mu) \mapsto \mu * \lambda$), *where $\lambda \in \mathscr{C}'(G)$, $\mu \in \mathscr{M}(G)$, defines on $\mathscr{M}(G)$ the structure of a left* (resp. *right*) *module over the algebra $\mathscr{C}'(G)$.*

This follows from Props. 1 and 3.

PROPOSITION 5. — *Let $\lambda$ be a left* (resp. *right*) *Haar measure on* G, *and $\mu \in \mathscr{M}^1(G)$. Then $\mu$ and $\lambda$* (resp. *$\lambda$ and $\mu$*) *are convolvable, and $\mu * \lambda = \mu(1)\lambda$* (resp. $\lambda * \mu = \mu(1)\lambda$).

We can suppose that $\mu \geqslant 0$. Let $f \in \mathscr{K}_+(G)$. When $\lambda$ is a left Haar measure,

$$\int^* d\mu(x) \int^* f(xy)\, d\lambda(y) = \int^* d\mu(x) \int f(y)\, d\lambda(y) = \lambda(f)\|\mu\| ,$$

therefore the function $(x, y) \mapsto f(xy)$ is $(\mu \otimes \lambda)$-integrable, and its integral for $\mu \otimes \lambda$ is $\lambda(f)\|\mu\|$. One argues similarly when $\lambda$ is a right Haar measure.

PROPOSITION 6. — *Let $\mu$ and $\nu$ be two convolvable measures on* G. *Let $\chi$ be a continuous representation of* G *in* $\mathbf{C}^*$. *Then $\chi \cdot \mu$ and $\chi \cdot \nu$ are convolvable and $(\chi \cdot \mu) * (\chi \cdot \nu) = \chi \cdot (\mu * \nu)$.*

Let $f \in \mathscr{K}(G)$. Then $f\chi \in \mathscr{K}(G)$, therefore the function

$$(x, y) \mapsto f(xy)\chi(xy) = f(xy)\chi(x)\chi(y)$$

on $G \times G$ is integrable for $\mu \otimes \nu$; therefore the function $(x, y) \mapsto f(xy)$ is integrable for $(\chi \cdot \mu) \otimes (\chi \cdot \nu)$; therefore $\chi \cdot \mu$ and $\chi \cdot \nu$ are convolvable. Moreover,

$$\begin{aligned} \langle \chi \cdot \mu * \chi \cdot \nu, f \rangle &= \int f(xy)\chi(x)\chi(y)\, d\mu(x)\, d\nu(y) \\ &= \int (f\chi)(xy)\, d\mu(x)\, d\nu(y) = \langle \mu * \nu, \chi f \rangle , \end{aligned}$$

whence $(\chi \cdot \mu) * (\chi \cdot \nu) = \chi \cdot (\mu * \nu)$.

PROPOSITION 7. — *Let* G *and* G′ *be two locally compact groups,* $u$ *a continuous representation of* G *in* G′, *and* $\mu_1, \dots, \mu_n$ *measures on* G, *all* $\neq 0$. *Then the following assertions are equivalent:*

(i) $u$ *is* $\mu_i$*-proper for all* $i$, *and the measures* $u(|\mu_i|)$ *are convolvable;*

(ii) *the* $\mu_i$ *are convolvable and* $u$ *is proper for* $\underset{i}{*}(|\mu_i|)$.

*When these conditions are satisfied,*

$$\underset{i}{\ast}\, u(\mu_i) = u\Big(\underset{i}{\ast}\, \mu_i\Big).$$

This follows from §1, No. 2, Cor. of Prop. 1.

COROLLARY. — *Let* G *be a locally compact group,* $\mu_1, \dots, \mu_n$ *measures on* G. *For the sequence* $(\mu_i)_{1 \leqslant i \leqslant n}$ *to be convolvable, it is necessary and sufficient that the sequence* $(\check{\mu}_{n-i})_{0 \leqslant i \leqslant n-1}$ *be so, in which case*

$$(\mu_1 * \cdots * \mu_n)^{\vee} = \check{\mu}_n * \cdots * \check{\mu}_1 .$$

This follows from Prop. 7 on considering the isomorphism $x \mapsto x^{-1}$ of G onto the opposite group.

## 2. The case of a group operating on a space

Let X be a locally compact space on which a locally compact group G operates on the left continuously by

$$(s, x) \mapsto s \cdot x .$$

If $\mu_1, \dots, \mu_n$ are measures on G and $\nu$ is a measure on X, these will be said to be convolvable if they are so for the mapping $(s_1, \dots, s_n, x) \mapsto s_1 \cdots s_n x$ of $\mathrm{G}^n \times \mathrm{X}$ into X, and their convolution product is to be understood in the sense of this mapping.

If $s \in \mathrm{G}$ and $x \in \mathrm{X}$, then

$$\varepsilon_s * \varepsilon_x = \varepsilon_{sx} . \tag{4}$$

If $s \in \mathrm{G}$ and $\mu \in \mathscr{M}(\mathrm{X})$, then

$$\varepsilon_s * \mu = \boldsymbol{\gamma}(s)\mu \tag{5}$$

by §1, No. 1, Example 3.

PROPOSITION 8. — *Let* $\mu$ *be a measure on* G, $\nu$ *a measure on* X.

(i) *If* $\mu$ *has compact support, then* $\mu$ *and* $\nu$ *are convolvable.*

(ii) *If* $\nu$ *has compact support, and if* G *operates properly in* X, *then* $\mu$ *and* $\nu$ *are convolvable.*

This follows from Prop. 4 of §1, No. 4.

PROPOSITION 9. — *For convolution,* $\mathscr{M}^1(\mathrm{X})$ *is a left module over* $\mathscr{M}^1(\mathrm{G})$, *while* $\mathscr{M}(\mathrm{X})$ *and* $\mathscr{C}'(\mathrm{X})$ *are left modules over* $\mathscr{C}'(\mathrm{G})$.

This follows from Prop. 8, and from §1, Props. 1, 3 and the Cor. of Prop. 5.

PROPOSITION 10. — *Let* $\mu$ *be a measure on* G, $\nu$ *a measure on* X, $\mu$ *and* $\nu$ *being convolvable. Suppose in addition that there exists a positive measure* $\beta$ *on* X *such that* $\boldsymbol{\gamma}(s)\nu$ *has base* $\beta$ *for every* $s \in \mathrm{G}$. *Then* $\mu * \nu$ *has base* $\beta$.

Let K be a $\beta$-negligible compact subset of X. Then K is $\boldsymbol{\gamma}(s)|\nu|$-negligible for every $s \in \mathrm{G}$. Now,

$$|\mu| * |\nu| = \int_{\mathrm{G}} (\varepsilon_s * |\nu|)\, d|\mu|(s)$$

(§1, No. 5, Prop. 7), and the mapping $s \mapsto \varepsilon_s * |\nu|$ is vaguely continuous (§2, Prop. 6). Therefore K is $|\mu| * |\nu|$-negligible by Ch. V, §3, No. 3, Th. 1. Therefore $|\mu| * |\nu|$ has base $\beta$ (Ch. V, §5, No. 5, Th. 2).

## 3. Convolution and linear representations

PROPOSITION 11. — *Let* G *be a locally compact group,* E *a quasi-complete locally convex space,* $U$ *a continuous representation of* G *in* E.

(i) *If* $\lambda \in \mathscr{C}'(\mathrm{G})$, $\mu \in \mathscr{C}'(\mathrm{G})$, *then* $U(\lambda * \mu) = U(\lambda)U(\mu)$.

(ii) *Suppose that* E *is a Banach space, and let* $\rho(s) = \|U(s)\|$ *for* $s \in \mathrm{G}$. *If* $\lambda \in \mathscr{M}^{\rho}(\mathrm{G})$, $\mu \in \mathscr{M}^{\rho}(\mathrm{G})$, *then* $U(\lambda * \mu) = U(\lambda)U(\mu)$.

Let $\lambda, \mu$ be in $\mathscr{C}'(\mathrm{G})$. For any $x \in \mathrm{E}$ one has, by applying notably Props. 1 and 4 of Ch. VI, §1, No. 1,

$$\begin{aligned} U(\lambda * \mu)x &= \int_{\mathrm{G}} U(s)x\, d(\lambda * \mu)(s) \\ &= \int_{\mathrm{G}\times\mathrm{G}} U(st)x\, d\lambda(s)\, d\mu(t) = \int_{\mathrm{G}\times\mathrm{G}} U(s)U(t)x\, d\lambda(s)\, d\mu(t) \\ &= \int_{\mathrm{G}} U(\lambda)U(t)x\, d\mu(t) = U(\lambda)\int_{\mathrm{G}} U(t)x\, d\mu(t) = U(\lambda)U(\mu)x, \end{aligned}$$

whence (i). An analogous argument may be applied in case (ii).

With G still a locally compact group, let us assume that G operates continuously on the left in a locally compact space X. This defines (§2, No. 4) a continuous linear representation $\boldsymbol{\gamma}$ of G in $\mathscr{M}(\mathrm{X})$ (equipped with the topology of compact convergence in $\mathscr{K}(\mathrm{X})$).

PROPOSITION 12. — *If* $\lambda \in \mathscr{C}'(\mathrm{G})$ *and* $\mu \in \mathscr{M}(\mathrm{X})$, *then*

$$\boldsymbol{\gamma}(\lambda)\mu = \lambda * \mu .$$

By Prop. 7 of §1, No. 5,

$$\lambda * \mu = \int_{\mathrm{G}} (\varepsilon_s * \mu)\, d\lambda(s) .$$

Now, $\varepsilon_s * \mu = \boldsymbol{\gamma}(s)\mu$ (No. 2, formula (5)) and

$$\int_{\mathrm{G}} (\boldsymbol{\gamma}(s)\mu) d\lambda(s) = \boldsymbol{\gamma}(\lambda)\mu$$

by the definition of $\boldsymbol{\gamma}(\lambda)$.

COROLLARY. — *The mapping* $(\lambda, \mu) \mapsto \lambda * \mu$ *of* $\mathscr{C}'(\mathrm{G}) \times \mathscr{M}(\mathrm{X})$ *into* $\mathscr{M}(\mathrm{X})$ *is hypocontinuous relative to the equicontinuous subsets of* $\mathscr{C}'(\mathrm{G})$ *and the compact subsets of* $\mathscr{M}(\mathrm{X})$ *(*$\mathscr{C}'(\mathrm{G})$ *and* $\mathscr{M}(\mathrm{X})$ *being equipped with the topology of compact convergence in* $\mathscr{C}(\mathrm{G})$ *and* $\mathscr{K}(\mathrm{X})$, *respectively).*

For, $\mathscr{M}(\mathrm{X})$, equipped with the topology of compact convergence in $\mathscr{K}(\mathrm{X})$, is quasi-complete. Therefore the mapping $(\lambda, \mu) \mapsto \boldsymbol{\gamma}(\lambda)\mu$ of $\mathscr{C}'(\mathrm{G}) \times \mathscr{M}(\mathrm{X})$ into $\mathscr{M}(\mathrm{X})$ is hypocontinuous relative to the equicontinuous subsets of $\mathscr{C}'(\mathrm{G})$ and the compact subsets of $\mathscr{M}(\mathrm{X})$ (§2, No. 6). It then suffices to apply Prop. 12.

*Remarks.* — 1) Let $\lambda_0 \in \mathscr{C}'(\mathrm{G})$. The mapping $\mu \mapsto \lambda_0 * \mu$ of $\mathscr{M}(\mathrm{X})$ into $\mathscr{M}(\mathrm{X})$ is vaguely continuous. For, let $f \in \mathscr{K}(\mathrm{X})$. One has

$$\langle \lambda_0 * \mu, f \rangle = \int f(sx)\, d\lambda_0(s)\, d\mu(x) = \langle \mu, g \rangle ,$$

where $g(x) = \int f(sx)\, d\lambda_0(s)$. Now, $g$ is continuous (Ch. VII, §1, No. 1, Lemma 1). On the other hand, let S be the support of $\lambda_0$ and K that of $f$. The conditions $sx \in \mathrm{K}$ and $s \in \mathrm{S}$ imply $x \in \mathrm{S}^{-1}\mathrm{K}$; therefore the support of $g$ is contained in $\mathrm{S}^{-1}\mathrm{K}$, so that $g \in \mathscr{K}(\mathrm{X})$. Then $\langle \lambda_0 * \mu, f \rangle = \langle \mu, g \rangle$ is a vaguely continuous function of $\mu$, which proves our assertion.

2) Let $\mu_0 \in \mathscr{M}(\mathrm{X})$. The mapping $\lambda \mapsto \lambda * \mu_0$ of $\mathscr{C}'(\mathrm{G})$ into $\mathscr{M}(\mathrm{X})$ is continuous for the topologies $\sigma(\mathscr{C}'(\mathrm{G}), \mathscr{C}(\mathrm{G}))$ and $\sigma(\mathscr{M}(\mathrm{X}), \mathscr{K}(\mathrm{X}))$. For,

let $f \in \mathscr{K}(X)$. Setting $h(s) = \int f(sx)\, d\mu_0(x)$, we have $\langle f, \lambda * \mu_0 \rangle = \langle h, \lambda \rangle$, and $h \in \mathscr{C}(G)$ (Ch. VII, §1, No. 1, Lemma 1).

PROPOSITION 13. — *The mapping* $(s, \mu) \mapsto \gamma(s)\mu$ *of* $G \times \mathscr{M}_+(X)$ *into* $\mathscr{M}_+(X)$ *is continuous when the set* $\mathscr{M}_+(X)$ *of positive measures on* X *is equipped with the vague topology.*

Since $\gamma(s)\mu = \gamma(ss_0^{-1})\gamma(s_0)\mu$, it follows from *Remark* 1 that it suffices to prove the continuity of the mapping under consideration at a point of the form $(e, \mu_0)$ with $\mu_0 \in \mathscr{M}_+(X)$. Given a function $f \in \mathscr{K}(X)$ and a number $\varepsilon > 0$, it is thus a matter of showing that there exist a neighborhood U of $e$ in G and a neighborhood W of $\mu_0$ in $\mathscr{M}_+(X)$ such that the relations $s \in U$, $\mu \in W$ imply

$$\left| \int f(sx)\, d\mu(x) - \int f(x)\, d\mu_0(x) \right| \leqslant \varepsilon . \tag{6}$$

Let V be a compact neighborhood of the support K of $f$ in X, and let $\varphi \in \mathscr{K}_+(X)$ be such that $\varphi(x) = 1$ on V; there exists a neighborhood $W_0$ of $\mu_0$ in $\mathscr{M}_+(X)$ such that $a = \sup_{\mu \in W_0} \mu(V)$ is finite: it suffices to take for $W_0$ the set of $\mu \in \mathscr{M}_+(X)$ such that $|\langle \varphi, \mu - \mu_0 \rangle| \leqslant 1$. Since the mapping $(s, x) \mapsto sx$ is continuous, there is, on the other hand, a compact neighborhood $U_0$ of $e$ in G such that $sK \subset V$ for all $s \in U_0$; the function $(s, x) \mapsto f(sx)$ is then uniformly continuous in $U_0 \times V$ and so there is a neighborhood $U \subset U_0$ of $e$ such that $|f(sx) - f(x)| \leqslant \varepsilon/2a$ for all $s \in U$ and $x \in V$. For $s \in U$ and $\mu \in W_0$, we therefore have

$$\left| \int f(sx)\, d\mu(x) - \int f(x)\, d\mu(x) \right| \leqslant \varepsilon/2 ;$$

if $W \subset W_0$ is the neighborhood of $\mu_0$ in $\mathscr{M}_+(X)$ formed by the measures $\mu \in W_0$ such that $\left| \int f(x)\, d\mu(x) - \int f(x)\, d\mu_0(x) \right| \leqslant \varepsilon/2$, U and W meet the requirements.

## §4. CONVOLUTION OF MEASURES AND FUNCTIONS

### 1. Convolution of a measure and a function

Let X be a locally compact space on which a locally compact group G operates on the left continuously. Let $\beta$ be a positive measure on X, quasi-invariant under G. Let $\chi$ be a function $> 0$ on $G \times X$, measurable for every measure on $G \times X$, and such that, for every $s \in G$, $\chi(s^{-1}, \cdot)$ is a density of $\gamma(s)\beta$ with respect to $\beta$:

$$\gamma(s)\beta = \chi(s^{-1}, \cdot) \cdot \beta, \tag{1}$$

which, with the conventions of Ch. VII, §1, No. 1, may be written:

$$d\beta(sx) = \chi(s, x)\, d\beta(x). \tag{1'}$$

These data will remain fixed in Nos. 1, 2, 3 (an exception being made in *Remark* 2 of No. 2).

Recall (§2, No. 5) that if $\chi$ is continuous and $\beta$ has support X, then $\chi$ is a multiplier.

Let $f$ be a locally $\beta$-integrable complex function on X, and let $\mu$ be a measure on G. For every $s \in G$, the measure $\gamma(s)(f \cdot \beta)$ has base $\beta$ since $\beta$ is quasi-invariant. Therefore, if $\mu$ and $f \cdot \beta$ are convolvable, then $\mu * (f \cdot \beta)$ *has base* $\beta$ (§3, No. 2, Prop. 10).

DEFINITION 1. — *If $\mu$ and $f \cdot \beta$ are convolvable, $\mu$ and $f$ are said to be convolvable relative to $\beta$. Every density of $\mu * (f \cdot \beta)$ with respect to $\beta$ is called a convolution product of $\mu$ and $f$ relative to $\beta$ and is denoted $\mu *^{\beta} f$.*

One omits $\beta$ when no confusion is possible. Convolution for several measures on G and a function on X is defined in an analogous manner.

The various convolution products of $\mu$ and $f$ are equal locally $\beta$-almost everywhere. If $\beta$ has support X and if there exists a convolution product of $\mu$ and $f$ that is continuous, then the latter is uniquely determined; it is then called *the* convolution product of $\mu$ and $f$ relative to $\beta$.

Let $s \in G$ and let $f$ be a locally $\beta$-integrable complex function on X. Then $\varepsilon_s$ and $f$ are convolvable, and

$$\varepsilon_s * (f \cdot \beta) = \gamma(s)(f \cdot \beta) = \big(\gamma(s)f\big) \cdot \big(\gamma(s)\beta\big) = \big(\gamma(s)f\big) \cdot \chi(s^{-1}, \cdot) \cdot \beta,$$

therefore

$$(\varepsilon_s * f)(x) = \chi(s^{-1}, x) f(s^{-1}x) = \big(\gamma_{\chi}(s) f\big)(x) \tag{2}$$

locally $\beta$-almost everywhere.

*Lemma* 1. — *Let $\mu$ be a measure on G. Then $\chi$ is locally $(\mu \otimes \beta)$-integrable, and the image of $\mu \otimes \beta$ under the homeomorphism $(s, x) \mapsto (s, s^{-1}x)$ of $G \times X$ onto $G \times X$ is $\chi \cdot (\mu \otimes \beta)$.*

We may suppose that $\mu \geqslant 0$. Let $\mathrm{F} \in \mathscr{K}_+(\mathrm{G} \times \mathrm{X})$. Then

$$\iint \mathrm{F}(s, s^{-1}x)\, d\mu(s)\, d\beta(x) = \int d\mu(s) \int \mathrm{F}(s, s^{-1}x)\, d\beta(x)$$
$$= \int d\mu(s) \int \mathrm{F}(s,x)\, d(\gamma(s^{-1})\beta)(x) = \int d\mu(s) \int \mathrm{F}(s,x)\chi(s,x)\, d\beta(x).$$

Now, the function $(s,x) \mapsto \mathrm{F}(s,x)\chi(s,x)$ has compact support and is $(\mu \otimes \beta)$-measurable. By Ch. V, §8, No. 3, Prop. 7, the preceding equality proves that this function is $(\mu \otimes \beta)$-integrable and that

$$\iint \mathrm{F}(s, s^{-1}x)\, d\mu(s)\, d\beta(x) = \iint \mathrm{F}(s,x)\chi(s,x)\, d\mu(s)\, d\beta(x).$$

This proves at the same time both assertions of Lemma 1.

PROPOSITION 1. — *Let $\mu$ be a measure on $\mathrm{G}$, $f$ a locally $\beta$-integrable complex function on $\mathrm{X}$. Suppose that the function $s \mapsto f(s^{-1}x)\chi(s^{-1},x)$ is essentially $\mu$-integrable except for a locally $\beta$-negligible set of values of $x$, and that the function $x \mapsto \int |f(s^{-1}x|\, \chi(s^{-1},x)\, d|\mu|(s)$, defined locally almost everywhere for $\beta$, is locally $\beta$-integrable. Then $\mu$ and $f$ are convolvable.*

We may assume that $f \geqslant 0$ and $\mu \geqslant 0$. Let $h \in \mathscr{K}_+(\mathrm{X})$. We are to prove that the function $(s,x) \mapsto h(sx)$ is essentially integrable for $\mu \otimes (f \cdot \beta) = (1 \otimes f) \cdot (\mu \otimes \beta)$ (Ch. V, §8, No. 5, Prop. 10), that is, that $\iint^{\bullet} h(sx)f(x)\, d\mu(s)\, d\beta(x) < +\infty$ (Ch. V, §5, No. 3, Prop. 3); it will clearly suffice to prove that there exists an $a > 0$ such that for every compact subset $\mathrm{K}$ of $\mathrm{G}$,

$$\iint^{\bullet} h(sx)f(x)\varphi_{\mathrm{K}}(s)\, d\mu(s)\, d\beta(x) \leqslant a.$$

By Lemma 1,

$$\iint^{\bullet} h(sx)f(x)\varphi_{\mathrm{K}}(s)\, d\mu(s)\, d\beta(x)$$
$$= \iint^{*} h(x)f(s^{-1}x)\varphi_{\mathrm{K}}(s)\chi(s^{-1},x)\, d\mu(s)\, d\beta(x).$$

Now, the function $(s,x) \mapsto h(x)f(s^{-1}x)\varphi_{\mathrm{K}}(s)\chi(s^{-1},x)$ is $(\mu \otimes \beta)$-measurable (Lemma 1) and has compact support. The preceding expression is therefore equal (Ch. V, §8, No. 3, Prop. 7) to

$$\int^{*} h(x)\, d\beta(x) \int^{*} f(s^{-1}x)\varphi_{\mathrm{K}}(s)\chi(s^{-1},x)\, d\mu(s)$$
$$\leqslant (\sup h) \int_{\mathrm{S}}^{*} d\beta(x) \int^{\bullet} f(s^{-1}x)\chi(s^{-1},x)\, d\mu(s),$$

where S denotes the support of $h$. Whence the proposition.

PROPOSITION 2. — *Let $\mu$ be a measure on* G, $f$ *a locally $\beta$-integrable complex function on* X. *Assume that one of the following conditions is satisfied:*

(i) $f$ *and* $\chi$ *are continuous;*

(ii) G *operates properly in* X *and* $f$ *is zero on the complement of a countable union of compact sets;*

(iii) $\mu$ *is carried by a countable union of compact sets.*

*If* $\mu$ *and* $f$ *are convolvable, then the function* $s \mapsto f(s^{-1}x)\chi(s^{-1},x)$ *is essentially $\mu$-integrable except for a locally $\beta$-negligible set of values of* $x$, *and one has, locally almost $\beta$-everywhere,*

$$(\mu *^{\beta} f)(x) = \int_{\mathrm{G}} f(s^{-1}x)\chi(s^{-1},x)\,d\mu(s) = \int_{\mathrm{G}} (\gamma_{\chi}(s)f)(x)\,d\mu(s). \tag{3}$$

Let $h \in \mathscr{K}(\mathrm{X})$. Since $\mu$ and $f$ are convolvable, the function $(s,x) \mapsto h(sx)f(x)$ is essentially $(\mu \otimes \beta)$-integrable. By Lemma 1, the function $(s,x) \mapsto h(x)f(s^{-1}x)\chi(s^{-1},x)$ is essentially $(\mu \otimes \beta)$-integrable. Under hypothesis (i) or (ii) of the statement, one then deduces that this function is $(\mu \otimes \beta)$-integrable; for, in the first case it is continuous and one applies Prop. 3 of Ch. V, §1, No. 1, and in the second case it is zero outside a countable union of compact sets, and one applies Prop. 7, 2) of No. 2, *loc. cit.* By the Lebesgue–Fubini theorem,

$$\begin{aligned}\iint h(sx)\,d\mu(s)\,d(f\cdot\beta)(x) &= \iint h(x)f(s^{-1}x)\chi(s^{-1},x)\,d\mu(s)\,d\beta(x)\\ &= \int h(x)\,d\beta(x)\int f(s^{-1}x)\chi(s^{-1},x)\,d\mu(s),\end{aligned}$$

the function $x \mapsto g(x) = \int f(s^{-1}x)\chi(s^{-1},x)\,d\mu(s)$ being moreover locally $\beta$-integrable. One thus sees that

$$\langle h, \mu * (f\cdot\beta)\rangle = \langle h, g\cdot\beta\rangle,$$

whence $g = \mu *^{\beta} f$.

Suppose now that $\mu$ is carried by the union S of a sequence of compact sets. The function

$$(s,x) \mapsto h(x)f(s^{-1}x)\chi(s^{-1},x)\varphi_{\mathrm{S}}(s)$$

is essentially $(\mu \otimes \beta)$-integrable, and zero outside a countable union of compact sets, hence $(\mu\otimes\beta)$-integrable. Since $\mu = \varphi_{\mathrm{S}}\cdot\mu$, the argument concludes as before.

*Remark.* — The hypothesis (iii) of Prop. 2 is satisfied notably when $\mu$ is *bounded*. For, for every $n > 0$, there then exists a compact subset $K_n$ of G such that

$$|\mu|(G - K_n) \leqslant \frac{1}{n}$$

(Ch. IV, §4, No. 7), and $\mu$ is carried by the union of the $K_n$. More generally, let $\rho$ be a lower semi-continuous finite function $> 0$ on G such that $\rho(st) \leqslant \rho(s)\rho(t)$; if $\mu \in \mathscr{M}^\rho$, the hypothesis (iii) is satisfied; for, $\rho \cdot \mu$ is bounded, and $\mu$ is carried by the same subsets as $\rho \cdot \mu$ since, on every compact subset of G, $\rho$ is bounded below by a constant $> 0$.

## 2. Examples of convolvable measures and functions

In Props. 3 and 4, $\mathscr{C}'(G)$ and $\mathscr{M}(G)$ are equipped with the topology of *compact convergence* in $\mathscr{C}(G)$ and $\mathscr{K}(G)$, respectively.

PROPOSITION 3. — *Assume $\chi$ continuous. Let $\mu \in \mathscr{C}'(G)$, $f \in \mathscr{C}(X)$. Then:*

(i) *$\mu$ and $f$ are convolvable relative to $\beta$.*

(ii) *Formula* (3) *of* No. 1 *defines for every $x \in X$ a convolution product $\mu *^\beta f$ that is continuous and is none other than the element $\boldsymbol{\gamma}_\chi(\mu)f$ defined by the continuous representation $\boldsymbol{\gamma}_\chi$ of G in $\mathscr{C}(X)$; moreover, the mapping $(\mu, f) \mapsto \mu *^\beta f$ is hypocontinuous relative to the equicontinuous subsets of $\mathscr{C}'(G)$ and the compact subsets of $\mathscr{C}(X)$.*

(iii) *If in addition $f \in \mathscr{K}(X)$, then the product $\mu *^\beta f$ of* (ii) *belongs to $\mathscr{K}(X)$ and the mapping $(\mu, f) \mapsto \mu *^\beta f$ is hypocontinuous relative to the equicontinuous subsets of $\mathscr{C}'(G)$ and the compact subsets of $\mathscr{K}(X)$.*

We know that $\mu$ and $f$ are convolvable (§3, No. 2, Prop. 8 (i)). On the other hand, with the notations of §2, we have

$$\boldsymbol{\gamma}_\chi(\mu)f = \int (\boldsymbol{\gamma}_\chi(s)f)d\mu(s) \in \mathscr{C}(X)$$

since $\mathscr{C}(X)$ is complete. In particular, for every $x \in X$,

$$(\boldsymbol{\gamma}_\chi(\mu)f)(x) = \int (\boldsymbol{\gamma}_\chi(s)f)(x)\, d\mu(s).$$

This, combined with Prop. 2 (i), and §2, No. 6, proves (ii). Finally, if $f \in \mathscr{K}(X)$ then $\mu*(f\cdot\beta)$ has compact support (§3, No. 2, Prop. 9), therefore $\mu*^\beta f \in \mathscr{K}(X)$. For, let us consider the continuous representation $U$ of G in the completion $\mathscr{K}(X)^\frown$ obtained by extending by continuity the continuous operators $\gamma_\chi(s)$ in $\mathscr{K}(\mathrm{X})$ (§2, No. 1, *Remark* 3). Let S be the support of $\mu$. The functions $\gamma_\chi(s)f$, for $s \in \mathrm{S}$, have their support contained in a fixed compact set K. The set $\mathscr{K}(\mathrm{X}, \mathrm{K})$ is a complete linear subspace of $\mathscr{K}(\mathrm{X})$. Therefore $U(\mu)f \in \mathscr{K}(\mathrm{X})$. One sees as before that $U(\mu)f = \mu *^\beta f$, and (iii) again follows from §2, No. 6.

PROPOSITION 4. — *Assume that* G *operates properly in* X *and that* $\chi$ *is continuous. Let* $\mu \in \mathscr{M}(\mathrm{G})$ *and* $f \in \mathscr{K}(\mathrm{X})$.

(i) $\mu$ *and* $f$ *are convolvable relative to* $\beta$.

(ii) *Formula* (3) *of* No. 1 *defines for every* $x \in \mathrm{X}$ *a convolution product* $\mu *^\beta f$ *that is continuous.*

(iii) *The mapping* $(\mu, f) \mapsto \mu *^\beta f$ *of* $\mathscr{M}(\mathrm{G}) \times \mathscr{K}(\mathrm{X})$ *into* $\mathscr{C}(\mathrm{X})$ *is hypocontinuous relative to the bounded subsets of* $\mathscr{M}(\mathrm{G})$ *and the compact subsets of* $\mathscr{K}(\mathrm{X})$ *that are contained in some subspace* $\mathscr{K}(\mathrm{X}, \mathrm{L})$ (*where* L *is a variable compact subset of* X).

We know that $\mu$ and $f$ are convolvable (§3, No. 2, Prop. 8 (ii)), and it is clear that the integrals occurring in (3) exist for every $x \in \mathrm{X}$. Let K and L be two compact subsets of X. There exists a compact subset H of G such that the relations $x \in \mathrm{K}$ and $s^{-1}x \in \mathrm{L}$ imply $s \in \mathrm{H}$; let $\varphi \in \mathscr{K}_+(\mathrm{G})$ with $\varphi(s) = 1$ for $s \in \mathrm{H}$. Then, for $f \in \mathscr{K}(\mathrm{X}, \mathrm{L})$ and $x \in \mathrm{K}$,

$$\begin{aligned}\int f(s^{-1}x)\chi(s^{-1}, x)\, d\mu(s) &= \int f(s^{-1}x)\chi(s^{-1}, x)\varphi(s)\, d\mu(s)\\ &= ((\varphi \cdot \mu) *^\beta f)(x).\end{aligned}$$

Consequently $\int f(s^{-1}x)\chi(s^{-1}, x)\, d\mu(s)$ is a continuous function of $x$ and defines a convolution product $\mu *^\beta f \in \mathscr{C}(\mathrm{X})$. Moreover, the mapping $\mu \mapsto \varphi \cdot \mu$ of $\mathscr{M}(\mathrm{G})$ into $\mathscr{C}'(\mathrm{G})$ is continuous for the topologies of compact convergence. Prop. 3 (iii) therefore implies that the mapping $(\mu, f) \mapsto \mu *^\beta f$ of $\mathscr{M}(\mathrm{G}) \times \mathscr{K}(\mathrm{X}, \mathrm{L})$ into $\mathscr{C}(\mathrm{X})$ is, for every compact subset L of X, hypocontinuous relative to the compact subsets of $\mathscr{K}(\mathrm{X}, \mathrm{L})$. In particular, the mapping $(\mu, f) \mapsto \mu *^\beta f$ of $\mathscr{M}(\mathrm{G}) \times \mathscr{K}(\mathrm{X})$ into $\mathscr{C}(\mathrm{X})$ is separately continuous. Since $\mathscr{K}(\mathrm{X})$ is barreled, this mapping is hypocontinuous relative to the bounded subsets of $\mathscr{M}(\mathrm{G})$ (TVS, III, §5, No. 3, Prop. 6).

*Remark* 1. — Under the hypotheses of Prop. 4, the mapping $\mu \mapsto \mu *^\beta f$ of $\mathscr{M}_+(\mathrm{G})$ into $\mathscr{C}(\mathrm{X})$ is continuous when $\mathscr{M}_+(\mathrm{G})$ is equipped with the *vague* topology, for every $f \in \mathscr{K}(\mathrm{X})$. For, let K be a compact subset of X, S the (compact) support of $f$; since G operates properly in X, the set of $s \in \mathrm{G}$ for which there exists an $x \in \mathrm{K}$ with $s^{-1}x \in \mathrm{S}$ is a compact subset L of G (GT, III, §4, No. 5, Th. 1). Let $\varepsilon$ be a number $> 0$, $\varphi$ a function in $\mathscr{K}_+(\mathrm{G})$ equal to 1 on the compact set L, $\mu_0$ an element of $\mathscr{M}_+(G)$; the set $W_0$ of measures $\mu \in \mathscr{M}_+(G)$ such that

$$\left|\int \varphi(s)\,d\mu(s) - \int \varphi(s)\,d\mu_0(s)\right| \leqslant \varepsilon$$

is a neighborhood of $\mu_0$ in $\mathscr{M}_+(G)$. On the other hand, the function $(s,x) \mapsto f(s^{-1}x)\chi(s^{-1},x)$ is uniformly continuous on $L \times K$, therefore there exists a finite number of points $x_i \in K$ $(1 \leqslant i \leqslant n)$ such that for every $x \in K$, there is an $i$ for which

$$|f(s^{-1}x)\chi(s^{-1},x) - f(s^{-1}x_i)\chi(s^{-1},x_i)| \leqslant \varepsilon$$

for all $s \in L$. Since $\mu(L) \leqslant \int \varphi(s)\,d\mu_0(s) + \varepsilon$ for all $\mu \in W_0$, also

$$\left|\int f(s^{-1}x)\chi(s^{-1},x)\,d\mu(s) - \int f(s^{-1}x_i)\chi(s^{-1},x_i)\,d\mu(s)\right| \leqslant \varepsilon\Big(\int \varphi(s)\,d\mu_0(s) + \varepsilon\Big)$$

for every $x$ satisfying the preceding inequality and every $\mu \in W_0$. Now let $W$ be the neighborhood of $\mu_0$ in $\mathscr{M}_+(G)$ formed by the measures $\mu \in W_0$ such that

$$\left|\int f(s^{-1}x_i)\chi(s^{-1},x_i)\,d\mu(s) - \int f(s^{-1}x_i)\chi(s^{-1},x_i)\,d\mu_0(s)\right| \leqslant \varepsilon$$

for $1 \leqslant i \leqslant n$. It is clear that for every measure $\mu \in W$ and every $x \in K$,

$$\left|\int f(s^{-1}x)\chi(s^{-1},x)\,d\mu(s) - \int f(s^{-1}x)\chi(s^{-1},x)\,d\mu_0(s)\right| \leqslant \varepsilon\Big(2\int \varphi(s)\,d\mu_0(s) + 2\varepsilon + 1\Big),$$

and since $\varepsilon$ is arbitrary, this proves our assertion.

PROPOSITION 5. — *Assume $\chi$ a continuous multiplier and each function $\chi(s,\cdot)$ bounded.*

(i) *The function $s \mapsto \rho(s) = \sup_{x\in X} \chi(s^{-1},x)$ on G is lower semi-continuous $> 0$ and satisfies $\rho(st) \leqslant \rho(s)\rho(t)$ for all $s,t$ in G.*

(ii) *Let $\mu \in \mathscr{M}^{\rho}(G)$ and $f \in L^{\infty}(X,\beta)$.*[1] *Then $\mu$ and $f$ are convolvable and $\mu *^{\beta} f$ is given locally almost everywhere by the formula* (3) *of* No. 1. *One has $\mu *^{\beta} f \in L^{\infty}(X,\beta)$, and $\|\mu *^{\beta} f\|_{\infty} \leqslant \|\mu\|_{\rho}\|f\|_{\infty}$.*

[1] For a function $f$, the expression "$f \in L^{\infty}(X,\beta)$" is an abuse of notation signifying that, depending on the context, the symbol $f$ is to be interpreted either as a function or the equivalence class of a function. In particular, the symbol $\mu *^{\beta} f$ can stand for either a function defined locally $\beta$-almost everywhere, or the equivalence class of such a function for the relation of equality locally $\beta$-almost everywhere.

(iii) *If, moreover,* $f \in \mathscr{C}^\infty(\mathrm{X})$ (resp. $\overline{\mathscr{K}(\mathrm{X})}$), *then formula* (3) *of* No. 1 *defines for every* $x$ *a convolution product* $\mu *^\beta f$ *that belongs to* $\mathscr{C}^\infty(\mathrm{X})$ (resp. $\overline{\mathscr{K}(\mathrm{X})}$).

(iv) *If* $f \in \overline{\mathscr{K}(\mathrm{X})}$, *then the convolution product* $\mu *^\beta f$ *defined by* (3) *is none other than the element* $\boldsymbol{\gamma}_\chi(\mu) f$ *defined by the continuous representation* $\boldsymbol{\gamma}_\chi$ *of* G *in* $\overline{\mathscr{K}(\mathrm{X})}$.

The identity $\chi(st, x) = \chi(s, tx)\chi(t, x)$ implies at once that $\rho(st) \leqslant \rho(s)\rho(t)$. On the other hand, $\rho$ is lower semi-continuous, being the upper envelope of continuous functions.

Let $\mu \in \mathscr{M}^\rho(\mathrm{G})$. By Prop. 1 of No. 1, $\mu$ and 1 are convolvable; Prop. 2 (i) shows that $(|\mu| *^\beta 1)(x) \leqslant \int_{\mathrm{G}} \rho(s)\, d|\mu|(s)$ locally $\beta$-almost everywhere. Therefore, if $f$ is $\beta$-measurable and $|f| \leqslant 1$, then $\mu$ and $f$ are convolvable and $\mathrm{N}_\infty(\mu *^\beta f) \leqslant \int \rho(s)\, d|\mu|(s)$. Moreover, $\mu *^\beta f$ is given locally almost everywhere by formula (3) of No. 1, because condition (iii) of Prop. 2 of No. 1 is satisfied. This implies (ii).

Suppose $f$ continuous and bounded by 1 in absolute value. It is clear that the integrals occurring in (3) exist for all $x \in \mathrm{X}$. Let us show that they depend continuously on $x$. We can suppose $\mu \geqslant 0$. Let $x_0 \in \mathrm{X}$ and $\varepsilon > 0$. Let K be a compact subset of G such that $\int_{\mathrm{G}-\mathrm{K}} \rho(s)\, d\mu(s) \leqslant \varepsilon$. There exists a neighborhood V of $x_0$ in X such that $x \in \mathrm{V}$ implies

$$|f(s^{-1}x)\chi(s^{-1}, x) - f(s^{-1}x_0)\chi(s^{-1}, x_0)| \leqslant \varepsilon/\mu(\mathrm{K})$$

for $s \in \mathrm{K}$. Then, for $x \in \mathrm{V}$,

$$\Big| \int f(s^{-1}x)\chi(s^{-1}, x)\, d\mu(s) - \int f(s^{-1}x_0)\chi(s^{-1}, x_0)\, d\mu(s) \Big| \leqslant 2\int_{\mathrm{G}-\mathrm{K}} \rho(s)\, d\mu(s) + \int_{\mathrm{K}} \frac{\varepsilon}{\mu(\mathrm{K})}\, d\mu(s) \leqslant 3\varepsilon,$$

whence our assertion. Suppose that in addition $f \in \overline{\mathscr{K}(\mathrm{X})}$. Let H be a compact subset of X such that $|f(y)| \leqslant \varepsilon$ for $y \notin \mathrm{H}$. Let $x \notin \mathrm{KH}$. Then $s^{-1}x \notin \mathrm{H}$ for $s \in \mathrm{K}$, therefore

$$\Big| \int_{\mathrm{G}} f(s^{-1}x)\chi(s^{-1}, x)\, d\mu(s) \Big| \leqslant \int_{\mathrm{G}-\mathrm{K}} \rho(s)\, d\mu(s) + \int_{\mathrm{K}} \varepsilon\, \rho(s)\, d\mu(s) \leqslant \varepsilon\Big(1 + \int_{\mathrm{G}} \rho(s)\, d\mu(s)\Big),$$

which completes the proof of (iii).

Finally, if $f \in \overline{\mathscr{K}(\mathrm{X})}$ then, since $\varepsilon_x \in \mathscr{M}^1(\mathrm{X})$ for all $x \in \mathrm{X}$, we have

$$\big(\boldsymbol{\gamma}_\chi(\mu) f\big)(x) = \int \big(\boldsymbol{\gamma}_\chi(s) f\big)(x) d\mu(s),$$

thus $\boldsymbol{\gamma}_\chi(\mu)f$ is the convolution product $\mu *^\beta f$ defined by (3).

PROPOSITION 6. — *Assume $\chi$ a continuous multiplier and each function $\chi(s,\cdot)$ bounded. Let $\rho(s) = \sup_{x\in X} \chi(s^{-1}, x)$. Let $p$ and $q$ be two conjugate exponents ($1 \leqslant p < +\infty$). Let $\mu \in \mathscr{M}^{\rho^{1/q}}(G)$ and $f \in L^p(X, \beta)$.*[2] *Then:*

(i) *$\mu$ and $f$ are convolvable;*

(ii) *the convolution product $\mu *^\beta f$ is given locally $\beta$-almost everywhere by the formula* (3), *and is equal locally $\beta$-almost everywhere to a function $g \in L^p(X, \beta)$ such that $\|g\|_p \leqslant \|\mu\|_{\rho^{1/q}} \|f\|_p$;*

(iii) *$g$ is equal to the element $\boldsymbol{\gamma}_\chi(\mu)f$ defined by the continuous representation $\boldsymbol{\gamma}_\chi$ of G in $L^p(X, \beta)$.*

We have

$$\int^* \|\gamma_\chi(s)f\|_p \, d|\mu|(s) \leqslant \Big(\int^* \rho(s)^{1/q} \, d|\mu|(s)\Big) \|f\|_p < +\infty$$

by §2, No. 5, formula (5). On the other hand, the mapping $s \mapsto \boldsymbol{\gamma}_\chi(s)f$ of G into $L^p(X, \beta)$ is continuous (§2, No. 5, Prop. 9). Therefore this mapping is $\mu$-integrable. Let

$$g = \int_G (\boldsymbol{\gamma}_\chi(s)f) d\mu(s) \in L^p(X, \beta).$$

We have $\|g\|_p \leqslant \big(\int \rho^{1/q}(s) \, d|\mu|(s)\big) \|f\|_p$. Applying the preceding remarks to $|f|$, one sees that the mapping $s \mapsto \varepsilon_s * |f|$ of G into $L^p(X, \beta)$ is $\mu$-integrable, therefore that, for every $h \in \mathscr{K}(X)$, the mapping $s \mapsto \langle h, \varepsilon_s * (|f| \cdot \beta)\rangle$ is $\mu$-integrable. Prop. 7 of §1, No. 5 then proves that $\mu$ and $f \cdot \beta$ are convolvable. Moreover,

$$\begin{aligned} \int_X g(x)h(x) \, d\beta(x) &= \int_G d\mu(s) \int_X (\boldsymbol{\gamma}_\chi(s)f)(x)h(x) \, d\beta(x) \\ &= \int_G \langle h, \varepsilon_s * (f \cdot \beta)\rangle \, d\mu(s), \end{aligned}$$

and this last integral is equal to $\langle h, \mu * (f \cdot \beta)\rangle$ by Prop. 7 of §1, No. 5. One therefore sees that $g$ is a convolution product of $\mu$ and $f$. This convolution product is given locally $\beta$-almost everywhere by (3), by Prop. 2 and the *Remark* that follows it.

[2] For a function $f$, the expression "$f \in L^p(X, \beta)$" is an abuse of notation signifying that, depending on the context, the symbol $f$ is to be interpreted either as a function defined $\beta$-almost everywhere, or as the equivalence class of such a function for the relation of equality $\beta$-almost everywhere. Thus $f \in L^p$ can symbolize either $f \in \mathscr{L}^p$ or $\dot{f} \in L^p$.

By an abuse of notation, it is often one of the functions $g$ of the statement that is denoted $\mu *^{\beta} f$, which permits writing

$$\|\mu *^{\beta} f\|_p \leqslant \|\mu\|_{\rho^{1/q}} \|f\|_p .$$

If X is countable at infinity, this style of notation is, moreover, entirely justified.

COROLLARY. — *Under the hypotheses of Prop. 6, the mapping* $(\mu, f) \mapsto \mu *^{\beta} f$ *defines on* $\mathrm{L}^p(\mathrm{X}, \beta)$ *the structure of a left module over* $\mathscr{M}^{\rho^{1/q}}(\mathrm{G})$ $(1 \leqslant p \leqslant +\infty)$.

This follows from Props. 5 and 6 and the associativity of the convolution product.

*Remark* 2. — Let X be a locally compact space on which a locally compact group G operates continuously on the right by $(x, s) \mapsto xs$. Let $\beta$ be a positive measure on X. Let $\chi$ be a function $> 0$ on $\mathrm{G} \times \mathrm{X}$, measurable for every measure on $\mathrm{G} \times \mathrm{X}$, such that $\boldsymbol{\delta}(s)\beta = \chi(s, \cdot) \cdot \beta$ for every $s \in \mathrm{G}$. Let $f$ be a locally $\beta$-integrable function on X and let $\mu$ be a measure on G. If $f \cdot \beta$ and $\mu$ are convolvable (for the mapping $(x, s) \mapsto xs$ of $\mathrm{X} \times \mathrm{G}$ into X), then $(f \cdot \beta) * \mu$ has base $\beta$. We then say that $f$ and $\mu$ are convolvable relative to $\beta$; every density of $(f \cdot \beta) * \mu$ with respect to $\beta$ is called a convolution product of $f$ and $\mu$ relative to $\beta$, and is denoted $f *^{\beta} \mu$ or simply $f * \mu$.

Let $\mathrm{G}^0$ be the group opposite to G. By $(s, x) \mapsto xs$, $\mathrm{G}^0$ operates continuously on the left in X. To say that $f$ and $\mu$ are convolvable in the foregoing sense is equivalent to saying that $\mu$ and $f$ are convolvable for $\mathrm{G}^0$ operating on the left in X; and the convolution products $f *^{\beta} \mu$ are none other than the convolution products $\mu *^{\beta} f$ for $\mathrm{G}^0$ operating on the left in X. On the other hand, for $s \in \mathrm{G}^0$ one has $\boldsymbol{\gamma}(s)\beta = \chi(s^{-1}, \cdot) \cdot \beta$. The results of Nos. 1 and 2 may then be translated immediately into results concerning the products $f *^{\beta} \mu$. In particular:

1) If $s \in \mathrm{G}$ and $f$ is locally $\beta$-integrable, then $f$ and $\varepsilon_s$ are convolvable and

$$(f * \varepsilon_s)(x) = \chi(s^{-1}, x) f(xs^{-1}) . \tag{4}$$

2) If $f$ and $\mu$ are convolvable and if one of the conditions (i), (ii), (iii) of Prop. 2 is fulfilled, then $f *^{\beta} \mu$ is given locally $\beta$-almost everywhere by

$$(f *^{\beta} \mu)(x) = \int_{\mathrm{G}} f(xs^{-1}) \chi(s^{-1}, x)\, d\mu(s) . \tag{5}$$

We leave to the reader the task of translating the other statements. Note that if $\chi$ is continuous and $\beta$ has support X, then

$$\chi(ts, x) = \chi(s, xt)\chi(t, x) \qquad (x \in \mathrm{X};\ s, t \text{ in } \mathrm{G}) . \tag{6}$$

## 3. Convolution and transposition

Let us return to the hypotheses and notations of the beginning of No. 1, but let us assume in addition that $\beta$ is *relatively invariant with multiplier* $\chi$; $\chi$ is therefore a continuous function on $G$.

PROPOSITION 7. — *Let $f$ be a locally $\beta$-integrable function on* $X$, $\nu$ *a measure on* $X$, *and* $\mu$ *a measure on* $G$. *Assume that:*

(i) $\mu$ *and* $f$ *are convolvable and formula* (3) *of No.* 1 *defines locally $\beta$-almost everywhere a convolution product* $\mu *^{\beta} f$.

(ii) $\chi \cdot \check{\mu}$ *and* $\nu$ *are convolvable.*

(iii) *The function* $g(s,x) = f(s^{-1}x)\chi(s^{-1})$ *is* $(\mu \otimes \nu)$*-integrable.*

*Then* $f$ *is essentially integrable for* $(\chi \cdot \check{\mu}) * \nu$, *the function* $\mu *^{\beta} f$ *defined by* (3) *is* $\nu$*-integrable, and*

$$\nu(\mu *^{\beta} f) = ((\chi \cdot \check{\mu}) * \nu)(f). \tag{7}$$

Since $g(s,x)$ is integrable for $\mu \otimes \nu$, the function $f(sx)$ is essentially integrable for $(\chi \cdot \check{\mu}) \otimes \nu$ and $f$ is essentially integrable for $(\chi \cdot \check{\mu}) * \nu$. By the Lebesgue–Fubini theorem, $\mu *^{\beta} f = \int g(s,x)\, d\mu(s)$ is $\nu$-integrable and

$$\begin{aligned}\nu(\mu *^{\beta} f) &= \iint f(s^{-1}x)\chi(s^{-1})\, d\mu(s)\, d\nu(x)\\ &= \iint f(sx)\chi(s)\, d\check{\mu}(s)\, d\nu(x) = ((\chi \cdot \check{\mu}) * \nu)(f).\end{aligned}$$

*Examples.* — 1) One can take $f \in \mathscr{C}(X)$, $\nu \in \mathscr{C}'(X)$ and $\mu \in \mathscr{C}'(G)$ by Prop. 3, and the Cor. of Prop. 5 of §1, No. 4. The formula (7) then means that the endomorphism $\nu \mapsto (\chi \cdot \check{\mu}) * \nu$ of $\mathscr{C}'(X)$ is the *transpose* of the endomorphism $f \mapsto \mu * f$ of $\mathscr{C}(X)$.

2) One can take $f \in \mathscr{K}(X)$, $\nu \in \mathscr{M}(X)$ and $\mu \in \mathscr{C}'(G)$ by Prop. 3, Prop. 8 of §3, No. 2, and the remark that the support of the continuous function $g(s,x)$ intersects the support of $\mu \otimes \nu$ in a compact set. The formula (7) then means that the endomorphism $\nu \mapsto (\chi \cdot \check{\mu}) * \nu$ of $\mathscr{M}(X)$ is the *transpose* of the endomorphism $f \mapsto \mu * f$ of $\mathscr{K}(X)$.

3) If $G$ operates properly on $X$, one can take $f \in \mathscr{K}(X)$, $\nu \in \mathscr{C}'(X)$ and $\mu \in \mathscr{M}(G)$ by Prop. 4, Prop. 8 of §3, No. 2, and the same remark as in *Example* 2.

PROPOSITION 8. — *Let $f$ and $g$ be two locally $\beta$-integrable functions on* $X$ *and let* $\mu \in \mathscr{M}(G)$. *Assume that:*

(i) $\mu$ *and* $f$ *are convolvable and the formula* (3) *of No.* 1 *defines locally* $\beta$*-almost everywhere a convolution product* $\mu *^{\beta} f$.

(ii) $\chi \cdot \check{\mu}$ *and* $g$ *are convolvable and the formula* (3) *of No.* 1 *(with* $\mu$ *replaced by* $\chi \cdot \check{\mu}$ *and* $f$ *by* $g$*) defines locally* $\beta$*-almost everywhere a convolution product* $(\chi \cdot \check{\mu}) *^{\beta} g$.

(iii) *There exists a function* $\psi$ *on* G, *equal locally* $\mu$*-almost everywhere to* 1, *such that the function*

$$h(s, x) = g(x) f(s^{-1}x)\chi(s^{-1})\psi(s)$$

*is* $(\mu \otimes \beta)$*-integrable.*

*Then the functions* $g(x)\big((\mu *^{\beta} f)(x)\big)$ *and* $f(x)\big(((\chi \cdot \check{\mu}) *^{\beta} g)(x)\big)$ *are essentially* $\beta$*-integrable, and*

$$\int f(x)\big(((\chi \cdot \check{\mu}) *^{\beta} g)(x)\big) d\beta(x) = \int g(x)\big((\mu *^{\beta} f)(x)\big) d\beta(x). \tag{8}$$

For, by (iii) and the Lebesgue–Fubini theorem, the function

$$x \mapsto g(x) \int f(s^{-1}x)\chi(s^{-1})\psi(s)\, d\mu(s)$$

is $\beta$-integrable, and

$$\begin{aligned} \mathrm{I} &= \iint f(s^{-1}x) g(x)\chi(s^{-1})\psi(s)\, d\mu(s)\, d\beta(x) \\ &= \int g(x)\, d\beta(x) \int f(s^{-1}x)\chi(s^{-1})\psi(s)\, d\mu(s). \end{aligned}$$

But $\psi \cdot \mu = \mu$, consequently

$$\int f(s^{-1}x)\chi(s^{-1})\psi(s)\, d\mu(s) = (\mu *^{\beta} f)(x)$$

locally $\beta$-almost everywhere. This shows that the function

$$x \mapsto g(x)\big((\mu *^{\beta} f)(x)\big)$$

is essentially $\beta$-integrable and that

$$\mathrm{I} = \int g(x)\big((\mu *^{\beta} f)(x)\big) d\beta(x).$$

On the other hand, Lemma 1 shows that the function

$$(s,x) \mapsto g(sx)f(x)\chi(s^{-1})\psi(s)$$

is integrable for $(\chi\cdot\mu)\otimes\beta$. Therefore the function $(s,x) \mapsto g(s^{-1}x)f(x)\psi(s^{-1})$ is integrable for $\check{\mu}\otimes\beta$, and

$$\begin{aligned} \mathrm{I} &= \iint g(s^{-1}x)f(x)\psi(s^{-1})\, d\check{\mu}(s)\, d\beta(x) \\ &= \int f(x)\, d\beta(x) \int g(s^{-1}x)\psi(s^{-1})\, d\check{\mu}(s). \end{aligned}$$

But $\check{\psi}\cdot\check{\mu} = \check{\mu}$ and so $\int g(s^{-1}x)\psi(s^{-1})\, d\check{\mu}(s) = \big((\chi\cdot\check{\mu}) *^{\beta} g\big)(x)$ locally $\beta$-almost everywhere. This shows that the function

$$x \mapsto f(x)\big(\big((\chi\cdot\check{\mu}) *^{\beta} g\big)(x)\big)$$

is essentially $\beta$-integrable and that

$$\mathrm{I} = \int f(x)\big(\big((\chi\cdot\check{\mu}) *^{\beta} g\big)(x)\big)\, d\beta(x).$$

This proves the proposition.

*Examples.* — 4) One can take $f \in \mathscr{C}(\mathrm{X})$, $g \in \mathscr{K}(\mathrm{X})$ and $\mu \in \mathscr{C}'(\mathrm{G})$ (with $\psi = 1$).

5) If G operates properly on X, one can take $f \in \mathscr{K}(\mathrm{X})$, $g \in \mathscr{K}(\mathrm{X})$ and $\mu \in \mathscr{M}(\mathrm{G})$ (with $\psi = 1$).

6) One can take $f \in \mathrm{L}^p(\mathrm{X},\beta)$, $g \in \mathrm{L}^q(\mathrm{X},\beta)$ and $\mu \in \mathscr{M}^{\rho}(\mathrm{G})$, where $1 \leqslant p < +\infty$, $\dfrac{1}{p} + \dfrac{1}{q} = 1$, $\rho = \chi^{-1/q}$. The conditions (i) and (ii) are satisfied by Props. 5 and 6. Let us prove (iii). We have seen that $\mu$ is carried by a set S that is a countable union of compact sets. Let us take for $\psi$ the characteristic function of S. The function $h$ is $(\mu\otimes\beta)$-measurable: for, the function $(s,x) \mapsto g(x)\chi(s^{-1})\psi(s)$ is so, as is the function $(s,x) \mapsto f(s^{-1}x)$ by Lemma 1. Moreover, $g$ being zero outside a countable union of $\beta$-integrable sets, $h$ is zero outside a countable union of $(\mu\otimes\beta)$-integrable sets. We then have (Ch. V, §8, No. 3, Prop. 7):

$$\begin{aligned} (9) \qquad \mathrm{J} &= \iint^{*} |g(x)f(s^{-1}x)|\, \chi(s^{-1})\psi(s)\, d|\mu|(s)\, d\beta(x) \\ &= \int^{*} |g(x)|\, d\beta(x) \int^{*} |f(s^{-1}x)|\, \chi(s^{-1})\psi(s)\, d|\mu|(s). \end{aligned}$$

But since $g$ (resp. $\psi$) is zero outside a countable union of integrable sets, the upper integrals of the second member of (9) are equal to the essential upper integrals (Ch. V, §1, No. 2, Prop. 7). Now (Ch. V, §5, No. 3, Prop. 3)

$$\int^{\bullet} |f(s^{-1}x)|\,\chi(s^{-1})\psi(s)\,d|\mu|(s) = \int^{\bullet} |f(s^{-1}x)|\,\chi(s^{-1})\,d|\mu|(s)$$

since $\mu = \psi \cdot \mu$. By Prop. 6, this last integral is finite and is equal to $(|\mu| *^{\beta} |f|)(x)$ locally $\beta$-almost everywhere. Therefore

$$\mathrm{J} = \int^{\bullet} |g(x)|\big(|\mu| *^{\beta} |f|\big)(x)\,d\beta(x)\,,$$

and J is finite since $g \in \mathrm{L}^q$ and $|\mu| *^{\beta} |f| \in \mathrm{L}^p$ (Prop. 6). Therefore $h$ is $(\mu \otimes \beta)$-integrable.

The formula (8) then means that the endomorphism $g \mapsto (\chi \cdot \check{\mu}) * g$ of $\mathrm{L}^q(\mathrm{X}, \beta)$ is, for $\mu \in \mathscr{M}^p(\mathrm{G})$, the *transpose* of the endomorphism $f \mapsto \mu * f$ of $\mathrm{L}^p(\mathrm{X}, \beta)$.

## 4. Convolution of a measure and a function on a group

Let G be a locally compact group. Throughout Nos. 4 and 5, we fix a relatively invariant positive measure $\beta \neq 0$ on G; let $\chi$ and $\chi'$ be its left and right multipliers (recall that $\chi' = \chi\Delta_{\mathrm{G}}$). If $\mu$ is a measure on G and $f$ is a locally $\beta$-integrable function on G, the convolvability of $\mu$ and $f$ and the products $\mu * f$ (resp. the convolvability of $f$ and $\mu$ and the products $f * \mu$) may be defined by regarding G as operating on itself on the left (resp. on the right) by translations. Let us make explicit in this situation some of the preceding results:

1) Let $\mu$ be a measure on G, $f$ a locally $\beta$-integrable complex function on G. Assume verified one of the following conditions:

(i) $f$ is continuous;

(ii) $f$ is zero on the complement of a countable union of compact sets;

(iii) $\mu$ is carried by a countable union of compact sets.

If $\mu$ and $f$ are convolvable then, locally $\beta$-almost everywhere,

$$(\mu * f)(x) = \int_{\mathrm{G}} f(s^{-1}x)\chi(s^{-1})\,d\mu(s)\,. \tag{10}$$

If $f$ and $\mu$ are convolvable then, locally $\beta$-almost everywhere,

$$(f * \mu)(x) = \int_{\mathrm{G}} f(xs^{-1})\chi'(s^{-1})\,d\mu(s)\,. \tag{11}$$

2) Let $p$ and $q$ be two conjugate exponents ($1 \leqslant p \leqslant +\infty$). If $\mu \in \mathscr{M}^{\chi^{-1/q}}(\mathrm{G})$ and $f \in \mathrm{L}^p(\mathrm{G}, \beta)$, then $\mu$ and $f$ are convolvable, and $\mu * f$ is equal locally $\beta$-almost everywhere to a function in $\mathrm{L}^p(\mathrm{G}, \beta)$; one has (with an abuse of notations already noted)

$$\|\mu * f\|_p \leqslant \|\mu\|_{\chi^{-1/q}} \|f\|_p .$$

If $\mu \in \mathscr{M}^{\chi'^{-1/q}}(\mathrm{G})$ and $f \in \mathrm{L}^p(\mathrm{G}, \beta)$, then $f$ and $\mu$ are convolvable, and $f * \mu$ is equal locally $\beta$-almost everywhere to a function in $\mathrm{L}^p(\mathrm{G}, \beta)$; one has $\|f * \mu\|_p \leqslant \|\mu\|_{\chi'^{-1/q}} \|f\|_p$.

3) The mappings $(\mu, f) \mapsto \mu * f$, $(f, \mu) \mapsto f * \mu$ define on $\mathrm{L}^p(\mathrm{G}, \beta)$ the structures of a left module over $\mathscr{M}^{\chi^{-1/q}}(\mathrm{G})$ and a right module over $\mathscr{M}^{\chi'^{-1/q}}(\mathrm{G})$. The two external laws on $\mathrm{L}^p(\mathrm{G}, \beta)$ are permutable by the associativity of convolution.

4) If $\mu * f$ is continuous and is given at every point by (10), then

$$(\mu * f)(e) = \int f(s^{-1}) \chi(s^{-1})\, d\mu(s) . \tag{12}$$

If $f * \mu$ is continuous and is given at every point by (11), then

$$(f * \mu)(e) = \int f(s^{-1}) \chi'(s^{-1})\, d\mu(s) . \tag{13}$$

## 5. Convolution of functions on a group

We conserve the notations $\mathrm{G}$, $\beta$, $\chi$, $\chi'$ of No. 4.

Recall that if $f$ is a complex function on $\mathrm{G}$, the property of being locally $\beta$-integrable is independent of the choice of $\beta$. Let $\mathscr{L}(\mathrm{G})$ be the set of functions having this property. If $f \in \mathscr{L}(\mathrm{G})$, $g \in \mathscr{L}(\mathrm{G})$, the relation

«$f \cdot \beta$ and $g \cdot \beta$ are convolvable»

is independent of the choice of $\beta$ (§3, No. 1, Prop. 6). We shall then say that $f$ and $g$ are *convolvable.* By No. 1, $(f \cdot \beta) * (g \cdot \beta)$ is of the form $h \cdot \beta$ with $h \in \mathscr{L}(\mathrm{G})$, $h$ being determined up to locally $\beta$-negligible sets. We shall write $h = f *^{\beta} g$ and we shall say that $h$ is a *convolution product* of $f$ and $g$ relative to $\beta$. (One omits $\beta$ when no confusion is possible.) If $\beta$ is replaced by $\psi \cdot \beta$, $\psi$ being a continuous representation of G in $\mathbf{R}_+^*$, $h$ does not change (§3, No. 1, Prop. 6); if $\beta$ is replaced by $a\beta$ ($a \in \mathbf{R}_+^*$), then $h$

is replaced by $ah$. The convolution product of several functions on G is defined in an analogous manner.

If one of the convolutions of $f$ and $g$ is continuous, it is uniquely determined since the support of $\beta$ is G. It is then called *the* convolution product of $f$ and $g$ relative to $\beta$.

It is clear that

$$f *^\beta g = (f \cdot \beta) *^\beta g = f *^\beta (g \cdot \beta). \tag{14}$$

PROPOSITION 9. — *Let $f, g$ be in $\mathscr{L}(\mathrm{G})$. Assume that the function $s \mapsto g(s^{-1}x)f(s)\chi(s^{-1})$ is essentially $\beta$-integrable except for a locally $\beta$-negligible set of values of $x$, and that the function*

$$x \mapsto \int |g(s^{-1}x)f(s)|\,\chi(s^{-1})\,d\beta(s),$$

*defined locally $\beta$-almost everywhere, is locally $\beta$-integrable. Then $f$ and $g$ are convolvable.*

This follows from Prop. 1 of No. 1.

PROPOSITION 10. — *Let $f, g$ be in $\mathscr{L}(\mathrm{G})$. Assume that one of these two functions is continuous or is zero on the complement of a countable union of compact sets. If $f$ and $g$ are convolvable, then the function $f * g$ is given locally $\beta$-almost everywhere by*

$$\begin{aligned}(f * g)(x) &= \int_{\mathrm{G}} g(s^{-1}x)f(s)\chi(s^{-1})\,d\beta(s) \\ &= \int_{\mathrm{G}} f(xs^{-1})g(s)\chi'(s^{-1})\,d\beta(s).\end{aligned} \tag{15}$$

This follows from Prop. 2 of No. 1, and the remarks in No. 4.

In particular, if $f * g$ is continuous and is given at every point by (15), then

$$(f * g)(e) = \int g(s^{-1})f(s)\chi(s^{-1})\,d\beta(s) = \int f(s^{-1})g(s)\chi'(s^{-1})\,d\beta(s). \tag{16}$$

Still more particularly, if $\beta$ is a left and right Haar measure, and if $f * g$ and $g * f$ are continuous and are given at every point by (15) and the analogous formula for $g * f$, then

$$(f * g)(e) = (g * f)(e) = \int f(s)g(s^{-1})\,d\beta(s). \tag{17}$$

PROPOSITION 11. — *Let $f, g$ be in $\mathscr{L}(\mathrm{G})$. Assume that one of the functions $f, g$ is continuous, and that one of the functions $f, g$ has compact support. Then $f$ and $g$ are convolvable. The formula* (15) *defines for all $x \in \mathrm{G}$ a product $f * g$ that is continuous. If $f \in \mathscr{K}(\mathrm{G})$ and $g \in \mathscr{K}(\mathrm{G})$, then $f * g \in \mathscr{K}(\mathrm{G})$.*

This follows from Props. 3 and 4 of No. 2.

PROPOSITION 12. — *Let $p$ and $q$ be two conjugate exponents $(1 \leqslant p \leqslant +\infty)$. If $f\chi^{-1/q} \in \mathrm{L}^1(\mathrm{G}, \beta)$ and $g \in \mathrm{L}^p(\mathrm{G}, \beta)$, then $f$ and $g$ are convolvable, $f * g$ is equal locally $\beta$-almost everywhere to a function in $\mathrm{L}^p(\mathrm{G}, \beta)$, and*

$$\|f * g\|_p \leqslant \|f\chi^{-1/q}\|_1 \, \|g\|_p \, .$$

*If $f \in \mathrm{L}^p(\mathrm{G}, \beta)$ and $g\chi'^{-1/q} \in \mathrm{L}^1(\mathrm{G}, \beta)$, then $f$ and $g$ are convolvable, $f * g$ is equal locally $\beta$-almost everywhere to a function in $\mathrm{L}^p(\mathrm{G}, \beta)$, and*

$$\|f * g\|_p \leqslant \|f\|_p \, \|g\chi'^{-1/q}\|_1 \, .$$

This follows from Props. 5 and 6 of No. 2 and the remarks in No. 4.

PROPOSITION 13. — *If $f\chi^{-1} \in \mathrm{L}^1(\mathrm{G}, \beta)$ and $g \in \overline{\mathscr{K}(\mathrm{G})}$, or if $f \in \overline{\mathscr{K}(\mathrm{G})}$ and $g\chi'^{-1} \in \mathrm{L}^1(\mathrm{G}, \beta)$, then $f$ and $g$ are convolvable, and* (15) *defines for every $x \in \mathrm{G}$ a product $f * g$ that belongs to $\overline{\mathscr{K}(\mathrm{G})}$.*

This follows from Prop. 5 of No. 2, and the remarks in No. 4.

PROPOSITION 14. — *If $f\chi^{-1} \in \mathrm{L}^1(\mathrm{G}, \beta)$ and $g \in \mathrm{L}^\infty(\mathrm{G}, \beta)$, then the formula* (15) *defines for every $x \in \mathrm{G}$ a product $f * g$ that is bounded and is uniformly continuous for the right uniform structure of $\mathrm{G}$.*

We already know that $f * g$ belongs to $\mathrm{L}^\infty(\mathrm{G}, \beta)$ (No. 2, Prop. 5); moreover, $(f * g)(x) = \int f(xs^{-1})g(s)\,d\nu(s)$, on setting $\nu = \chi'^{-1} \cdot \beta$; $\nu$ is a right Haar measure. Therefore

$$\begin{aligned} |(f * g)(x) - (f * g)(x')| &\leqslant \|g\|_\infty \int |f(xs^{-1}) - f(x's^{-1})|\,d\nu(s) \\ &= \|g\|_\infty \int |(f(s^{-1}) - f(x'x^{-1}s^{-1})\,d\nu(s) \end{aligned}$$

and the latter integral is arbitrarily small provided $x'x^{-1}$ is in a suitable neighborhood of $e$ (§2, No. 5, Prop. 8).

PROPOSITION 15. — *Let $p$ and $q$ be two conjugate exponents $(1 < p < +\infty)$. Assume that $\beta$ is left-invariant. Let $f \in \mathrm{L}^p(\mathrm{G}, \beta)$, $g \in \mathrm{L}^q(\mathrm{G}, \check{\beta})$. Then $f$ and $g$ are convolvable. The formula* (15) *defines, for every $x \in \mathrm{G}$, a product $f * g$ that belongs to $\overline{\mathscr{K}(\mathrm{G})}$ and is such that $\|f * g\|_\infty \leqslant \|f\|_p \, \|\check{g}\|_q$.*

For, we have $\check{g} \in \mathrm{L}^q(\mathrm{G},\beta)$, therefore the function $s \mapsto g(s^{-1}x)f(s)$ is $\beta$-integrable for every $x \in \mathrm{G}$. Moreover,

$$\int |g(s^{-1}x)f(s)|\, d\beta(s) \leqslant \left(\int |f(s)|^p\, d\beta(s)\right)^{1/p} \left(\int |g(s^{-1}x)|^q\, d\beta(s)\right)^{1/q}$$
$$= \|f\|_p \left(\int |\check{g}(x^{-1}s)|^q\, d\beta(s)\right)^{1/q} = \|f\|_p\, \|\check{g}\|_q\,,$$

therefore $f$ and $g$ are convolvable (Prop. 9). One sees at the same time that (15) defines for every $x$ a product $f * g$ such that

$$|(f * g)(x)| \leqslant \|f\|_p\, \|\check{g}\|_q\,.$$

For $f, g$ in $\mathscr{K}(\mathrm{G})$, we have $f * g \in \mathscr{K}(\mathrm{G})$ (Prop. 11); therefore, for $f \in \mathrm{L}^p(\mathrm{G},\beta)$ and $g \in \mathrm{L}^q(\mathrm{G},\check{\beta})$, the product $f * g$ furnished by (15) is the uniform limit of functions in $\mathscr{K}(\mathrm{G})$, hence belongs to $\overline{\mathscr{K}(\mathrm{G})}$.

COROLLARY. — *Let $f \in \mathrm{L}^2(\mathrm{G},\beta)$, $g \in \mathrm{L}^2(\mathrm{G},\beta)$. Then $f$ and $\check{\bar{g}}$ are convolvable. One of the convolutions $f * \check{\bar{g}}$ belongs to $\overline{\mathscr{K}(\mathrm{G})}$ and its value at $e$ is $\int_{\mathrm{G}} f(s)\overline{g(s)}\, d\beta(s)$.*

It suffices to take $p = q = 2$ in Prop. 15 and to apply (16).

We no longer assume $\beta$ to be left-invariant. Let $\rho$ be a lower semi-continuous finite function $> 0$ on G, such that $\rho(st) \leqslant \rho(s)\rho(t)$ for all $s, t$ in G. We denote by $\mathrm{L}^\rho(\mathrm{G},\beta)$ the set of equivalence classes of the complex functions on G that are integrable for $\rho \cdot \beta$. By the mapping $f \mapsto f \cdot \beta$, $\mathrm{L}^\rho(\mathrm{G},\beta)$ may be identified with the set of elements of $\mathscr{M}^\rho(\mathrm{G})$ that have base $\beta$ (a set that is independent of the choice of $\beta$). If one sets

$$\|f\|_\rho = \int_{\mathrm{G}} |f(s)|\, \rho(s)\, d\beta(s)$$

for $f \in \mathrm{L}^\rho(\mathrm{G},\beta)$, this identification is compatible with the norms, thus $\mathrm{L}^\rho(\mathrm{G},\beta)$ appears as a complete normed subalgebra of $\mathscr{M}^\rho(\mathrm{G})$. It is even a two-sided ideal of $\mathscr{M}^\rho(\mathrm{G})$ by Prop. 10 of §3, No. 2. (For $\rho = 1$, one recovers one of the assertions of No. 4.) In particular, $\mathrm{L}^1(\mathrm{G},\beta)$ may be identified with a closed two-sided ideal of $\mathscr{M}^1(\mathrm{G})$.

PROPOSITION 16. — *Let $U$ be a continuous representation of G in a Banach space E. Set $\rho(s) = \|U(s)\|$ for all $s \in \mathrm{G}$. For every $f \in \mathrm{L}^\rho(\mathrm{G},\beta)$, set $U(f) = U(f \cdot \beta)$. Then $f \mapsto U(f)$ is a linear representation of the algebra $\mathrm{L}^\rho(\mathrm{G},\beta)$ in E, such that $\|U(f)\| \leqslant \|f\|_\rho$.*

This follows from §2, No. 6 and §3, No. 3, Prop. 11.

## 6. Applications

PROPOSITION 17. — *Let* G *be a locally compact group,* A *a subset of* G, *measurable and not locally negligible for a Haar measure. Then* $A \cdot A^{-1}$ *is a neighborhood of* $e$.

Let $\beta$ be a left Haar measure. There exists a compact subset K of G such that $B = A \cap K$ is integrable with measure $> 0$ for $\beta$. Let us apply the Cor. of Prop. 15 with $f = g = \varphi_B$. The function $F = \varphi_B * \check{\varphi}_B$ is continuous and $> 0$ at $e$. Therefore there exists a neighborhood V of $e$ such that $F(x) > 0$ for $x \in V$. Now,

$$F(x) = \int \varphi_B(s)\varphi_B(x^{-1}s)\, d\beta(s) = \beta(B \cap xB).$$

Therefore, for $x \in V$, one has $B \cap xB \neq \varnothing$, whence $x \in B \cdot B^{-1}$. Thus $V \subset B \cdot B^{-1} \subset A \cdot A^{-1}$.

COROLLARY 1. — *Let* H *be a subgroup of* G *measurable for a Haar measure* $\beta$. *Then* H *is either open or locally* $\beta$*-negligible.*

For, $H = H \cdot H^{-1}$, therefore if H is not locally $\beta$-negligible, then H contains a neighborhood of $e$ (Prop. 17) hence is open (GT, III, §2, No. 1, Cor. of Prop. 4).

COROLLARY 2. — *Let* L *be a subset of* G *stable for multiplication and whose complement is locally negligible for a Haar measure* $\beta$. *Then* $L = G$.

For, $L^{-1}$ and $L \cap L^{-1}$ have locally $\beta$-negligible complements. Now, $L \cap L^{-1}$ is a subgroup, hence is open (Cor. 1) and therefore closed. Therefore $G - (L \cap L^{-1})$, which is open and locally $\beta$-negligible, is empty. Thus $G = L \cap L^{-1}$.

PROPOSITION 18. — *Let* G *be a locally compact group,* $\Gamma$ *a set equipped with a multiplication* $(u, v) \mapsto uv$ *and a Hausdorff topology such that:*

1) *the topology of* $\Gamma$ *is invariant under the translations;*

2) *the restriction of the multiplication to every compact subset of* $\Gamma \times \Gamma$ *is continuous.*

*Let* $f : G \to \Gamma$ *be a mapping of* G *into* $\Gamma$ *such that* $f(xy) = f(x)f(y)$ *for* $x, y$ *in* G, *and measurable for a Haar measure* $\beta$ *on* G. *Then* $f$ *is continuous.*

Set $g(x) = f(x^{-1})$ for $x \in G$. Since $f$ and $g$ are $\beta$-measurable, there exists a non $\beta$-negligible compact subset K of G such that the restrictions of $f$ and $g$ to K are continuous. The mapping $(x, y) \mapsto f(xy^{-1}) = f(x)g(y)$ of $K \times K$ into $\Gamma$ is continuous because the multiplication of $\Gamma$ is continuous on $f(K) \times g(K)$; now, this mapping may be written as $\varphi \circ \psi$, where $\psi$ is the mapping $(x, y) \mapsto xy^{-1}$ of $K \times K$ onto $K \cdot K^{-1}$, and $\varphi$ is the restriction of $f$ to $K \cdot K^{-1}$. Let R be the equivalence relation defined on $K \times K$ by $\psi$. The mapping $\psi'$ of $(K \times K)/R$ onto $K \cdot K^{-1}$ deduced from $\psi$ by passage to the quotient is continuous, therefore $(K \times K)/R$ is Hausdorff and $\psi'$ is a homeomorphism. Since $\varphi \circ \psi$ is continuous, one sees that the restriction of $f$ to $K \cdot K^{-1}$ is continuous. Now, $K \cdot K^{-1}$ is a neighborhood of $e$ (Prop. 17), therefore $f$ is continuous at $e$. For every $x_0 \in G$, $f(x_0 x) = f(x_0)f(x)$, thus $f$ is continuous at $x_0$ because the topology of $\Gamma$ is invariant under translations.

COROLLARY 1. — *Let* G *be a locally compact group,* $\beta$ *a Haar measure on* G, E *a Hausdorff barreled locally convex space,* $U$ *a linear representation of* G *in* E, *such that* $U(s) \in \mathscr{L}(E;E)$ *for all* $s \in G$, $\beta$*-measurable when* $\mathscr{L}(E;E)$ *is equipped with the topology of pointwise convergence. Then* $U$ *is a continuous linear representation.*

Let $\Gamma$ be the group of automorphisms of E, equipped with the topology of pointwise convergence. This topology is Hausdorff and is invariant under translations. Let K be a compact subset of $\Gamma$. Then K is bounded in $\mathscr{L}(E;E)$ equipped with the topology of pointwise convergence, hence is equicontinuous (TVS, III, §4, No. 2, Th. 1); therefore the mapping $(u,v) \mapsto v \circ u$ of $K \times K$ into $\mathscr{L}(E;E)$ is continuous (*loc. cit.*, §5, No. 5, Cor. 1 of Prop. 9). Therefore, for every $x \in E$, the mapping $s \mapsto U(s)x$ of G into E is continuous (Prop. 18). Since E is barreled, $U$ is continuous (§2, No. 1, Prop. 1).

COROLLARY 2. — *Let* G *be a locally compact group,* $\beta$ *a Haar measure on* G, E *a separable Banach space, and* $U$ *a linear representation of* G *in* E *such that* $U(s) \in \mathscr{L}(E;E)$ *for all* $s \in G$. *Let* $(a_m)$ *be a total sequence in* E, *and let* $(a'_n)$ *be a dense sequence in the unit ball* B′ *of the dual* E′ *of* E, *equipped with the weak topology. Assume that the functions* $s \mapsto \langle U(s)a_m, a'_n\rangle$ *on* G *are* $\beta$*-measurable. Then* $U$ *is a continuous linear representation.*

Let us first show that for every $z' \in E'$, the scalar functions

$$s \mapsto \langle U(s)a_m, z'\rangle$$

are $\beta$-measurable; we may restrict ourselves to the case that $\|z'\| \leqslant 1$, and, since B′ is metrizable for the weak topology (TVS, III, §3, No. 4, Cor. 2 of Prop. 6), there exists a subsequence $(a'_{n_k})$ of $(a'_n)$ that converges weakly to $z'$; the function

$$s \mapsto \langle U(s)a_m, z'\rangle$$

is thus the limit of a sequence of $\beta$-measurable functions, whence our assertion. It follows that the mapping $s \mapsto U(s)a_m$ of G into E is $\beta$-measurable for every $m$ (Ch. IV, §5, No. 5, Prop. 10). On the other hand, there exists a sequence $(b_m)$ of elements of E, linear combinations of the $a_i$, that is dense in the unit ball of E. For every $s \in G$, $\|U(s)\| = \sup_m \|U(s)b_m\|$, therefore $s \mapsto \|U(s)\|$ is measurable. Let K be a compact subset of G and let $\varepsilon > 0$. There exists a compact subset $K_0$ of K such that $\beta(K - K_0) \leqslant \varepsilon$ and such that the restrictions to $K_0$ of the functions $s \mapsto U(s)a_m$ and $s \mapsto \|U(s)\|$ are continuous. Then the $U(s)$ for $s \in K_0$ are equicontinuous, and the topology of pointwise convergence induces on $U(K_0)$ the topology of pointwise convergence in the set of $a_m$ (TVS, III, §3, No. 4, Prop. 5). Consequently the mapping $s \mapsto U(s)$ of $K_0$ into $\mathscr{L}_s(E;E)$ is continuous. It then suffices to apply Cor. 1.

## 7. Regularization

PROPOSITION 19. — *Let* G *be a locally compact group,* $\beta$ *a relatively invariant positive measure* $\neq 0$ *on* G, $\mathfrak{B}$ *a base for the filter of neighborhoods of* $e$ *in* G, *consisting of compact neighborhoods. For every* $V \in \mathfrak{B}$, *let* $f_V$ *be a continuous function* $\geqslant 0$ *on* G, *with support contained in* V, *such that* $\int f_V \, d\beta = 1$. *If* $\mu$ *is a measure on* G *then, in* $\mathscr{M}(G)$ *equipped with the topology of compact convergence in* $\mathscr{K}(G)$,

$$\mu = \lim_V (\mu * f_V) \cdot \beta = \lim_V (f_V * \mu) \cdot \beta,$$

*the limit being taken with respect to the section filter of* $\mathfrak{B}$.

For the topology of compact convergence in $\mathscr{C}(G)$, $f_V \cdot \beta$ tends to $\varepsilon_e$ with respect to the section filter of $\mathfrak{B}$ (§2, No. 7, Cor. 1 of Lemma 4). Therefore $\mu = \lim_V \mu * (f_V \cdot \beta) = \lim_V (f_V \cdot \beta) * \mu$ in $\mathscr{M}(G)$ equipped with the topology of compact convergence in $\mathscr{K}(G)$ (§3, No. 3, Cor. of Prop. 12).

*Remarks.* — 1) We thus see that every measure on G is the limit of measures admitting a *continuous density* with respect to every Haar measure (for the topology indicated in Prop. 19 and *a fortiori* for the vague topology).

2) If G is metrizable, $\mathfrak{B}$ can be taken to be a *sequence* $(V_n)$ of neighborhoods. Then $\mu$ is the limit of the sequence of measures $(\mu * f_{V_n}) \cdot \beta$ with continuous densities. *If G is a real Lie group, the $f_{V_n}$ can be taken to be infinitely differentiable; we shall see later on that the densities $\mu * f_{V_n}$ are then infinitely differentiable.*

PROPOSITION 20. — *We conserve the hypotheses and notations of Prop.* 19. *Let* $p \in [1, +\infty[$ *and* $g \in L^p(G, \beta)$. *Then*

$$g = \lim_V g *^{\beta} f_V = \lim_V f_V *^{\beta} g$$

*in the sense of the norm* $N_p$, *the limit being taken with respect to the section filter of* $\mathfrak{B}$.

It suffices to apply Prop. 6 (iii), and §2, No. 7, Cor. 3 of Lemma 4.

*Remark* 3. — By Prop. 15, the functions $g*f_V$, $f_V*g$ belong to $\overline{\mathscr{K}(G)}$.

COROLLARY. — *Let* W *be a closed linear subspace of* $L^1(G,\beta)$. *For* W *to be a left* (resp. *right*) *ideal of* $L^1(G,\beta)$, *it is necessary and sufficient that* W *be invariant under the left* (resp. *right*) *translations of* G.

Suppose that W is a left ideal. Let $s \in G$ and $g \in W$. We have $\varepsilon_s * g = \lim\limits_{V} f_V * (\varepsilon_s * g) = \lim\limits_{V} (f_V * \varepsilon_s) * g$, and $(f_V * \varepsilon_s) * g \in W$, therefore $\varepsilon_s * g \in W$, thus $\gamma(s)g \in W$. Conversely, if W is invariant under the left translations, then $\mu *^{\beta} g \in W$ for $\mu \in \mathscr{M}^1(G)$ and $g \in W$, therefore W is *a fortiori* a left ideal of $L^1(G,\beta)$. One argues similarly for right ideals.

*Example.* — We take $G = \mathbf{R}$. Let us define a function $F_n \in \mathscr{K}(\mathbf{R})$ by

$$\begin{aligned} F_n(x) &= (1-x^2)^n \quad &&\text{if } x \in [-1,1] \\ F_n(x) &= 0 &&\text{if } x \notin [-1,1]. \end{aligned}$$

Let $A_n = \int_{-1}^{+1} F_n(x)\,dx$, and $G_n = A_n^{-1} F_n$. It is immediate that the measures $G_n(x)\,dx$ satisfy the conditions of §2, No. 7, Cor. 1 of Lemma 4. Let $\mu$ be a measure on $\mathbf{R}$ whose support is contained in $[-1/2, 1/2]$. Then

$$\begin{aligned} (\mu * G_n)(x) &= \int_{\mathbf{R}} G_n(x-y)\,d\mu(y) \\ &= A_n^{-1} \int_{-1/2}^{1/2} F_n(x-y)\,d\mu(y). \end{aligned}$$

If $-1/2 \leqslant x \leqslant 1/2$, then

$$(\mu * G_n)(x) = A_n^{-1} \int_{-1/2}^{1/2} [1-(x-y)^2]^n\,d\mu(y),$$

therefore $\mu * G_n$ coincides on $[-1/2, 1/2]$ with a polynomial. In particular, if $f$ is a continuous function with support contained in $[-1/2, 1/2]$, then $f*G_n$ coincides in $[-1/2, 1/2]$ with a polynomial; moreover, by Prop. 5 (iv), and §2, No. 7, Cor. 3 of Lemma 4, $f * G_n$ converges uniformly to $f$. *If $f$ is of class $C^r$, the derivatives $D^s(f * G_n)$ tend uniformly to $D^s f$ for $0 \leqslant s \leqslant r$.*

## §5. THE SPACE OF CLOSED SUBGROUPS

*Throughout this section,* G *denotes a locally compact group and* $\mu$ *a right Haar measure on* G.

### 1. The space of Haar measures on the closed subgroups of G

*Lemma* 1. — *Let* $\alpha$ *be a positive measure* $\neq 0$ *on* G, S *its support; the following two conditions are equivalent:*

a) S *is a closed subgroup of* G *and the measure induced by* $\alpha$ *on* S *is a right Haar measure on* S.

b) $\boldsymbol{\delta}(s)\alpha = \alpha$ *for every* $s \in \mathrm{S}$.

*Moreover, when these conditions are satisfied, the set of* $t \in \mathrm{G}$ *such that* $\boldsymbol{\delta}(t)\alpha = \alpha$ *is equal to* S.

It is clear that a) implies b); conversely, the relation b) implies that $\mathrm{S}x = \mathrm{S}$ for every $x \in \mathrm{S}$; in other words, the relations $x \in \mathrm{S}$ and $y \in \mathrm{S}$ imply that $y \in \mathrm{S}x$, or again that $yx^{-1} \in \mathrm{S}$, and since S is nonempty, S is a closed subgroup of G. The set of $t \in \mathrm{G}$ such that $\mathrm{S}t = \mathrm{S}$ is then equal to S itself, whence the last assertion.

*For the rest of the section, we denote by* $\Gamma$ *the set of positive measures* $\neq 0$ *on* G *satisfying the conditions of Lemma* 1, *and for every* $\alpha \in \Gamma$ *we denote by* $\mathrm{H}_\alpha$ *the closed subgroup of* G *that is the support of* $\alpha$.

PROPOSITION 1. — *The set* $\Gamma$ *is closed in the space* $\mathscr{M}_+(\mathrm{G}) - \{0\}$ *equipped with the vague topology.*

We first prove the following lemmas:

*Lemma* 2. — *Let* X *be a locally compact space and for every measure* $\alpha \in \mathscr{M}_+(\mathrm{X}) - \{0\}$, *let* $\mathrm{S}_\alpha$ *be the support of* $\alpha$. *Let* $\Phi$ *be a filter on* $\mathscr{M}_+(\mathrm{X}) - \{0\}$ *that converges vaguely to a measure* $\alpha_0 \neq 0$. *Then, for every neighborhood* V *of a point* s *of the support of* $\alpha_0$, *there exists a set* $\mathrm{M} \in \Phi$ *such that, for every* $\alpha \in \mathrm{M}$, *one has* $\mathrm{V} \cap \mathrm{S}_\alpha \neq \varnothing$.

For, if $\varphi \in \mathscr{K}_+(\mathrm{X})$ is a function with support contained in V and such that $\int \varphi(x)\, d\alpha_0(x) > 0$, by definition there exists a set $\mathrm{M} \in \Phi$ such that $\int \varphi(x)\, d\alpha(x) > 0$ for all $\alpha \in \mathrm{M}$, which implies $\mathrm{V} \cap \mathrm{S}_\alpha \neq \varnothing$.

*Lemma* 3. — *Let* E *be a set filtered by a filter* $\Phi$, *and let* $\xi \mapsto \alpha(\xi)$ *be a mapping of* E *into* $\Gamma$ *that converges vaguely with respect to* $\Phi$ *to a measure* $\alpha_0 \neq 0$. *On the other hand, let* $\xi \mapsto t_\xi$ *be a mapping of* E *into* G *such that* $t_\xi \in \mathrm{H}_{\alpha(\xi)}$ *for every* $\xi \in \mathrm{E}$. *If* $s$ *is a cluster point of the mapping* $\xi \mapsto t_\xi$ *with respect to* $\Phi$, *then* $\boldsymbol{\delta}(s)\alpha_0 = \alpha_0$.

Replacing if necessary $\Phi$ by a finer filter, we can suppose that $s$ is a limit of $\xi \mapsto t_\xi$ with respect to $\Phi$; by Lemma 1, $\boldsymbol{\delta}(t_\xi)\alpha(\xi) = \alpha(\xi)$ for every $\xi \in \mathrm{E}$, and the conclusion follows from the continuity of the mapping $(u, \lambda) \mapsto \boldsymbol{\delta}(u)\lambda$ on $\mathrm{G} \times \mathscr{M}_+(\mathrm{G})$ (§3, No. 3, Prop. 13).

To prove Prop. 1 it suffices, by Lemma 1, to show that if a filter $\Psi$ on $\Gamma$ converges vaguely to a measure $\alpha_0 \neq 0$ and if $s$ belongs to the support of $\alpha_0$, then $\boldsymbol{\delta}(s)\alpha_0 = \alpha_0$. Now, for every neighborhood V of $s$ in G, there exists an $\mathrm{M} \in \Psi$ such that, for every $\alpha \in \mathrm{M}$, one has $\mathrm{V} \cap \mathrm{H}_\alpha \neq \varnothing$, by Lemma 2. For every neighborhood V of $s$ and every $\alpha \in \Gamma$, let $t_{\mathrm{V},\alpha}$ be a point of $\mathrm{V} \cap \mathrm{H}_\alpha$ if $\mathrm{V} \cap \mathrm{H}_\alpha \neq \varnothing$, and any point of $\mathrm{H}_\alpha$ in the contrary case; if $\Theta$ is the section filter of the filter of neighborhoods of $s$, and $\Phi$ is the product filter $\Theta \times \Psi$, then $s$ is, by the foregoing, a cluster point of $(\mathrm{V}, \alpha) \mapsto t_{\mathrm{V},\alpha}$ with respect to $\Phi$. Since, on the other hand, the mapping $(\mathrm{V}, \alpha) \mapsto \alpha$ has $\alpha_0$ as limit with respect to $\Phi$, the proposition follows from Lemma 3.

PROPOSITION 2. — *Let* $\varphi$ *be a function in* $\mathscr{K}_+(\mathrm{G})$ *such that* $\varphi(e) > 0$. *Then the set* $\Gamma_\varphi$ *of measures* $\alpha \in \Gamma$ *such that* $\int \varphi(x)\, d\alpha(x) = 1$ *is compact for the vague topology.*

The set $\Gamma_\varphi$ is the intersection of $\Gamma$ with the hyperplane of $\mathscr{M}(\mathrm{G})$ formed by the $\alpha$ such that $\int \varphi(x)\, d\alpha(x) = 1$; since this hyperplane is vaguely closed in $\mathscr{M}(\mathrm{G})$ and does not contain 0, it follows from Prop. 1 that $\Gamma_\varphi$ is vaguely closed in $\mathscr{M}(\mathrm{G})$. It therefore suffices to show that for every compact subset K of G, one has $\sup_{\alpha \in \Gamma_\varphi} \alpha(\mathrm{K}) < +\infty$ (Ch. III, §1, No. 9, Prop. 15). Now, let U be the open neighborhood of $e$ in G defined by the inequality $\varphi(x) > \varphi(e)/2$; since $1 = \int \varphi(x)\, d\alpha(x) \geqslant \int_{\mathrm{U}} \varphi(x)\, d\alpha(x)$ for $\alpha \in \Gamma_\varphi$, one sees that, on setting $c = 2/\varphi(e)$, one has $\alpha(\mathrm{U}) \leqslant c$ for every $\alpha \in \Gamma_\varphi$. Let V be a symmetric open neighborhood of $e$ in G such that $\mathrm{V}^2 \subset \mathrm{U}$; let us show that $\alpha(\mathrm{V}x) \leqslant c$ for every $x \in \mathrm{G}$ and every $\alpha \in \Gamma_\varphi$. Indeed, this relation is trivial if $\mathrm{V}x$ does not intersect the support $\mathrm{H}_\alpha$ of $\alpha$; if, on the contrary, there exists an $h \in \mathrm{V}x \cap \mathrm{H}_\alpha$, then $h = vx$ for some $v \in \mathrm{V}$, whence

$$\mathrm{V}x = \mathrm{V}v^{-1}h \subset \mathrm{V}^2 h \subset \mathrm{U}h,$$

and since $\boldsymbol{\delta}(h)\alpha = \alpha$, it follows that $\alpha(\mathrm{V}x) \leqslant \alpha(\mathrm{U}h) = \alpha(\mathrm{U}) \leqslant c$. Now let $(x_i)_{1 \leqslant i \leqslant n}$ be a sequence of points of K such that the $\mathrm{V}x_i$ form a covering of K; it follows from the foregoing that $\alpha(\mathrm{K}) \leqslant \sum_{i=1}^{n} \alpha(\mathrm{V}x_i) \leqslant nc$ for every $\alpha \in \Gamma_\varphi$; Q.E.D.

PROPOSITION 3. — *Under the hypotheses of Prop. 2, the mapping* $\alpha \mapsto \left(\langle\varphi,\alpha\rangle, \dfrac{\alpha}{\langle\varphi,\alpha\rangle}\right)$ *is a homeomorphism of* $\Gamma$ *onto the product space* $\mathbf{R}_+^* \times \Gamma_\varphi$.

Since the mapping $\alpha \mapsto \langle\varphi,\alpha\rangle$ is vaguely continuous, it suffices to observe that $\langle\varphi,\alpha\rangle \neq 0$ for every measure $\alpha \in \Gamma$, since $e$ belongs to the support $\mathrm{H}_\alpha$ of $\alpha$ and $\varphi(e) > 0$.

## 2. Semi-continuity of the volume of the homogeneous space

In this No., for every measure $\alpha \in \Gamma$ we set

$$\mathrm{Q}_\alpha = \mathrm{G}/\mathrm{H}_\alpha, \tag{1}$$

and we denote by $\pi_\alpha$ the canonical mapping $\mathrm{G} \to \mathrm{Q}_\alpha$.

Let $\Gamma^0$ be the subset of $\Gamma$ formed by the measures $\alpha$ such that the subgroup $\mathrm{H}_\alpha$ of G is *unimodular*; the elements of $\Gamma^0$ are characterized by the fact that $\alpha(f) = \alpha(\check{f})$ for every function $f \in \mathscr{K}(\mathrm{G})$ (every function of $\mathscr{K}(\mathrm{H}_\alpha)$ being extendible to a function of $\mathscr{K}(\mathrm{G})$ by Urysohn's theorem); it follows that $\Gamma^0$ is a *closed* subset of $\Gamma$. Recall that for every $\alpha \in \Gamma^0$, the quotient measure $\mu_\alpha = \mu/\alpha$ on $\mathrm{Q}_\alpha$ is defined and is relatively invariant under G (Ch. VII, §2, No. 6, Th. 3); also recall that for every function $f \in \mathscr{K}(\mathrm{G})$,

$$\int_{\mathrm{G}} f(x)\,d\mu(x) = \int_{\mathrm{Q}_\alpha} d\mu_\alpha(\dot{x}) \int_{\mathrm{H}_\alpha} f(xs)\,d\alpha(s), \tag{2}$$

where $\dot{x} = \pi_\alpha(x)$ is the canonical image of $x \in \mathrm{G}$ in $\mathrm{Q}_\alpha$.

PROPOSITION 4. — *Let* $\Gamma^0$ *be the set of measures* $\alpha \in \Gamma$ *such that* $\mathrm{H}_\alpha$ *is unimodular, and for every* $\alpha \in \Gamma^0$ *set* $\mu_\alpha = \mu/\alpha$; *then the mapping* $\alpha \mapsto \|\mu_\alpha\|$ *of* $\Gamma^0$ *into* $\overline{\mathbf{R}}$ *is lower semi-continuous for the vague topology.*

For every $\alpha \in \Gamma^0$ and every function $f \in \mathscr{K}(\mathrm{G})$, set

$$f_\alpha(\dot{x}) = \int_{\mathrm{H}_\alpha} f(xs)\,d\alpha(s) = (f * \alpha)(x),$$

where the convolution product is taken relative to the right Haar measure $\mu$ and where one makes use of the fact that $\check{\alpha} = \alpha$ (§4, No. 4, formula (11)). We know (Ch. VII, §2, No. 1, Prop. 2) that the mapping $f \mapsto f_\alpha$ of $\mathscr{K}_+(\mathrm{G})$ into $\mathscr{K}_+(\mathrm{Q}_\alpha)$ is *surjective*; therefore, by (2),

$$\|\mu_\alpha\| = \sup_{f \in \mathscr{K}_+(\mathrm{G}),\, f \neq 0} \mu_\alpha(f_\alpha)/\|f_\alpha\| = \sup_{f \in \mathscr{K}_+(\mathrm{G}),\, f \neq 0} \mu(f)/\|f_\alpha\|,$$

where one has set

$$\|f_\alpha\| = \sup_{\dot{x}\in Q_\alpha} |f_\alpha(\dot{x})| = \sup_{x\in G} |(f*\alpha)(x)|\,. \tag{3}$$

To establish the proposition, it will suffice to show that, given $f \in \mathscr{K}_+(G)$, the mapping $\alpha \mapsto \|f_\alpha\|$ is vaguely continuous. Now, let K be the support of $f$; the function $f*\alpha$ has its support contained in $KH_\alpha$ and is invariant on the right under $H_\alpha$; consequently

$$\|f_\alpha\| = \sup_{x\in K} |(f*\alpha)(x)|\,.$$

The conclusion therefore follows from the fact that the mapping $\alpha \mapsto f*\alpha$ of $\mathscr{M}_+(G)$ equipped with the vague topology, into $\mathscr{C}(G)$ equipped with the topology of compact convergence, is continuous (§4, No. 2, *Remark* 1).

Recall that if, for a measure $\alpha \in \Gamma^0$, $\|\mu_\alpha\|$ is *finite*, then G is necessarily *unimodular* (Ch. VII, §2, No. 6, Cor. 3 of Th. 3).

PROPOSITION 5. — *Let $g$ be a $\mu$-integrable positive numerical function and let $\Gamma^0(g)$ be the set of measures $\alpha \in \Gamma^0$ such that $\int^* g(xs)\,d\alpha(s) \geqslant 1$ for all $x \in G$. Then the mapping $\alpha \mapsto \|\mu_\alpha\|$ of $\Gamma^0(g)$ into $\overline{\mathbf{R}}$ is vaguely continuous.*

For every measure $\alpha \in \Gamma^0(G)$, recall (Ch. VII, §2, No. 3, Prop. 5) that the function

$$g_\alpha(\dot{x}) = \int_{H_\alpha} g(xs)\,d\alpha(s)$$

is defined $\mu_\alpha$-almost everywhere on $Q_\alpha$, is $\mu_\alpha$-integrable, and

$$\int_G g(x)\,d\mu(x) = \int_{Q_\alpha} g_\alpha(\dot{x})\,d\mu_\alpha(\dot{x})\,. \tag{4}$$

In view of Prop. 4, it suffices to prove that, in $\Gamma^0(g)$, $\alpha \mapsto \|\mu_\alpha\|$ is *upper* semi-continuous. Fix a measure $\alpha \in \Gamma^0(g)$, and let K be a compact subset of G. There exists on $Q_\alpha$ a continuous function with compact support, taking its values in $[0,1]$, equal to 1 on the compact set $\pi_\alpha(K)$; since the mapping $f \mapsto f_\alpha$ of $\mathscr{K}_+(G)$ into $\mathscr{K}_+(Q_\alpha)$ is surjective (Ch. VII, §2, No. 1, Prop. 2), one sees that there exists a function $f \in \mathscr{K}_+(G)$ such that

$$(f*\alpha)(x) = \int_G f(xs)\,d\alpha(s) \begin{cases} \leqslant 1 & \text{for all } x \in G \\ = 1 & \text{for all } x \in K\,. \end{cases}$$

Since $\beta \mapsto f * \beta$ is a continuous mapping of $\mathscr{M}_+(\mathrm{G})$, equipped with the vague topology, into $\mathscr{C}(\mathrm{G})$ equipped with the topology of compact convergence (§4, No. 2, *Remark* 1), one sees that for every $\varepsilon > 0$, the set $\mathrm{U}_\varepsilon$ of $\beta \in \Gamma^0(\mathrm{G})$ such that

$$f_\beta(\dot{x}) = \int_{\mathrm{G}} f(xs)\, d\beta(s) > 1 - \varepsilon \quad \text{for all } x \in \mathrm{K}$$

is an open neighborhood of $\alpha$ in $\Gamma^0(g)$; for every $\beta \in \mathrm{U}_\varepsilon$, we then have, by virtue of the formula (2),

$$(5) \quad \|\mu_\alpha\| \geqslant \int_{\mathrm{G}} f(x)\, d\mu(x) = \int_{\mathrm{Q}_\beta} f_\beta(\dot{x})\, d\mu_\beta(\dot{x}) \geqslant (1-\varepsilon)\, \mu_\beta(\pi_\beta(\mathrm{K})).$$

Given a number $\varepsilon > 0$, let us choose a function $h \in \mathscr{K}_+(\mathrm{G})$ such that $\int_{\mathrm{G}} |g(x) - h(x)|\, d\mu(x) \leqslant \varepsilon$, and let us take $\mathrm{K} = \mathrm{Supp}(h)$ in the foregoing. For every $\beta \in \Gamma^0(g)$, by hypothesis $g_\beta(\dot{x}) \geqslant 1$ almost everywhere (for $\mu_\beta$) in $\mathrm{Q}_\beta$, therefore

$$\mu_\beta(\mathrm{Q}_\beta - \pi_\beta(\mathrm{K})) \leqslant \int_{\mathrm{Q}_\beta - \pi_\beta(\mathrm{K})} g_\beta(\dot{x})\, d\mu_\beta(\dot{x}) = \int_{\mathrm{G}-\mathrm{KH}_\beta} g(x)\, d\mu(x)$$

by virtue of (4); since $h$ is zero outside $\mathrm{K}$, and *a fortiori* outside $\mathrm{KH}_\beta$, it follows that

$$\begin{aligned}\mu_\beta(\mathrm{Q}_\beta - \pi_\beta(\mathrm{K})) &\leqslant \int_{\mathrm{G}-\mathrm{KH}_\beta} |g(x) - h(x)|\, d\mu(x) \\ &\leqslant \int_{\mathrm{G}} |g(x) - h(x)|\, d\mu(x) \leqslant \varepsilon;\end{aligned}$$

combining this result with (5), one sees that

$$\|\mu_\beta\| \leqslant \varepsilon + \|\mu_\alpha\|/(1-\varepsilon)$$

when $\beta \in \mathrm{U}_\varepsilon$, which completes the proof.

COROLLARY 1. — *Let* $\mathrm{K}$ *be a compact subset of* $\mathrm{G}$, $\mathrm{V}$ *a symmetric compact neighborhood of* $e$ *in* $\mathrm{G}$, $c$ *a real number* $> 0$. *The restriction of the mapping* $\alpha \mapsto \|\mu_\alpha\|$ *to the set of* $\alpha \in \Gamma^0$ *such that* $\mathrm{G} = \mathrm{KH}_\alpha$ *and* $\alpha(\mathrm{V}) \geqslant c$ *is vaguely continuous.*

For, let $g \in \mathscr{K}_+(\mathrm{G})$ be a function such that $g(x) \geqslant 1/c$ for $x \in \mathrm{KV}$. For every $x \in \mathrm{K}$,

$$\int g(xs)\, d\alpha(s) \geqslant \int_{\mathrm{V}} g(xs)\, d\alpha(s) \geqslant 1$$

for $\alpha$ satisfying the conditions of the statement; since, moreover, $\pi_\alpha(\mathrm{K}) = \mathrm{Q}_\alpha$, one has $\alpha \in \Gamma^0(g)$, whence the corollary.

COROLLARY 2. — *Let* A *be a* $\mu$*-integrable subset of* G. *The restriction of the mapping* $\alpha \mapsto \|\mu_\alpha\|$ *to the set* $\mathrm{N_A}$ *of normalized Haar measures of the discrete subgroups* H *of* G *such that* $\mathrm{G} = \mathrm{AH}$, *is vaguely continuous.*

For $a \in \mathrm{A}$ and $\alpha \in \mathrm{N_A}$,

$$\int \varphi_{\mathrm{A}}(as)\, d\alpha(s) \geqslant \varphi_{\mathrm{A}}(a) = 1,$$

and since $\pi_\alpha(\mathrm{A}) = \mathrm{Q}_\alpha$, one has $\mathrm{N_A} \subset \Gamma^0(\varphi_{\mathrm{A}})$, and the corollary therefore follows from Prop. 5.

## 3. The space of closed subgroups of G

Let us denote by $\Sigma$ the set of *closed subgroups* of G; if one associates to each measure $\alpha \in \Gamma$ the subgroup $\mathrm{H}_\alpha$ that is the support of $\alpha$, one obtains a mapping (called canonical) of $\Gamma$ into $\Sigma$, which is clearly surjective and permits canonically identifying $\Sigma$ with the set of orbits of the group of homotheties in $\Gamma$ with ratio $> 0$. The set $\Sigma$, equipped with the quotient topology of the vague topology on $\Gamma$, is called *the space of closed subgroups* of G.

THEOREM 1. — *Let* G *be a locally compact group. The space* $\Sigma$ *of closed subgroups of* G *is compact. Moreover, one has the following properties:*

(i) *The set* $\Sigma^0$ *of unimodular closed subgroups of* G *is closed in* $\Sigma$ *(hence is compact).*

(ii) *If* G *is generated by a compact neighborhood of* $e$, *then the set* $\Sigma^0_c$ *of unimodular closed subgroups* H *of* G *such that the quotient space* G/H *is compact, is open in* $\Sigma^0$ *(hence is locally compact).*

(iii) *For every relatively compact open neighborhood* U *of* $e$ *in* G, *the set* $\mathrm{D_U}$ *of discrete subgroups* H *of* G *such that* $\mathrm{H} \cap \mathrm{U} = \{e\}$ *is closed in* $\Sigma^0$ *(hence is compact).*

It follows from Prop. 3 of No. 1 that $\Sigma$ is homeomorphic to $\Gamma_\varphi$, hence is compact by Prop. 2 of No. 1. Moreover, it was noted at the beginning of No. 2 that the set $\Gamma^0$ of measures $\alpha \in \Gamma$ such that $\mathrm{H}_\alpha$ is unimodular is closed in $\Gamma$; since $\Gamma^0$ is stable under the homotheties with ratio $> 0$, the image $\Sigma^0$ of $\Gamma^0$ in $\Sigma$ is a closed subset of $\Sigma$, which proves (i).

Property (ii) will be a consequence of the following proposition:

PROPOSITION 6. — *Suppose that the locally compact group* G *is generated by a compact neighborhood of* $e$. *Then the set* $\Gamma^0_c$ *of measures* $\alpha \in \Gamma^0$

*such that* $G/H_\alpha$ *is compact is open in* $\Gamma^0$, *and the restriction to* $\Gamma^0_c$ *of the mapping* $\alpha \mapsto \|\mu_\alpha\|$ *is vaguely continuous.*

With the notations of Prop. 5 of No. 2, we have, for $g \in \mathscr{K}_+(G)$,

$$\Gamma^0(g) \subset \Gamma^0_c. \tag{6}$$

For, if K is the support of $g$, the relation $\int g(xs)\,d\alpha(s) \geqslant 1$ for all $x \in G$ implies $KH_\alpha = G$, the integral obviously being zero on the complement of $KH_\alpha$, therefore $G/H_\alpha = \pi_\alpha(K)$ is compact. Given a measure $\alpha \in \Gamma^0_c$, it will therefore suffice to define a function $g \in \mathscr{K}_+(G)$ such that $\Gamma^0(g)$ is a neighborhood of $\alpha$ in $\Gamma^0$. Since $G/H_\alpha$ is compact and the canonical mapping $f \mapsto f_\alpha$ of $\mathscr{K}_+(G)$ into $\mathscr{K}_+(G/H_\alpha)$ is surjective (Ch. VII, §2, No. 2), there exists a function $g \in \mathscr{K}_+(G)$ such that $\int g(xs)\,d\alpha(s) = 2$ for *every* $x \in G$. Let K be the (compact) support of $g$, L a symmetric compact neighborhood of $e$ in G that generates G; the mapping $\beta \mapsto g*\beta$ of $\mathscr{M}_+(G)$ into $\mathscr{C}(G)$ being vaguely continuous (§4, No. 2, *Remark* 1), there exists a neighborhood W of $\alpha$ in $\Gamma^0$ such that

$$(g * \beta)(x) = \int g(xs)\,d\beta(s) \geqslant 1 \tag{7}$$

for all $\beta \in W$ and $x \in LK$. The first member of (7) being equal to zero outside $KH_\beta$, the relation $\beta \in W$ implies

$$LK \subset KH_\beta,$$

from which one deduces, by induction on $n$, that $L^nK \subset KH_\beta$ for every integer $n > 0$; since L generates G, we therefore have $G = KH_\beta$ for every measure $\beta \in W$, which proves that $W \subset \Gamma^0_c$. On the other hand, the first member of (7) being invariant on the right under $H_\beta$, the inequality (7) is also valid for $x \in LKH_\beta = G$; therefore $W \subset \Gamma^0(g)$, which proves the proposition.

Finally, (iii) will be a consequence of the following proposition:

PROPOSITION 7. — *Let* $N \subset \Gamma^0$ *be the subspace of normalized Haar measures on the discrete subgroups of* G, *and for every relatively compact open neighborhood* U *of* $e$ *in* G, *let* $N_U$ *be the subset of* N *formed by the* $\alpha$ *such that* $H_\alpha \cap U = \{e\}$. *Then:*

a) $N_U$ *is compact.*

b) *The interiors of the sets* $N_U$ *in* N *form a covering of* N, *as* U *runs over the set of relatively compact open neighborhoods of* $e$ *in* G.

c) *For a subset* M *of* N *to be relatively compact in* N, *it is necessary and sufficient that there exist a relatively compact open neighborhood* U *of* $e$ *in* G *such that* $M \subset N_U$.

Since $D_U$ is the image of $N_U$ under the canonical continuous mapping $\Gamma \to \Sigma$, the assertion (iii) of Th. 1 will result at once from Prop. 7 *a*).

To prove Prop. 7, we observe that $N_U$ can be defined as the subset of $\Gamma^0$ formed by the $\alpha$ such that both

$$\alpha(\{e\}) \geqslant 1 \quad \text{and} \quad \alpha(U) \leqslant 1 .$$

Now, if A is compact (resp. open and relatively compact) in G, then the mapping $\alpha \mapsto \alpha(A)$ of $\mathscr{M}_+(G)$ into $\mathbf{R}$ is upper (resp. lower) semi-continuous for the vague topology (Ch. IV, §4, No. 4, Cor. 3 of Prop. 5 and *loc. cit.*, §1, No. 1, Prop. 4); we thus see that $N_U$ is a *closed* subset of $\Gamma^0$. Moreover, let $\varphi \in \mathscr{K}_+(G)$ be a function such that $\varphi(e) = 1$ and $\varphi(x) = 0$ on $G - U$; it is clear that $\int \varphi(x)\, d\alpha(x) = 1$ for all $\alpha \in N_U$; Prop. 2 of No. 1 therefore shows that $N_U$ is a *compact* set, which proves a). On the other hand let V be a relatively compact open neighborhood of $e$ in G such that $\overline{V} \subset U$, and let $\varphi \in \mathscr{K}_+(G)$, with support contained in U and such that $\varphi(x) = 1$ on V. Then $\alpha(\varphi) = 1$ for $\alpha \in N_U$, therefore there exists a neighborhood W of $\alpha$ in N such that $\beta(\varphi) < 2$ for $\beta \in W$; it is then clear that $W \subset N_V$, therefore $N_V$ is a neighborhood of $N_U$. Since the $N_U$ cover N, this proves b). Finally, every compact subset M of N is contained in a finite union of sets $N_{U_i}$ $(1 \leqslant i \leqslant n)$, and since $\bigcup_i N_{U_i} \subset N_U$, where $U = \bigcap_i U_i$, this proves c).

COROLLARY. — *The subspace* N *of* $\Gamma^0$ *is locally compact.*

## 4. The case of groups without arbitrarily small finite subgroups

THEOREM 2. — *Let* G *be a locally compact group satisfying the following condition:*

(L) *There exists a neighborhood of* $e$ *in* G *that contains no finite subgroup of* G *not reduced to* $e$.

*The following properties then hold:*

(i) *The set* D *of discrete subgroups of* G *is locally closed in* $\Sigma$ *(which is equivalent to saying that it is locally compact).*

(ii) *For a closed subset* A *of* D *to be compact, it is necessary and sufficient that there exist a neighborhood* U *of* $e$ *in* G *such that* $H \cap U = \{e\}$ *for every subgroup* $H \in A$.

(iii) *If in addition* G *is generated by a compact neighborhood of* $e$, *then the set* $D_c$ *of discrete subgroups* H *of* G *such that* G/H *is compact is locally closed in* $\Sigma$ *(hence is locally compact).*

We have $D_c = D \cap \Sigma_c^0$, therefore (iii) is a consequence of (i), and Th. 1 (ii) of No. 3.

With the notations of No. 3, Prop. 7, it suffices, for proving (i) and (ii), to prove that:

PROPOSITION 8. — *The canonical bijection of* N *onto* D *is a homeomorphism.*

Now, if $\Gamma_d$ is the set of Haar measures on the discrete subgroups of G, then D is canonically homeomorphic to the space of orbits of the group of homotheties in $\Gamma_d$ with ratio $> 0$ (GT, I, §5, No. 2, Prop. 4). It therefore suffices to prove that the canonical mapping $\alpha \mapsto \big(\alpha(\{e\}), \alpha/\alpha(\{e\})\big)$ of $\Gamma_d$ onto $\mathbf{R}_+^* \times \mathrm{N}$ is a *homeomorphism*, which will result from the following lemma:

*Lemma* 4. — *If the group* G *satisfies the condition* (L), *the mapping* $\alpha \mapsto \alpha(\{e\})$ *of* $\Gamma_d$ *into* $\mathbf{R}_+^*$ *is vaguely continuous.*

Let us consider a measure $\alpha \in \Gamma_d$; let $V_0$ be a relatively compact open neighborhood of $e$ in G such that $H_\alpha \cap V_0 = \{e\}$ and such that there exists no finite subgroup of G contained in $V_0$ and not reduced to $e$. Let V be a symmetric compact neighborhood of $e$ such that $V^3 \subset V_0$, and let U be a symmetric neighborhood of $e$ such that $U^2 \subset V$. Let $\varphi$ (resp. $\psi$) be a function in $\mathscr{K}_+(G)$, with values in $[0,1]$, equal to 1 on $V^3$ (resp. at the point $e$) and with support contained in $V_0$ (resp. in U). The set of measures $\beta \in \Gamma_d$ such that $|\beta(\varphi) - \alpha(\varphi)| \leqslant \varepsilon$ and $|\beta(\psi) - \alpha(\psi)| \leqslant \varepsilon$ is a neighborhood W of $\alpha$. We propose to show that, provided $\varepsilon$ is taken to be sufficiently small, $H_\beta \cap V = \{e\}$ *for every* $\beta \in W$; it will then follow that $\beta(\psi) = \beta(\{e\})$, hence that $|\beta(\{e\}) - \alpha(\{e\})| \leqslant \varepsilon$, which will prove the lemma.

It will suffice to show that, for $\beta \in W$,

$$(V^2 - V) \cap H_\beta = \varnothing. \tag{8}$$

For, suppose that this point is established: then, for $x$ and $y$ in $V \cap H_\beta$, one has $xy^{-1} \in V^2 \cap H_\beta$; but, by virtue of (8), this implies $xy^{-1} \in V \cap H_\beta$; in other words, $V \cap H_\beta$ is a *subgroup* of G, which is obviously discrete and compact, hence finite; but then, by the choice of $V_0$, this implies that indeed $V \cap H_\beta = \{e\}$.

Let us argue by contradiction and so assume that there exists a point $z$ of $V^2 - V$ that belongs to $H_\beta$; by the choice of U and V, we have $\psi(sz^{-1}) + \psi(s) \leqslant \varphi(s)$ in G, the relation $z \notin U^2$ implying $Uz \cap U = \varnothing$. Since

$$\int \psi(sz^{-1})\, d\beta(s) = \int \psi(s)\, d\beta(s),$$

it follows that $2\beta(\psi) \leqslant \beta(\varphi) \leqslant \alpha(\varphi) + \varepsilon$; but we also have

$$\beta(\psi) \geqslant \alpha(\psi) - \varepsilon,$$

and by construction $\alpha(\varphi) = \alpha(\psi) = \alpha(\{e\})$. We thus arrive at a contradiction by taking $\varepsilon < \alpha(\{e\})/3$. Q.E.D.

In graphic terms, a group G satisfying the condition (L) is said to *not have arbitrarily small finite subgroups.* *It can be shown that every Lie group satisfies the condition (L); but this condition is not characteristic of Lie groups; for example, the multiplicative group of $p$-adic integers congruent to 1 mod $p$ satisfies (L).$_*$

## 5. The case of abelian groups

Let G be a locally compact group, $\mathrm{N} \subset \Gamma^0$ the subspace of normalized Haar measures on the discrete subgroups of G, and $\mathrm{N}_c$ the subset of N corresponding to the discrete subgroups H of G such that G/H is *compact*; thus $\mathrm{N}_c = \mathrm{N} \cap \Gamma_c^0$ in the notations of No. 3, Prop. 6; and if the group G is generated by a compact neighborhood of $e$, it follows from No. 3, Prop. 6 that $\mathrm{N}_c$ is *open* in N (hence is *locally compact* by No. 3, Cor. of Prop. 7) and that the restriction to $\mathrm{N}_c$ of the mapping $\alpha \mapsto \|\mu_\alpha\|$ is *vaguely continuous.*

PROPOSITION 9. — *Let* G *be a locally compact* abelian *group, generated by a compact neighborhood of* $e$. *For a subset* A *of* $\mathrm{N}_c$ *to be relatively compact in* $\mathrm{N}_c$, *it is necessary and sufficient that it satisfy the following two conditions*:

(i) *There exists an open neighborhood* U *of* $e$ *in* G *such that* $\mathrm{H}_\alpha \cap \mathrm{U} = \{e\}$ *for all* $\alpha \in \mathrm{A}$.

(ii) *There exists a constant* $k$ *such that* $\mu_\alpha(\mathrm{G}/\mathrm{H}_\alpha) \leqslant k$ *for all* $\alpha \in \mathrm{A}$.

If $\mathrm{A} \subset \mathrm{N}_c$ is relatively compact in $\mathrm{N}_c$, it is *a fortiori* so in N, and the necessity of the conditions (i) and (ii) therefore follows from No. 3, Props. 6 and 7 (without assuming G to be abelian). Conversely, suppose that $\mathrm{A} \subset \mathrm{N}_c$ satisfies these conditions; if $\overline{\mathrm{A}}$ is the closure of A *in* N, then $\overline{\mathrm{A}}$ is compact by virtue of No. 3, Prop. 7; moreover, since $\alpha \mapsto \|\mu_\alpha\|$ is lower semi-continuous on $\Gamma^0$ for the vague topology (No. 2, Prop. 4), the condition (ii) implies that one also has $\|\mu_\alpha\| \leqslant k$ for all $\alpha \in \overline{\mathrm{A}}$. Now, since G is abelian, $\mu_\alpha = \mu/\alpha$ is a Haar measure on the group $\mathrm{G}/\mathrm{H}_\alpha$, and $\mathrm{G}/\mathrm{H}_\alpha$ is therefore compact for every $\alpha \in \overline{\mathrm{A}}$ (Ch. VII, §1, No. 2, Prop. 2). This means that $\overline{\mathrm{A}} \subset \mathrm{N}_c$, thus A is relatively compact in $\mathrm{N}_c$.

COROLLARY. — *Let* G *be a locally compact abelian group, generated by a compact neighborhood of* $e$ *and satisfying the condition* (L) *of No.* 4.

*Let* $D_c$ *be the set of discrete subgroups* H *of* G *such that* G/H *is compact, and, for every* $H \in D_c$, *let* $v(H)$ *be the total mass* $\mu_\alpha(G/H)$, *where* $\mu_\alpha$ *is the quotient measure of* $\mu$ *by the normalized Haar measure* $\alpha$ *of* H. *For a subset* A *of the space* $D_c$ *to be relatively compact in* $D_c$, *it is necessary and sufficient that it satisfy the following two conditions*:

(i) *There exists an open neighborhood* U *of* $e$ *in* G *such that* $H \cap U = \{e\}$ *for all* $H \in A$.

(ii) *There exists a constant* $k$ *such that* $v(H) \leqslant k$ *for all* $H \in A$.

Taking into account Prop. 9, this follows at once from the fact that $D_c$ is the image of $N_c$ under the canonical bijection of N onto D, and the fact that, under the hypotheses made, this bijection is a homeomorphism (No. 4, Prop. 8).

*Example.* — Let us take $G = \mathbf{R}^n$ and for $\mu$ the Lebesgue measure; all of the hypotheses of the Cor. of Prop. 9 are satisfied. The discrete subgroups H of G such that G/H is compact are none other than the discrete subgroups of *rank* $n$ (GT, VII, §1, No. 1, Th. 1); such a subgroup H is generated by a basis $(a_i)_{1\leqslant i\leqslant n}$ of $\mathbf{R}^n$, and

$$v(H) = |\det(a_1, \dots, a_n)|$$

(the determinant being taken with respect to the canonical basis of $\mathbf{R}^n$) (Ch. VII, §2, No. 10, Th. 4). The space $D_c$ can be interpreted here in the following way: every subgroup $H \in D_c$ is the transform $g \cdot \mathbf{Z}^n$ of the subgroup $\mathbf{Z}^n$ by an element $g \in \mathbf{GL}(n, \mathbf{R})$, and the subgroup of $\mathbf{GL}(n, \mathbf{R})$ leaving $\mathbf{Z}^n$ stable may be identified with $\mathbf{GL}(n, \mathbf{Z})$. Consequently $D_c$ may be canonically identified, as a (non-topological) homogeneous space, with $\mathbf{GL}(n, \mathbf{R})/\mathbf{GL}(n, \mathbf{Z})$. On the other hand, $\mathbf{GL}(n, \mathbf{R})$ operates continuously in $\mathbf{R}^n$, hence also in $\mathscr{M}_+(\mathbf{R}^n)$ for the vague topology (§3, No. 3, Prop. 13), hence in the subspace $N_c$ of $\mathscr{M}_+(\mathbf{R}^n)$; moreover, the canonical homeomorphism (No. 4, Prop. 8) of $N_c$ onto $D_c$ is compatible with the laws of operation of $\mathbf{GL}(n, \mathbf{R})$. Since $\mathbf{GL}(n, \mathbf{R})$ is countable at infinity and $D_c$ is locally compact, the bijection of $\mathbf{GL}(n, \mathbf{R})/\mathbf{GL}(n, \mathbf{Z})$ onto $D_c$ defined above is a *homeomorphism* (Ch. VII, App. I, Lemma 2). The Cor. of Prop. 9 therefore gives a criterion for compactness in the homogeneous space $\mathbf{GL}(n, \mathbf{R})/\mathbf{GL}(n, \mathbf{Z})$.

## 6. Another interpretation of the topology of the space of closed subgroups

Let $\mathfrak{F}$ be the set of closed subsets of G; one defines a *Hausdorff uniform structure* on $\mathfrak{F}$ in the following way: for every compact subset K of G and every neighborhood V of $e$ in G, let P(K, V) be the set of pairs (X, Y) of elements of $\mathfrak{F}$ such that both

$$\text{(9)} \qquad \mathrm{X} \cap \mathrm{K} \subset \mathrm{VY} \quad \text{and} \quad \mathrm{Y} \cap \mathrm{K} \subset \mathrm{VX}.$$

Let us show that the set of P(K, V) is a fundamental system of entourages for a Hausdorff uniform structure $\mathscr{U}$ on $\mathfrak{F}$. The axioms $(\mathrm{U}'_{\mathrm{I}})$ and $(\mathrm{U}'_{\mathrm{II}})$ of GT, II, §1, No. 1 are obviously satisfied; moreover, the relations $\mathrm{K} \subset \mathrm{K}'$ and $\mathrm{V}' \subset \mathrm{V}$ imply $\mathrm{P}(\mathrm{K}', \mathrm{V}') \subset \mathrm{P}(\mathrm{K}, \mathrm{V})$; to verify $(\mathrm{U}'_{\mathrm{III}})$, one can therefore limit oneself to the case that V is a symmetric compact neighborhood of $e$, so that VK is compact. Suppose that $(\mathrm{X}, \mathrm{Y}) \in \mathrm{P}(\mathrm{VK}, \mathrm{V})$ and $(\mathrm{Y}, \mathrm{Z}) \in \mathrm{P}(\mathrm{VK}, \mathrm{V})$; then $\mathrm{X} \cap \mathrm{K} \subset \mathrm{X} \cap \mathrm{VK} \subset \mathrm{VY}$, and if $y \in \mathrm{Y}$ is such that $vy \in \mathrm{K}$ for some $v \in \mathrm{V}$, then necessarily $y \in \mathrm{VK}$, therefore

$$\mathrm{X} \cap \mathrm{K} \subset \mathrm{V}(\mathrm{Y} \cap \mathrm{VK});$$

on the other hand, $\mathrm{Y} \cap \mathrm{VK} \subset \mathrm{VZ}$, whence $\mathrm{X} \cap \mathrm{K} \subset \mathrm{V}^2\mathrm{Z}$, and one shows similarly that $\mathrm{Z} \cap \mathrm{K} \subset \mathrm{V}^2\mathrm{X}$, which proves $(\mathrm{U}'_{\mathrm{III}})$. Finally, if X, Y are two distinct elements of $\mathfrak{F}$, there exists for example a point $a \in \mathrm{X}$ such that $a \notin \mathrm{Y}$, hence a symmetric compact neighborhood V of $e$ such that $\mathrm{V}a \cap \mathrm{Y} = \varnothing$, that is, $a \notin \mathrm{VY}$; *a fortiori* $(\mathrm{X}, \mathrm{Y}) \notin \mathrm{P}(\mathrm{V}a, \mathrm{V})$, which completes the proof of our assertion.

This established, let us consider on the set $\Sigma$ of closed subgroups of G the topology $\mathscr{T}$ induced by the topology of the uniform space $\mathfrak{F}$ just defined. We shall see that this topology is *identical to the topology defined in No.* 3. It will suffice to prove that the mapping $\alpha \mapsto \mathrm{H}_\alpha$ of $\Gamma$ into $\Sigma$ is *continuous* when $\Sigma$ is equipped with the topology $\mathscr{T}$: for, the same will then be true of the restriction of this mapping to $\Gamma_\varphi$ (with notations as in No. 1, Prop. 2), which is bijective; but since $\Gamma_\varphi$ is compact and the topology $\mathscr{T}$ is separated, the mapping $\alpha \mapsto \mathrm{H}_\alpha$ of $\Gamma_\varphi$ into $\Sigma$ will then be a homeomorphism.

Thus let $\alpha_0$ be a point of $\Gamma$ and let $\Phi$ be a filter on $\Gamma$ that converges to $\alpha_0$; we are to show that, with respect to $\Phi$, $\mathrm{H}_\alpha$ tends to $\mathrm{H}_{\alpha_0}$ for the topology $\mathscr{T}$. Let K be a compact subset of G, V a symmetric compact neighborhood of $e$ in G; for every $x \in \mathrm{H}_{\alpha_0} \cap \mathrm{K}$, there exists a set $\mathrm{M}(x) \in \Phi$ such that for every $\alpha \in \mathrm{M}(x)$, one has $\mathrm{V}x \cap \mathrm{H}_\alpha \neq \varnothing$ (No. 1, Lemma 2), whence $\mathrm{V}x \subset \mathrm{V}^2\mathrm{H}_\alpha$; on covering $\mathrm{H}_{\alpha_0} \cap \mathrm{K}$ by a finite number of sets $\mathrm{V}x_i$, one sees that if $\mathrm{M} = \bigcap_i \mathrm{M}(x_i)$, then $\mathrm{H}_{\alpha_0} \cap \mathrm{K} \subset \mathrm{V}^2\mathrm{H}_\alpha$ for every $\alpha \in \mathrm{M}$. Conversely, suppose that there existed an open neighborhood U of $e$ in G such that, for every set $\mathrm{L} \in \Phi$, there is at least one $\alpha \in \mathrm{L}$ for which $\mathrm{H}_\alpha \cap \mathrm{K} \not\subset \mathrm{UH}_{\alpha_0}$; if $\omega(\mathrm{L})$ is the set of $\alpha \in \mathrm{L}$ having this property, the $\omega(\mathrm{L})$ would form a base of a filter $\Phi'$ on $\Gamma$ finer than $\Phi$, and, for every $\alpha$ belonging to the union E of the $\omega(\mathrm{L})$ for $\mathrm{L} \in \Phi$, there would exist a

$t_\alpha \in H_\alpha \cap K$ not belonging to $UH_{\alpha_0}$; for $\alpha \notin E$, take for $t_\alpha$ any point of $H_\alpha$. Since $K \cap \complement(UH_{\alpha_0})$ is compact, there would exist a cluster point $s$ of $\alpha \mapsto t_\alpha$ with respect to $\Phi'$, belonging to $K \cap \complement(UH_{\alpha_0})$; but since $\Phi'$ converges to $\alpha_0$ in $\Gamma$, this contradicts Lemma 3 of No. 1.

# Exercises

## §1

1) Let $\Gamma$ be a proper[1] closed convex cone in $\mathbf{R}^n$. Show that the mapping $(x,y) \mapsto x+y$ of $\Gamma \times \Gamma$ into $\Gamma$ is proper. Deduce from this that any two measures on $\Gamma$ are convolvable for the mapping $(x,y) \mapsto x+y$.

2) Let G be a locally compact group and $\Gamma$ the compact space obtained by adjoining to G a point at infinity $\omega$. One extends the law of composition of G to $\Gamma$ by setting $x\omega = \omega x = \omega$ for every $x \in \Gamma$. To every measure $\mu$ on $\Gamma$ there corresponds, on the one hand a bounded measure $\mu_1$ on G, on the other a complex number $\mu(\{\omega\})$. Show that if $*$ (resp. $\widehat{*}$) denotes the convolution defined by the multiplication in G (resp. $\Gamma$), then $(\mu \mathbin{\widehat{*}} \nu)_1 = \mu_1 * \nu_1$ and

$$(\mu \mathbin{\widehat{*}} \nu)(\omega) = \mu(\omega)\nu_1(\mathrm{G}) + \nu(\omega)\mu_1(\mathrm{G}) + \mu(\omega)\nu(\omega)$$

for any two measures $\mu$ and $\nu$ on $\Gamma$.

## §2

1) Let $(\mathrm{G}_\iota)_{\iota\in\mathrm{I}}$ be a family of locally compact groups, all but finitely many of them compact. Let $U_\iota$ be a continuous linear representation of $\mathrm{G}_\iota$ in a locally convex space $\mathrm{E}_\iota$. For every $s = (s_\iota) \in \mathrm{G} = \prod_\iota \mathrm{G}_\iota$, let $U(s)$ be the endomorphism $(x_\iota) \mapsto (U_\iota(s)x_\iota)$ of $\mathrm{E} = \prod_\iota \mathrm{E}_\iota$. Show that $U$ is a continuous linear representation of G in E. Let E′ be the topological direct sum of the $\mathrm{E}_\iota$. Let $V(s)$ be the restriction of $U(s)$ to E′. Show that $V$ is a continuous linear representation of G in E′.

[1] Translation of *saillant* (TVS, II, §2, No. 4).

2) Let $U_1$ (resp. $U_2$) be a continuous linear representation of a locally compact group G (resp. H) in a locally convex space E (resp. F). For $u \in \mathscr{L}(\mathrm{E};\mathrm{F})$, $x \in \mathrm{G}$, $y \in \mathrm{H}$, set

$$V(x,y)\cdot u = U_2(y)\circ u \circ U_1(x).$$

Show that the mapping $(x,y) \mapsto V(x,y)$ is a continuous linear representation of the group $\mathrm{G}^0 \times \mathrm{H}$ in the space $\mathscr{L}(\mathrm{E};\mathrm{F})$ equipped with the topology of compact convergence. (Use Prop. 9 of TVS, III, §4, No. 4 and the fact that, for K compact in G, $U_1(\mathrm{K})$ is equicontinuous.)

¶ 3) Let G be a locally compact group, $U$ a continuous linear representation of G in a locally convex space E, and E′ the dual of E equipped with the strong topology.

*a*) Show that for every compact subset K of G, ${}^tU(\mathrm{K})$ is equicontinuous.

*b*) Let F be the set of $a' \in \mathrm{E}'$ such that the mapping $s \mapsto {}^tU(s)a'$ of G into E′ is continuous. Show that F is a closed linear subspace of E′ stable for ${}^tU(\mathrm{G})$ and that the representation deduced by restricting to F the contragredient representation of $U$ is continuous.

*c*) Assume that E is quasi-complete. Let $\alpha$ be a left Haar measure on G. Show that $f \mapsto U(f\cdot\alpha)$ is a continuous mapping of $\mathscr{K}(\mathrm{G})$ into $\mathscr{L}(\mathrm{E};\mathrm{E})$ equipped with the topology of bounded convergence. (Use Prop. 17 of Ch. VI, §1, No. 7.) Show that F is weakly dense in E′. (Prove that ${}^tU(f)a' \in \mathrm{F}$ for all $a' \in \mathrm{E}'$ and $f \in \mathscr{K}(\mathrm{G})$, then use Cor. 3 of Lemma 4.) Deduce from this that if E is semi-reflexive, then the contragredient representation of $U$ in E′ equipped with the strong topology is continuous.

*d*) Show that if $U$ is taken to be the left regular representation of G in $\mathrm{L}^1(\mathrm{G},\alpha)$ ($\alpha$ still being a left Haar measure on G), then F is the subspace of $\mathrm{E}' = \mathrm{L}^\infty(\mathrm{G},\alpha)$ formed by the uniformly continuous functions.

4) Let H be a Hilbert space. A continuous representation $U$ of G in H is said to be *unitary* if the endomorphisms $U(s)$ are unitary for all $s \in \mathrm{G}$. For every $\mu \in \mathscr{M}(\mathrm{G})$, let $\mu^*$ be the conjugate measure of $\check{\mu}$. Show that if $\mu \in \mathscr{M}^1(\mathrm{G})$ then $U(\mu^*) = U(\mu)^*$.

5) Let G be a locally compact group, H a closed subgroup of G, and $U$ a continuous linear representation of H in a locally convex space E. Let K be a compact subset of G. Let $\mathscr{K}^U(\mathrm{K})$ be the space of continuous functions on G, with values in E, with support contained in KH, and satisfying $f(xh) = U(h)^{-1}f(x)$ ($x \in \mathrm{G}$, $h \in \mathrm{H}$). Let $\mathscr{K}^U$ be the union of the $\mathscr{K}^U(\mathrm{K})$, equipped with the direct limit topology of the topologies of uniform convergence in K on each of the spaces $\mathscr{K}^U(\mathrm{K})$. For $f \in \mathscr{K}^U$ and $s \in \mathrm{G}$, define $V(s)f \in \mathscr{K}^U$ by

$$(V(s)f)(t) = f(s^{-1}t).$$

Show that $V$ is a continuous linear representation of G in $\mathscr{K}^U$.

6) Let G be a locally compact group, $\beta$ a relatively invariant, nonzero positive measure on G, $\chi$ and $\chi'$ its left and right multipliers. For $f \in \mathrm{L}^p_{\mathbf{C}}(\mathrm{G},\beta)$ and $s \in \mathrm{G}$, set

$$(U(s)f)(x) = \chi(s)^{-1/p}f(s^{-1}x)$$
$$(V(s)f)(x) = \chi'(s)^{-1/p}f(xs)$$
$$(Sf)(x) = (\chi\chi')(x)^{-1/p}\,\overline{f(x^{-1})}.$$

Then $U$ and $V$ are linear representations of G, and

$$S^2 = 1,\quad \|U(s)\| = \|V(s)\| = 1$$
$$U(s)V(t) = V(t)U(s),\quad SU(s)S = V(s)$$

for all $s, t$ in G.

7) Let E be a Hilbert space having an orthonormal basis $(e_s)_{s\in\mathbf{R}}$ equipotent to $\mathbf{R}$. For every $s \in \mathbf{R}$, denote by $U(s)$ the isometry of E such that $U(s)\cdot e_t = e_{s+t}$ for all $t \in \mathbf{R}$; the linear representation $s \mapsto U(s)$ of $\mathbf{R}$ in E is not continuous, even though the set of $U(s)$ is equicontinuous.

8) Let G be a locally compact abelian group, $\mu$ a Haar measure on G, $f$ a $\mu$-measurable finite numerical function on G. Assume that for every $s \in G$, the numerical function

$$x \mapsto f(sx) - f(x)$$

is continuous on G. Show that $f$ is then continuous. (Argue by contradiction; assuming that the function $f$ is not continuous at a point $x_0 \in G$, show first that there exists on G a filter $\mathfrak{F}$ with limit $e$ such that

$$\lim_{\mathfrak{F},s} |f(sx_0)| = +\infty,$$

and deduce from this that also $\lim_{\mathfrak{F},s} |f(sx)| = +\infty$ for *every* $x \in G$. If $g = |f|/(1+|f|)$, deduce from the last result a contradiction to the fact that for every compact subset $K \subset G$,

$$\lim_{\mathfrak{F},s} \int_K |g(sx) - g(x)|\, d\mu(x) = 0.)$$

9) Let G be a locally compact group, $\mu$ a left Haar measure on G, $f$ a $\mu$-integrable function. Let $\mathfrak{B}$ be a filter base formed of $\mu$-integrable sets of measure $> 0$, having $e$ as limit. For every $B \in \mathfrak{B}$, set $f_B(t) = \dfrac{1}{\mu(B)} \displaystyle\int_B f(st)\, d\mu(s)$. Show that for every integrable subset A of G, $\lim_{\mathfrak{B}} \int_A f_B(t)\, d\mu(t) = \int_A f(t)\, d\mu(t)$.

10) Let G be a locally compact group, E a Hausdorff locally convex space, E′ its dual, and $U$ a linear representation of G in E, continuous for the weakened topology $\sigma(E, E')$ on E. Assume that E is quasi-complete for $\sigma(E, E')$, so that $U(\mu)$ is defined for every measure $\mu \in \mathscr{C}'(G)$. Show that the bilinear mapping $(\mu, x) \mapsto U(\mu)\cdot x$ is hypocontinuous relative to the equicontinuous subsets of $\mathscr{C}'(G)$.

## §3

1) In $\mathbf{R}^2$, let $\lambda$ and $\mu$ be the positive measures

$$f \mapsto \int_0^{+\infty} f(x, 0)\, dx, \quad f \mapsto \int_0^{+\infty} f(-x, x)\, dx \qquad (f \in \mathscr{K}(\mathbf{R}^2)).$$

Show that $\lambda$ and $\mu$ are convolvable. Let $u$ be the homomorphism $(x, y) \mapsto x$ of $\mathbf{R}^2$ onto $\mathbf{R}$. Show that $u$ is $\lambda$-proper and $\mu$-proper, but that $u(\lambda)$ and $u(\mu)$ are not convolvable.

2) Let X be a locally compact space in which a locally compact group G operates on the left continuously. Let E be a linear subspace of $\mathscr{M}(X)$, stable for the $\gamma(s)$ ($s \in G$), equipped with a quasi-complete locally convex topology finer than the topology of compact convergence in $\mathscr{K}(X)$. For every $s \in G$, let $\gamma_E(s)$ be the restriction of $\gamma(s)$ to E. Assume that $\mu \in E$ implies $|\mu| \in E$, and that the representation $\gamma_E$ of G in E

is equicontinuous. Then, if $\mu \in \mathrm{E}$ and $\nu \in \mathscr{M}^1(\mathrm{G})$, $\nu$ and $\mu$ are convolvable and $\nu * \mu = \boldsymbol{\gamma}_{\mathrm{E}}(\nu)\mu \in \mathrm{E}$. (Make use notably of Prop. 17 of Ch. VI, §1, No. 7.)

3) For $x = (x_1, \dots, x_n) \in \mathbf{R}^n$ one sets $|x|^2 = x_1^2 + \dots + x_n^2$. Let $\mathscr{M}_1$ be the set of measures $\mu$ on $\mathbf{R}^n$ for which there exists a real number $k$ such that the function $(1 + |x|^2)^k$ is $\mu$-integrable. Let $\mathscr{M}_2$ be the set of measures $\nu$ on $\mathbf{R}^n$ for which the function $(1 + |x|^2)^k$ is $\nu$-integrable for every $k$. Show that if $\mu \in \mathscr{M}_1$ and $\nu \in \mathscr{M}_2$, then $\mu$ and $\nu$ are convolvable, and $\mu * \nu \in \mathscr{M}_1$; if $\mu \in \mathscr{M}_2$ and $\nu \in \mathscr{M}_2$, then $\mu * \nu \in \mathscr{M}_2$. (One first shows that if $u \geqslant 0$, $v \geqslant 0$, then

$$(1 + u^2)(1 + v^2) \geqslant \frac{1}{3}(1 + (u + v)^2);$$

from this, one deduces that for any $x, y$ in $\mathbf{R}^n$,

$$1 + |x|^2 \leqslant 3(1 + |y|^2)(1 + |x + y|^2).$$

Then let $\mu \in \mathscr{M}_1$, $\nu \in \mathscr{M}_2$ with $\mu \geqslant 0$, $\nu \geqslant 0$. Let $f$ be a continuous function $\geqslant 0$ on $\mathbf{R}^n$. There exists a $k$ such that $\mu = (1 + |x|^2)^k \cdot \mu_1$ with $\mu_1$ bounded; let $\nu_1 = (1 + |x|^2)^k \cdot \nu$, which is bounded. Then

$$\int^* f(x + y)\, d\mu(x)\, d\nu(y) \leqslant 3^k \int^* (1 + |x|^2)^k f(x)\, d(\mu_1 * \nu_1)(x).$$

Whence the convolvability of $\mu$ and $\nu$ and the fact that $\mu * \nu \in \mathscr{M}_1$. Argue in an analogous manner for $\mu, \nu$ in $\mathscr{M}_2$.)

4) Let $\mu$ be Lebesgue measure on $\mathbf{R}$, and $\nu$ Lebesgue measure on $[0, +\infty[$. Let $x_1, x_2$ be in $\mathbf{R}$. Show that the convolution product $((\varepsilon_{x_1} - \varepsilon_{x_2}) * \nu) * \mu$ is defined, but that $\mu$ and $\nu$ are not convolvable. Show that the convolution products $\nu * ((\varepsilon_{x_1} - \varepsilon_{x_2}) * \mu)$ and $(\nu * (\varepsilon_{x_1} - \varepsilon_{x_2})) * \mu$ are defined, but are distinct for $x_1 \neq x_2$.

5) Let G be a compact group, and $\mu$ a positive measure on G, with support G, such that $\mu * \mu = \mu$. Show that $\mu$ is the normalized Haar measure of G. (First show that $\|\mu\| = 1$. Then, if $\mu$ is not the Haar measure of G, there exists a function $f \in \mathscr{K}_+(\mathrm{G})$ such that

$$\int f(s)\, d\mu(s) \geqslant \int f(st)\, d\mu(s)$$

for all $t \in \mathrm{G}$, and such that $\int f(s)\, d\mu(s) > \int f(st)\, d\mu(s)$ for some $t$. Show that one then has $(\mu * \mu)(f) > \mu(f)$.)

6) *a*) Let $\mathrm{I} = [0, 1]$. Let $f$ be a continuous function $\geqslant 0$ on $\mathbf{R}$, with support contained in $[-1, 0]$. Show that the set of functions $\boldsymbol{\gamma}(s)f|\mathrm{I}$ $(s \in \mathrm{I})$ has infinite rank in $\mathscr{K}(\mathrm{I})$.

*b*) Let $f_1, \dots, f_n$ be in $\mathscr{K}(\mathbf{R})$. Let M be the set of measures $\mu \in \mathscr{M}(\mathrm{I})$ such that $\mu(f_1) = \dots = \mu(f_n) = 0$. Show that the set of functions $y \mapsto \int f(x + y)\, d\mu(x)$ $(y \in \mathrm{I})$, where $\mu$ runs over M, has infinite rank in $\mathscr{K}(\mathrm{I})$. (Make use of *a*).)

*c*) Let $g_1, \ldots, g_p$ be in $\mathscr{K}(\mathbf{R})$. Deduce from *b*) that there exist $\mu \in \mathrm{M}$ and $\nu \in \mathscr{M}(\mathrm{I})$ such that $\nu(g_1) = \ldots = \nu(g_p) = 0$, $(\nu * \mu)(f) \neq 0$.

*d*) Deduce from *c*) that the mapping $(\mu, \nu) \mapsto \nu * \mu$ of $\mathscr{M}(\mathrm{T}) \times \mathscr{M}(\mathrm{T})$ into $\mathscr{M}(\mathrm{T})$ is not vaguely continuous.

7) Let G be a locally compact group.

*a*) Show that if a positive measure $\mu \in \mathscr{C}'(\mathrm{G})$ admits an inverse in $\mathscr{C}'(\mathrm{G})$ that is a positive measure, then $\mu$ is necessarily a point measure.

*b*) Taking G to be the finite group $\mathbf{Z}/2\mathbf{Z}$, give an example of a non point measure $\mu \in \mathscr{M}(\mathrm{G})$ that is invertible in $\mathscr{M}(\mathrm{G})$.

8) Let G be a locally compact group; denote by $\mathscr{C}'_+(\mathrm{G})$ the set of positive measures on G with compact support. Show that the mapping $(\mu, \nu) \mapsto \mu * \nu$ of $\mathscr{M}_+(\mathrm{G}) \times \mathscr{C}'_+(\mathrm{G})$ into $\mathscr{M}_+(\mathrm{G})$ is continuous when $\mathscr{C}'(\mathrm{G})$ is equipped with the weak topology $\sigma(\mathscr{C}'(\mathrm{G}), \mathscr{C}(\mathrm{G}))$, and $\mathscr{M}(\mathrm{G})$ with the vague topology $\sigma(\mathscr{M}(\mathrm{G}), \mathscr{K}(\mathrm{G}))$. (Make use of Exer. 10 of Ch. III, §1 and the fact that G is paracompact.)

9) *a*) Let G be a locally compact group, B a bounded subset of $\mathscr{M}(\mathrm{G})$, C an equicontinuous subset of $\mathscr{C}'(\mathrm{G})$; show that if $\mathscr{C}'(\mathrm{G})$ is equipped with the weak topology $\sigma(\mathscr{C}'(\mathrm{G}), \mathscr{C}(\mathrm{G}))$, and $\mathscr{M}(\mathrm{G})$ with the vague topology $\sigma(\mathscr{M}(\mathrm{G}), \mathscr{K}(\mathrm{G}))$, then the mapping $(\mu, \nu) \mapsto \mu * \nu$ of $\mathrm{B} \times \mathrm{C}$ into $\mathscr{M}(\mathrm{G})$ is continuous. (Observe that all of the measures $\nu \in \mathrm{C}$ have their support in a common compact set, and make use of Ch. III, §4, No. 3, remarks following Prop. 6.)

*b*) If, in $\mathscr{M}^1(\mathbf{R})$, one sets $\mu_n = \varepsilon_n$, $\nu_n = \varepsilon_{-n}$, the sequences $(\mu_n)$ and $(\nu_n)$ tend vaguely to 0 for the weak topology $\sigma(\mathscr{M}^1(\mathbf{R}), \overline{\mathscr{K}(\mathbf{R})})$, but the sequence $(\mu_n * \nu_n)$ does not tend to 0 for the vague topology.

¶ 10) For a locally compact space T, one denotes by $\mathscr{C}^\infty(\mathrm{T})$ the Banach space of continuous bounded numerical functions on T. A subset H of $\mathscr{M}^1(\mathrm{T})$ is said to be *cramped* if, for every $\varepsilon > 0$, there exists a compact subset K of T such that $|\mu|(\mathrm{T} - \mathrm{K}) \leqslant \varepsilon$ for all $\mu \in \mathrm{H}$.

*a*) Show that if a subset H of $\mathscr{M}^1(\mathrm{G})$ is cramped and is bounded for the topology defined by the norm of $\mathscr{M}^1(\mathrm{T})$, then it is relatively compact for the topology $\sigma(\mathscr{M}^1(\mathrm{T}), \mathscr{C}^\infty(\mathrm{T}))$ (observe that H is relatively compact for $\sigma(\mathscr{M}^1(\mathrm{T}), \mathscr{K}(\mathrm{T}))$).

*b*) Assuming in addition that T is *paracompact*, show that, conversely, if H is a subset of $\mathscr{M}^1(\mathrm{T})$ relatively compact for $\sigma(\mathscr{M}^1(\mathrm{T}), \mathscr{C}^\infty(\mathrm{T}))$, then H is bounded for the norm of $\mathscr{M}^1(\mathrm{T})$ and is cramped. (Consider first the case that $\mathrm{T} = \mathbf{N}$, applying Exer. 15 of Ch. V, §5. Then contemplate the case that T is countable at infinity, the union of a sequence $(\mathrm{U}_n)$ of relatively compact open sets such that $\overline{\mathrm{U}}_n \subset \mathrm{U}_{n+1}$. Arguing by contradiction, show that one can reduce to the case that there would exist for each $n$ a continuous numerical function $f_n$ defined on T, with support contained in $\mathrm{U}_{n+1} - \overline{\mathrm{U}}_n$, such that $\|f_n\| \leqslant 1$, and a sequence $(\mu_n)$ of measures belonging to H and for which $\mu_n(f_n) \geqslant a > 0$ for all $n$. Consider then the continuous mapping $u : \mathrm{L}^1(\mathbf{N}) \to \mathscr{C}^\infty(\mathrm{T})$ such that $u((\xi_n)) = \sum_{n=0}^{\infty} \xi_n f_n$ and obtain a contradiction to what was proved for $\mathrm{T} = \mathbf{N}$, by considering the transpose of $u$. Finally, when T is an arbitrary paracompact locally compact space, T is the topological sum of a family $(\mathrm{T}_\alpha)$ of locally compact spaces that are countable at infinity; for every $\alpha$, let $m_\alpha = \sup_{\mu \in \mathrm{H}} |\mu|(\mathrm{T}_\alpha)$; show, arguing by contradiction and making use of the preceding case, that necessarily $m_\alpha = 0$ except for a countable infinity of indices $\alpha$.)

*c*) Show that the conclusion of *b*) is not valid for the non-paracompact locally compact space defined in Exer. 16 *h*) of Ch. IV, §4.

¶ 11) Let G be a locally compact group; on $\mathscr{M}^1(G)$, let us denote by $\mathscr{T}_I$ the topology $\sigma(\mathscr{M}^1(G), \overline{\mathscr{K}(G)})$, by $\mathscr{T}_{III}$ the topology $\sigma(\mathscr{M}^1(G), \mathscr{C}^\infty(G))$, and let us write $\mathscr{M}_I$, $\mathscr{M}_{III}$ for the space $\mathscr{M}^1(G)$ equipped respectively with $\mathscr{T}_I$ and $\mathscr{T}_{III}$.

*a*) Let A be a bounded subset of $\mathscr{M}_I$, and B a relatively compact subset of $\mathscr{M}_{III}$. Show that the restriction to $A \times B$ of the mapping $(\mu, \nu) \mapsto \mu * \nu$ of $\mathscr{M}_I \times \mathscr{M}_{III}$ into $\mathscr{M}_I$ is continuous (use Exer. 10 to reduce to evaluating an integral $\iint f(st)\,d\mu(s)\,d\nu(t)$ when $f(st) = \sum_i u_i(s) v_i(t)$ with $u_i$ and $v_i$ in $\mathscr{K}(G)$).

*b*) Give an example where G is compact (hence $\mathscr{T}_I = \mathscr{T}_{III}$) showing that $(\mu, \nu) \mapsto \mu * \nu$, regarded as a mapping of $\mathscr{M}_I \times \mathscr{M}_{III}$ into $\mathscr{M}_I$, is not hypocontinuous with respect to the relatively compact subsets of $\mathscr{M}_I$, nor with respect to the relatively compact subsets of $\mathscr{M}_{III}$ (cf. Exer. 6).

*c*) Let A, B be two relatively compact subsets of $\mathscr{M}_{III}$. Show that the restriction to $A \times B$ of the mapping $(\mu, \nu) \mapsto \mu * \nu$ of $\mathscr{M}_{III} \times \mathscr{M}_{III}$ into $\mathscr{M}_{III}$ is continuous (same method as in *a*)).

*d*) Denote by E the subspace of $\mathbf{R}^G$ formed by the linear combinations of characteristic functions of open subsets of G, by $\mathscr{T}_{IV}$ the topology $\sigma(\mathscr{M}^1(G), E)$, and by $\mathscr{M}_{IV}$ the space $\mathscr{M}^1(G)$ equipped with $\mathscr{T}_{IV}$. Recall that the topologies induced by $\mathscr{T}_{III}$ and $\mathscr{T}_{IV}$ on a bounded subset of $\mathscr{M}^1(G)$ consisting of positive measures, are in general distinct (Ch. V, §5, Exer. 16 *c*)). Recall also that the compact subsets of $\mathscr{M}_{IV}$ are the same as those of $\mathscr{M}^1(G)$ for the weakened topology $\sigma(\mathscr{M}^1(G), (\mathscr{M}^1(G))')$ on the Banach space $\mathscr{M}^1(G)$ (Ch. VI, §2, Exer. 12). Show that if A, B are two relatively compact subsets of $\mathscr{M}_{IV}$, the restriction to $A \times B$ of the mapping $(\mu, \nu) \mapsto \mu * \nu$ of $\mathscr{M}_{IV} \times \mathscr{M}_{IV}$ into $\mathscr{M}_{IV}$ is continuous. (Restrict attention to the case that A and B are compact, and begin by proving that the image of $A \times B$ under the preceding mapping is then compact for $\sigma(\mathscr{M}^1(G), (\mathscr{M}^1(G))')$; for this, apply the theorems of Eberlein and Šmulian (TVS, IV, §5, No. 3), and thus reduce to proving that if $(\mu_n)$, $(\nu_n)$ are two sequences that converge to 0 in $\mathscr{M}_{IV}$, then the same is true of the sequence $(\mu_n * \nu_n)$; make use of Prop. 12 and Exer. 15 of Ch. V, §5. Finally, to prove that $(\mu, \nu) \mapsto \mu * \nu$ is continuous for $\mathscr{T}_{IV}$, use *c*) and the fact that $\mathscr{T}_{III}$ is coarser than $\mathscr{T}_{IV}$.)

*e*) Take $G = \mathbf{R}^2$; let $a, b$ be the vectors of the canonical basis of G over $\mathbf{R}$, $\rho_n$ the measure on the interval $I = [0, \pi]$ of $\mathbf{R}$ having as density with respect to Lebesgue measure the function $\sin(2^n x)$, and $\mu_n$ the measure $\rho_n \otimes \varepsilon_0$ on $\mathbf{R} \times \mathbf{R}$; on the other hand let $\nu_n = \varepsilon_{b/2^n} - \varepsilon_0$ on G. Show that the sequence $(\mu_n)$ tends to 0 in $\mathscr{M}_{IV}$ and that the sequence $(\nu_n)$ tends to 0 in $\mathscr{M}_{III}$, but that the sequence $(\mu_n * \nu_n)$ does not tend to 0 in $\mathscr{M}_{IV}$ (cf. Ch. V, §5, Exer. 16).

*f*) Give an example where G is compact and where $(\mu, \nu) \mapsto \mu * \nu$, regarded as a mapping of $\mathscr{M}_{IV} \times \mathscr{M}_{IV}$ into $\mathscr{M}_{IV}$, is not hypocontinuous relative to the compact subsets of $\mathscr{M}_{IV}$ (same method as in Exer. 6, on observing that for a given $f \in E$ there exist compact subsets $H \subset \mathscr{M}_{IV}$ such that the set of functions $t \mapsto \int f(st)\,d\mu(s)$, where $\mu$ runs over H, has finite rank over $\mathbf{R}$).

12) Let G be a locally compact group that is not unimodular.

*a*) Show that there exists a bounded positive measure $\mu$ on G such that $\Delta_G \cdot \mu$ is not bounded (take $\mu$ to be discrete).

*b*) Let $\mu'$ be a left Haar measure on G. Then $\mu$ and $\mu'$ are convolvable (Prop. 5). Show that $\mu'$ and $\mu$ are not convolvable.

13) Let $r$ be a number such that $0 < r < 1$; for every integer $n \geqslant 1$, denote by $\lambda_{n,r}$ the measure $(\varepsilon_{r^n} + \varepsilon_{-r^n})/2$ on $\mathbf{R}$, and set $\mu_{n,r} = \lambda_{1,r} * \lambda_{2,r} * \cdots * \lambda_{n,r}$.

*a*) Show that the sequence $(\mu_{n,r})$ converges vaguely to a measure $\mu_r$ on $\mathbf{R}$, with support contained in $I = [-1, +1]$ (prove that for every interval U of $\mathbf{R}$, the sequence $(\mu_{n,r}(U))$ is convergent).

*b*) Show that for $r < 1/2$, the measure $\mu_r$ is *alien* to the Lebesgue measure on $\mathbf{R}$, but that $\mu_{1/2}$ is the measure induced on I by Lebesgue measure.

*c*) Let $\nu_{1/4}$ be the image of $\mu_{1/4}$ under the homothety $t \mapsto 2t$ in $\mathbf{R}$. Show that $\mu_{1/4} * \nu_{1/4} = \mu_{1/2}$, even though $\mu_{1/4}$ and $\nu_{1/4}$ are both alien to Lebesgue measure (use Exer. 11 *a*)).

## §4

¶ 1) Let G be a locally compact group, $\beta$ a left Haar measure on G, A a $\beta$-measurable subset of G, and $\nu$ a nonzero positive measure on G. Assume that $s\mathrm{A}$ is locally $\nu$-negligible for every $s \in \mathrm{G}$. Show that A is locally $\beta$-negligible. (Reduce to the case that A is relatively compact and $\nu$ is bounded. Prove that $\nu$ and $\varphi_\mathrm{A} \cdot \beta$ are convolvable and that $\nu *^\beta \varphi_\mathrm{A} = 0$, whence $0 = \|\nu * \varphi_\mathrm{A} \cdot \beta\| = \|\nu\| \cdot \|\varphi_\mathrm{A} \cdot \beta\|$.) Show, on admitting the continuum hypothesis, that this result may fail to hold if A is not assumed to be $\beta$-measurable. (Take $\mathrm{G} = \mathbf{R}^2$, take for $\nu$ the Haar measure on the subgroup $\mathbf{R} \times \{0\}$, and apply Exer. 7 *c*) of Ch. V, §8.)

2) Let H be the additive group $\mathbf{R}$ equipped with the discrete topology. Let G be the locally compact group $\mathbf{R} \times \mathbf{R} \times \mathrm{H}$. Let $\alpha$ be a Haar measure on G, $\beta$ a Haar measure on $\mathbf{R} \times \mathrm{H}$, and $\mu = \varepsilon_0 \otimes \beta \in \mathscr{M}(\mathrm{G})$. Construct a function $f \geqslant 0$ on G, locally $\alpha$-negligible (hence such that $\mu$ and $f$ are convolvable) but such that no translate of $f$ is $\mu$-measurable. (Imitate the construction of Ch. V, §3, Exer. 4.)

¶ 3) Let G be a locally compact group.

*a*) Let $\mu$ be a nonzero bounded positive measure on G such that $\mu * \mu = \mu$. Show that the support S of $\mu$ is compact. (Let $f \in \mathscr{K}_+(\mathrm{G})$ with $f \neq 0$; choose a left Haar measure on G, with respect to which convolution products will be taken; we have $\mu * f \in \overline{\mathscr{K}(\mathrm{G})}$; let $x \in \mathrm{S}$ be such that $(\mu * f)(x) = \sup_{y \in \mathrm{S}} (\mu * f)(y)$; show that $(\mu * f)(y) = (\mu * f)(x)$ for every $y \in \mathrm{S}$.)

*b*) Show that S is a compact subgroup of G and that $\mu$ is the normalized Haar measure of S. (Make use of *a*), Exer. 21 of GT, III, §4, and Exer. 5 of §3.)

¶ 4) Let G be a locally compact group. For every $t \in \mathbf{R}_+^*$, let $\mu_t$ be a nonzero bounded positive measure on G. Assume that the mapping $t \mapsto \mu_t$ is continuous for the topology $\sigma(\mathscr{M}^1(\mathrm{G}), \overline{\mathscr{K}(\mathrm{G})})$, and that $\mu_{s+t} = \mu_s * \mu_t$ ($s, t$ in $\mathbf{R}_+^*$).

*a*) Show that there exists a number $c \in \mathbf{R}$ such that $\|\mu_t\| = \exp(ct)$. (Observe that $t \mapsto \|\mu_t\|$ is lower semi-continuous and that $\|\mu_{s+t}\| = \|\mu_s\| \cdot \|\mu_t\|$. Make use of Prop. 18.)

*b*) Suppose that $c = 0$. Show that $\mu_t$ converges weakly as $t \to 0$ to the normalized Haar measure of a compact subgroup of G. (Make use of the weak compactness of the unit ball of $\mathscr{M}^1(\mathrm{G})$, and show that for every weak cluster point $\mu$ of $t \mapsto \mu_t$ with respect to the filter of neighborhoods of 0 in $\mathbf{R}_+^*$, one has $\mu_t * \mu = \mu_t$ for every $t$, then $\mu * \mu = \mu$; next, apply Exer. 3.)

¶ 5) Let G be a locally compact group countable at infinity, $\beta$ a left Haar measure on G, $a \in \mathbf{R}_+^*$, and $(\nu_u)_{0 \leqslant u \leqslant a}$ a family of positive measures on G satisfying the following conditions:

(i) $\nu_0 = \varepsilon_e$; $\nu_{u+v} = \nu_u * \nu_v$ if $u + v \leqslant a$; $\nu_u = \check{\nu}_u$;

(ii) for $0 < u \leqslant a$, $\nu_u = f_u \cdot \beta$ with $f_u \geqslant 0$ lower semi-continuous;

(iii) $\nu_u$ is a vaguely continuous function of $u$.

*a*) Let $f \in \mathscr{K}_+(\mathrm{G})$. Show that, for $0 < u \leqslant a$,

$$\int (f_{u/2} * f)(z)^2 \, d\beta(z) = \int f(x)(f_u * f)(x) \, d\beta(x). \tag{1}$$

*b*) Let $f \in \mathscr{K}(\mathrm{G})$. Show that $f_{u/2} * f$ has $\beta$-integrable square for $0 < u \leqslant a$, and that (1) again holds.

*c*) Let $f \in \mathscr{K}(\mathrm{G})$ be such that $\nu_a * f = 0$. Show that $f = 0$. (One has $\nu_{a/2^n} * f = 0$ for every integer $n > 0$ by *b*), therefore $f = 0$ by §3, No. 3, *Remark* 1).)

*d*) Let $\nu \in \mathscr{C}'(\mathrm{G})$ be such that $\nu_a * \nu = 0$. Show that $\nu = 0$. (Regularize $\nu$ by functions in $\mathscr{K}(\mathrm{G})$ and apply *c*).)

6) Let G be a locally compact group. Show that if $\mathscr{K}(\mathrm{G})$ is commutative for convolution, then G is abelian. (Show by regularization that $\mathscr{C}'(\mathrm{G})$ is commutative, and apply this to the measures $\varepsilon_s$, where $s \in \mathrm{G}$.)

7) Let G be a locally compact group and $\beta$ a left Haar measure on G. Show that the algebra $\mathrm{L}^1(\mathrm{G}, \beta)$ has a unity element if and only if G is discrete. (Suppose G non-discrete, and let $f_0 \in \mathrm{L}^1(\mathrm{G}, \beta)$. There exists a compact neighborhood V of $e$ such that

$$\int_{\mathrm{V}} |f_0(x)|\, d\beta(x) < 1 .$$

Let U be a symmetric compact neighborhood of $e$ such that $\mathrm{U}^2 \subset \mathrm{V}$. Then, for almost every $x \in \mathrm{U}$,

$$|(\varphi_{\mathrm{U}} * f_0)(x)| = \int_{\mathrm{U}} |f_0(y^{-1}x)|\, d\beta(y) \leqslant \int_{\mathrm{V}} |f_0(x)|\, d\beta(x) < 1 ,$$

thus $f_0$ is not a unity element for $\mathrm{L}^1(\mathrm{G}, \beta)$.)

¶ 8) Let G be a locally compact group countable at infinity, operating continuously on the left in a Polish locally compact space T. Let $\nu$ be a positive measure on T that is quasi-invariant under G. Let R be a $\nu$-measurable equivalence relation on T, compatible with G. There then exists (Ch. VI, §3, No. 4, Prop. 2) a Polish locally compact space B, and a $\nu$-measurable mapping $p$ of T into B, such that $\mathrm{R}\{x, y\}$ is equivalent to $p(x)) = p(y)$. Let $\nu'$ be a pseudo-image measure of $\nu$ under $p$, and let $b \mapsto \lambda_b$ $(b \in \mathrm{B})$ be a disintegration of $\nu$ by R. Show that the $\lambda_b$ are, for almost every $b \in \mathrm{B}$, quasi-invariant under G.

(Let $\chi$ be a function on $\mathrm{G} \times \mathrm{T}$ satisfying the conditions of Exer. 13 of Ch. VII, §1. Show that for every $s \in \mathrm{G}$, there exists a $\nu'$-negligible subset $\mathrm{N}(s)$ of B such that $\chi(s^{-1}, \cdot)$ is locally $\lambda_b$-integrable for $b \notin \mathrm{N}(s)$; for this, observe that for every $\psi \in \mathscr{K}(\mathrm{T})$, the function $x \mapsto \psi(x)\chi(s^{-1}, x)$ is $\lambda_b$-integrable except for $b$ belonging to a $\nu'$-negligible set $\mathrm{N}(s, \psi)$, and make use of Lemma 1 of Ch. VI, §3, No. 1. Set $\lambda'_{b,s} = 0$ if $b \in \mathrm{N}(s)$, and $\lambda'_{b,s} = \chi(s^{-1}, \cdot) \cdot \lambda_b$ if $b \notin \mathrm{N}(s)$. Show that the mapping $b \mapsto \lambda'_{b,s}$ is $\nu'$-adequate (use Lemma 3 of Ch. VI, §3, No. 1) and that $\gamma(s)\beta = \int \lambda'_{b,s}\, d\nu'(b)$. Show that on the other hand $\gamma(s)\beta = \int \gamma(s)\lambda_b\, d\nu'(b)$ and deduce from this that, for every $s \in \mathrm{G}$, one has $\gamma(s)\lambda_b = \chi(s^{-1}, \cdot) \cdot \lambda_b$ for almost every $b$, therefore for almost every $b$ one has $\gamma(s)\lambda_b = \chi(s^{-1}, \cdot) \cdot \lambda_b$ for almost every $s$. Then use Cor. 2 of Prop. 17 of §4 to infer that, for almost every $b$, $\gamma(s)\lambda_b$ is equivalent to $\lambda_b$ for all $s \in \mathrm{G}$.)

Show that if $\nu$ is relatively invariant under G with multiplier $\chi$, then the $\lambda_b$ are, for almost every $b$, relatively invariant with multiplier $\chi$.

9) Let G be a locally compact group, $\beta$ a left Haar measure on G, A and B two $\beta$-integrable sets such that $\beta(\mathrm{A}) \leqslant \beta(\mathrm{B})$ and $\beta^*(\mathrm{A}^{-1}) < +\infty$. Show that there exist

1) disjoint sets N, $\mathrm{K}_1$, $\mathrm{K}_2$, ... covering A, with N $\beta$-negligible and the $\mathrm{K}_n$ compact;

2) disjoint sets $N'$, $K'_1$, $K'_2$, ... covering $B$, with the $K'_n$ compact;
3) elements $s_n \in G$ such that $K'_n = s_n K_n$.
(Using the fact that $\beta(xA \cap B)$ depends continuously on $x$ and that

$$\int \beta(xA \cap B)\, d\beta(x) = \beta(A^{-1})\beta(B),$$

show that if $\beta(A) \neq 0$ then there exists an $x \in G$ such that $\beta(xA \cap B) \neq 0$.)

10) Let $G$ be a locally compact group, $\beta$ a left Haar measure on $G$. For any two $\beta$-integrable sets $A, B$ set $\rho(A, B) = \beta(A \cup B) - \beta(A \cap B)$.

*a*) Let $A$ be a $\beta$-integrable set. Show that $x \mapsto \rho(xA, A)$ is a continuous function.

*b*) Let $U$ be a neighborhood of $e$. Show that there exist a compact set $A$ and a number $\varepsilon > 0$ such that $\rho(xA, A) < \varepsilon$ implies $x \in U$. (Take for $A$ a neighborhood of $e$ such that $A \cdot A^{-1} \subset U$.)

*c*) For a subset $C$ of $G$ to be relatively compact, it is necessary and sufficient that there exist a $\beta$-integrable set $A$ and a number $a$ $(0 < a < 2\beta(A))$ such that $x \in C$ implies $\rho(xA, A) \leqslant a$.

11) *a*) Let $G$ be a locally compact group generated by a compact neighborhood of $e$. Let $\varphi$ be a non-surjective continuous endomorphism of $G$ belonging to the closure, for the topology of compact convergence, of the group $\mathscr{G}$ of (bicontinuous) automorphisms of $G$. Then $\lim_{\psi \in \mathscr{G},\, \psi \to \varphi} \operatorname{mod} \psi = 0$. (Let $K$ be a compact neighborhood of $e$ that generates $G$. Let $\mu$ be a left Haar measure on $G$. If $\mu(\varphi(K)) > 0$, then $\varphi(K) \cdot \varphi(K)^{-1}$ is a neighborhood of $e$ in $G$, therefore $\varphi(G)$ is an open subgroup of $G$; for $\psi \in \mathscr{G}$ sufficiently near $\varphi$, one has $\psi(K) \subset \varphi(G)$, therefore $\psi(G) \neq G$, which is absurd. Thus $\mu(\varphi(K)) = 0$. As $\psi \in \mathscr{G}$ tends to $\varphi$, $\mu(\psi(K))$ tends to $0$.)

*b*) Let $G$ be a free abelian group, a direct sum $G_1 \oplus G_2 \oplus \cdots$, where each $G_i$ is isomorphic to $\mathbf{Z}$. Consider $G$ as being discrete. Let $\varphi$ be the non-surjective endomorphism $(x_1, x_2, \ldots) \mapsto (0, x_1, x_2, \ldots)$ of $G$. Then $\varphi$ is the limit, for the topology of compact convergence, of automorphisms of $G$, and every automorphism of $G$ has modulus $1$.

12) For $t > 0$ and $x \in \mathbf{R}$, let $F_t(x) = te^{-\pi t^2 x^2}$. Let $f \in \overline{\mathscr{K}(\mathbf{R})}$. Show that $f * F_t$ tends to $f$ uniformly as $t$ tends to $+\infty$. (Show that the measures $F_t(x)\,dx$ satisfy the conditions of §2, No. 7, Lemma 4, with $a = 0$.)

¶ 13) Let $G$ be a locally compact group operating continuously on the left in a Polish locally compact space $T$. Let $\beta$ be a left Haar measure on $G$, $\nu$ a quasi-invariant positive measure on $T$, and $\chi(s, x)$ a function $> 0$ on $G \times T$ satisfying the conditions of Exer. 13 of Ch. VII, §1.

For $f \in L^p(T, \nu)$ $(1 \leqslant p < +\infty)$, set

$$(\gamma_{\chi,p}(s)f)(x) = \chi(s^{-1}, x)^{1/p} f(s^{-1}x).$$

*a*) Show that for every $s \in G$, $\gamma_{\chi,p}(s)$ is an isometric endomorphism of $L^p(T, \nu)$. Show that the mapping $s \mapsto \gamma_{\chi,p}(s)$ is a linear representation of $G$ in $L^p(T, \nu)$ (argue as in §2, No. 5).

*b*) Let $f \in \mathscr{L}^p(T, \nu)$, and let $h \in \mathscr{K}(G)$. Show that the function

$$(x, s) \mapsto f(s^{-1}x) h(s) \chi(s^{-1}, x)^{1/p}$$

is $p$-th power integrable for $\beta \otimes \nu$ (begin with the case $p = 1$, and use Lemma 1 of §4, No. 1). Show that if $q$ is the exponent conjugate to $p$, and if $g \in \mathscr{L}^q(T, \nu)$, then

$f(s^{-1}x)h(s)g(x)\chi(s^{-1},x)^{1/p}$ is integrable for $\beta\otimes\nu$ (write $h$ in the form $h_1h_2$ with $h_1,h_2$ in $\mathscr{K}(\mathrm{G})$ ). Then deduce from the Lebesgue–Fubini theorem that, for $f\in\mathrm{L}^p(\mathrm{T},\nu)$ and $g\in\mathrm{L}^q(\mathrm{T},\nu)$, the function

$$s\mapsto\int g(x)(\gamma_{\chi,p}(s)f)(x)\,d\nu(x)$$

is $\beta$-measurable.

*c*) Show, using Cor. 2 of Prop. 18 of §4, No. 6 and Lemma 1 of Ch. VI, §3, No. 1, that the representation $s\mapsto\gamma_{\chi,p}(s)$ of G in $\mathrm{L}^p(\mathrm{T},\nu)$ is *continuous* for $1\leqslant p<+\infty$.

*d*) Assume in addition that, for every $s\in\mathrm{G}$, the function $x\mapsto\chi(s^{-1},x)$ is bounded. For $f\in\mathrm{L}^p(\mathrm{T},\nu)$, set

$$(\gamma_\chi(s)f)(x)=\chi(s^{-1},x)f(s^{-1}x)\,.$$

Show that $s\mapsto\gamma_\chi(s)$ is a continuous representation of G in $\mathrm{L}^p(\mathrm{T},\nu)$ for $1\leqslant p<+\infty$. (One shows, as in the proof of Prop. 9 of §2, No. 5, that $s\mapsto\gamma_\chi(s)$ is a representation of G by endomorphisms of $\mathrm{L}^p(\mathrm{T},\nu)$. One then observes that if $f\in\mathrm{L}^p(\mathrm{T},\nu)$ and $h\in\mathscr{K}(\mathrm{G})$, Lemma 1 of §4, No. 1 shows that $h(s)f(s^{-1}x)\chi(s^{-1},x)$ is $p$-th power integrable for $\beta\otimes\nu$. The proof is concluded as in *c*).)

14) *a*) Let G be a locally compact group, $f$ a lower semi-continuous positive function on G, $\mu$ a positive measure on G. Show that the function

$$x\mapsto\int^* f(s^{-1}x)\,d\mu(s)=g(x)$$

is lower semi-continuous on G (cf. Ch. IV, §1, No. 1, Th. 1); for $\mu$ and $f$ to be convolvable, it is necessary and sufficient that $g$ be integrable for a left Haar measure on G.

*b*) On the group $\mathrm{G}=\mathbf{R}\times\mathbf{R}$, let $\mu$ be the measure $\varepsilon_0\otimes\lambda$, where $\lambda$ is Lebesgue measure on $\mathbf{R}$; let $f(x,y)=(1-|xy-2|)^+$; show that the function

$$g(x,y)=\int f(x-s,y-t)\,d\mu(s,t)$$

is everywhere finite on G, but is not continuous and is not integrable for Lebesgue measure on G.

¶ 15) Let G be a locally compact group, $\beta$ a left Haar measure on G; by an abuse of language, in what follows one identifies $\beta$-integrable numerical functions with their classes in $\mathrm{L}^1(\mathrm{G},\beta)$; same abuse for the $\mathrm{L}^p(\mathrm{G},\beta)$.

*a*) Let $A$ be a continuous endomorphism of the Banach space $\mathrm{L}^1(\mathrm{G},\beta)$ such that, for every $s\in\mathrm{G}$, one has $A(f*\varepsilon_s)=A(f)*\varepsilon_s$ for all $f\in\mathrm{L}^1(\mathrm{G},\beta)$. Show that for every function $g\in\mathscr{K}(\mathrm{G})$, one then has $A(f*g)=A(f)*g$ (observe that $s\mapsto g(s)A(f*\varepsilon_s)$ is a $\beta$-integrable mapping of G into $\mathrm{L}^1(\mathrm{G},\beta)$ ); converse. From this, deduce that there exists one and only one bounded measure $\mu$ on G such that $A(f)=\mu*f$ for all $f\in\mathrm{L}^1(\mathrm{G},\beta)$, and that $\|A\|=\|\mu\|$. (With the notations of Prop. 19 of No. 7, consider the limit of $A(f_V*g)$ with respect to an ultrafilter finer than the section filter of $\mathfrak{B}$, making use of the compactness of the unit ball of $\mathscr{M}^1(\mathrm{G})$ for the weak topology $\sigma(\mathscr{M}^1(\mathrm{G}),\overline{\mathscr{K}(\mathrm{G})})$.)

*b*) Let $\mu$ be a bounded measure on G; show directly that the norm of the continuous endomorphism $\gamma(\mu):f\mapsto\mu*f$ of $\mathrm{L}^\infty(\mathrm{G},\beta)$ is equal to $\|\mu\|$. (Reduce to the case that $\mu$ has compact support and has a continuous density with respect to $|\mu|$.) From this, deduce anew that the continuous endomorphism $\gamma(\mu):f\mapsto\mu*f$ of $\mathrm{L}^1(\mathrm{G},\beta)$ has norm equal to $\|\mu\|$.

c) Assume that G is *compact* and $\mu$ is *positive*. Show that for $1 < p < +\infty$, the norm of the continuous endomorphism $\boldsymbol{\gamma}(\mu) : f \mapsto \mu * f$ of $L^p(G, \beta)$ is equal to $\|\mu\|$.

d) Take for G the cyclic group of order 3. Give an example of a measure $\mu$ on G such that the norm of the endomorphism $\boldsymbol{\gamma}(\mu)$ of $L^p(G, \beta)$ is strictly less than $\|\mu\|$ for $1 < p < +\infty$.

¶ 16) Notations and conventions are those of Exer. 15.

a) Show that, for a bounded measure $\mu$ on G to be such that $\|\mu * f\|_1 = \|f\|_1$ for all $f \in L^1(G, \beta)$, it is necessary and sufficient that $\mu$ be a point measure of norm 1. (Using the fact that the endomorphism $\boldsymbol{\gamma}(|\mu|)$ of $L^1(G, \beta)$ has norm $\|\mu\|$ (Exer. 15), show that one necessarily has, for every function $f \in \mathscr{K}(G)$,

$$\left| \int f \, d\mu \right| = \int |f| \, d|\mu| ;$$

from this, first deduce that $\mu = c|\mu|$, where $c$ is a constant of absolute value 1, then that $\mu$ is a point measure.)

b) Take for G the cyclic group of order 3. Give an example of a measure $\mu$ on G, not a point measure, such that $\|\mu * f\|_2 = \|f\|_2$ for every numerical function $f$ defined on G.

17) Notations and conventions are those of Exer. 15.

a) Let $\mu$ be a bounded measure on G; for the endomorphism $\boldsymbol{\gamma}(\mu) : f \mapsto \mu * f$ of $L^p(G, \beta)$ to be surjective ($1 \leqslant p < +\infty$), it is necessary and sufficient that there exist a $c_p > 0$ such that $\|\check{\mu} * g\|_q \geqslant c_p \|g\|_q$ for all $g \in L^q(G, \beta)$ (where $\dfrac{1}{p} + \dfrac{1}{q} = 1$) (cf. TVS, IV, §4, No. 2, remarks following Cor. 3 of Th. 1).

b) If $\mu$ is a measure with base $\beta$, and G is not discrete, show that $\boldsymbol{\gamma}(\mu)$ is never surjective for $1 \leqslant p \leqslant +\infty$ (for $p < +\infty$, use a), reducing to the case that the density of $\mu$ with respect to $\beta$ belongs to $\mathscr{K}(G)$; for $p = +\infty$, argue directly by observing that for $f \in L^1(G, \beta)$ and $g \in L^\infty(G, \beta)$, $f * g$ is uniformly continuous for the right uniform structure).

c) If G is abelian, $1 \leqslant p \leqslant 2$ and the endomorphism $\boldsymbol{\gamma}(\mu)$ of $L^p(G, \beta)$ is surjective, show that it is bijective (using regularization, show that if $\boldsymbol{\gamma}(\mu)$ is not injective then the endomorphism $\boldsymbol{\gamma}(\check{\mu})$ of $L^q(G, \beta)$ is not injective: one makes use of Ch. IV, §6, No. 5, Cor. of Prop. 4).

d) Take $G = \mathbf{Z}$ and $\mu = \varepsilon_1 - \varepsilon_0$; show that $\boldsymbol{\gamma}(\mu)$ is injective in the $L^p(\mathbf{Z}, \beta)$ for which $p \neq +\infty$, but not in $L^\infty(\mathbf{Z}, \beta)$, and it is not surjective for any value of $p$.

¶ 18) Notations and conventions are those of Exer. 15.

a) Let $\mu$ be a bounded measure on G, whose support contains at least two distinct points. Show that there exists a compact set K, and a number $k$ such that $0 < k < 1$, for which $\|\varphi_{sK} \cdot \mu\| \leqslant k\|\mu\|$ for all $s \in G$. (If $t, t'$ are two distinct points in the support of $\mu$, take for K a neighborhood of $e$ sufficiently small that $sK$ cannot intersect both $tK$ and $t'K$.)

b) Let $\mu$ be a bounded measure $\geqslant 0$ on G, whose support contains at least two distinct points. Show that there exists a function $f \in L^\infty(G, \beta)$, not equivalent to a function $\geqslant 0$, such that $\mu * f$ is $\geqslant 0$ locally almost everywhere for $\beta$. (Use a), taking $f$ to be equal to $-1$ on K and to a suitable positive constant on $\complement K$.) If, in addition, the support of $\mu$ is compact, there exists a function $f$ having the preceding properties and having compact support.

c) Let $\mu_1, \mu_2$ be two nonzero, bounded positive measures on G, permutable for convolution, such that the support K of $\mu_1$ is compact and contains $e$. Set $\mu = \mu_1 + \mu_2$. Let V be a symmetric compact neighborhood of $e$ containing K, $g$ the function equal to $\mu * \varphi_V$ on $\complement V$, and to 0 on V; show that the function $\mu * (g - (\mu_1 * \varphi_V))$ is $\geqslant 0$ locally almost everywhere in $\complement(V^2)$. On the other hand, show that there exists a compact set H such that the function $\mu_2 * \varphi_H$ is $\geqslant 0$ almost everywhere in $V^2$.

*d*) Deduce from *c*) that if $\mu$ is a bounded positive measure on G whose support contains at least two distinct points, there exists a function $f$ that belongs to every $L^p(G,\beta)$ $(1 \leqslant p \leqslant +\infty)$, is not equivalent to a function $\geqslant 0$, and is such that $\mu * f$ is $\geqslant 0$ locally almost everywhere, when one makes in addition one of the following hypotheses: $\alpha$) G is abelian; $\beta$) there is a point $a \in G$ such that $\mu(\{a\}) > 0$. (Suitably decompose $\mu$ as a sum of two permutable measures $\geqslant 0$.)

¶ 19) Notations and conventions are those of Exer. 15.

*a*) Show that if a positive bounded measure $\mu$ on G is such that for every function $f \in L^p(G,\beta)$ ($p$ given, $1 \leqslant p < \infty$) the relation $\mu * f \geqslant 0$ locally almost everywhere implies $f \geqslant 0$ locally almost everywhere, then one can conclude that $\mu$ is a point measure in each of the following cases: $\alpha$) G is compact; $\beta$) G is discrete; $\gamma$) G is abelian (use Exer. 18) (*). For $p = +\infty$, the same conclusion is valid without a supplementary hypothesis on G.

*b*) Let $\mu$ be a positive bounded measure on G such that: 1° $\gamma(\mu)$ is a surjective endomorphism of $L^1(G,\beta)$; 2° for every function $f \in L^1(G,\beta)$, the relation $\mu * f \geqslant 0$ locally almost everywhere implies $f \geqslant 0$ locally almost everywhere. Show that $\mu$ is then a point measure. (Observe that for every function $g \in L^\infty(G,\beta)$, the relation $\check{\mu} * g \geqslant 0$ locally almost everywhere implies $g \geqslant 0$ locally almost everywhere.)

20) Let G be a locally compact group, $\beta$ a left Haar measure on G.

*a*) Let $f$ be a bounded numerical function on G, uniformly continuous for the left uniform structure. Show that for every bounded measure $\mu$ on G, $\mu * f$ is uniformly continuous for the left uniform structure; moreover, with the notations of No. 7, Prop. 19, $f$ is the limit, for the topology of uniform convergence in G, of the functions $f_V * f$ with respect to the section filter of $\mathfrak{B}$.

*b*) Take $G = \mathbf{T}$; give an example of a continuous function $h$ on $\mathbf{T}$ that is not of the form $f * g$, where $f$ and $g$ belong to $L^2(\mathbf{T},\beta)$. (Use Exers. 15 *b*) and 16 of Ch. IV, §6.)

¶ 21) Let G be a locally compact group, $\beta$ a left Haar measure on G; one canonically identifies $L^1(G,\beta)$ with a subspace of $\mathscr{M}^1(G)$.

*a*) With the notations of §3, Exer. 11, let A be a relatively compact subset of $\mathscr{M}_{\mathrm{III}}$, B a subset of $L^1(G,\beta)$, relatively compact in $\mathscr{M}_{\mathrm{IV}}$. Show that the restriction to $A \times B$ of the mapping $(\mu,\nu) \mapsto \mu * \nu$ of $\mathscr{M}_{\mathrm{III}} \times \mathscr{M}_{\mathrm{IV}}$ into $\mathscr{M}_{\mathrm{IV}}$ is continuous. (First prove that if A and B are compact, the image of $A \times B$ under this mapping is compact; use Exer. 10 of §3, as well as the criterion $\alpha$) of Ch. V, §5, Exer. 15. To show next that $(\mu,\nu) \mapsto \mu * \nu$ is continuous, make use of Exer. 11 *c*) of §3.)

*b*) Let $\mathscr{U}_s^\infty(G)$ be the set of bounded numerical functions on G uniformly continuous for the left uniform structure. Denote by $\mathscr{T}_{\mathrm{II}}$ the topology $\sigma(\mathscr{M}^1(G), \mathscr{U}_s^\infty(G))$, and by $\mathscr{M}_{\mathrm{II}}$ the space $\mathscr{M}^1(G)$ equipped with $\mathscr{T}_{\mathrm{II}}$. Taking $G = \mathbf{R}$, give an example of a sequence $(\mu_n)$ of measures on G tending to 0 for $\mathscr{T}_{\mathrm{II}}$ and a sequence $(f_n)$ of functions in $L^1(G,\beta)$, tending to 0 for $\mathscr{T}_{\mathrm{IV}}$ (or, what comes to the same, for the topology $\sigma(L^1(G,\beta), L^\infty(G,\beta))$), such that the sequence $(\mu_n * f_n)$ does not tend to 0 for $\mathscr{T}_{\mathrm{IV}}$ (take $f_n(t) = \sin nt$ in the interval $[0,\pi]$, $f_n(t) = 0$ elsewhere).

*c*) Let A be a relatively compact subset of $\mathscr{M}_{\mathrm{II}}$, B a subset of $L^1(G,\beta)$, relatively compact for the topology of the norm $\|f\|_1$. Show that the restriction to $A \times B$ of the mapping $(\mu, f) \mapsto \mu * f$ of $\mathscr{M}_{\mathrm{III}} \times L^1(G,\beta)$ into $L^1(G,\beta)$ is continuous ($L^1(G,\beta)$ being equipped with its normed space topology). (Reduce to proving that for $f \in \mathscr{K}(G)$, the mapping $\mu \mapsto \mu * f$ of $\mathscr{M}_{\mathrm{II}}$ into $L^1(G,\beta)$ is continuous in A. Using the fact that for $g \in L^\infty(G,\beta)$ and $f \in \mathscr{K}(G)$, $g * f$ is uniformly continuous for the left uniform structure, first show that the mapping $\mu \mapsto \mu * f$ of $\mathscr{M}_{\mathrm{II}}$ into $\mathscr{M}_{\mathrm{IV}}$ is continuous. Next, using Exer. 20 *a*), reduce to proving that if C is a subset of $L^1(G,\beta)$ compact for the topology

(*) For an example where $\mu$ is not a point measure and is such that $\mu * f \geqslant 0$ implies $f \geqslant 0$, see J.H. Williamson, *Proc. Edinburgh Math. Soc.*, **11** (1958/59), 71–77.

$\sigma(L^1(G,\beta), L^\infty(G,\beta))$ and $h \in \mathscr{K}(G)$, the mapping $\mu \mapsto \mu * h$ of C into $L^1(G,\beta)$ is continuous when $L^1(G,\beta)$ is equipped with its normed space topology. By the same reasoning as in $a$), show that for this it suffices to prove that the image of C under this mapping is compact for the normed space topology. To this end, use Šmulian's theorem, the fact that C is cramped (§3, Exer. 10) and Lebesgue's theorem.)

*d*) Let A be a relatively compact subset of $\mathscr{M}_{\mathrm{II}}$ containing 0, B a relatively compact subset of $\mathscr{M}_{\mathrm{I}}$; show that for every $\nu_0 \in$ B, the restriction to A × B of the mapping $(\mu,\nu) \mapsto \mu * \nu$ of $\mathscr{M}_{\mathrm{II}} \times \mathscr{M}_{\mathrm{I}}$ into $\mathscr{M}_{\mathrm{II}}$ is continuous at the point $(0,\nu_0)$. (Make use of Exer. 20 $a$), and Exer. 21 $c$).)

*e*) Let A be a bounded subset of $\mathscr{M}^1(G)$, $f$ a function in $L^1(G,\beta)$; show that the restriction to A of the mapping $\mu \mapsto \mu * f$ of $\mathscr{M}_{\mathrm{III}}$ into $\mathscr{M}_{\mathrm{IV}}$ is continuous.

*f*) Let $\mathscr{M}^1_+(G)$ be the set of bounded positive measures on G. Show that if a subset A of $\mathscr{M}^1_+(G)$ is compact for $\mathscr{T}_{\mathrm{II}}$, it is also compact for $\mathscr{T}_{\mathrm{III}}$. (Reduce to proving that A is a cramped set. Argue by contradiction, using the fact that for $f \in \mathscr{K}(G)$, the image of A under the mapping $\mu \mapsto \mu * f$ is a cramped set by virtue of $c$).)

*g*) Show that for $G = \mathbf{R}$, the topologies induced on $\mathscr{M}^1_+(G)$ by $\mathscr{T}_{\mathrm{II}}$ and $\mathscr{T}_{\mathrm{III}}$ are distinct.

¶ 22) Notations are those of Exer. 21 of §4 and Exer. 11 of §3.

*a*) Let $(\mu_n)$ be a sequence of bounded measures on G. Assume that for every function $f \in L^1(G,\beta)$, the sequence $(\mu_n * f)$ converges to 0 in $\mathscr{M}_{\mathrm{I}}$. Show that the sequence of norms $(\|\mu_n\|)$ is bounded (use the Banach–Steinhaus theorem for the family of mappings $f \mapsto \mu_n * f$ of $L^1(G,\beta)$ into itself, as well as Exer. 15). Deduce from this that the sequence $(\mu_n)$ tends to 0 in $\mathscr{M}_{\mathrm{I}}$ (use Exer. 20).

*b*) Let $(\mu_n)$ be a sequence of bounded measures on G such that the sequence $(\mu_n * f)$ tends vaguely to 0 for every function $f \in L^1(G,\beta)$; show that the sequence $(\mu_n)$ tends vaguely to 0 (for every compact subset K of G, show, arguing as in $a$), that the sequence $(|\mu_n|(K))$ is bounded). Give an example (with $G = \mathbf{Z}$) where the sequence $(\|\mu_n\|)$ is not bounded.

*c*) Let $(\mu_n)$ be a sequence of bounded measures on G such that for every function $f \in L^1(G,\beta)$, the sequence $(\mu_n * f)$ converges to 0 in $\mathscr{M}_{\mathrm{II}}$; show that the sequence $(\mu_n)$ converges to 0 in $\mathscr{M}_{\mathrm{II}}$ (make use of $a$)).

23) Let G be a locally compact group, $\beta$ a left Haar measure on G, $f$ and $g$ two *positive* locally $\beta$-integrable functions. Assume that $f$ and $g$ are convolvable and that one of these functions is zero on the complement of a countable union of compact sets. Show that $f * g$ is equal locally almost everywhere to a lower semi-continuous function (use Prop. 15 of No. 5).

24) Show that the inequalities of Props. 12 and 15 of No. 5 cannot be improved by inserting a constant factor $c < 1$ in their right members.

25) Let G be a locally compact group, $\beta$ a left Haar measure on G, $\mu$ a positive measure on G. If A is a $\mu$-integrable set and B is a Borel set in G, the function

$$u : s \mapsto \mu(A \cap sB)$$

is $\beta$-measurable in G; if, in addition, $B^{-1}$ is $\beta$-integrable, then so is $u$, and

$$\int \mu(A \cap sB)\, d\beta(s) = \mu(A)\beta(B^{-1}).$$

Give an example of a measure $\mu$ such that $s \mapsto \mu(A \cap sB)$ is not continuous; if $\mu$ is a measure with base $\beta$, and if A is $\mu$-integrable and B is a Borel set, then the function $s \mapsto \mu(A \cap sB)$ is continuous on G.

¶ 26) Let G be a locally compact group, $\beta$ a left Haar measure on G. For a subset H of $L^p(G,\beta)$ $(1 \leqslant p < +\infty)$ to be relatively compact (for the topology of convergence in mean of order $p$), it is necessary and sufficient that the following conditions be satisfied: 1° H is bounded in $L^p(G,\beta)$; 2° for every $\varepsilon > 0$, there exists a compact subset K of G such that $\|f\varphi_{G-K}\|_p \leqslant \varepsilon$ for every function $f \in H$; 3° for every $\varepsilon > 0$, there exists a neighborhood V of $e$ in G such that $\|\boldsymbol{\gamma}(s)f - f\|_p \leqslant \varepsilon$ for all $f \in H$ and $s \in V$. (To prove that the conditions are sufficient, observe that if $g \in \mathscr{K}(G)$ and if L is a compact subset of G, then the image, under the mapping $f \mapsto g * f$, of the set $\varphi_L \cdot H$, is an equicontinuous subset of $\mathscr{K}(G)$.)

¶ 27) Let G be a locally compact group, $\beta$ a left Haar measure on G. If G is not reduced to $e$, then $L^1(G,\beta)$ is an algebra (for convolution) admitting divisors of zero other than 0. One may proceed as follows to form two nonzero elements $f, g$ of $L^1(G,\beta)$ such that $f * g = 0$:

1° The case that G contains a compact subgroup H, not reduced to $e$, in which $\Delta(x) = 1$. Take $f$ to be the characteristic function of a set A, and $g$ to be a difference $\varphi_{sB} - \varphi_B$ of two characteristic functions, with A and B suitably chosen.

2° The case that $G = \mathbf{Z}$. Show that one can then take

$$f(n) = \frac{1}{2n-1} - \frac{1}{2n+1}$$

for all $n \in \mathbf{Z}$, and $g(n) = f(-n)$.

3° The general case. First prove that there exists an $a \neq e$ in G such that $\Delta(a) = 1$. The closure H in G of the subgroup generated by $a$ is then either a compact subgroup or a subgroup isomorphic to $\mathbf{Z}$ (GT, V, §1, Exer. 2). In the first case, use the result of 1°; in the second, take

$$f(t) = \sum_{n=-\infty}^{+\infty} \alpha_n \varphi_{Ua^{-n}}(t), \quad g(t) = \sum_{n=-\infty}^{+\infty} \beta_n \varphi_{a^n U}(t),$$

suitably choosing U and the sequences $(\alpha_n)$, $(\beta_n)$ with the help of 2°.

¶ 28) Let $G_1, G_2$ be two locally compact groups, $\beta_1, \beta_2$ left Haar measures on $G_1, G_2$, and $A_1$ (resp. $A_2$) the topological algebra (over $\mathbf{R}$) $L^1(G_1,\beta_1)$ (resp. $L^1(G_2,\beta_2)$). Let T be an algebra isomorphism of $A_1$ onto $A_2$, such that the relation $f \geqslant 0$ almost everywhere is equivalent to $T(f) \geqslant 0$ almost everywhere.

*a*) Show that T is an isomorphism of topological algebras. (First note that if a decreasing sequence $(f_n)$ of elements of $A_1$ tends to 0 almost everywhere, then the same is true of the sequence of the $T(f_n)$; deduce from this that for every sequence $(f_n)$ that is bounded above in $A_1$, $T(\sup(f_n)) = \sup(T(f_n))$, and conclude with the help of Fatou's lemma.)

*b*) Show that T may be extended, in only one way, to an isomorphism of the topological algebra $\mathscr{M}^1(G_1)$ onto the topological algebra $\mathscr{M}^1(G_2)$, and that there exist a topological isomorphism $u$ of $G_1$ onto $G_2$ and a continuous homomorphism $\chi$ of $G_1$ into $\mathbf{R}_+^*$ such that $T(\varepsilon_s) = \chi(s)\varepsilon_{u(s)}$ (make use of Exer. 15 *a*) and Exer. 19 *b*); to prove that $u$ and $\chi$ are continuous, observe that T defines an isomorphism of the algebra $\mathscr{L}(A_1)$ of continuous endomorphisms of the topological vector space $A_1$ onto the analogous algebra $\mathscr{L}(A_2)$, and that this isomorphism is bicontinuous for the topology of pointwise convergence; on the other hand, observe that $s \mapsto \boldsymbol{\delta}(s^{-1})$ is an isomorphism of $G_1$ onto a multiplicative subgroup of $\mathscr{L}(A_1)$).

*c*) Deduce from *b*) that, for every $f \in A_1$,

$$(T(f))(t) = \chi(u^{-1}(t))f(u^{-1}(t))$$

for all $t \in G_2$.

Recall (A, III, §2, Exer. 9) that there exist finite groups $G_1, G_2$ such that the algebras $L^1(G_1, \beta_1)$ and $L^1(G_2, \beta_2)$ are isomorphic without $G_1$ and $G_2$ being so.

## §5

1) Let X be a locally compact space, $\mathfrak{F}$ (or $\mathfrak{F}(X)$) the set of closed subsets of X. For every compact subset K of X, every compact neighborhood L of K, and every entourage V of the unique uniform structure of L, let $Q(K, L, V)$ be the set of pairs $(A, B)$ of elements of $\mathfrak{F}$ satisfying the two conditions

$$A \cap K \subset V(B \cap L) \quad \text{and} \quad B \cap K \subset V(A \cap L).$$

*a*) Show that the sets $Q(K, L, V)$ form a fundamental system of entourages for a Hausdorff uniform structure on $\mathfrak{F}$.

*b*) Let $\mathscr{U}$ be a uniform structure compatible with the topology of X. For every compact subset K of X and every entourage W of $\mathscr{U}$, let $P(K, W)$ be the set of pairs $(A, B)$ of $\mathfrak{F}$ satisfying the two conditions

$$A \cap K \subset W(B) \quad \text{and} \quad B \cap K \subset W(A).$$

Show that the sets $P(K, W)$ form a fundamental system of entourages for the uniform structure defined in *a*).

*c*) Let X′ be the Alexandroff compactification of X, $\omega$ the point at infinity of X′; for every closed subset A of X, let $A' = A \cup \{\omega\}$. Show that the mapping $A \mapsto A'$ is an isomorphism of the uniform space $\mathfrak{F}(X)$ defined in *a*) onto the subspace of $\mathfrak{F}(X')$ formed by the closed subsets containing $\omega$ (make use of *b*)). Deduce from this that $\mathfrak{F}(X)$ is compact (cf. GT, II, §4, Exer. 15).

2) Let G be a locally compact group, $\mathfrak{F}(G)$ the uniform space of closed subsets of G (defined in No. 6 or in Exer. 1). Prove directly that the set $\Sigma$ of closed subgroups of G is closed in $\mathfrak{F}(G)$ (without using Prop. 3 of No. 1).

3) Let G be a locally compact group, $\chi$ a continuous representation of G in $\mathbf{R}_+^*$. Generalize Props. 4, 5 and 6 to the set $\Gamma_\chi$ of measures $\alpha \in \Gamma$ such that the restriction of $\chi$ to $H_\alpha$ is the modulus of $\alpha$. (Replace the measure $\mu$ by $\chi \cdot \mu$ and consider the quotient measures $(\chi \cdot \mu)/\check{\alpha}$.)

4) Let G be a locally compact group that is not generated by any compact subset of G, and let H be a discrete subgroup of G such that G/H is compact. Show that H cannot be generated by a finite number of elements, but the normalized Haar measure $\alpha_0$ of H belongs to the closure in $\Gamma^0$ of the set of $\alpha$ such that $\|\mu_\alpha\| = +\infty$ (consider the normalized Haar measures of the subgroups of H that are generated by a finite number of elements).

5) Let G be the discrete group that is the direct sum of an infinite sequence $(G_n)$ of subgroups with two elements, and let $p_n$ be the projection of G onto $G_n$. Set $H_n = \overset{-1}{p_n}(e)$, and let $\alpha_n$ be the normalized Haar measure of $H_n$. If $\alpha$ is the normalized Haar measure of G, show that in $\Gamma_c^0$ the sequence $(\alpha_n)$ converges to $\alpha$ but that $\|\mu_{\alpha_n}\| = 2$ and $\|\mu_\alpha\| = 1$.

6) Let G be a locally compact group that does not satisfy the condition (L) of No. 4.

*a*) Show (with the notations of No. 4) that the canonical bijection of N onto D is not a homeomorphism. (For every neighborhood W of $e$ in G, let A(W) be the set of finite subgroups $\neq \{e\}$ contained in W; show that the A(W) form the base of a filter that converges in D to the subgroup $\{e\}$, but that if $\alpha_H$ denotes the normalized Haar measure of H, the measures $\alpha_H$ do not converge to $\varepsilon_e$.)

*b*) If in addition G is metrizable, show that there exists a compact subset B of the space D of discrete subgroups of G, that is not contained in any of the sets $D_U$ defined in Th. 1 of No. 3.

# HISTORICAL NOTE

## (Chapters VII and VIII)

(N.B. — The Roman numerals refer to the bibliography at the end of this note.)

The concepts of length, area and volume are, with the Greeks, essentially based on their *invariance* under displacements: « Things that coincide ($\varepsilon\varphi\alpha\rho\mu\acute{o}\xi o\nu\tau\alpha$) are equal » (*Eucl. El.*, Book I, 'Common notion' 4); and it is by an ingenious use of this principle that all of the formulas giving the areas or volumes of the classical 'figures' (polygons, conic sections, polyhedra, spheres, etc.) are obtained, sometimes by methods of finite decomposition, sometimes by 'exhaustion' (*). In modern language, one can say that what the Greek geometers did was to prove the existence of 'set functions', additive and invariant under displacements, but defined only for sets of a very special type. The integral calculus may be regarded as responding to the need for enlarging the domain of definition of these set functions, and, from Cavalieri to H. Lebesgue, it is this preoccupation that was to be at the forefront of the research of analysts; as for the property of invariance under displacements, it passed to a secondary status, having become a trivial consequence of the general formula for change of variables in double or triple integrals and the fact that an orthogonal transformation has determinant equal to $\pm 1$. Even in non-euclidean geometries (though the group of displacements is different there), the point of view remains the same: in a general way, Riemann de-

(*) It can be shown that if two plane polygons $P, P'$ have the same area, there are two polygons $R \supset P$, $R' \supset P'$ each of which can be decomposed into a finite number of polygons $R_i$ (resp. $R'_i$) $(1 \leqslant i \leqslant m)$ without common interior point, such that $R_i$ and $R'_i$ can be deduced from each other by means of a displacement (depending on $i$) and such that $R$ (resp. $R'$) is the union of a finite family of polygons $S_j$ (resp. $S'_j$) $(0 \leqslant j \leqslant n)$, without common interior point, with $S_0 = P$, $S'_0 = P'$, and $S'_j$ obtainable from $S_j$ by a displacement for $1 \leqslant j \leqslant n$. However, M. DEHN proved (Ueber den Rauminhalt, *Math. Ann.*, **55** (1902), 465–478) that this property is no longer valid for the volume of polyhedra, and that the exhaustion methods employed from EUDOXUS onward were therefore unavoidable.

fined the infinitesimal elements of area or volume (or their analogues for dimensions $\geqslant 3$) beginning with a $ds^2$ by the classical euclidean formulas, and their invariance under the transformations that leave the $ds^2$ invariant is therefore almost a tautology.

It is only around 1890 that there appeared other, less immediate extensions of the concept of measure invariant under a group, with the development of the theory of *integral invariants*, notably by H. Poincaré and E. Cartan; H. Poincaré considered only one-parameter groups operating in a portion of space, whereas E. Cartan was above all interested in groups of displacements, but operating in spaces other than the one where they are defined. For example, he thus determined among other things (II) the invariant (under the group of displacements) measure on the space of lines of $\mathbf{R}^2$ or of $\mathbf{R}^3$ (*); moreover, he noted that in a general way the integral invariants for a Lie group are none other than particular differential invariants and that it is therefore possible to determine them all by the methods of Lie. However, it does not seem that anyone had thought of considering nor of using an invariant measure on the group itself, prior to the fundamental work of A. Hurwitz in 1897 (V). Seeking to form polynomials (on $\mathbf{R}^n$) invariant under the orthogonal group, Hurwitz starts from the remark that for a finite group of linear transformations, the problem is immediately solved by taking the *average* of the transforms $s \cdot \mathrm{P}$ of any polynomial P by all of the elements $s$ of the group—which gave him the idea, for the orthogonal group, of replacing the average by an integral with respect to an invariant measure; he gave explicitly the expression of the latter with the help of the parametric representation by means of the Euler angles, but immediately observed (independently of E. Cartan) that the methods of Lie yielded the existence of an invariant measure for every Lie group. Perhaps due to the decline of invariant theory at the beginning of the 20th century, Hurwitz's ideas received scarcely any immediate echo, and were not exploited until 1924 onward, with the extension to compact groups, by I. Schur and H. Weyl, of the classical theory of Frobenius on the linear representations of finite groups. The former restricted himself to the case of the orthogonal group, and showed how Hurwitz's method permitted extending the classical orthogonality relations of the characters—an idea that H. Weyl combined with the work of E. Cartan on semi-simple Lie algebras, to obtain explicit expressions for the characters of the irreducible representations of compact Lie groups and the theorem on complete reducibility (XI *a*)), then, by a bold extension of the concept of 'regular representation', the celebrated Peter–Weyl theorem, a perfect analogue of the decomposition of the regular representation into its

(*) The invariant measure on the space of lines of the plane had already been essentially determined in connection with problems in 'geometric probability', notably by CROFTON, whose works were probably not known to E. CARTAN at the time.

irreducible components in the theory of finite groups (XI, *b*)).

A year earlier, O. Schreier had founded the general theory of topological groups, and from then on it was clear that the arguments in the Peter–Weyl memoir would remain valid unchanged for every topological group on which an 'invariant measure' could be defined. Actually, the general concepts of topology and measure were at the time still in rapid development, and neither the category of topological groups on which one could hope to define an invariant measure, nor the sets for which this 'measure' was to be defined, seemed to be clearly delineated. The only obvious point was that one could not hope to extend to the general case the infinitesimal methods proving the existence of an invariant measure on a Lie group. Now, another current of ideas, growing out of work on Lebesgue measure, led precisely to more direct methods of attack. Hausdorff had proved, in 1914, that there does not exist an additive set function, not identically zero, that is defined for *all* subsets of $\mathbf{R}^3$ and is invariant under displacements, and it was natural to investigate whether this result was also valid for $\mathbf{R}$ and $\mathbf{R}^2$: a problem that was solved by S. Banach in 1923 in a surprising way, by showing that, on the contrary, such a 'measure' did indeed exist (I); his method, highly ingenious, already rested on a construction by transfinite induction and on the consideration of the 'means' $\frac{1}{n}\sum_{k=1}^{n} f(x + \alpha_k)$ of the translates of a function by elements of the group (*). It was analogous ideas that enabled A. Haar, in 1933 (IV), to take the decisive step, by proving the existence of an invariant measure for locally compact groups with a countable base for open sets: guided by the method of approximating a volume, in classical integral Calculus, by a juxtaposition of arbitrarily small congruent cubes, he obtained, with the aid of the diagonal method, the invariant measure as a limit of a sequence of 'approximate measures', a procedure which is essentially the one we have used in Ch. VII, §1. This discovery had a very great impact, in particular because it immediately allowed J. von Neumann to solve, for compact groups, the famous "5th problem" of Hilbert on the characterization of Lie groups by purely topological properties (excluding all differential structure given in advance). However, it was immediately perceived that to make efficient use of the invariant measure, it was necessary to know not only its existence, but to also know that it was unique up to a constant factor; this point was first proved by J. von Neumann for compact groups, using a method of defining Haar measure via 'means' of continuous functions, analogous to those of Banach (VII *a*)); then J. von Neumann

(*) J. von NEUMANN showed, in 1929, that the underlying reason for the difference in behavior between $\mathbf{R}$ and $\mathbf{R}^2$ on the one hand, and the $\mathbf{R}^n$ for $n \geqslant 3$ on the other, was to be found in the commutativity of the group of rotations of the space $\mathbf{R}^2$.

(VII *b*)) and A. Weil (X), by different methods, simultaneously obtained uniqueness for the case of locally compact groups, with A. Weil indicating at the same time how Haar's method could be extended to general locally compact groups. It was also A. Weil (*loc. cit.*) who obtained the condition for the existence of a relatively invariant measure on a homogeneous space, and showed, finally, that the existence of a 'measure' (endowed with reasonable properties) on a Hausdorff topological group, implied *ipso facto* that the group is locally precompact. This work essentially completed the general theory of Haar measure; the only recent addition to be cited is the concept of quasi-invariant measure, which was scarcely identified before around 1950, in connection with the theory of representations of locally compact groups in Hilbert spaces.

The history of the convolution product is more complex. From the beginning of the 19th century, it was observed that if, for example, $\mathrm{F}(x,t)$ is a solution of a partial differential equation in $x$ and $t$, linear and with constant coefficients, then

$$\int_{-\infty}^{+\infty} \mathrm{F}(x-s,t)f(s)\,ds$$

is also a solution of the same equation; since before 1820, Poisson, among others, had used this idea to write the solutions of the heat equation in the form

$$\text{(1)} \qquad \int_{-\infty}^{+\infty} \exp\Big(-\frac{(x-s)^2}{4t}\Big)f(s)\,ds.$$

A little later, the expression

$$\text{(2)} \qquad \frac{1}{2\pi}\int_{-\pi}^{+\pi} \frac{\sin\frac{2n+1}{2}(x-t)}{\sin\frac{x-t}{2}}f(t)\,dt$$

for the partial sum of a Fourier series, and the study, by Dirichlet, of the limit of this integral as $n$ tends to $+\infty$, provided the first example of a 'regularization' $f \mapsto \rho_n * f$ on the torus $\mathbf{T}$ (actually, by a sequence of non-positive 'kernels', which greatly complicates the study); under the name of 'singular integrals', the analogous integral expressions were a subject of choice among analysts at the end of the 19th century and the beginning of the 20th, from P. du Bois-Reymond to H. Lebesgue. On $\mathbf{R}$, Weierstrass made use of the integral (1) in the proof of his theorem on approximation by polynomials, and gave in this connection the general principle of regularization by a sequence of positive 'kernels' $\rho_n$ of the form $x \mapsto c_n\rho(x/n)$. On $\mathbf{T}$, the most famous example of regularization by positive kernels was given a little later by Fejér, and from this moment on, it is the standard procedure that was to be the basis of most of the 'summation methods' for series of functions.

However, these works, due to the dissymmetry of the roles played by the 'kernel' and the function regularized, scarcely revealed the algebraic properties of the convolution product. We are indebted above all to Volterra for having placed the emphasis on this point. He made a general study of the 'composition' $\mathrm{F} * \mathrm{G}$ of two functions of two variables

$$(\mathrm{F} * \mathrm{G})(x, y) = \int_x^y \mathrm{F}(x, t)\mathrm{G}(t, y)\, dt\,,$$

which he viewed as a generalization, 'by passage from finite to infinite', of the product of two matrices (IX). Very early he singled out the case (called 'of closed cycle' because of its interpretation in the theory of heredity) where $\mathrm{F}$ and $\mathrm{G}$ depend only on $y - x$; the same is then true of $\mathrm{H} = \mathrm{F} * \mathrm{G}$, and if one sets $\mathrm{F}(x, y) = f(y - x)$, $\mathrm{G}(x, y) = g(y - x)$, then

$$\mathrm{H}(x, y) = h(y - x)\,,$$

where

$$h(t) = \int_0^t f(t - s)g(s)\, ds\,,$$

so that, for $t \geqslant 0$, $h$ coincides with the convolution of the functions $f_1, g_1$ equal, respectively, to $f$ and $g$ when $t \geqslant 0$, and to $0$ when $t < 0$.

Nevertheless, the algebraic formalism developed by Volterra did not reveal the connections with the group structure of $\mathbf{R}$ and the Fourier transformation. This is not the place to relate the history of the latter; but it is appropriate to note that from Cauchy on, the analysts who treated the Fourier integral devoted themselves above all to finding ever wider conditions for the validity of various 'inversion' formulas, and somewhat neglected its algebraic properties. One could certainly not say the same regarding this of the works of Fourier himself (or of those of Laplace on the analogous integral $\int_0^{+\infty} e^{-st} f(t)\, dt$); but these transformations had been introduced essentially in connection with *linear* problems, and it is therefore not very surprising that it was a long time before anyone thought of considering the product of two Fourier transforms (with exception made for products of trigonometric series or of power series, but the connection with the convolution of discrete measures obviously could not have been perceived in the 19th century). The first mention of this product and of convolution over $\mathbf{R}$ is probably to be found in a memoir of Tchebychef (VIII), in connection with questions in probability theory. In fact, in this theory the convolution $\mu * \nu$

of two 'laws of probability' on **R** (positive measures of total mass 1) is none other than the 'composed' probability law of $\mu$ and $\nu$ (for the addition of the corresponding 'random variables'). To be sure, for Tchebychef it is still only a question of the convolution of probability laws having a density (with respect to Lebesgue measure), hence of the convolution of functions; moreover, it only comes up in his work in an episodic way, and it was to be so in the several rare works in which it appeared before the period 1920–1930. In 1920, P. J. Daniell, in a note (III) little noticed at the time, defined the convolution of two arbitrary measures on **R** and the Fourier transform of such a measure, and observed explicitly that the Fourier transform carried convolution over to an ordinary product—a formalism that, from 1925 on, was to be used intensively by probabilists, especially following P. Lévy. But the fundamental importance of convolution in the theory of groups was only fully recognized by H. Weyl in 1927; he noticed that for a compact group, the convolution of functions plays the role of multiplication in the algebra of a finite group, allowing him to subsequently define the 'regular representation'; at the same time, he found in regularization the equivalent of the unity element of the algebra of a finite group. It remained to make the synthesis of all of these points of view, accomplished in the book of A. Weil (X), preparing the way for the later generalizations which were to constitute, on the one hand I. Gelfand's theory of normed algebras, and on the other the convolution of distributions.

Haar measure and convolution have rapidly become essential tools in the tendency towards algebraization that so strongly marks modern Analysis; we shall have occasion to develop numerous applications of them in later Books. The only one that we have treated in these chapters concerns the 'variation' of the closed subgroups (and notably of the discrete subgroups) of a locally compact group. This theory, starting from a result of K. Mahler in the Geometry of numbers, was inaugurated in 1950 by C. Chabauty, and has just been considerably developed and deepened by Macbeath and Swierczkowski (VII), whose principal results we have reproduced here.

# BIBLIOGRAPHY


(I) S. BANACH, Sur le problème de la mesure, *Fund. Math.*, **4** (1923), pp. 7–33.

(II) E. CARTAN, Le principe de dualité et certaines intégrales multiples de l'espace tangentiel et de l'espace réglé, *Bull. Soc. Math. France,* **24** (1896), pp. 140–177 ( = *Œuvres complètes*, v. $\mathrm{II}_1$, pp. 265–302).

(III) P. J. DANIELL, Stieltjes–Volterra products, *Congr. Intern. des Math.*, Strasbourg, 1920, pp. 130–136.

(IV) A. HAAR, Der Massbegriff in der Theorie der kontinuierlichen Gruppen, *Ann. of Math.*, (2), **34** (1933), pp. 147–169 ( = *Gesammelte Arbeiten*, pp. 600–622).

(V) A. HURWITZ, Ueber die Erzeugung der Invarianten durch Integration, *Gött. Nachr.*, 1897, pp. 71–90 ( = *Math. Werke*, v. II, pp. 546–564).

(VI) A. M. MACBEATH, S. SWIERCZKOWSKI, Limits of lattices in a compactly generated group, *Canad. J. Math.*, **12** (1960), pp. 427–437.

(VII) J. von NEUMANN, *a*) Zum Haarschen Mass in topologischen Gruppen, *Comp. Math.*, **1** (1934), pp. 106–114 ( = *Collected Works*, v. II, n° 22); *b*) The uniqueness of Haar's measure, *Mat. Sbornik*, **1 (43)** (1936), pp. 721–734 ( = *Collected Works*, v. IV, n° 6).

(VIII) P. TCHEBYCHEF, Sur deux théorèmes relatifs aux probabilités, *Acta. Math.*, **14** (1890), pp. 305–315) ( = *Œuvres*, v. II, pp. 481–491).

(IX) V. VOLTERRA, *Leçons sur les fonctions de lignes*, Paris (Gauthier-Villars), 1913.

(X) A. WEIL, L'intégration dans les groupes topologiques et ses applications, *Actual. Scient. et Ind.*, n° 869, Paris, Hermann, 1940 (2ᵉ éd., *ibid.*, n° 869-1145, Paris, Hermann, 1953).

(XI) H. WEYL, *a*) Theorie des Darstellung kontinuierlicher halbeinfacher Gruppen durch lineare Transformationen, *Math. Zeit.*, **23** (1925), pp. 271–309, **24** (1926), pp. 328–395 and 789–791 ( = *Selecta*, Basel–Stuttgart (Birkhäuser), 1956, pp. 262–366); *b*) (with F. PETER) Die Vollständigkeit der primitiven Darstellungen einer geschlossenen kontinuierlichen Gruppe, *Math. Ann.*, **97** (1927), pp. 737–755 ( = *Selecta*, pp. 387–404).

CHAPTER IX

# Measures on Hausdorff topological spaces

*If* T *is a set, and* A *is a subset of* T, *we denote by* $\varphi_A$ *the characteristic function of* A, *provided this does not lead to any confusion. The set* $\overline{\mathbf{R}}_+^T$ *of numerical functions* $\geqslant 0$ *(finite or not) defined on* T *will be denoted by* $\mathscr{F}_+(T)$, *or simply* $\mathscr{F}_+$ *if there is no ambiguity as to* T; *this set will always be equipped with its natural order structure. Recall that the product of two elements of* $\mathscr{F}_+$ *is always defined, thanks to the convention* $0 \cdot (+\infty) = 0$. *If* A *is a subset of* T, *and* $f$ *is a function defined on* T, *the restriction* $f|A$ *of* $f$ *to* A *may be denoted* $f_A$ *in this chapter, if this creates no confusion; an analogous notation will be employed for induced measures. On the other hand, if* $f \in \mathscr{F}_+(A)$ *we shall denote by* $f^0$ *the extension by* 0 *of* $f$ *to* T, *that is, the function defined on* T *that coincides with* $f$ *on* A *and with* 0 *on* T − A.

*All topological spaces considered in this chapter are assumed to be Hausdorff, absent express mention to the contrary. From* §1, *No.* 4 *on, except for* §5, *all measures will be assumed to be positive, absent express mention to the contrary.*

## §1. PREMEASURES AND MEASURES ON A TOPOLOGICAL SPACE

### 1. Encumbrances

DEFINITION 1. — *Let* T *be a set. One calls encumbrance on* T *any mapping* $p$ *of* $\mathscr{F}_+(T)$ *into* $\overline{\mathbf{R}}_+$ *that has the following properties:*

*a) If* $f$ *and* $g$ *are two elements of* $\mathscr{F}_+$ *such that* $f \leqslant g$, *then* $p(f) \leqslant p(g)$.

*b) If* $f$ *is an element of* $\mathscr{F}_+$, *and* $t$ *is a number* $\geqslant 0$, *then* $p(tf) = tp(f)$.

*c) If $f$ and $g$ are two elements of $\mathscr{F}_+$, then $p(f+g) \leqslant p(f)+p(g)$.*

*d) If $(f_n)$ is an increasing sequence of elements of $\mathscr{F}_+$, and if $f = \lim_{n\to\infty} f_n$, then $p(f) = \lim_{n\to\infty} p(f_n)$.*

*If A is a subset of T, we write $p(\mathrm{A})$ instead of $p(\varphi_{\mathrm{A}})$.*

The condition *b)* implies that $p(0) = 0$. On the other hand, let $(f_n)$ be a sequence of elements of $\mathscr{F}_+$; the conditions *c)* and *d)* imply the inequality

$$p\Big(\sum_n f_n\Big) \leqslant \sum_n p(f_n)$$

(the *inequality of countable convexity*).

For example, let T be a locally compact space, $\mu$ a positive measure on T; then $\mu^*$ and $\mu^\bullet$ are encumbrances on T. This follows from Props. 10, 11, 12 and Th. 3 of Ch. IV, §1, No. 3 for $\mu^*$, and from Prop. 1 of Ch. V, §1, No. 1 for $\mu^\bullet$.

PROPOSITION 1. — *Let $(p_\alpha)_{\alpha\in\mathrm{A}}$ be a family of encumbrances on T. The sum and upper envelope of the family $(p_\alpha)$ (in $\mathscr{F}_+(\mathscr{F}_+(\mathrm{T}))$) are then encumbrances.*

The sum of a finite family of encumbrances obviously being an encumbrance, it suffices to treat the case of the upper envelope. The properties *a)*, *b)*, *c)* of Definition 1 being obviously satisfied, it remains to establish *d)*. Set $p = \sup_\alpha p_\alpha$; then, with the notations of Definition 1 *d)*,

$$p(f) = \sup_\alpha p_\alpha(f) = \sup_\alpha \sup_n p_\alpha(f_n) = \sup_n \sup_\alpha p_\alpha(f_n) = \sup_n p(f_n).$$

DEFINITION 2. — *Let $p$ be an encumbrance on a set T. One says that $p$ is bounded if $p(\mathrm{T}) < +\infty$. If T is a topological space, $p$ is said to be locally bounded provided that every $x \in \mathrm{T}$ admits a neighborhood V such that $p(\mathrm{V}) < +\infty$.*

It then follows from the properties *a)* and *c)* of Def. 1 that $p(\mathrm{K}) < +\infty$ for every compact subset K of T. In particular, if T is compact, then every locally bounded encumbrance on T is bounded.

Let $p$ be an encumbrance on a set T, and A a subset of T. For every function $f \in \mathscr{F}_+(\mathrm{A})$, let $f^0$ be the extension by 0 of $f$ to T; the mapping $f \mapsto p(f^0)$ on $\mathscr{F}_+(\mathrm{A})$ is then an encumbrance, called *the encumbrance induced by $p$ on* A, and is denoted $p|\mathrm{A}$ or $p_{\mathrm{A}}$.

Let T and U be two sets, $\pi$ a mapping of T into U, and $p$ an encumbrance on T. The encumbrance $\pi(p)$ on U, whose value for $f \in \mathscr{F}_+(\mathrm{U})$ is given by

$$\big(\pi(p)\big)(f) = p(f \circ \pi),$$

is called the *image encumbrance* of $p$ under $\pi$.

Let $p$ be an encumbrance on a set T; $p$ is said to be *concentrated* on a subset A of T if $p(\mathrm{T}-\mathrm{A})=0$.

*Lemma 1. — If the encumbrance $p$ is concentrated on $\mathrm{A}\subset\mathrm{T}$, then $p(f)=p(f\varphi_{\mathrm{A}})$ for every $f\in\mathscr{F}_+(\mathrm{T})$.*

For, set $\mathrm{T}-\mathrm{A}=\mathrm{B}$, so that $p(\varphi_{\mathrm{B}})=0$; then

$$f\varphi_{\mathrm{B}}\leqslant(+\infty)\cdot\varphi_{\mathrm{B}}=\sup_{n\in\mathbf{N}} n\varphi_{\mathrm{B}},$$

therefore $p(f\varphi_{\mathrm{B}})=0$ by properties *a*), *b*), *d*) of Def. 1. It follows from *c*) that $p(f)\leqslant p(f\varphi_{\mathrm{A}})+p(f\varphi_{\mathrm{B}})=p(f\varphi_{\mathrm{A}})$, and finally $p(f)=p(f\varphi_{\mathrm{A}})$ by *a*).

## 2. Premeasures and measures

Let T be a topological space, and let $\mathfrak{K}$ be the set of compact subsets of T, ordered by inclusion. For every $\mathrm{K}\in\mathfrak{K}$, let $\mathscr{M}(\mathrm{K};\mathbf{C})$ be the set of complex measures on K. For every pair $(\mathrm{K},\mathrm{L})$ of elements of $\mathfrak{K}$ such that $\mathrm{K}\subset\mathrm{L}$, let $\iota_{\mathrm{KL}}$ be the mapping of $\mathscr{M}(\mathrm{L};\mathbf{C})$ into $\mathscr{M}(\mathrm{K};\mathbf{C})$ that associates to each measure $\mu$ on L the measure $\mu_{\mathrm{K}}$ induced by $\mu$ on K (Ch. IV, §5, No. 7, Def. 4). Then $\iota_{\mathrm{KM}}=\iota_{\mathrm{KL}}\circ\iota_{\mathrm{LM}}$ when K, L and M are compact subsets of T such that $\mathrm{K}\subset\mathrm{L}\subset\mathrm{M}$; this follows from the transitivity of induced measures (Ch. V, §7, No. 2, Prop. 4). The elements of the *inverse limit* of the family $(\mathscr{M}(\mathrm{K};\mathbf{C}))_{\mathrm{K}\in\mathfrak{K}}$ for the mappings $\iota_{\mathrm{KL}}$ will be called *premeasures* on T. In other words:

DEFINITION 3. — *One calls premeasure on a topological space* T *every mapping $w$ that associates, to every compact subset* K *of* T, *a measure $w_{\mathrm{K}}$ on* K, *and that has the following property*:

*If* K *and* L *are compact subsets of* T *such that* $\mathrm{K}\subset\mathrm{L}$, *the measure* $(w_{\mathrm{L}})_{\mathrm{K}}$ *induced by* $w_{\mathrm{L}}$ *on* K *is equal to* $w_{\mathrm{K}}$.

*The premeasure $w$ is said to be real* (resp. *positive*) *if all of the measures $w_{\mathrm{K}}$ are real* (resp. *positive*).

Let $w$ and $w'$ be two premeasures on T, $t$ a complex number; the premeasures $w+w'$ and $tw$ are defined by the formulas $(w+w')_{\mathrm{K}}=w_{\mathrm{K}}+w'_{\mathrm{K}}$, $(tw)_{\mathrm{K}}=tw_{\mathrm{K}}$ for every compact subset K of T. The premeasures on T obviously form a vector space, which is denoted $\mathscr{P}(\mathrm{T};\mathbf{C})$; the space of real premeasures will be denoted $\mathscr{P}(\mathrm{T};\mathbf{R})$, or more often $\mathscr{P}(\mathrm{T})$, and the convex cone of positive premeasures will be denoted $\mathscr{P}_+(\mathrm{T})$. Let $w$ be a premeasure; the mapping $\mathrm{K}\mapsto|w_{\mathrm{K}}|$ is then a premeasure on T (Ch. IV, §5, No. 7, Lemma 3), which will be denoted $|w|$. If $w$ is real, one sets

$w^+ = \frac{1}{2}(|w| + w)$, $w^- = \frac{1}{2}(|w| - w)$; these two premeasures being positive, one sees that every real premeasure is the difference of two positive premeasures. Clearly $(w^+)_K = (w_K)^+$, $(w^-)_K = (w_K)^-$ for every compact subset K of T.

The vector space $\mathscr{P}(T)$ is ordered by the cone $\mathscr{P}_+(T)$. It is clear that $w^+ = \sup(w, 0)$, $w^- = \sup(-w, 0)$; consequently, $\mathscr{P}(T)$ is lattice-ordered and $\sup(w, w') = w + (w' - w)^+$, $\inf(w, w') = w - (w' - w)^-$. Moreover, clearly

$$(\sup(w, w'))_K = \sup(w_K, w'_K), \quad (\inf(w, w'))_K = \inf(w_K, w'_K)$$

for every compact subset K of T.

DEFINITION 4. — *Let $w$ be a positive premeasure on* T. *We shall set, for every function $f \in \mathscr{F}_+(T)$,*

$$w^\bullet(f) = \sup_K (w_K)^\bullet(f_K), \tag{1}$$

*where* K *runs over the set of compact subsets of* T.

For each compact set K, let $p^K$ be the image encumbrance of the encumbrance $(w_K)^\bullet$ under the canonical injection of K into T; $w^\bullet$ is the upper envelope of the encumbrances $p^K$, hence is an encumbrance (No. 1, Prop. 1). One says that $w^\bullet$ is the *essential upper integral* associated with the positive premeasure $w$. One often writes $\int^\bullet f\,dw$ or $\int^\bullet f(t)\,dw(t)$ instead of $w^\bullet(f)$.

*Remark* 1). — If $v$ and $w$ are two positive premeasures, then $(v + w)^\bullet = v^\bullet + w^\bullet$ (Ch. V, §1, No. 1, Prop. 3). If $v$ and $w$ are two complex premeasures, then $|v + w|^\bullet \leqslant |v|^\bullet + |w|^\bullet$.

PROPOSITION 2. — *a) Let $w$ be a positive premeasure. For every compact subset* K *of* T, *the encumbrance $(w^\bullet)_K$ induced by $w^\bullet$ on* K *is equal to $(w_K)^\bullet$. For every function $f \in \mathscr{F}_+(T)$, one has the relations $(w_K)^\bullet(f_K) = w^\bullet(f\varphi_K)$ and*

$$w^\bullet(f) = \sup_K w^\bullet(f\varphi_K). \tag{2}$$

*b) Conversely, let $p$ be an encumbrance on* T *satisfying the following conditions:*

1) *For every compact subset* K *of* T, *there exists a positive measure $w_K$ on* K *such that $p_K = (w_K)^\bullet$.*

2) *For every function $f \in \mathscr{F}_+(T)$, $p(f) = \sup_K p(f\varphi_K)$.*

*The mapping $w : K \mapsto w_K$ is then a positive premeasure on* T, *and $p = w^\bullet$.*

Let us prove *a*): let $g \in \mathscr{F}_+(\mathrm{K})$ and let $g^0$ be the extension by zero of $g$ to T; then, by the definition of induced encumbrances,

$$(w^\bullet)_\mathrm{K}(g) = w^\bullet(g^0) = \sup_\mathrm{L} (w_\mathrm{L})^\bullet(g^0|\mathrm{L}),$$

where L runs over the set of compact subsets of T, or merely the set of those that contain K. But if L contains K, then $(w_\mathrm{L})^\bullet(g^0|\mathrm{L}) = (w_\mathrm{K})^\bullet(g)$ from the fact that $g^0|\mathrm{L}$ is zero outside K (Ch. V, §7, No. 1, Prop. 1), which proves the first assertion. Therefore

$$(w_\mathrm{K})^\bullet(f_\mathrm{K}) = (w^\bullet)_\mathrm{K}(f_\mathrm{K}) = w^\bullet((f_\mathrm{K})^0) = w^\bullet(f\varphi_\mathrm{K})$$

for all $f \in \mathscr{F}_+(\mathrm{T})$, and (2) merely translates the formula (1).

Let us pass to *b*): the measure $w_\mathrm{K}$ considered in 1) is unique (Ch. V, §1, No. 1). Let us show that the mapping $\mathrm{K} \mapsto w_\mathrm{K}$ is a premeasure: let K and L be two compact subsets such that $\mathrm{K} \subset \mathrm{L}$, and let $\lambda$ be the measure induced by $w_\mathrm{L}$ on K; everything comes down to showing that $\lambda^\bullet = (w_\mathrm{K})^\bullet$. Now, $\lambda^\bullet = ((w_\mathrm{L})^\bullet)_\mathrm{K}$ (Ch. V, §7, No. 1, Prop. 1); since $(w_\mathrm{L})^\bullet = p_\mathrm{L}$, we have $\lambda^\bullet = (p_\mathrm{L})_\mathrm{K} = p_\mathrm{K} = (w_\mathrm{K})^\bullet$.

Let us denote by $w$ the premeasure $\mathrm{K} \mapsto w_\mathrm{K}$; since $p_\mathrm{K} = (w_\mathrm{K})^\bullet = (w^\bullet)_\mathrm{K}$, we have $p(f\varphi_\mathrm{K}) = p_\mathrm{K}(f_\mathrm{K}) = (w^\bullet)_\mathrm{K}(f_\mathrm{K}) = w^\bullet(f\varphi_\mathrm{K})$. The two encumbrances $p$ and $w^\bullet$ are therefore equal by virtue of the formula (2) and the hypothesis 2) on $p$.

Q.E.D.

Since the induced encumbrance $(w^\bullet)_\mathrm{K}$ is equal to $(w_\mathrm{K})^\bullet$, there is no ambiguity in simply writing $w^\bullet_\mathrm{K}$. We shall employ this notation from now on.

COROLLARY. — *Let $v$ and $w$ be two positive premeasures on* T, *such that $v^\bullet(\mathrm{L}) = w^\bullet(\mathrm{L})$ for every compact subset* L *of* T; *then $v = w$. In particular, the relation $v^\bullet = w^\bullet$ implies $v = w$.*

For, let K be a compact set in T; for every compact set $\mathrm{L} \subset \mathrm{K}$, one has the relation

$$w_\mathrm{K}(\mathrm{L}) = w^\bullet_\mathrm{K}(\mathrm{L}) = w^\bullet(\mathrm{L}) = v^\bullet(\mathrm{L}) = v^\bullet_\mathrm{K}(\mathrm{L}) = v_\mathrm{K}(\mathrm{L})$$

by Prop. 2; therefore $w_\mathrm{K} = v_\mathrm{K}$ (Ch. IV, §4, No. 10, Cor. 3 of Prop. 19), and finally $w = v$ by the definition of premeasures.

DEFINITION 5. — *Let $w$ be a premeasure on a topological space* T. *One says that $w$ is a measure* (resp. *a bounded measure*) *if the encumbrance $|w|^\bullet$ is locally bounded* (resp. *bounded*) (cf. No. 1, Def. 2).

The set of complex measures on T is obviously a vector space (*Remark* 1), which will be denoted $\mathscr{M}(\mathrm{T};\mathbf{C})$. The space of real measures will be denoted $\mathscr{M}(T;\mathbf{R})$ or more often $\mathscr{M}(T)$, and the cone of positive measures will be denoted $\mathscr{M}_+(T)$.

If $w$ is a complex measure, its real part and its imaginary part are real measures. If $w$ is a real measure, $w^+$ and $w^-$ are positive measures. Every complex (resp. real) measure is thus a linear combination (resp. difference) of positive measures.

*Remarks.* — 2) If T is *locally compact*, then every premeasure $w$ on T is a measure. For, every $x \in T$ admits a compact neighborhood K, and $|w|^\bullet(K) = \|w_K\| < +\infty$, so that the encumbrance $|w|^\bullet$ is locally bounded.

3) For every Borel subset A of T (in particular for $A = T$) and every positive measure $\mu$ on T, the number $\mu^\bullet(A)$ is the supremum of the measures $\mu^\bullet(K)$ of the compact subsets of A. Indeed, for every compact subset K of A, one has $\mu^\bullet(K) \leqslant \mu^\bullet(A)$; on the other hand, if $\mathfrak{K}$ is the set of compact subsets of T, then

$$\mu^\bullet(A) = \sup_{K\in\mathfrak{K}} \mu_K^\bullet(A\cap K) = \sup_{K\in\mathfrak{K}} \sup_{\substack{L\in\mathfrak{K}\\ L\subset A\cap K}} \mu_K^\bullet(L) \leqslant \sup_{\substack{L\in\mathfrak{K}\\ L\subset A}} \mu^\bullet(L)$$

by Cor. 1 of Th. 4 of Ch. IV, §4, No. 6.

## 3. Examples of measures

*Example* 1. — *Measures on a locally compact space.*

The following proposition shows that the theory of this chapter contains that of Ch. IV. In the statement, the word 'measure' and the notation $\mathscr{M}(T;\mathbf{C})$ are taken in the sense of the earlier chapters.

PROPOSITION 3. — *Let* T *be a locally compact space, and let* $\mu$ *be a measure on* T. *Denote by* $W(\mu)$ *the mapping that associates to each compact subset* K *of* T *the induced measure* $\mu_K$. *Then* $W(\mu)$ *is a premeasure on* T, *one has* $W(|\mu|) = |W(\mu)|$, *and the linear mapping* $W : \mu \mapsto W(\mu)$ *is a bijection of the space* $\mathscr{M}(T;\mathbf{C})$ *onto the space* $\mathscr{P}(T;\mathbf{C})$ *of premeasures on* T. *Moreover, if* $\mu$ *is positive then* $\mu^\bullet = \big(W(\mu)\big)^\bullet$.

It is clear that $W(\mu)$ is a premeasure (Ch. V, §7, No. 2, Prop. 4) and that the mapping W is linear. The relation $W(\mu) = 0$ means that $\mu$ induces the measure 0 on every compact set in T; then $\mu(f) = 0$ for $f \in \mathscr{K}(T;\mathbf{C})$, thus $\mu = 0$, which proves that W is injective. It remains to prove that W is surjective. Since every premeasure is a linear combination of positive premeasures, it will suffice to construct, for every *positive* premeasure $w$, a positive measure $\mu$ such that $w = W(\mu)$. Let $f \in \mathscr{K}(T)$ be a given function, and let L be a compact set containing the support of $f$; the number $w_L(f_L)$ is independent of the choice of L, by the definition of induced measures, so that one can set $\mu(f) = w_L(f_L)$; then $\mu$ is a positive linear form on $\mathscr{K}(T)$, that is, a positive measure. Let us verify that $w = \mathrm{W}(\mu)$; first, the relation $\mu^\bullet(f) = w_\mathrm{L}^\bullet(f_\mathrm{L})$ extends to the case that $f$ is a finite upper semi-continuous function that is positive and is zero outside $\mathrm{L}$. For, let $\mathrm{M}$ be a compact neighborhood of $\mathrm{L}$, and $\mathscr{H}$ the (decreasing directed) set of continuous functions on $\mathrm{T}$, with support contained in $\mathrm{M}$, that are $\geqslant f$. Then (Ch. IV, §4, No. 4, Cor. 2 of Prop. 5)

$$\mu^\bullet(f) = \inf_{h\in\mathscr{H}} \mu(h) = \inf_{h\in\mathscr{H}} w_\mathrm{M}(h_\mathrm{M}) = w_\mathrm{M}^\bullet(f_\mathrm{M}),$$

and on the other hand $w_\mathrm{M}^\bullet(f_\mathrm{M}) = w_\mathrm{L}^\bullet(f_\mathrm{L})$ since $f_\mathrm{M}$ is zero on $\mathrm{M} - \mathrm{L}$ (Ch. V, §7, No. 1, Prop. 1). In particular, if $f$ is taken to be the extension by 0 of an element of $\mathscr{K}_+(\mathrm{L})$, this formula shows that $\mu_\mathrm{L} = w_\mathrm{L}$ by the definition of induced measures, thus indeed $\mathrm{W}(\mu) = w$.

If $\mu$ is positive, then

$$\mu^\bullet(f) = \sup_\mathrm{K} \mu^\bullet(f\varphi_\mathrm{K}) = \sup_\mathrm{K} \mu_\mathrm{K}^\bullet(f_\mathrm{K}) = \big(\mathrm{W}(\mu)\big)^\bullet(f)$$

for every $f \in \mathscr{F}_+(\mathrm{T})$ (Ch. V, §1, Def. 1 and §7, Prop. 1). The relation $|\mathrm{W}(\mu)| = \mathrm{W}(|\mu|)$ is obvious (Ch. IV, §5, No. 7, Lemma 3).

Q.E.D.

When $\mathrm{T}$ is *locally compact*, we shall from now on *identify* the spaces $\mathscr{M}(\mathrm{T};\mathbf{C})$ and $\mathscr{P}(\mathrm{T};\mathbf{C})$ by means of the bijection $\mathrm{W}$.

*Example 2. — Measures with compact support on a topological space.*

*Lemma 2. — Let* $\mathrm{T}$ *be a topological space,* $\mathrm{L}$ *a compact subset of* $\mathrm{T}$, *and* $\lambda$ *a positive measure on* $\mathrm{L}$. *There exists a unique positive measure* $\mu$ *on* $\mathrm{T}$ *such that, for every function* $f \in \mathscr{F}_+(\mathrm{T})$,

$$\mu^\bullet(f) = \lambda^\bullet(f_\mathrm{L}). \tag{2}$$

Let us set $p(f) = \lambda^\bullet(f_\mathrm{L})$ for every $f \in \mathscr{F}_+(\mathrm{T})$, and let us show that the conditions 1) and 2) of Prop. 2 *b*) are satisfied. The second is obviously satisfied: indeed, $p(f) = p(f\varphi_\mathrm{K})$ if $\mathrm{K}$ contains $\mathrm{L}$. If $\mathrm{K} \subset \mathrm{T}$ is compact, and if $h \in \mathscr{F}_+(\mathrm{K})$, then

$$p_\mathrm{K}(h) = p(h^0) = \lambda^\bullet(h^0|\mathrm{L}).$$

But $h^0|\mathrm{L}$ is the extension by 0 of $h_{\mathrm{K}\cap\mathrm{L}}$ to $\mathrm{L}$: the last expression is therefore equal to $(\mu_\mathrm{K})^\bullet(h)$, where $\mu_\mathrm{K}$ is the image of $\lambda|\mathrm{K}\cap\mathrm{L}$ under the injection of $\mathrm{K}\cap\mathrm{L}$ into $\mathrm{K}$ (Ch. V, §6, No. 2, Prop. 2 and §7, No. 1, Prop. 1), and $p_\mathrm{K} = (\mu_\mathrm{K})^\bullet$. Condition 1) of Prop. 2 *b*) is therefore also satisfied, and the existence of $\mu$ follows at once.

Q.E.D.

We shall say that $\mu$ is the measure on T *defined* by $\lambda$. In particular, for every point $x$ of T one can define the measure $\varepsilon_x$; it is characterized by $(\varepsilon_x)^\bullet(f) = f(x)$ for $f \in \mathscr{F}_+(\mathrm{T})$.

*Remarks.* — 1) When T is locally compact, $\mu$ is the image of $\lambda$ under the injection of L into T. We shall see in §2, No. 3, *Example*, when image measures will have been treated, that this interpretation remains valid for arbitrary spaces.

2) We shall also see that the measures defined in *Example* 2 are positive measures on T with compact support (No. 6, *Remark* 2)).

*We shall henceforth consider only positive measures, absent express mention to the contrary. For the rest of this section,* T *will denote a topological space and* $\mu$ *a positive measure on* T.

Numerous results in the following subsections may be extended to positive premeasures. This extension is left to the reader.

## 4. Locally negligible sets and functions

DEFINITION 6. — *A function* $f \in \mathscr{F}_+$ (resp. *a subset* A *of* T) *is said to be locally negligible for the measure* $\mu$ *if* $\mu^\bullet(f) = 0$ (resp. $\mu^\bullet(\mathrm{A}) = 0$). *One says that* $\mu$ *is concentrated on a subset* A *of* T *if* T − A *is locally* $\mu$*-negligible.*

*Remarks.* — 1) The concepts so defined coincide, when T is locally compact, with the usual concepts.

2) After we have defined *negligible* sets, we will see that the locally negligible sets are indeed those whose germ, at every point of T, is the germ of a negligible set (No. 9, Cor. 2 of Prop. 14).

3) As in Chs. IV and V, the expression 'locally almost everywhere' will be synonymous with 'except on a locally negligible set'.

4) If $\theta$ is a complex measure, we shall say that a function (resp. a subset of T) is locally negligible for $\theta$ if it is so for the positive measure $|\theta|$.

*Example.* — Let L be a compact subset of T, $\lambda$ a measure on L, and $\mu$ the measure on T defined by $\lambda$ (No. 3, *Example* 2). The formula (3) implies at once that a function $f \in \mathscr{F}_+(\mathrm{T})$ is locally $\mu$-negligible if and only if $f_{\mathrm{L}}$ is $\lambda$-negligible.

It follows immediately from formula (1) that a function $f \in \mathscr{F}_+(\mathrm{T})$ is locally $\mu$-negligible if and only if $f_{\mathrm{K}}$ is $\mu_{\mathrm{K}}$-negligible for every compact subset K of T. Thus the properties of locally negligible sets reduce at once to those of negligible sets in compact spaces, treated in Ch. IV. Here are some results that will be used henceforth without further reference.

— For a function $f \geqslant 0$ to be locally negligible, it is necessary and sufficient that $f(t) = 0$ locally almost everywhere (Ch. IV, §2, No. 3, Th. 1). If $\mathbf{f}$ is a function with values in a Banach space, it is therefore equivalent to say that $\mathbf{f} = 0$ locally almost everywhere or that $\mu^\bullet(|\mathbf{f}|) = 0$; in this case we shall again say that $\mathbf{f}$ is locally negligible.

— The sum and upper envelope of a sequence of locally negligible functions $\geqslant 0$ are locally negligible (*loc. cit.*, No. 1, Prop. 2).

— If $f$ and $g$ are two functions $\geqslant 0$ that are equal locally almost everywhere, then $\mu^\bullet(f) = \mu^\bullet(g)$ (*loc. cit.*, No. 3, Prop. 6).

## 5. Measurable sets and functions

DEFINITION 7. — *A function $f$ defined on* T, *with values in a topological space* F *(Hausdorff or not) is said to be measurable for the measure $\mu$ (or to be $\mu$-measurable) if, for every compact subset* K *of* T, *the function $f_{\mathrm{K}}$ is $\mu_{\mathrm{K}}$-measurable.*

This amounts to saying that there exists, for every compact set K, a partition of K into a $\mu_{\mathrm{K}}$-negligible set N and a sequence $(\mathrm{K}_n)$ of compact sets, such that the restriction of $f$ to each $\mathrm{K}_n$ is continuous. Since it is equivalent to say that N is $\mu_{\mathrm{K}}$-negligible or that it is locally $\mu$-negligible (No. 4), one sees that $f$ is $\mu$-measurable if and only if, for every compact set K, there exists a partition of K into a locally $\mu$-negligible set N and a sequence $(\mathrm{K}_n)$ of compact sets such that $f_{\mathrm{K}_n}$ is continuous for all $n$. This definition is identical to that of Def. 1 of Ch. IV, §5, No. 1, and one thus recovers the usual concept of measurable function when T is locally compact.

A subset A of T is said to be measurable if its characteristic function is measurable. When A is $\mu$-measurable and $\mu^\bullet(\mathrm{A}) < +\infty$, this number is denoted simply $\mu(\mathrm{A})$ and is called the *measure* of A. One similarly writes $\mu(f)$ for $\mu^\bullet(f)$ when $f$ is $\geqslant 0$, $\mu$-measurable, and $\mu^\bullet(f) < +\infty$.

If $\theta$ is a complex measure on T, a function $f$ (resp. a subset of T) is said to be $\theta$-measurable if it is measurable for the positive measure $|\theta|$. The results below may be extended to complex measures.

*Example.* — Let L be a compact subset of T, $\lambda$ a measure on L, and $\mu$ the measure on T defined by $\lambda$ (No. 3, *Example* 2). A function $f$ defined on T is $\mu$-measurable if and only if $f_{\mathrm{L}}$ is $\lambda$-measurable. For, this condition is obviously necessary. Conversely, if it is satisfied, there exists a partition of L into a $\lambda$-negligible set N and a sequence $(\mathrm{L}_n)$ of compact sets, such that $f_{\mathrm{L}_n}$ is continuous for all $n$. If K is a compact subset of T, the set $\mathrm{K} - \bigcup_n (\mathrm{K} \cap \mathrm{L}_n)$ has an intersection with L that is $\lambda$-negligible, hence this set is $\mu$-negligible by formula (3) of No. 3, and the restriction of $f$ to $\mathrm{K} \cap \mathrm{L}_n$ is continuous for all $n$.

Def. 7 permits extending, without a new proof, a number of results on measurable functions to the case of spaces not locally compact. Here are some of them, which we shall use henceforth without further reference: the open sets and the closed sets of T are $\mu$-measurable; the $\mu$-measurable sets form a tribe (Ch. IV, §5, No. 4, Cor. 2 of Th. 2), that contains the Borel sets of T (*loc. cit.*, Cor. 3), and the Souslin sets (Ch. IV, §5, No. 1, Cor. 2 of Prop. 3)[1]. The usual algebraic operations on numerical functions preserve measurability (Ch. IV, §5, No. 3), as do the operations of countable passage to the limit (*loc. cit.*, No. 4, Th. 2 and Cor. 1). The following property merits more explicit mention:

PROPOSITION 4. — *Let $f$ be a positive function and $(g_n)_{n\geqslant 1}$ a sequence of $\mu$-measurable positive functions on* T. *Setting $g = \sum_{n\geqslant 1} g_n$, one has*

$$\mu^\bullet(fg) = \sum_{n\geqslant 1} \mu^\bullet(fg_n)\,. \tag{4}$$

Set $h_n = \sum_{i=1}^{n} g_i$ for all $n \geqslant 1$. For every compact subset K of T,

$$\mu^\bullet_{\mathrm{K}}((fh_n)_{\mathrm{K}}) = \sum_{i=1}^{n} \mu^\bullet_{\mathrm{K}}((fg_i)_{\mathrm{K}})$$

by Prop. 2 of Ch. V, §1, No. 1 applied to the compact space K. Passing to the limit with respect to the increasing directed set of compact subsets of T, one obtains

$$\mu^\bullet(fh_n) = \sum_{i=1}^{n} \mu^\bullet(fg_i)\,.$$

Now, $fg$ is the limit of the increasing sequence $(fh_n)_{n\geqslant 1}$, whence $\mu^\bullet(fg) = \lim_{n\to\infty} \mu^\bullet(fh_n)$; the preceding formula then immediately implies (4).

COROLLARY. — *Let $(\mathrm{A}_n)$ be a sequence of pairwise disjoint measurable subsets, with union* A. *For every subset* B *of* T,

$$\mu^\bullet(\mathrm{A}\cap\mathrm{B}) = \sum_{n} \mu^\bullet(\mathrm{A}_n\cap\mathrm{B})$$

*and in particular*

$$\mu^\bullet(\mathrm{A}) = \sum_{n} \mu^\bullet(\mathrm{A}_n)\,.$$

Among the properties of measurable functions or sets that extend as above to Hausdorff spaces, we cite also Prop. 12 of Ch. IV, §5, No. 8 ($\mu$-dense families of compact sets). Thus, a function $f$ with values in a topological

---

[1] The proof of this corollary is valid without modification for Souslin sets in a nonmetrizable locally compact space (GT, IX, §6, No. 9, Th. 5).

space (Hausdorff or not) is $\mu$-measurable if and only if the set of compact subsets K of T, such that $f_K$ is continuous, is $\mu$-dense (*loc. cit.*, No. 10, Prop. 15).

## 6. Directed families; support of a measure

PROPOSITION 5. — *a) Let* H *be an increasing directed set of functions* $\geqslant 0$ *that are lower semi-continuous on every compact subset of* T. *Then*

$$\mu^\bullet\Big(\sup_{h\in H} h\Big) = \sup_{h\in H} \mu^\bullet(h). \tag{5}$$

*b) Let* H *be a decreasing directed set of functions* $\geqslant 0$ *that are upper semi-continuous on every compact subset of* T. *If there exists in* H *a function* $h_0$ *such that* $\mu^\bullet(h_0) < +\infty$, *then*

$$\mu^\bullet\Big(\inf_{h\in H} h\Big) = \inf_{h\in H} \mu^\bullet(h). \tag{6}$$

For every compact set $K \subset T$, we have in case *a*)

$$\mu^\bullet\Big(\sup_{h\in H} h\varphi_K\Big) = \mu_K^\bullet\Big(\sup_{h\in H} h_K\Big) = \sup_{h\in H} \mu_K^\bullet(h_K) = \sup_{h\in H} \mu^\bullet(h\varphi_K),$$

and in case *b*)

$$\mu^\bullet\Big(\inf_{h\in H} h\varphi_K\Big) = \mu_K^\bullet\Big(\inf_{h\in H} h_K\Big) = \inf_{h\in H} \mu_K^\bullet(h_K) = \inf_{h\in H} \mu^\bullet(h\varphi_K),$$

by Prop. 2 of No. 2, and Prop. 8 of Ch. V, §1, No. 2. Case *a*) follows at once, by passage to the supremum with respect to K (No. 2, Prop. 2). To treat case *b*), denote by $\varepsilon$ a number $> 0$, and choose a compact set K such that $\mu^\bullet(h_0\varphi_K) \geqslant \mu^\bullet(h_0) - \varepsilon$. We then have (No. 5, Prop. 4) $\mu^\bullet(h_0\varphi_{\complement K}) \leqslant \varepsilon$; for every function $h \in H$ that is $\leqslant h_0$, we therefore have $\mu^\bullet(h\varphi_{\complement K}) \leqslant \varepsilon$, and finally $\mu^\bullet(h\varphi_K) \geqslant \mu^\bullet(h) - \varepsilon$ by Prop. 4 of No. 5. Therefore

$$\mu^\bullet\Big(\inf_{h\in H} h\Big) \geqslant \mu^\bullet\Big(\inf_{h\in H} h\varphi_K\Big) = \inf_{h\in H,\, h\leqslant h_0} \mu^\bullet(h\varphi_K) \geqslant \inf_{h\in H,\, h\leqslant h_0} \mu^\bullet(h) - \varepsilon.$$

Consequently the left side of (6) is $\geqslant$ the right side; the reverse inequality being obvious, the proposition is established.

COROLLARY. — *a) Let* $(U_\alpha)_{\alpha\in I}$ *be an increasing directed family of open subsets of* T, *with union* U. *Then* $\mu^\bullet(U) = \sup_{\alpha\in I} \mu^\bullet(U_\alpha)$.

*b) Let* $(F_\alpha)_{\alpha\in I}$ *be a decreasing directed family of closed subsets of* T, *with intersection* F. *If there exists an* $\alpha \in I$ *such that* $\mu^\bullet(F_\alpha)$ *is finite, then* $\mu^\bullet(F) = \inf_{\alpha\in I} \mu^\bullet(F_\alpha)$.

By the preceding corollary, there exists a largest locally negligible open set; this justifies the following definition:

DEFINITION 8. — *The support of a measure* $\mu$ *on* T *is defined to be the complement of the largest locally* $\mu$*-negligible open set in* T.

The support of $\mu$ is denoted $\operatorname{Supp}(\mu)$.

*Remarks.* — 1) If $\mu$ is a complex measure, the support of $\mu$ is defined to be the support of the positive measure $|\mu|$; it is again the complement of the largest locally $\mu$-negligible open set.

2) Let us show that the measures introduced in *Example* 2 of No. 3 are measures with compact support in T. Let $\mu$ be a positive measure on T whose support is a compact set K, and let $\nu$ be the measure defined by $\mu_K$ (in the sense of No. 3). Let $f \in \mathscr{F}_+(T)$; then

$$\nu^\bullet(f) = \mu_K^\bullet(f_K) \qquad \text{(No. 3, formula (3)).}$$

The encumbrance $\mu^\bullet$ being concentrated on K, we also have

$$\mu^\bullet(f) = \mu^\bullet(f\varphi_K) = \mu^\bullet((f_K)^0) = \mu_K^\bullet(f_K),$$

whence $\mu^\bullet = \nu^\bullet$, and finally $\mu = \nu$. Conversely, if K is a compact set in T and $\lambda$ a measure on K, and if $\mu$ is the measure on T defined by $\lambda$, then $\mu^\bullet(\complement K) = 0$ (No. 3, formula (3)); consequently, the support of $\mu$ is contained in K, hence is compact.

## 7. Upper envelopes and sums of measures

PROPOSITION 6. — *Let* $(\lambda_\alpha)_{\alpha\in A}$ *be an increasing directed family of measures on* T, *and let* $p = \sup_\alpha \lambda_\alpha^\bullet$. *For the family* $(\lambda_\alpha)$ *to be bounded above in* $\mathscr{M}(T)$, *it is necessary and sufficient that the encumbrance* $p$ *be locally bounded. The family* $(\lambda_\alpha)$ *then admits a supremum* $\lambda$ *in* $\mathscr{M}(T)$, *and* $\lambda^\bullet = p$. *For every compact set* K, *the measure* $\lambda_K$ *is the supremum of the measures* $(\lambda_\alpha)_K$ *in* $\mathscr{M}(K)$.

If the family $(\lambda_\alpha)$ is bounded above in $\mathscr{M}(T)$, then $p$ is obviously locally bounded. Conversely, let us assume $p$ to be locally bounded, and let us show that it satisfies the conditions 1) and 2) of Prop. 2 *b*) of No. 2. For 2), this results from the following equalities:

$$p(f) = \sup_\alpha \lambda_\alpha^\bullet(f) = \sup_\alpha \sup_K \lambda_\alpha^\bullet(f\varphi_K) = \sup_K \sup_\alpha \lambda_\alpha^\bullet(f\varphi_K) = \sup_K p(f\varphi_K).$$

On the other hand, let K be a compact set; the encumbrance $p_K$ is equal to the upper envelope of the encumbrances $(\lambda_\alpha^\bullet)_K$ and it is bounded since $p$ is locally bounded. The measures $(\lambda_\alpha)_K$ therefore admit a supremum $\lambda_K$ in $\mathscr{M}(K)$, and $\lambda_K^\bullet = p_K$ (Ch. V, §1, No. 4, Prop. 11). The condition 1) of Prop. 2 *b*) of No. 2 is thus satisfied, therefore there exists a measure $\lambda$ on T such that $\lambda^\bullet = p$; it is clear that $\lambda$ is the supremum of the measures $\lambda_\alpha$.

DEFINITION 9. — *Let $(\mu_i)_{i\in I}$ be a family of measures on* T. *Let* A *be the set of finite subsets of* I; *for every $\alpha \in A$ let $\lambda_\alpha = \sum_{i\in\alpha} \mu_i$. If the family $(\lambda_\alpha)$ admits a supremum $\mu$ in $\mathscr{M}(T)$, the family $(\mu_i)$ is said to be summable, $\mu$ is called the sum of the family $(\mu_i)$, and one writes $\mu = \sum_{i\in I} \mu_i$.*

This definition extends the definition of Ch. V, §2, No. 1.

PROPOSITION 7. — *For the family $(\mu_i)_{i\in I}$ to be summable, with sum $\mu$, it is necessary and sufficient that the encumbrance $p = \sum_{i\in I} \mu_i^\bullet$ be locally bounded, in which case $p = \mu^\bullet$. For every compact subset* K *of* T, *the family $\big((\mu_i)_K\big)_{i\in I}$ is then summable in $\mathscr{M}(K)$, and $\mu_K = \sum_{i\in I} (\mu_i)_K$.*

With notations as in Def. 9, $\lambda_\alpha^\bullet = \sum_{i\in\alpha} \mu_i^\bullet$ for every finite subset $\alpha$ of I (No. 2, *Remark* 1). The statement is then an immediate consequence of Prop. 6.

The relation $\mu_K = \sum_{i\in I} (\mu_i)_K$ and Prop. 2 of Ch. V, §2, No. 2 yields the following result:

PROPOSITION 8. — *Let $\mu$ be the sum of a summable family $(\mu_i)_{i\in I}$ of measures on* T. *In order that a mapping $f$ of* T *into a topological space* F *(Hausdorff or not) be $\mu$-measurable, it is necessary and sufficient that $f$ be $\mu_i$-measurable for every $i \in I$.*

## 8. Crushings

DEFINITION 10. — *One calls crushing of* T *for $\mu$, or $\mu$-crushing, any locally countable family $(K_\alpha)_{\alpha\in A}$ of pairwise disjoint compact subsets of* T *such that the set $N = T - \bigcup_{\alpha\in A} K_\alpha$ is locally $\mu$-negligible.*

PROPOSITION 9. — *a) There exists a crushing $(K_\alpha)_{\alpha\in A}$ of* T *for $\mu$.*

*b) Let $(K_\alpha)_{\alpha\in A}$ be a crushing of* T *for $\mu$. If $\mu_\alpha$ is the measure on* T *defined by $\mu_{K_\alpha}$* (No. 3, *Example* 2), *then the family $(\mu_\alpha)_{\alpha\in A}$ is summable,* *its sum is equal to* $\mu$, *and, for every function* $f \in \mathscr{F}_+(\mathrm{T})$,

$$\mu^\bullet(f) = \sum_{\alpha\in\mathrm{A}} \mu^\bullet_\alpha(f) = \sum_{\alpha\in\mathrm{A}} \mu^\bullet_{\mathrm{K}_\alpha}(f_{\mathrm{K}_\alpha}).\,^{(1)} \tag{7}$$

*For a mapping* $g$ *of* T *into a topological space* G *(Hausdorff or not) to be* $\mu$*-measurable, it is necessary and sufficient that* $g_{\mathrm{K}_\alpha}$ *be* $\mu_{\mathrm{K}_\alpha}$*-measurable for every* $\alpha \in \mathrm{A}$.

A) *Existence of a crushing*:

The proof is a repetition of that of Prop. 14 of Ch. IV, §5, No. 9, with slight modifications. Let $\mathfrak{K}$ be the set of compact subsets K of T such that $\mathrm{Supp}(\mu_\mathrm{K}) = \mathrm{K}$, and let $\mathscr{H}$ be the set (ordered by inclusion) of subsets $\mathfrak{L}$ of $\mathfrak{K}$ consisting of pairwise disjoint sets. Let us first show that every element $\mathfrak{L}$ of $\mathscr{H}$ is *locally countable*. Let $x$ be a point of T, and V an open neighborhood of $x$ such that $\mu^\bullet(\mathrm{V}) < +\infty$; let $\mathfrak{L}_\mathrm{V}$ be the set of $\mathrm{K} \in \mathfrak{L}$ that intersect V. If $(\mathrm{K}_i)_{1\leqslant i\leqslant n}$ is a finite sequence of distinct elements of $\mathfrak{L}_\mathrm{V}$, we have, by the Cor. of Prop. 4,

$$\sum_{i=1}^{n} \mu^\bullet(\mathrm{K}_i \cap \mathrm{V}) = \mu^\bullet\Big(\mathrm{V} \cap \Big(\bigcup_{i=1}^{n} \mathrm{K}_i\Big)\Big) \leqslant \mu^\bullet(\mathrm{V}),$$

because the $\mathrm{K}_i$ are pairwise disjoint. Thus,

$$\sum_{\mathrm{K}\in\mathfrak{L}_\mathrm{V}} \mu^\bullet(\mathrm{K}\cap\mathrm{V}) < +\infty .$$

Now, $\mu^\bullet(\mathrm{K}\cap\mathrm{V}) = \mu^\bullet_\mathrm{K}(\mathrm{K}\cap\mathrm{V}) > 0$ for every $\mathrm{K} \in \mathfrak{L}_\mathrm{V}$, because $\mathrm{K}\cap\mathrm{V}$ is nonempty, open in K, and the support of $\mu_\mathrm{K}$ is all of K; $\mathfrak{L}_\mathrm{V}$ is therefore countable, and $\mathfrak{L}$ is indeed locally countable. It is immediate that $\mathscr{H}$ is inductive, and nonempty (one has $\varnothing \in \mathscr{H}$). Thus, let $\mathfrak{H}$ be a maximal element of $\mathscr{H}$. We are going to show that the set $\mathrm{N} = \mathrm{T} - \bigcup_{\mathrm{K}\in\mathfrak{H}} \mathrm{K}$ is locally negligible. By Prop. 2, it suffices to verify that $\mu^\bullet(\mathrm{N}\cap\mathrm{L}) = 0$ for every compact set L, or again that $\mu^\bullet_\mathrm{L}(\mathrm{N}\cap\mathrm{L}) = 0$. We shall argue by contradiction. Thus, suppose that $\mu^\bullet_\mathrm{L}(\mathrm{N}\cap\mathrm{L}) > 0$. Since the set of $\mathrm{K} \in \mathfrak{H}$ that intersect L is countable, $\mathrm{N}\cap\mathrm{L}$ is $\mu_\mathrm{L}$-measurable; therefore there exists a compact set J contained in $\mathrm{N}\cap\mathrm{L}$ such that $\mu^\bullet_\mathrm{L}(\mathrm{J}) > 0$. Let S be the support of the nonzero measure $(\mu_\mathrm{L})_\mathrm{J} = \mu_\mathrm{J}$; it is contained in N, the measure $\mu_\mathrm{S}$ is nonzero, and $\mathrm{Supp}(\mu_\mathrm{S}) = \mathrm{S}$ (Ch. IV, §5, No. 7, Lemma 2). The set $\mathfrak{H} \cup \{\mathrm{S}\}$ therefore belongs to $\mathscr{H}$, in contradiction with the maximal character of $\mathfrak{H}$. This proves the existence of a crushing.

---

(1) We shall see later on (§2, No. 2) that $\mu_\alpha$ is the measure $\varphi_{\mathrm{K}_\alpha}\cdot\mu$.

B) *Proof of* (7):

For every $\alpha \in \mathrm{A}$, $\mu^\bullet_\alpha(f) = \mu^\bullet_{\mathrm{K}_\alpha}(f_{\mathrm{K}_\alpha}) = \mu^\bullet(f\varphi_{\mathrm{K}_\alpha})$ by formula (3) of No. 3 and Prop. 2 *a*) of No. 2; these formulas show that the encumbrance $\sum_{\alpha\in\mathrm{A}} \mu^\bullet_\alpha$ is $\leqslant \mu^\bullet$, hence that the family $(\mu_\alpha)_{\alpha\in\mathrm{A}}$ is summable (No. 7, Prop. 7). It thus suffices to show that $\mu = \sum_{\alpha\in\mathrm{A}} \mu_\alpha$, that is, to establish the formula

$$\mu^\bullet_{\mathrm{K}} = \sum_{\alpha\in\mathrm{A}} (\mu_\alpha)^\bullet_{\mathrm{K}} \tag{8}$$

for every compact subset K of T. Now, K being fixed, the set A′ of $\alpha \in \mathrm{A}$ such that $\mathrm{K}_\alpha$ intersects K is countable. Let $g \in \mathscr{F}_+(\mathrm{K})$; then $g^0 = g^0\varphi_{\mathrm{N}} + \sum_{\alpha\in\mathrm{A}} g^0\varphi_{\mathrm{K}_\alpha}$, and $g^0\varphi_{\mathrm{K}_\alpha} = 0$ for $\alpha \in \mathrm{A} - \mathrm{A}'$; by Prop. 4 of No. 5, it follows that $\mu^\bullet(g^0) = \sum_{\alpha\in\mathrm{A}} \mu^\bullet(g^0\varphi_{\mathrm{K}_\alpha})$, whence

$$\mu^\bullet_{\mathrm{K}}(g) = \mu^\bullet(g^0) = \sum_{\alpha\in\mathrm{A}} \mu^\bullet(g^0\varphi_{\mathrm{K}_\alpha}) = \sum_{\alpha\in\mathrm{A}} \mu^\bullet_\alpha(g^0) = \sum_{\alpha\in\mathrm{A}} (\mu_\alpha)^\bullet_{\mathrm{K}}(g)\,;$$

thus (8) is established.

C) *Measurability*:

For a function $g$ defined on T to be $\mu$-measurable, it is necessary and sufficient that it be $\mu_\alpha$-measurable for every $\alpha \in \mathrm{A}$ (No. 7, Prop. 8); but this amounts to saying that $g_{\mathrm{K}_\alpha}$ is $\mu_{\mathrm{K}_\alpha}$-measurable for all $\alpha \in \mathrm{A}$ (No. 5, *Example*). Q.E.D.

As in Prop. 14 of Ch. IV, §5, No. 9, one can require the compact sets $\mathrm{K}_\alpha$ to belong to a $\mu$-dense set of compact subsets of T, given in advance. We will only need the following result, which we shall establish directly:

PROPOSITION 10. — *If $g$ is a $\mu$-measurable mapping with values in a topological space* G *(Hausdorff or not), there exists a $\mu$-crushing $(\mathrm{L}_\beta)_{\beta\in\mathrm{B}}$ of* T *such that the restrictions $g_{\mathrm{L}_\beta}$ are continuous for all $\beta \in \mathrm{B}$.*

Consider a crushing $(\mathrm{K}_\alpha)_{\alpha\in\mathrm{A}}$ of T for $\mu$. Since the mapping $g$ is measurable, there exists for each $\alpha \in \mathrm{A}$ a partition of $\mathrm{K}_\alpha$ into a sequence $(\mathrm{K}_{\alpha n})$ of compact sets and a locally negligible set $\mathrm{N}_\alpha$, such that the restriction of $g$ to each of the sets $\mathrm{K}_{\alpha n}$ is continuous. The family $(\mathrm{K}_{\alpha n})_{(\alpha,n)\in\mathrm{A}\times\mathbf{N}}$ is then the sought-for crushing. For, it is locally countable, and the set $\mathrm{N}' = \mathrm{N} \cup \big(\bigcup_\alpha \mathrm{N}_\alpha\big)$ is locally negligible, because a compact set intersects at most a countable infinity of sets $\mathrm{N}_\alpha$.

SCHOLIUM. — Let $(\mathrm{K}_\alpha)_{\alpha\in\mathrm{A}}$ be a crushing of T, and let $\mathrm{N} = \mathrm{T} - \bigcup_\alpha \mathrm{K}_\alpha$. We denote by T′ the locally compact space obtained by equipping T with the sum topology of the topologies of the subspaces $K_\alpha$ and any locally compact topology on N (unless expressly mentioned to the contrary, N will always be equipped with the discrete topology). For each $\alpha \in A$, let $i_\alpha$ be the canonical injection of $K_\alpha$ into $T'$, and let $\mu'_\alpha$ be the measure on $T'$ that is the image of $\mu_{K_\alpha}$ under $i_\alpha$. The family $(\mu'_\alpha)$ is summable: for, if $f$ is a continuous function on $T'$ with compact support, then $\mathrm{Supp}(f)$ intersects $K_\alpha$ for only a finite number of indices $\alpha$. We set $\mu' = \sum_{\alpha \in A} \mu'_\alpha$. The set N being locally negligible for $\mu'$, since it is so for each $\mu'_\alpha$ (Prop. 9), the family $(K_\alpha)_{\alpha \in A}$ is a $\mu'$-crushing of $T'$; now, the measure induced by $\mu'$ on $K_\alpha$ is obviously $\mu_{K_\alpha}$ and the formula (7), applied to $\mu$ and to $\mu'$, shows that $\mu^\bullet = \mu'^\bullet$. Similarly, the last assertion of the statement of Prop. 9, applied to $\mu$ and to $\mu'$, shows that *the measurable mappings are the same for the two measures* $\mu$ *and* $\mu'$.

These two properties permit reducing nearly all of the theory of integration with respect to $\mu$ to the theory elaborated for locally compact spaces. These considerations will be developed in No. 10.

Here is another application of the concept of crushing:

PROPOSITION 11. — *Let* X *be a* $\mu$*-measurable subset of* T. *There exists a locally countable family* $(L_\alpha)_{\alpha \in A}$ *of compact subsets of* X, *pairwise disjoint, such that* $X - \bigcup_{\alpha \in A} L_\alpha$ *is locally* $\mu$*-negligible. If, in addition,* X *is the union of a sequence* $(X_n)$ *of measurable sets such that* $\mu^\bullet(X_n) < +\infty$, *then the set* B *of* $\alpha \in A$ *such that* $\mu^\bullet(L_\alpha) \neq 0$ *is countable, and* $X - \bigcup_{\beta \in B} L_\beta$ *is locally* $\mu$*-negligible.*

Let $f$ be the characteristic function of X, and let $(K_\alpha)_{\alpha \in A}$ be a crushing of T such that the restriction of $f$ to each of the $K_\alpha$ is continuous (Prop. 10). The set $L_\alpha = K_\alpha \cap X$ is then compact for every $\alpha \in A$, and $(L_\alpha)_{\alpha \in A}$ is the desired family. Let us pass to the second assertion; the measurable sets $X_n$ may clearly be assumed to be disjoint, and it suffices to establish the assertion for each of them. In other words, changing notation if necessary, we can suppose that $\mu^\bullet(X) < +\infty$. The set B of $\alpha \in A$ such that $\mu^\bullet(L_\alpha) > 0$ is then countable, and it only remains to prove that the set $N = \bigcup_{\alpha \in A - B} L_\alpha$ is locally negligible. But K is a compact set; the family $(L_\alpha)_{\alpha \in A}$ being locally countable, the set $K \cap N$ is the union of a *countable* subfamily of the family $(K \cap L_\alpha)_{\alpha \in A - B}$, and this set is therefore locally negligible. The same is then true of N (No. 2, Prop. 2) and the proposition is established.

## 9. Upper integral

DEFINITION 11. — *For every function* $f \in \mathscr{F}_+(\mathrm{T})$, *one defines the upper integral of* $f$ *(with respect to the measure* $\mu$*) to be the finite or infinite positive number*

$$\mu^*(f) = \inf_g \mu^\bullet(g), \tag{9}$$

*where* $g$ *runs over the set of lower semi-continuous functions that are* $\geqslant f$.

The notations $\int^* f(t)\,d\mu(t)$ and $\int^* f\,d\mu$ are also used. When T is locally compact, this definition coincides with the usual definition (Ch. V, §1, No. 1, Prop. 4). Clearly $\mu^\bullet(f) \leqslant \mu^*(f)$, with equality when $f$ is lower semi-continuous. If A is a subset of T, one writes $\mu^*(\mathrm{A})$ instead of $\mu^*(\varphi_\mathrm{A})$, and this number is called the *outer measure* of A. The measurable sets with finite outer measure are called *integrable sets*, as in the case of locally compact spaces.

A function $\mathbf{f}$ with values in a Banach space or in $\overline{\mathbf{R}}$ such that $\mu^*(|\mathbf{f}|) = 0$ is said to be *negligible*; a set $\mathrm{A} \subset \mathrm{T}$ is said to be negligible if $\varphi_\mathrm{A}$ is negligible, that is, if $\mu^*(\mathrm{A}) = 0$. The expression *almost everywhere* is introduced as in Ch. IV, §2, No. 3.

PROPOSITION 12. — *The function* $\mu^*$ *is an encumbrance on* T.

The properties *a*), *b*), *c*) of Def. 1 of No. 1 are obvious. The proof of the property *d*) is identical to that of Th. 3 of Ch. IV, §1, No. 3, on taking into account Props. 4 and 5 *a*).

COROLLARY. — *A function* $\mathbf{f}$, *with values in a Banach space or in* $\overline{\mathbf{R}}$, *is negligible if and only if* $\mathbf{f}(t) = 0$ *almost everywhere.*

One reduces immediately to the case of a positive function. The proof is then identical to that of Th. 1 of Ch. IV, §2, No. 3.

PROPOSITION 13. — *For every subset* A *of* T, $\mu^*(\mathrm{A})$ *is the infimum of the outer measures of the open sets containing* A.

The proof is identical to that of Prop. 19 of Ch. IV, §1, No. 4.

DEFINITION 12. — *Let* $\mathbf{f}$ *be a function defined on* T, *with values in a Banach space or in* $\overline{\mathbf{R}}$. *One says that* $\mathbf{f}$ *is moderated for the measure* $\mu$, *or* $\mu$*-moderated, if* $\mathbf{f}$ *is zero on the complement of a countable union of integrable open sets. A subset* A *of* T *is said to be moderated if the function* $\varphi_\mathrm{A}$ *is moderated. The measure* $\mu$ *is said to be moderated if the function* 1 *is* $\mu$*-moderated.*

For example, since the encumbrance $\mu^\bullet$ is locally bounded, every compact subset K of T is contained in an open set V such that $\mu^\bullet(\mathrm{V}) < +\infty$; a function that is zero outside a compact set is therefore moderated. A negligible function is moderated. The remarks following Def. 2 of Ch. V, §1, No. 2 can immediately be extended to the present context. In particular, the sum of a sequence of moderated positive functions is moderated.

*Remarks.* — 1) On a Lindelöf space T (TG, IX, Appendix I, Def. 1),[1] and in particular on a Souslin space (*ibid.*, Cor. of Prop. 1), every measure is moderated. For, the open sets of finite measure form a covering of T, from which one can extract a countable covering of T.

2) Beware, however, that the existence of a sequence of Borel sets of finite measure for $\mu$, with union T, does not necessarily imply the existence of a sequence of *open* sets of finite measure with union T (in other words, does not imply that $\mu$ is moderated). See Exer. 8.

PROPOSITION 14. — *Let $f \in \mathscr{F}_+(\mathrm{T})$. If $f$ is $\mu$-moderated, then $\mu^*(f) = \mu^\bullet(f)$; if $f$ is not $\mu$-moderated, then $\mu^*(f) = +\infty$.*

If $\mu^*(f) < +\infty$, there exists a lower semi-continuous function $g \geqslant f$ such that $\mu^\bullet(g) < +\infty$. For every $n \in \mathbf{N}$, let $\mathrm{G}_n$ be the set of $t \in \mathrm{T}$ such that $g(t) > 1/n$; the set $\mathrm{G}_n$ is open, one has $\mu^\bullet(\mathrm{G}_n) \leqslant n\mu^\bullet(g) < +\infty$, and $f$ is zero outside the union of the $\mathrm{G}_n$: the function $f$ is therefore moderated.

Next, let us show that $\mu^*$ and $\mu^\bullet$ have the same value for moderated functions. Since $\mu^*$ and $\mu^\bullet$ are encumbrances, it suffices to establish the relation $\mu^*(f) = \mu^\bullet(f)$ when $f$ is a positive function, bounded above by a constant M, and zero outside an open set G of finite measure, which we shall now do.

The measure $\mu$ is the supremum, in $\mathscr{M}(\mathrm{T})$, of an increasing directed family $(\mu_i)_{i\in\mathrm{I}}$ of measures with compact support: this follows at once from Prop. 9 of No. 8. Let $g$ be a lower semi-continuous function on T, between $f$ and the lower semi-continuous function $\mathrm{M}\varphi_\mathrm{G}$. Set $\nu_i = \mu - \mu_i$; then $\mu^\bullet = \mu_i^\bullet + \nu_i^\bullet$ (No. 2, *Remark* 1), consequently

$$\begin{aligned}\mu^\bullet(g) - \mu^\bullet(f) &= \big(\mu_i^\bullet(g) - \mu_i^\bullet(f)\big) + \big(\nu_i^\bullet(g) - \nu_i^\bullet(f)\big)\\ &\leqslant \big(\mu_i^\bullet(g) - \mu_i^\bullet(f)\big) + \nu_i^\bullet(\mathrm{M}\varphi_\mathrm{G}).\end{aligned}$$

One has $\nu_i^\bullet(\mathrm{M}\varphi_\mathrm{G}) = \mu^\bullet(\mathrm{M}\varphi_\mathrm{G}) - \mu_i^\bullet(\mathrm{M}\varphi_\mathrm{G})$ and $\mu^\bullet(\mathrm{M}\varphi_\mathrm{G}) = \sup \mu_i^\bullet(\mathrm{M}\varphi_\mathrm{G})$ (No. 7, Prop. 6); the number $\nu_i^\bullet(\mathrm{M}\varphi_\mathrm{G})$ may therefore be made arbitrarily small by a suitable choice of $i$. Thus everything comes down to showing that one can find, for any number $c > 0$ and any index $i \in \mathrm{I}$, a lower semi-continuous function $g$ between $f$ and $\mathrm{M}\varphi_\mathrm{G}$, such that $\mu_i^\bullet(g) - \mu_i^\bullet(f) \leqslant c$. Now, let L be the compact support of the measure $\mu_i$, and let $\lambda$ be the measure $(\mu_i)_\mathrm{L}$; since $\mu_i$ is concentrated on L, one has $\mu_i^\bullet(h) = \mu_i^\bullet(h\varphi_\mathrm{L}) = \lambda^\bullet(h_\mathrm{L})$ for every function $h \in \mathscr{F}_+(\mathrm{T})$ (No. 1, Lemma 1 and No. 2, Prop. 2); therefore

$$\mu_i^\bullet(g) - \mu_i^\bullet(f) = \lambda^\bullet(g_\mathrm{L}) - \lambda^\bullet(f_\mathrm{L}).$$

[1] The cited appendix in TG does not appear in GT (which translated an earlier edition of Ch. IX). Lindelöf spaces are defined in GT, I, §9, Exer. 14; this exercise and the definition of Souslin space (GT, IX, §6, No. 2, Def. 2) cover the material needed here.

But L is compact; therefore $\lambda^\bullet = \lambda^*$, consequently there exists a lower semi-continuous function $h$ defined on L, that is $\geqslant f_{\rm L}$ and is such that $\lambda^\bullet(h) \leqslant \lambda^\bullet(f_{\rm L})+c$. Since the set L is closed in T, the function $k$ equal to $h$ on L, and to $+\infty$ on T $-$ L, is lower semi-continuous on T and is $\geqslant f$, and $\lambda^\bullet(k_{\rm L}) = \lambda^\bullet(h) \leqslant \lambda^\bullet(f_{\rm L}) + c$. It remains only to set $g = \inf(k, {\rm M}\varphi_{\rm G})$: $g$ is lower semi-continuous, $g \geqslant f$ and

$$\mu_i^\bullet(g) \leqslant \mu_i^\bullet(k) = \lambda^\bullet(k_{\rm L}) \leqslant \lambda^\bullet(f_{\rm L}) + c = \mu_i^\bullet(f) + c.$$

COROLLARY 1. — *For a function to be negligible, it is necessary and sufficient that it be locally negligible and moderated.*

COROLLARY 2. — *For a function $f$ to be locally negligible, it is necessary and sufficient that every $x \in$ T possess a neighborhood V such that $f\varphi_{\rm V}$ is negligible.*

For, if this property is satisfied, $f\varphi_{\rm K}$ is negligible for every compact set K, and $f$ is therefore locally negligible (No. 2, Prop. 2). Conversely, suppose that $f$ is locally negligible, and let $x$ be a point of T; $x$ admits an open neighborhood V of finite measure. The function $f\varphi_{\rm V}$ is then locally negligible and moderated, hence is negligible.

COROLLARY 3. — *Let* $\mathbf{f}$ *be a moderated function defined on* T. *There exists a sequence* $({\rm K}_n)$ *of pairwise disjoint compact sets, and a negligible set* H, *such that* $\mathbf{f} = \mathbf{f}\varphi_{\rm H} + \sum\limits_n \mathbf{f}\varphi_{{\rm K}_n}$.

For, let G be a set that is a countable union of integrable open sets, such that $\mathbf{f}$ is zero outside G; then G is the union of a sequence $({\rm K}_n)$ of pairwise disjoint compact sets and a locally negligible set H (No. 8, Prop. 11); but H is moderated, therefore negligible.

COROLLARY 4. — *If $\mu$ and $\nu$ are two measures on* T *such that* $\mu^* = \nu^*$, *then* $\mu = \nu$.

For, the equality $\mu^* = \nu^*$ implies that $\mu^\bullet(f) = \nu^\bullet(f)$ for every positive function $f$ that is moderated for $\mu$ and $\nu$, hence for every positive function with compact support; it follows that $\mu^\bullet = \nu^\bullet$ (No. 2, Prop. 2), then $\mu = \nu$ (No. 2, Cor. of Prop. 2).

COROLLARY 5. — *If $\mu$ is a moderated measure on* T, *there exists a sequence $(\mu_n)_{n\in\mathbf{N}}$ of measures with compact support such that* $\mu = \sum\limits_{n\in\mathbf{N}} \mu_n$.

By hypothesis, the constant function 1 is $\mu$-moderated. Let us apply Cor. 3 to the case $f = 1$; thus, there exists a sequence $({\rm K}_n)_{n\in\mathbf{N}}$ of pairwise disjoint compact subsets of T such that $1 = \sum\limits_{n\in\mathbf{N}} \varphi_{{\rm K}_n}$ $\mu$-almost everywhere. Let $\mu_n$ be the measure on T defined by the measure $\mu_{{\rm K}_n}$ on ${\rm K}_n$ (No. 3, *Example* 2). One knows (No. 6, *Remark* 2) that $\mu_n$ has compact support, and that $\mu_n^\bullet(f) = \mu^\bullet(f\varphi_{K_n})$ for $f \in \mathscr{F}_+(T)$. Now, $f$ is equal to $\sum_{n\in\mathbf{N}} f\varphi_{K_n}$ $\mu$-almost everywhere, whence

$$\mu^\bullet(f) = \sum_{n\in\mathbf{N}} \mu^\bullet(f\varphi_{K_n}) = \sum_{n\in\mathbf{N}} \mu_n^\bullet(f).$$

It follows that $\mu = \sum_{n\in\mathbf{N}} \mu_n$ (No. 7, Prop. 7).

## 10. Integration theory

DEFINITION 13. — *Let $p \in [1, +\infty[$; one denotes by $\overline{\mathscr{L}}^p(T, \mu)$ (resp. $\overline{\mathscr{L}}^p_F(T, \mu)$ if F is a Banach space) the set of mappings* $\mathbf{f}$ *of* T *into* $\overline{\mathbf{R}}$ (resp. *into* F) *that are $\mu$-measurable and satisfy $\mu^\bullet(|\mathbf{f}|^p) < +\infty$. One denotes by $\mathscr{L}^p(T, \mu)$* (resp. $\mathscr{L}^p_F(T, \mu)$) *the set of $\mu$-moderated elements of $\overline{\mathscr{L}}^p(T, \mu)$* (resp. $\overline{\mathscr{L}}^p_F(T, \mu)$).

We will write $\overline{N}_p(\mathbf{f}) = \left(\mu^\bullet(|\mathbf{f}|^p)\right)^{1/p}$, $N_p(\mathbf{f}) = \left(\mu^*(|\mathbf{f}|^p)\right)^{1/p}$. We denote by $\overline{N}_\infty(\mathbf{f})$ the infimum of the numbers $k \geqslant 0$ such that $|\mathbf{f}| \leqslant k$ locally $\mu$-almost everywhere; if $\overline{N}_\infty(\mathbf{f}) < +\infty$, $\mathbf{f}$ is said to be essentially bounded. The set of measurable and essentially bounded mappings of T into $\overline{\mathbf{R}}$ (resp. into F) is denoted $\overline{\mathscr{L}}^\infty(T, \mu)$ (resp. $\overline{\mathscr{L}}^\infty_F(T, \mu)$). The elements of $\overline{\mathscr{L}}^1_F(T, \mu)$ (resp. $\mathscr{L}^1_F(T, \mu)$) are called essentially integrable functions (resp. integrable functions) with values in F.

If $\mu$ is a complex measure, one sets

$$\overline{\mathscr{L}}^p_F(T, \mu) = \overline{\mathscr{L}}^p_F(T, |\mu|) \quad \text{and} \quad \mathscr{L}^p_F(T, \mu) = \mathscr{L}^p_F(T, |\mu|).$$

The above notations are often abbreviated to $\overline{\mathscr{L}}^p_F(\mu)$, $\overline{\mathscr{L}}^p_F$ or $\mathscr{L}^p(\mu)$, $\mathscr{L}^p$, if this does not lead to any confusion.

We saw in No. 8 (*Scholium*) that one can construct a locally compact space T′, having the same underlying set as T and a topology finer than that of T, and equip T′ with a measure $\mu'$ such that the $\mu$-measurable functions and the $\mu'$-measurable functions are the same, and such that the essential upper integrals of positive functions for $\mu$ and $\mu'$ are equal. It follows that the sets $\overline{\mathscr{L}}^p_F(\mu)$ and $\overline{\mathscr{L}}^p_F(\mu')$ are identical for $1 \leqslant p \leqslant +\infty$.[1] This also implies without new proof that $\overline{\mathscr{L}}^p_F$ is a vector space, and that the function $\overline{N}_p$ is a semi-norm on $\overline{\mathscr{L}}^p_F(\mu)$, for which this space is complete.

Let $\mathbf{f}$ be an element of $\overline{\mathscr{L}}^p_F$ $(1 \leqslant p < +\infty)$; since one has $\mu^\bullet(|\mathbf{f}|^p) = \mu'^\bullet(|\mathbf{f}|^p) < +\infty$, Prop. 7 of Ch. V, §1, No. 2 implies that $\mathbf{f}$ is zero outside

(1) Note that the space $\mathscr{L}^p_F(\mu)$ is contained in $\mathscr{L}^p_F(\mu')$, but is in general distinct from it.

the union of a sequence of compact subsets of $T'$ and a locally $\mu'$-negligible set; the latter set being locally $\mu$-negligible, and every compact subset of $T'$ being compact in $T$, it follows that $\mathbf{f}$ is equal locally $\mu$-almost everywhere to a $\mu$-moderated function. Let us denote by $\overline{\mathscr{N}}_F$ (resp. $\mathscr{N}_F$) the space of locally $\mu$-negligible (resp. $\mu$-negligible) functions; we thus have $\overline{\mathscr{L}}^p_F = \mathscr{L}^p_F + \overline{\mathscr{N}}_F$, and $\mathscr{N}_F = \mathscr{L}^p_F \cap \overline{\mathscr{N}}_F$ (No. 9, Cor. 1 of Prop. 14). The space $\overline{\mathscr{L}}^p_F / \overline{\mathscr{N}}_F$ may therefore be canonically identified with $\mathscr{L}^p_F / \mathscr{N}_F$, and one verifies immediately that this identification preserves norm; this quotient space is denoted $L^p_F(\mu)$. It can be interpreted as the normed space associated with each of the semi-normed spaces $\overline{\mathscr{L}}^p_F(\mu)$ and $\mathscr{L}^p_F(\mu)$; since $\overline{\mathscr{L}}^p_F$ is complete, the same is true of $L^p_F$ and $\mathscr{L}^p_F$.

The set of functions $\mathbf{f}$ with values in $F$, continuous on $T'$ with compact support, is dense in $\overline{\mathscr{L}}^p_F(\mu') = \overline{\mathscr{L}}^p_F(\mu)$ (Ch. IV, §3, No. 4, Def. 2). Let us take up again the notations of the *Scholium* of No. 8. Since a compact subset of $T'$ intersects only a finite number of the compact sets $K_\alpha$, every continuous function $\mathbf{f}$ on $T'$ with compact support may be written as a sum

$$\mathbf{f} = \sum_{\alpha \in A} \mathbf{f}_\alpha + \mathbf{g},$$

where $\mathbf{f}_\alpha$ is, for every $\alpha$, the extension by 0 of a continuous function on $K_\alpha$, where $\mathbf{f}_\alpha = 0$ except for a finite number of indices, and where $\mathbf{g}$ is locally $\mu$-negligible. We thus have the following result:

PROPOSITION 15. — *The set of functions* $\mathbf{f}$ *with values in* $F$, *such that* $\mathrm{Supp}(\mathbf{f})$ *is compact and such that the restriction of* $\mathbf{f}$ *to* $\mathrm{Supp}(\mathbf{f})$ *is continuous, is dense in* $\overline{\mathscr{L}}^p_F(\mu)$ *and in* $\mathscr{L}^p_F(\mu)$, *for* $1 \leqslant p < +\infty$.

Note that these functions *are not* continuous functions *on* $T$ with compact support.

Let us pass to the definition of the integral.

PROPOSITION 16. — *There exists one and only one continuous linear mapping* $\mathbf{f} \mapsto \int \mathbf{f}\, d\mu$, *of the space* $\overline{\mathscr{L}}^1_F(\mu)$ *into* $F$, *having the following property*:

*If* $\mathbf{f}$ *is of the form* $t \mapsto g(t)\mathbf{a}$, *where* $\mathbf{a} \in F$, *and where* $g$ *is a positive function, finite,* $\mu$*-measurable and such that* $\mu^\bullet(g) < +\infty$, *then* $\int \mathbf{f}\, d\mu = \mu^\bullet(g) \cdot \mathbf{a}$.

For, the semi-normed spaces $\overline{\mathscr{L}}^1_F(\mu)$ and $\overline{\mathscr{L}}^1_F(\mu')$ are identical. Since $\mu^\bullet = \mu'^\bullet$, the mapping $\mathbf{f} \mapsto \int \mathbf{f}\, d\mu'$ satisfies the conditions of the statement. On the other hand, the set of functions of the form $\mathbf{f} = g \cdot \mathbf{a}$ considered in the statement is *total* in $\overline{\mathscr{L}}^1_F(\mu')$ (Ch. IV, §3, No. 5, Prop. 10), whence uniqueness.

*One says that* $\int \mathbf{f}\, d\mu$ *is the integral of* $\mathbf{f}$ *with respect to* $\mu$, *and this vector is also denoted* $\mu(\mathbf{f})$ *or* $\int \mathbf{f}(t)\, d\mu(t)$.

Since $\int \mathbf{f}\, d\mu = \int \mathbf{f}\, d\mu'$ for every essentially integrable function $\mathbf{f}$ with values in F, all of the theory of the essential integral extends to measures on Hausdorff spaces, without new proofs; from it, one deduces results relative to the ordinary integral by restricting oneself to moderated functions. We cite in particular the following results:

- Th. 3 of Ch. IV, §3, No. 4, its extension to $\overline{\mathscr{L}}^p_{\mathrm{F}}$, and its two corollaries.
- Th. 4 of Ch. IV, §3, No. 5 (composition with a continuous linear mapping) and its corollaries; Props. 9, 11 and 12 of the same No.
- All of the results of Ch. IV, §3, No. 6, relative to the ordered vector space structure of $\mathrm{L}^p$.
- All of the results of Ch. IV, §3, No. 7, and in particular Lebesgue's theorem.
- All of the results of Ch. IV, §3, No. 8, on the relations between the spaces $\mathrm{L}^p_{\mathrm{F}}$.
- Theorem 2 of Ch. IV, §4, No. 3 (the statement of Lebesgue's theorem specific to $\mathrm{L}^1_{\mathrm{F}}$).
- Hölder's inequality (Ch. IV, §6, No. 4, Th. 2) and its corollaries.
- The relations between the spaces $\mathrm{L}^p_{\mathrm{F}}$ established in Ch. IV, §6, No. 5.
- The results on the duality of the spaces $\mathrm{L}^p$ established in Ch. V, §5, No. 8.
- The Dunford–Pettis theorem (Ch. VI, §2, No. 5, Th. 1), its Corollaries 1 and 2, and Prop. 10 of Ch. VI, §2, No. 6 (dual of $\mathrm{L}^1_{\mathrm{F}}$).

## §2. OPERATIONS ON MEASURES

*As in the preceding section,* T *denotes a Hausdorff topological space, and* $\mu$ *a measure on* T. *Recall that all measures are assumed positive, absent mention to the contrary.*

### 1. Induced measure on a measurable subspace

Let X be a subset of T, and let $\nu$ be the restriction of the mapping $\mu : \mathrm{K} \mapsto \mu_{\mathrm{K}}$ to the set of compact subsets of X; it is clear that $\nu$ is a premeasure on X. On the other hand, let $x \in \mathrm{X}$ and let V be an open neighborhood of $x$ in T such that $\mu^\bullet(V) < +\infty$; then

$$\nu^\bullet(X \cap V) = \sup_{\substack{K \text{ compact} \\ K \subset X \cap V}} \mu^\bullet(K) \leqslant \mu^\bullet(V) < +\infty$$

by *Remark* 3 of §1, No. 2, so that $\nu$ is a measure.

When X is not $\mu$-measurable, the encumbrances $\nu^\bullet$ and $(\mu^\bullet)_X$ are not necessarily equal and the measure $\nu$ presents no interest.

DEFINITION 1. — *Let* X *be a* $\mu$*-measurable subset of* T. *The restriction of* $\mu : K \mapsto \mu_K$ *to the set of compact subsets of* X *is called the measure induced by* $\mu$ *on the subspace* X, *and is denoted by* $\mu_X$ *or* $\mu|X$.

PROPOSITION 1. — *Let* X *be a* $\mu$*-measurable subset of* T. *The encumbrance* $(\mu_X)^\bullet$ *is equal to the encumbrance* $(\mu^\bullet)_X$ *induced by* $\mu^\bullet$ *on* X (§1, No. 1). *In other words,* $(\mu_X)^\bullet(g) = \mu^\bullet(g^0)$ *for every function* $g \in \mathscr{F}_+(X)$.

Let $f \in \mathscr{F}_+(X)$ and let $f^0$ be the extension by 0 of $f$ to T. One has $(\mu^\bullet)_X(f) = \mu^\bullet(f^0) = \sup_L \mu^\bullet(f^0\varphi_L)$, where L runs over the set of compact subsets of T (§1, No. 2, Prop. 2); similarly $(\mu_X)^\bullet(f) = \sup_K \mu_K^\bullet(f_K) = \sup_K \mu^\bullet(f^0\varphi_K)$, where K runs over the set of compact subsets of X. Thus it all comes down to showing that $\mu^\bullet(f^0\varphi_L) = \sup_K \mu^\bullet(f^0\varphi_K)$ for every compact subset L of T, where K runs over the set of compact subsets of $L \cap X$. Now, let $(K_n)$ be an increasing sequence of compact sets contained in $L \cap X$, such that $(L \cap X) - \bigcup_n K_n$ is locally $\mu$-negligible (§1, No. 8, Prop. 11); $f^0$ being zero outside X, $f^0\varphi_L$ is zero outside $L \cap X$ hence is equal locally almost everywhere to the upper envelope of the sequence $(f^0\varphi_{K_n})$. This implies that $\mu^\bullet(f^0\varphi_L) = \sup_n \mu^\bullet(f^0\varphi_{K_n})$, whence the desired result.

*Remarks.* — 1) The relation $(\mu_X)^\bullet = (\mu^\bullet)_X$ permits using the notation $\mu_X^\bullet$ without ambiguity; we shall do so henceforth. The preceding Prop. 1 and Prop. 2 of §1, No. 2 show that the measures denoted $\mu_K$ until now, for K compact, are indeed induced measures in the sense of Def. 1. Similarly, if T is locally compact, and if X is a locally compact subspace of T, the above Prop. 1 and Prop. 1 of Ch. V, §7, No. 1 show that Def. 1 coincides with the definition of Ch. IV, §5, No. 7.

2) Def. 1 may be extended to the case that $\mu$ is a complex measure on T. To show in this case that the premeasure $\mu_X$ is a measure, it suffices to observe that $|\mu_K| = |\mu|_K$ for every compact subset K of X (§1, No. 2).

By Prop. 1, a subset Y of X is $\mu_X$-measurable (resp. locally $\mu_X$-negligible) if and only if it is $\mu$-measurable (resp. locally $\mu$-negligible). If Y is $\mu_X$-measurable, hence $\mu$-measurable, the induced measures $(\mu_X)_Y$ and $\mu_Y$ are obviously equal by virtue of Prop. 1 (*transitivity of induced measures*).

*Remark* 3). — Let X be a $\mu$-measurable subset of T. By Prop. 10 of §1, No. 8, applied to $g = \varphi_X$, there exists a crushing $(K_\alpha)_{\alpha \in A}$ of T such that for each $\alpha \in A$, either $K_\alpha \subset X$ or $K_\alpha \subset \complement X$. If one modifies the topology of T by the procedure of the Scholium of §1, No. 8, the space X′ obtained by equipping X with the topology induced by T′ is locally compact, and one knows how to associate to $\mu$ (resp. to $\mu_X$) a measure $\mu'$ (resp. $\nu$) on T′ (resp. on X′) that admits the same essential upper integral as $\mu$ (resp. $\mu_X$): this implies that $\mu'_{X'} = \nu$. Since the locally negligible sets, measurable mappings, and essentially integrable functions with values in a Banach space, are the same for $\mu$ and $\mu'$, and for $\mu_X$ and $(\mu_X)' = \nu = \mu'_{X'}$, the theory of integration with respect to an induced measure reduces to that which was treated in Ch. V, §7 in the special case of locally compact spaces. We leave to the reader the task of transcribing the results.

## 2. Measures defined by numerical densities

DEFINITION 2. — *A function* **f** *defined on* T, *with values in* $\overline{\mathbf{R}}$ *or in a Banach space, is said to be locally* $\mu$*-integrable if* **f** *is* $\mu$*-measurable and every point* $x \in T$ *admits a neighborhood* V *such that* $\mu^\bullet(|\mathbf{f}|\varphi_V) < +\infty$.

This definition coincides with the one given in Ch. V, §5, No. 1, in case T is locally compact.

Let $f$ be a locally $\mu$-integrable positive function; the mapping $K \mapsto f_K \cdot \mu_K$ is a premeasure (Ch. V, §7, No. 1, Cor. 2 of Th. 1), which will be denoted $f \cdot \mu$.

PROPOSITION 2. — *If* $f$ *is a positive locally* $\mu$*-integrable function, then, for every function* $g \in \mathscr{F}_+(T)$, *one has the relation*

$$(f \cdot \mu)^\bullet(g) = \mu^\bullet(fg). \tag{1}$$

Indeed, for every compact set K in T,

$$(f \cdot \mu)^\bullet_K(g_K) = (f_K \cdot \mu_K)^\bullet(g_K) = \mu^\bullet_K(f_K g_K) = \mu^\bullet_K((fg)_K),$$

on using the definition of $f \cdot \mu$ and Prop. 3 of Ch. V, §5, No. 3. Prop. 2 follows from this on passing to the supremum over K.

Now let $x \in T$ and let V be a neighborhood of $x$ such that $\mu^\bullet(f\varphi_V) < +\infty$ (Def. 2); then $(f \cdot \mu)^\bullet(V) = \mu^\bullet(f\varphi_V) < +\infty$, therefore $f \cdot \mu$ is a measure.

DEFINITION 3. — *Let* $f$ *be a locally* $\mu$*-integrable positive function. The measure* $f \cdot \mu : K \mapsto f_K \cdot \mu_K$ *is called the measure with density* $f$ *with respect* *to $\mu$, or the product measure of $\mu$ by the function $f$. Every measure of the form $f \cdot \mu$, where $f$ is positive and locally $\mu$-integrable, is called a measure with base $\mu$.*

*Remarks.* — 1) The definition of $f \cdot \mu$ extends to the case that $f$ is a complex locally integrable function; one then has $|f \cdot \mu| = |f| \cdot \mu$, which implies at once that $f \cdot \mu$ is a measure, not just a premeasure. We retain the expression 'measures with base $\mu$' to designate the complex measures so defined.

2) Similarly, if $\theta$ is a complex measure, $f$ is said to be locally $\theta$-integrable if it is locally $|\theta|$-integrable, and one defines the measure $f \cdot \theta : \mathrm{K} \mapsto f_{\mathrm{K}} \cdot \theta_{\mathrm{K}}$. One has $|f \cdot \theta| = |f| \cdot |\theta|$ (Ch. V, §5, No. 2, Prop. 2). In this No., we shall leave aside everything concerning non-positive measures.

PROPOSITION 3. — *Let $\nu$ be a measure on $\mathrm{T}$. For $\nu$ to be of the form $f \cdot \mu$, where $f$ is a locally $\mu$-integrable positive function, it is necessary and sufficient that every $\mu$-negligible compact set be $\nu$-negligible. If $f'$ is a second locally $\mu$-integrable function such that $\nu = f' \cdot \mu$, then $f = f'$ locally $\mu$-almost everywhere.*

The condition is obviously necessary (Prop. 2). Conversely, suppose that every $\mu$-negligible compact set is $\nu$-negligible. Let us introduce a crushing $(\mathrm{K}_\alpha)_{\alpha \in \mathrm{A}}$ of $\mathrm{T}$ for the measure $\mu + \nu$ and let us set $\mathrm{N} = \mathrm{T} - \bigcup_{\alpha \in \mathrm{A}} \mathrm{K}_\alpha$. It is clear that $(\mathrm{K}_\alpha)_{\alpha \in \mathrm{A}}$ is a crushing for $\mu$ and for $\nu$, and Prop. 9 of §1, No. 3 therefore implies the following relations for every $g \in \mathscr{F}_+$:

$$\mu^\bullet(g) = \sum_{\alpha \in \mathrm{A}} \mu^\bullet_{\mathrm{K}_\alpha}(g_{\mathrm{K}_\alpha}), \quad \nu^\bullet(g) = \sum_{\alpha \in \mathrm{A}} \nu^\bullet_{\mathrm{K}_\alpha}(g_{\mathrm{K}_\alpha}).$$

Consider a compact set $\mathrm{C} \subset \mathrm{K}_\alpha$ that is $\mu_{\mathrm{K}_\alpha}$-negligible; then $\mathrm{C}$ is locally $\mu$-negligible, hence locally $\nu$-negligible, and finally $\nu_{\mathrm{K}_\alpha}$-negligible by the definition of $\nu$. It then follows from the Lebesgue–Nikodym theorem (Ch. V, §5, No. 5, Th. 2) that $\nu_{\mathrm{K}_\alpha}$ admits a density $f_\alpha$ with respect to $\mu_{\mathrm{K}_\alpha}$. Let $f$ be the function that coincides with $f_\alpha$ on each of the sets $\mathrm{K}_\alpha$, and with 0 on $\mathrm{N}$; the function $f$ is $\mu$-measurable (Ch. IV, §5, No. 10, Prop. 16), and for every function $g \in \mathscr{F}_+$ one has, by the above relations and Prop. 3 of Ch. V, §5, No. 3,

$$\nu^\bullet(g) = \sum_{\alpha \in \mathrm{A}} \nu^\bullet_{\mathrm{K}_\alpha}(g_{\mathrm{K}_\alpha}) = \sum_{\alpha \in \mathrm{A}} \mu^\bullet_{\mathrm{K}_\alpha}(f_\alpha g_{\mathrm{K}_\alpha}) = \sum_{\alpha \in \mathrm{A}} \mu^\bullet_{\mathrm{K}_\alpha}((fg)_{\mathrm{K}_\alpha}) = \mu^\bullet(fg).$$

It then follows first of all that $f$ is locally $\mu$-integrable: if $x$ is a point of $\mathrm{T}$, and if $\mathrm{V}$ is a neighborhood of $x$ such that $\nu^\bullet(\mathrm{V}) < +\infty$, then $\mu^\bullet(f\varphi_{\mathrm{V}}) < +\infty$. Next, Prop. 2 shows that the measures $\nu$ and $f \cdot \mu$ have the same essential upper integral. They are therefore equal (§1, No. 2, Cor. of Prop. 2). The uniqueness of $f$ being obvious based on the case of compact spaces, the proposition is established.

*Remark* 3). — The theory of integration with respect to a measure $\nu = f \cdot \mu$ reduces at once to the theory treated in Ch. V. For, let $(\mathrm{K}_\alpha)_{\alpha \in \mathrm{A}}$ be a crushing of T for $\mu$, hence for $\nu$, and let T′ be the locally compact space defined in the *Scholium* of §1, No. 8; we can associate to $\mu$ (resp. to $\nu$) a measure $\mu'$ (resp. $\nu'$) on T′, in such a way that the measurable functions, the essentially integrable functions with values in a Banach space, and the essential upper integrals of positive functions, are the same for $\mu$ and $\mu'$ (resp. for $\nu$ and $\nu'$). The function $f$ is therefore $\mu'$-measurable; it is locally $\mu'$-integrable, because T′ is locally compact, and a compact subset of T′ intersects only a finite number of the compact sets $\mathrm{K}_\alpha$ $(\alpha \in \mathrm{A})$. Finally, the relation $\nu'^\bullet(g) = \nu^\bullet(g) = \mu^\bullet(fg) = \mu'^\bullet(fg)$ proves that $\nu' = f \cdot \mu'$ (Ch. V, §5, No. 3, Prop. 3). We leave to the reader the task of transcribing the results of Ch. V, §5.

## 3. Image of a measure

DEFINITION 4. — *Let* $\pi$ *be a mapping of* T *into a topological space* X. *One says that* $\pi$ *is* $\mu$*-proper if* $\pi$ *is* $\mu$*-measurable, and if every point* $x$ *of* X *admits a neighborhood* V *such that* $\mu^\bullet(\overset{-1}{\pi}(\mathrm{V})) < +\infty$.

*Remarks.* — 1) When T and X are locally compact, this definition is equivalent to that of Ch. V, §6, No. 1.

2) A proper continuous mapping (GT, I, §10, No. 1, Def. 1) of T into X is $\mu$-proper for every measure $\mu$. For, let $x \in \mathrm{X}$; since $\overset{-1}{\pi}(x)$ is compact (*loc. cit.*, No. 2, Th. 1), the set $\overset{-1}{\pi}(x)$ has an open neighborhood H such that $\mu^\bullet(\mathrm{H}) < +\infty$. Set $\mathrm{V} = \mathrm{X} - \pi(\mathrm{T} - \mathrm{H})$; since $\pi$ is closed, V is open in X, contains $x$, and satisfies $\overset{-1}{\pi}(\mathrm{V}) \subset \mathrm{H}$, whence $\mu^\bullet(\overset{-1}{\pi}(\mathrm{V})) \leqslant \mu^\bullet(\mathrm{H}) < +\infty$.

3) If $\mu$ is bounded, every $\mu$-measurable mapping of T into X is $\mu$-proper.

4) If $\theta$ is a complex measure on T, $\pi$ is said to be $\theta$-proper if $\pi$ is proper for the positive measure $|\theta|$.

PROPOSITION 4. — *Let* $\pi$ *be a* $\mu$*-proper mapping of* T *into a topological space* X. *There exists one and only one measure* $\nu$ *on* X *such that* $\nu^\bullet$ *is equal to the image encumbrance* $\pi(\mu^\bullet)$ (§1, No. 1), *in other words, such that* $\nu^\bullet(g) = \mu^\bullet(g \circ \pi)$ *for all* $g \in \mathscr{F}_+(\mathrm{X})$.

Uniqueness is obvious (§1, No. 2, Cor. of Prop. 2). To establish existence, we shall first treat the case that $\mu$ is carried by a compact set K, such that the restriction of $\pi$ to K is continuous. Then $\mathrm{L} = \pi(\mathrm{K})$ is compact; let $\pi'$ be the continuous mapping of K into L induced by $\pi$, and let $\nu'$ be the image measure $\pi'(\mu_\mathrm{K})$ on L, $\nu$ the measure on X defined by $\nu'$ (§1, No. 3, *Example* 2). For every $g \in \mathscr{F}_+(\mathrm{X})$,

$$\nu^\bullet(g) = \nu'^\bullet(g_\mathrm{L}) = \mu_\mathrm{K}^\bullet(g_\mathrm{L} \circ \pi') = \mu_\mathrm{K}^\bullet((g \circ \pi)_\mathrm{K}) = \mu^\bullet((g \circ \pi)_\mathrm{K}^0) = \mu^\bullet(g \circ \pi)$$

(we have used successively formula (3) of §1, No. 3; Prop. 2 of Ch. V, §6, No. 2; the definition of $\mu_K^\bullet$ ; and the fact that $\mu$ is carried by K). In other words, $\nu^\bullet = \pi(\mu^\bullet)$.

Let us now pass to the general case; by Props. 10 and 9 of §1, No. 8, $\mu$ is the sum of a summable family $(\mu_\alpha)_{\alpha\in A}$ of measures with compact support, such that the restriction of $\pi$ to the support $K_\alpha$ of $\mu_\alpha$ is continuous for every $\alpha \in A$. The special case treated above permits associating to each measure $\mu_\alpha$ on T a measure $\nu_\alpha$ on X such that $\nu_\alpha^\bullet = \pi(\mu_\alpha^\bullet)$. Then, for $g \in \mathscr{F}_+(X)$,

$$\sum_{\alpha\in A} \nu_\alpha^\bullet(g) = \sum_{\alpha\in A} \mu_\alpha^\bullet(g \circ \pi) = \mu^\bullet(g \circ \pi).$$

The encumbrance $\pi(\mu^\bullet)$ is locally bounded, since $\pi$ is $\mu$-proper; the family $(\nu_\alpha)_{\alpha\in A}$ is therefore summable (§1, No. 7, Prop. 7), and its sum $\nu$ satisfies the statement.

DEFINITION 5. — *If $\pi$ is a $\mu$-proper mapping of* T *into a topological space* X, *the unique measure $\nu$ on* X *such that $\nu^\bullet(g) = \mu^\bullet(g \circ \pi)$ for all $g \in \mathscr{F}_+(X)$ is called the image measure of $\mu$ under $\pi$, and is denoted $\pi(\mu)$.*

*Example.* — Let K be a compact subspace of T, $i$ the canonical injection of K into T, and $\lambda$ a measure on K; since $i$ is continuous and $\lambda$ is bounded, $i$ is $\lambda$-proper. Formula (3) of §1, No. 3 shows that the "measure on T defined by $\lambda$" is the image measure $i(\lambda)$.

*Remark* 5). — If $\theta$ is a real measure and $\pi$ is $\theta$-proper, then $\pi$ is proper for the measures $\theta^+$ and $\theta^-$; one then sets $\pi(\theta) = \pi(\theta^+) - \pi(\theta^-)$. If $\theta$ is a complex measure and $\pi$ is $\theta$-proper, then $\pi$ is proper for the real measures $\mathscr{R}(\theta)$ and $\mathscr{I}(\theta)$; one then sets

$$\pi(\theta) = \pi(\mathscr{R}(\theta)) + i\pi(\mathscr{I}(\theta)).$$

PROPOSITION 5. — *Let $\pi$ be a $\mu$-proper mapping of* T *into a topological space* X, *and let $f$ be a mapping of* X *into a topological space* F (*Hausdorff or not*). *For $f$ to be $\pi(\mu)$-measurable, it is necessary and sufficient that $f\circ\pi$ be $\mu$-measurable.*

Let us take up again the proof of Prop. 4, and commence with the special case treated at the beginning, with the same notations; $g$ is measurable for the measure $\pi(\mu) = \nu$ if and only if $g_L$ is $\nu'$-measurable (§1, No. 5, *Example*); now, this is equivalent to saying that $g_L \circ \pi' = (g \circ \pi)_K$ is $\mu_K$-measurable (Ch. V, §6, No. 2, Prop. 3), and finally that $g\circ\pi$ is $\mu$-measurable (§1, No. 5, *Example*). Let us now pass to the general case, with the same notations as in the proof of Prop. 4; $f$ is $\nu$-measurable if and only if $f$ is $\nu_\alpha$-measurable for every $\alpha \in A$ (§1, No. 7, Prop. 8), hence if and only if $f\circ\pi$ is $\mu_\alpha$-measurable for every $\alpha \in A$ (the preceding special case) and finally if and only if $f \circ \pi$ is $\mu$-measurable (§1, No. 7, Prop. 8).

COROLLARY. — *Let* X *and* Y *be two topological spaces,* $\pi$ *a* $\mu$*-proper mapping of* T *into* X, *and* $\pi'$ *a* $\pi(\mu)$*-proper mapping of* X *into* Y. *The mapping* $\pi'' = \pi' \circ \pi$ *is then* $\mu$*-proper, and* $\pi''(\mu) = \pi'(\pi(\mu))$ ('transitivity of images of measures').

For, $\pi''$ is $\mu$-measurable (Prop. 5). Set $\mu' = \pi(\mu)$; the image encumbrance $\pi'(\mu'^{\bullet}) = \pi'(\pi(\mu^{\bullet}))$ is obviously equal to $\pi''(\mu^{\bullet})$. Since it is locally bounded, $\pi''$ is $\mu$-proper. The measures $\pi''(\mu)$ and $\pi'(\mu')$ then have the same essential upper integral, hence are equal.

PROPOSITION 6. — *Let* $\pi$ *be a* $\mu$*-proper mapping of* T *into a topological space* X, *and let* B *be a* $\pi(\mu)$*-measurable subset of* X. *Set* $A = \overset{-1}{\pi}(B)$, *and denote by* $\pi'$ *the mapping of* A *into* B *that coincides with* $\pi$ *on* A. *The set* A *is then* $\mu$*-measurable,* $\pi_A$ *and* $\pi'$ *are* $\mu_A$*-proper, and*

$$(2) \qquad (\pi(\mu))_B = (\pi_A(\mu_A))_B = \pi'(\mu_A).$$

The set A is $\mu$-measurable by Prop. 5 applied to $\varphi_B$; the mapping $\pi_A$ is clearly $\mu_A$-measurable by the definition of induced measures (No. 1), and it follows that $\pi'$ is measurable. Let $f$ be an element of $\mathscr{F}_+(B)$; denoting by zero exponents the extensions by 0 in X and in T, we have

$$(\pi(\mu)_B)^{\bullet}(f) = \pi(\mu)^{\bullet}(f^0) = \mu^{\bullet}(f^0 \circ \pi) = \mu^{\bullet}((f \circ \pi')^0) = \mu_A^{\bullet}(f \circ \pi'),$$

whence $(\pi(\mu)_B)^{\bullet} = \pi'(\mu_A^{\bullet})$. Since the encumbrance $(\pi(\mu)_B)^{\bullet}$ is locally bounded, the same is true of $\pi'(\mu_A^{\bullet})$ and therefore $\pi'$ is $\mu_A$-proper. The measures $\pi'(\mu_A)$ and $(\pi(\mu))_B$ have the same essential upper integral, hence are equal. The other relation may be established in an analogous manner.

PROPOSITION 7. — *Let* T *be a subspace of a topological space* X, *and let* $i$ *be the injection of* T *into* X.

*a) If* $\mu$ *is a measure on* T, *and if* $i$ *is* $\mu$*-proper, then the measure* $i(\mu)$ *is concentrated on* T, *and one has* $(i(\mu))_T = \mu$.

*b) If* $\lambda$ *is a measure on* X *such that* T *is* $\lambda$*-measurable, then* $i$ *is* $\lambda_T$*-proper, and* $i(\lambda_T) = \varphi_T \cdot \lambda$.

*a)* Set $\nu = i(\mu)$; the relation $\nu^{\bullet}(A) = \mu^{\bullet}(A \cap T)$, applied to $A = X - T$, shows that $\nu$ is concentrated on T. The relation $\nu_T = \mu$ is a special case of the relation (2), on taking $B = T = A$.

*b)* Let $f$ be a positive function defined on X; setting $\mu = \lambda_T$, one has $\mu^{\bullet}(f \circ i) = \lambda_T^{\bullet}(f_T) = \lambda^{\bullet}(f\varphi_T) \leqslant \lambda^{\bullet}(f)$ (Prop. 1); it follows that $i$ is $\mu$-proper. On the other hand, $\mu^{\bullet}(f \circ i)$ (resp. $\lambda^{\bullet}(f\varphi_T)$) is the essential upper integral of $f$ with respect to $i(\mu)$ (resp. $\varphi_{\mathrm{T}} \cdot \lambda$). These two measures are therefore equal.

*Remark* 6). — Let $\pi$ be a $\mu$-proper mapping of T into a topological space X. One reduces the theory of integration with respect to the image measure $\nu = \pi(\mu)$ to the theory treated in Ch. V, §6, in the following way. Let $(\mathrm{K}_\alpha)_{\alpha \in \mathrm{A}}$ (resp. $(\mathrm{L}_\beta)_{\beta \in \mathrm{B}}$) be a crushing of T (resp. of X) for $\mu$ (resp. for $\nu$), and set $\mathrm{N} = \mathrm{T} - \bigcup_{\alpha \in \mathrm{A}} \mathrm{K}_\alpha$, $\mathrm{P} = \mathrm{X} - \bigcup_{\beta \in \mathrm{B}} \mathrm{L}_\beta$. We can suppose that the restriction of $\pi$ to each $\mathrm{K}_\alpha$ is continuous (§1, No. 8, Prop. 10). Let $\mathrm{T}', \mathrm{X}'$ be the locally compact spaces constructed as in the Scholium of §1, No. 8 and let $\mu'$ and $\nu'$ be the measures on these spaces associated with $\mu$ and $\nu$. The topology of $\mathrm{T}'$ being the sum of the topologies of the subspaces $\mathrm{K}_\alpha$ and the discrete topology on N, $\pi$ is a continuous mapping of $\mathrm{T}'$ into X and the relation $\mu'^\bullet(g \circ \pi) = \mu^\bullet(g \circ \pi) = \nu^\bullet(g)$ (for $g \in \mathscr{F}_+(\mathrm{X})$) shows that $\pi$ is $\mu'$-proper and that $\pi(\mu') = \nu$. On the other hand, the identity mapping $i$ of X onto $\mathrm{X}'$ is $\nu$-proper, and $i(\nu) = \nu'$. It follows that $\pi$ is a $\mu'$-proper mapping of $\mathrm{T}'$ into $\mathrm{X}'$, and that the image of $\mu'$ under $\pi$ is $\nu'$ (Cor. of Prop. 5). We leave to the reader the task of transcribing the results of Ch. V, §6.

### 4. Lifting of measures

PROPOSITION 8. — *Let* T *and* X *be two topological spaces,* $\pi$ *a mapping of* T *into* X.

*a) Let* $\nu$ *be a bounded measure on* X. *In order that there exist a measure* $\mu$ *on* T *such that* $\pi$ *is* $\mu$*-proper and* $\pi(\mu) = \nu$, *it is necessary and sufficient that there exist, for every number* $\varepsilon > 0$, *a compact set* $\mathrm{K}_\varepsilon \subset \mathrm{T}$ *such that the restriction of* $\pi$ *to* $\mathrm{K}_\varepsilon$ *is continuous and* $\nu^\bullet\big(\mathrm{X} - \pi(\mathrm{K}_\varepsilon)\big) < \varepsilon$.

*b) Suppose that* $\pi$ *is injective; let* $\mu$ *and* $\mu'$ *be two measures on* T, *such that* $\pi$ *is proper for* $\mu$ *and* $\mu'$, *and such that* $\pi(\mu) = \pi(\mu')$. *Then* $\mu = \mu'$.

The condition stated in *a*) is necessary. For, if $\pi$ is $\mu$-proper and $\pi(\mu) = \nu$, the relation $\mu^\bullet(1) = \nu^\bullet(1) < +\infty$ implies that $\mu$ is bounded. Prop. 2 of §1, No. 2, applied to the function 1, implies the existence of a compact subset K of T such that $\mu^\bullet(\mathrm{T} - \mathrm{K}) < \varepsilon/2$. Since $\pi$ is $\mu$-measurable, there exists a compact set $\mathrm{K}_\varepsilon \subset \mathrm{K}$ such that the restriction of $\pi$ to $\mathrm{K}_\varepsilon$ is continuous, and such that $\mu^\bullet(\mathrm{K} - \mathrm{K}_\varepsilon) < \varepsilon/2$. Then (No. 3, Prop. 4)

$$\nu^\bullet\big(\mathrm{X} - \pi(\mathrm{K}_\varepsilon)\big) = \mu^\bullet\big(\mathrm{T} - \overset{-1}{\pi}(\pi(\mathrm{K}_\varepsilon))\big) < \varepsilon .$$

To show that the condition is sufficient, we first treat a special case.

*Lemma* 1. — *Let* U *and* V *be two compact spaces,* $h$ *a continuous mapping of* U *onto* V. *The mapping* $\lambda \mapsto h(\lambda)$ *of* $\mathscr{M}_+(\mathrm{U})$ *into* $\mathscr{M}_+(\mathrm{V})$ *is then surjective.*

For, let $a$ be the linear mapping $f \mapsto f \circ h$ of $\mathscr{C}(\mathrm{V})$ into $\mathscr{C}(\mathrm{U})$; since $h$ is surjective, $a$ is an isometry of $\mathscr{C}(\mathrm{V})$ onto a subspace H of $\mathscr{C}(\mathrm{U})$. Let $\theta$ be a positive measure on V; then $\theta \circ a^{-1}$ is a continuous linear form on H, which is extendible to a linear form $\eta$ on $\mathscr{C}(\mathrm{U})$ with the same norm, by virtue of the Hahn–Banach theorem (TVS, II, §3, No. 2, Cor. 3 of Th. 1); $\eta$ is then a measure on U, and $\theta(f) = \eta(f \circ h)$ for all $f \in \mathscr{C}(\mathrm{V})$, so that $\theta = h(\eta)$. Finally, $\theta(1) = \|\theta\| = \|\eta\|$, and $\theta(1) = \eta(1)$, so that $\eta$ is positive (Ch. V, §5, No. 5, Prop. 9).

Let us now prove the sufficiency of the condition stated in *a*). The condition implies the existence of a sequence $(\mathrm{K}_n)_{n\geqslant 1}$ of compact subsets of T, such that the restriction of $\pi$ to each $\mathrm{K}_n$ is continuous, and such that, for every $n$, $\nu^\bullet\big(\mathrm{X} - \pi(\mathrm{K}_n)\big) < 1/n$. The sequence $(\mathrm{K}_n)$ can be assumed to be increasing. Set $\mathrm{L}_n = \pi(\mathrm{K}_n)$ and denote by $\nu'_n$ the measure $\varphi_{\mathrm{L}_n - \mathrm{L}_{n-1}} \cdot \nu_{\mathrm{L}_n}$ on $\mathrm{L}_n$, with the convention $\mathrm{L}_0 = \varnothing$.

The restriction $\pi_{\mathrm{K}_n}$ being continuous, there exists a measure $\mu'_n$ on $\mathrm{K}_n$ such that $\pi_{\mathrm{K}_n}(\mu'_n) = \nu'_n$ (Lemma 1). Let $\mu_n$ be the image of $\mu'_n$ under the canonical injection of $\mathrm{K}_n$ into T, and let $g$ be an element of $\mathscr{F}_+(\mathrm{X})$. Using successively the fact that $\nu$ is concentrated on $\bigcup_n \mathrm{L}_n$, Prop. 4 of §1, No. 5; Prop. 2 of §1, No. 2; Prop. 4 of No. 3, and finally Prop. 7 of No. 3, we have

$$\begin{aligned}\nu^\bullet(g) &= \sum_n \nu^\bullet(\varphi_{\mathrm{L}_n - \mathrm{L}_{n-1}} g) = \sum_n \nu'^\bullet_n(g_{\mathrm{L}_n}) = \sum_n \mu'^\bullet_n(g_{\mathrm{L}_n} \circ \pi_{\mathrm{K}_n}) \\ &= \sum_n \mu'^\bullet_n\big((g \circ \pi)_{\mathrm{K}_n}\big) = \sum_n \mu^\bullet_n(g \circ \pi).\end{aligned}$$

Taking $g = 1$ in this formula, one sees that the family $(\mu_n)$ is summable and that its sum is a bounded measure $\mu$ (§1, No. 7, Prop. 7). By Prop. 5 of No. 3, the mapping $\pi$ is $\mu_n$-measurable for all $n$, because $\pi_{\mathrm{K}_n}$ is continuous, hence $\mu'_n$-measurable; it follows that $\pi$ is $\mu$-measurable (§1, No. 7, Prop. 8), hence $\mu$-proper since $\mu$ is bounded. The above relations then prove that the measures $\pi(\mu)$ and $\nu$ have the same essential upper integral, hence are equal (§1, No. 2, Cor. of Prop. 2).

Finally, let us assume that $\pi$ is injective, and let us prove *b*). Let $f$ be an element of $\mathscr{F}_+(\mathrm{T})$; since $\pi$ is injective, there exists a function $g \in \mathscr{F}_+(\mathrm{X})$ such that $f = g \circ \pi$ and, setting $\nu = \pi(\mu) = \pi(\mu')$, by Prop. 4 of No. 3 we have

$$\mu^\bullet(f) = \mu^\bullet(g \circ \pi) = \nu^\bullet(g) = \mu'^\bullet(g \circ \pi) = \mu'^\bullet(f).$$

The two measures $\mu$ and $\mu'$ thus have the same essential upper integral, which implies their equality (§1, No. 2, Cor. of Prop. 2).

*Remark.* — Suppose that $\pi$ is injective. Let $\theta$ be a complex measure such that $\pi$ is $\theta$-proper and $\pi(\theta) = 0$; then $\theta = 0$. Indeed, by separating $\theta$ into its real and imaginary parts, one can reduce to the case that $\theta$ is real. We then have $\pi(\theta^+) = \pi(\theta^-)$, therefore $\theta^+ = \theta^-$ (Prop. 8), and finally $\theta = 0$.

Here is an important case where condition *a*) of Prop. 8 is always satisfied.

PROPOSITION 9. — *Let* T *be a Souslin space* (GT, IX, §6, No. 2, Def. 2), X *a Hausdorff space,* $\pi$ *a continuous mapping of* T *onto* X, *and* $\nu$ *a bounded measure on* X. *Then there exists a bounded measure* $\mu$ *on* T *such that* $\pi(\mu) = \nu$.

The hypotheses obviously imply that X is a Souslin space.

Let us consider the set function $c : \mathrm{A} \mapsto \nu^\bullet(\pi(\mathrm{A}))$ on $\mathfrak{P}(\mathrm{T})$. The relation $\mathrm{A} \subset \mathrm{B}$ implies $c(\mathrm{A}) \leqslant c(\mathrm{B})$; if $(\mathrm{A}_n)$ is an increasing sequence of subsets of T, and if $\mathrm{A} = \bigcup_{n \in \mathbf{N}} \mathrm{A}_n$, then $c(\mathrm{A}) = \sup_n c(\mathrm{A}_n)$ from the fact that $\nu^\bullet$ is an encumbrance. Finally, let $\mathrm{A} \subset \mathrm{T}$ and let $\varepsilon$ be a number $> 0$; choose an open subset G of X containing $\pi(\mathrm{A})$, such that $\nu^\bullet(\mathrm{G}) \leqslant \nu^\bullet(\pi(\mathrm{A})) + \varepsilon$ (§1, No. 9, Prop. 13); the open subset $\mathrm{H} = \overset{-1}{\pi}(\mathrm{G})$ of T contains A, and $c(\mathrm{H}) \leqslant c(\mathrm{A}) + \varepsilon$. The function $c$ is therefore a right-continuous capacity on T (TG, IX, §6, No. 10, Def. 9)[1] and the theorem on capacitability (*loc. cit.*, Th. 6) implies the equality $c(\mathrm{T}) = \sup_{\mathrm{K}} c(\mathrm{K})$, where K runs over the set of compact subsets of T. Prop. 8 then implies the existence of the desired measure $\mu$.

## 5. Product of two measures

Let S and T be two topological spaces, equipped respectively with two (positive) premeasures $\lambda$ and $\mu$, and let X be the product space $\mathrm{S} \times \mathrm{T}$. Let K be a compact subset of X; let us denote by A and B the projections of K on S and T respectively, and set

$$\nu_{\mathrm{K}} = (\lambda_{\mathrm{A}} \otimes \mu_{\mathrm{B}})_{\mathrm{K}} . \tag{3}$$

We thus define a premeasure on X. For, let L be a compact subset of X containing K, and let C and D be its two projections; then $\mathrm{A} \subset \mathrm{C}$, $\mathrm{B} \subset \mathrm{D}$,

[1] A capacity $f$ on T is said to be right-continuous if, for every compact set K in T, $f(\mathrm{K}) = \inf_{\mathrm{U}} f(\mathrm{U})$ as U runs over the open sets $\mathrm{U} \supset \mathrm{K}$. This concept is not defined in GT, translated from an earlier edition of Ch. IX.

consequently, using the transitivity of induced measures, and Prop. 12 of Ch. V, §8, No. 5, we have

$$\begin{aligned}(\nu_{\mathrm{L}})_{\mathrm{K}} &= \big((\lambda_{\mathrm{C}} \otimes \mu_{\mathrm{D}})_{\mathrm{L}}\big)_{\mathrm{K}} = (\lambda_{\mathrm{C}} \otimes \mu_{\mathrm{D}})_{\mathrm{K}} \\ &= \big((\lambda_{\mathrm{C}} \otimes \mu_{\mathrm{D}})_{\mathrm{A}\times\mathrm{B}}\big)_{\mathrm{K}} = (\lambda_{\mathrm{A}} \otimes \mu_{\mathrm{B}})_{\mathrm{K}} = \nu_{\mathrm{K}} .\end{aligned}$$

DEFINITION 6. — *The premeasure* $\nu$ *defined by* (3) *is called the product premeasure of* $\lambda$ *and* $\mu$, *and is denoted* $\lambda \otimes \mu$.

This definition obviously extends to the case that $\lambda$ and $\mu$ are complex premeasures, and one then has $|\lambda\otimes\mu| = |\lambda|\otimes|\mu|$ (Ch. III, §4, No. 2, Prop. 3 and Ch. IV, §5, No. 7, Lemma 3).

We conserve the notations of Ch. III, §4 and Ch. V, §8 relative to products of measures and to iterated integrals. In particular, if $f$ and $g$ are two functions defined respectively on S and T, with values in $\overline{\mathbf{R}}_+$ or in $\mathbf{C}$, the function $(s,t) \mapsto f(s)g(t)$ on $\mathrm{S}\times\mathrm{T}$ will be denoted $f\otimes g$.

PROPOSITION 10. — *Let* $\nu$ *be the product premeasure of* $\lambda$ *and* $\mu$; *for every function* $f \in \mathscr{F}_+(\mathrm{S})$ *and every function* $g \in \mathscr{F}_+(\mathrm{T})$,

$$\nu^\bullet(f \otimes g) = \lambda^\bullet(f)\mu^\bullet(g) . \tag{4}$$

*The premeasure* $\nu$ *is the only premeasure on* $\mathrm{S}\times\mathrm{T}$ *that satisfies* (4).

As K (resp. L) runs over the set of compact subsets of S (resp. of T), we have

$$\begin{aligned}\nu^\bullet(f\otimes g) &= \sup_{\mathrm{K,L}} \nu^\bullet_{\mathrm{K}\times\mathrm{L}}\big((f\otimes g)_{\mathrm{K}\times\mathrm{L}}\big) = \sup_{\mathrm{K,L}} (\lambda_{\mathrm{K}}\otimes\mu_{\mathrm{L}})^\bullet(f_{\mathrm{K}}\otimes g_{\mathrm{L}}) \\ &= \sup_{\mathrm{K,L}} \lambda^\bullet_{\mathrm{K}}(f_{\mathrm{K}})\cdot\mu^\bullet_{\mathrm{L}}(g_{\mathrm{L}}) = \big(\sup_{\mathrm{K}} \lambda^\bullet_{\mathrm{K}}(f_{\mathrm{K}})\big)\big(\sup_{\mathrm{L}} \mu^\bullet_{\mathrm{L}}(g_{\mathrm{L}})\big) \\ &= \lambda^\bullet(f)\mu^\bullet(g)\end{aligned}$$

by Prop. 8 of Ch. V, §8, No. 3.

Let $\eta$ be a second premeasure on $\mathrm{S}\times\mathrm{T}$ satisfying (4), and let K and L be compact subsets of S and T respectively, $f$ and $g$ elements of $\mathscr{F}_+(\mathrm{K})$ and $\mathscr{F}_+(\mathrm{L})$ respectively. One has the relation $(f\otimes g)^0 = f^0\otimes g^0$ between the extensions by $0$, therefore (§1, No. 2, Prop. 2)

$$\begin{aligned}\eta^\bullet_{\mathrm{K}\times\mathrm{L}}(f\otimes g) = \eta^\bullet\big((f\otimes g)^0\big) &= \eta^\bullet(f^0\otimes g^0) \\ &= \lambda^\bullet(f^0)\mu^\bullet(g^0) = \lambda^\bullet_{\mathrm{K}}(f)\mu^\bullet_{\mathrm{L}}(g) .\end{aligned}$$

In particular, if one takes $f \in \mathscr{K}_+(\mathrm{K})$, $g \in \mathscr{K}_+(\mathrm{L})$, one sees that $\eta_{\mathrm{K}\times\mathrm{L}}$ has the characteristic property of the product measure $\lambda_{\mathrm{K}}\otimes\mu_{\mathrm{L}}$ (Ch. III, §4, No. 1, Th. 1). Therefore $\eta_{\mathrm{K}\times\mathrm{L}} = \nu_{\mathrm{K}\times\mathrm{L}}$; since every compact subset of $\mathrm{S}\times\mathrm{T}$

is contained in a set of the form $K \times L$, the transitivity of induced measures implies that $\eta = \nu$.

COROLLARY 1. — *If* $\lambda$ *and* $\mu$ *are measures, then* $\nu$ *is a measure.*

For, let $x = (s, t)$ be a point of X, and let U and V be neighborhoods of $s, t$ respectively, such that $\lambda^\bullet(U) < +\infty$, $\mu^\bullet(V) < +\infty$; the set $U \times V$ is a neighborhood of $x$, and $\nu^\bullet(U \times V) = \lambda^\bullet(U)\mu^\bullet(V) < +\infty$ by (4); the encumbrance $\nu^\bullet$ is thus locally bounded, and the premeasure $\nu$ is a measure.

This result extends at once to complex measures.

COROLLARY 2. — *If* A *is a subset of* S *locally negligible for* $\lambda$, *then* $A \times T$ *is locally* $\nu$*-negligible.*

COROLLARY 3. — *Suppose that* $\lambda$ (resp. $\mu$) *is the sum of a summable family* $(\lambda_\alpha)_{\alpha \in A}$ (resp. $(\mu_\beta)_{\beta \in B}$) *of measures on* S (resp. T). *The family* $(\lambda_\alpha \otimes \mu_\beta)_{(\alpha,\beta) \in A \times B}$ *is then summable, and its sum is* $\lambda \otimes \mu$.

For, let $p$ be the encumbrance $\sum\limits_{\alpha,\beta}(\lambda_\alpha \otimes \mu_\beta)^\bullet$; if $f \in \mathscr{F}_+(S)$ and $g \in \mathscr{F}_+(T)$, then obviously $p(f \otimes g) = \lambda^\bullet(f)\mu^\bullet(g)$. The proof of Cor. 1 then shows that $p$ is locally bounded, so that the family $(\lambda_\alpha \otimes \mu_\beta)$ is summable (§1, No. 7, Prop. 7). Its sum $\eta$ then satisfies $\eta^\bullet = p$ (§1, No. 7, Prop. 7), and Prop. 10 implies $\eta = \nu$.

## 6. Integration with respect to the product of two measures

*Throughout this No.,* $\lambda$ *and* $\mu$ *denote measures on* S *and* T, *respectively, and* $\nu$ *denotes the product measure* $\lambda \otimes \mu$ *on* $S \times T$. *In addition, if* $f$ *is a positive function on* $S \times T$, *for every* $s \in S$ *we denote by* $f_s$ *the function* $t \mapsto f(s,t)$ *on* T, *and by* $I_f$ *the function* $s \mapsto \mu^\bullet(f_s)$ *on* S.

*Lemma* 2. — *Let* $f$ *be a* $\nu$*-measurable positive function on* $S \times T$; *for every compact subset* L *of* T, *let* $I_f^L$ *be the function* $s \mapsto \mu^\bullet(f_s\varphi_L)$ *on* S. *Then the function* $I_f^L$ *is* $\lambda$*-measurable, and*

$$I_f = \sup_L I_f^L \tag{5}$$

$$\nu^\bullet(f) = \sup_L \lambda^\bullet(I_f^L), \tag{6}$$

*where* L *runs over the set of compact subsets of* T.

We first note that the inclusion $L \subset L'$ implies $I_f^L \leqslant I_f^{L'}$; on the other hand, $I_f^L(s) = \mu_L^\bullet((f_s)_L)$ for all $s \in S$. Formula (5) is therefore an immediate consequence of the definition of the encumbrance $\mu^\bullet$ given in

§1, No. 2. If K is a compact subset of S, and L a compact subset of T, then $\nu_{K\times L} = \lambda_K \otimes \mu_L$ by construction, and Prop. 7 of Ch. V, §8, No. 3 implies the relation

$$\nu^\bullet(f\varphi_{K\times L}) = \lambda_K^\bullet\big((I_f^L)_K\big). \tag{7}$$

Moreover, every compact subset of $S \times T$ is contained in a compact set of the form $K \times L$; passing to the upper envelope over all K and L in the preceding formula, we therefore obtain

$$\nu^\bullet(f) = \sup_L \sup_K \lambda_K^\bullet\big((I_f^L)_K\big) = \sup_L \lambda^\bullet(I_f^L), \tag{8}$$

namely (6).

Finally, Prop. 7 of Ch. V, §8, No. 3 implies that the restriction of $I_f^L$ to every compact subset K of S is $\lambda_K$-measurable; this is equivalent to saying that $I_f^L$ is $\lambda$-measurable.

PROPOSITION 11. — *Let $f$ be a lower semi-continuous function $\geqslant 0$ defined on $X = S \times T$.*

*a) The function $f_s : t \mapsto f(s,t)$ is lower semi-continuous on T for every $s \in S$.*

*b) The function $I_f : s \mapsto \int^\bullet f(s,t)\,d\mu(t)$ is lower semi-continuous on S, and*

$$\iint_X^\bullet f(s,t)\,d\nu(s,t) = \int_S^\bullet d\lambda(s) \int_T^\bullet f(s,t)\,d\mu(t). \tag{9}$$

The property *a*) is obvious, since the mapping $t \mapsto f(s,t)$ of T into $\overline{\mathbf{R}}$ is the composition of $f$ with the continuous mapping $t \mapsto (s,t)$ of T into X. To establish *b*), we will make use of a lemma:

*Lemma 3. — Let X be a topological space (Hausdorff or not), $f$ a lower semi-continuous function $\geqslant 0$ defined on X; then $f$ is the limit of an increasing sequence $(f_n)_{n\in\mathbf{N}}$ of lower semi-continuous functions on X, such that each function $f_n$ is a linear combination, with positive coefficients, of characteristic functions of open sets.*

Given two integers $k \geqslant 1$ and $n \geqslant 1$, let us denote by $J_{kn}$ the characteristic function of the interval $]k/2^n, +\infty]$ of $\overline{\mathbf{R}}$. For every $x \in \overline{\mathbf{R}}_+$, set $u_n(x) = 2^{-n} \sum_{k=1}^{n\cdot 2^n} J_{kn}(x)$; it is immediate that the sequence $\big(u_n(x)\big)_{n\geqslant 1}$ is increasing and admits $x$ as limit. The sequence of functions $f_n = u_n \circ f$ is therefore increasing and converges to $f$, and one has $f_n = 2^{-n} \sum_{k=1}^{n\cdot 2^n} \varphi_{U(k,n)}$, where $U(k,n)$ is the open set $\overset{-1}{f}(]k/2^n, +\infty])$ of X.

Let us pass to the proof of *b*). The function $I_f$ being the upper envelope of the increasing directed family of functions $I_f^L$, where L runs over the set of compact subsets of T (Lemma 2), it will suffice to show that the functions $I_f^L$ are lower semi-continuous; the formula (9) may then be deduced from (6) by passing to the upper envelope over L (§1, No. 6, Prop. 5).

Thus let $\mathscr{H}$ be the set of positive lower semi-continuous functions $f$ on $S \times T$ such that $I_f^L$ is lower semi-continuous for every compact subset L of T. By Prop. 5 of §1, No. 6, the supremum of every increasing directed set of elements of $\mathscr{H}$ belongs to $\mathscr{H}$. By Lemma 3, it will therefore suffice to prove that the characteristic function of an open set W of $S \times T$ belongs to $\mathscr{H}$. Moreover, by the definition of the product topology on $S \times T$, the open set W is the union of an increasing directed family $(W_\alpha)_{\alpha \in A}$ of open sets of the form

$$W = \bigcup_{1 \leqslant i \leqslant n} (U_i \times V_i),$$

where the $U_i$ are open in S and the $V_i$ are open in T; by the remarks made above, it will suffice to show that the characteristic function of such an open set belongs to $\mathscr{H}$. Let then $s \in S$, and let U be the intersection of the family (possibly empty) formed by the open sets $U_i$ containing $s$; one sees immediately that $\varphi_W(s,t) \leqslant \varphi_W(s',t)$ for all $s' \in U$ and $t \in T$, whence, by integration, $I_{\varphi_W}^L(s) \leqslant I_{\varphi_W}^L(s')$ for all $s' \in U$. Consequently $I_{\varphi_W}^L$ is lower semi-continuous, and the proposition is established.

COROLLARY 1. — *Let $f$ be a positive numerical function defined on* $X = S \times T$; *then*

$$\iint_X^* f(s,t)\, d\nu(s,t) \geqslant \int_S^* d\lambda(s) \int_T^* f(s,t)\, d\mu(t). \tag{10}$$

For, let $g$ be a lower semi-continuous function on X such that $g \geqslant f$; by Prop. 11,

$$\begin{aligned}\iint_X^* g(s,t)\, d\nu(s,t) &= \iint^\bullet g(s,t)\, d\nu(s,t) = \int^\bullet d\lambda(s) \int^\bullet g(s,t)\, d\mu(t) \\ &= \int^* d\lambda(s) \int^* g(s,t)\, d\mu(t) \geqslant \int^* d\lambda(s) \int^* f(s,t)\, d\mu(t).\end{aligned}$$

The inequality (10) is obtained by passing to the lower envelope over $g$.

COROLLARY 2. — *Let $f$ be a numerical function defined on* $\mathrm{S} \times \mathrm{T}$ *and* $\nu$*-negligible. Then the function* $f_s : t \mapsto f(s,t)$ *is* $\mu$*-negligible for* $\lambda$*-almost every* $s \in \mathrm{S}$.

PROPOSITION 12. — *Let $f$ be a* $\nu$*-measurable positive function defined on* $\mathrm{X} = \mathrm{S} \times \mathrm{T}$. *Assume that $f$ is* $\nu$*-moderated* (resp. *that* $\mu$ *is moderated*). *Then:*

*a) The set* $\mathrm{N}$ *of* $s \in \mathrm{S}$ *such that the function* $f_s : t \mapsto f(s,t)$ *is not* $\mu$*-measurable is negligible* (resp. *locally negligible*) *for* $\lambda$.

*b) The mapping* $s \mapsto \int^{\bullet} f(s,t)\, d\mu(t)$ *is* $\lambda$*-measurable, and*

$$\iint_{\mathrm{X}}^{\bullet} f(s,t)\, d\nu(s,t) = \int_{\mathrm{S}}^{\bullet} d\lambda(s) \int_{\mathrm{T}}^{\bullet} f(s,t)\, d\mu(t)\,. \tag{11}$$

We begin by establishing *b*) when $f$ is $\nu$-moderated. By Lemma 2, this part of the statement is valid when there exists a compact subset $\mathrm{L}$ of $\mathrm{T}$ such that $f$ is zero outside $\mathrm{S} \times \mathrm{L}$; for, in this case $\mathrm{I}_f = \mathrm{I}_f^{\mathrm{L}'}$ for every compact subset $\mathrm{L}'$ of $\mathrm{T}$ containing $\mathrm{L}$, and formula (11) reduces to (6). In particular, *b*) is established for a function $f$ that is zero outside a compact subset of $\mathrm{S} \times \mathrm{T}$. On the other hand, Cor. 1 of Prop. 11 implies that *b*) is true when $f$ is $\nu$-negligible. Since every $\nu$-moderated function is the sum of a $\nu$-negligible function and a sequence of functions with compact support (§1, No. 9, Cor. 3 of Prop. 14), the assertion *b*) is true when $f$ is $\nu$-moderated.

Similarly, the assertion *b*) is obvious when $\mu$ is carried by a compact subset $\mathrm{L}$ of $\mathrm{T}$ (Lemma 2). Suppose that $\mu$ is moderated; then there exists a sequence $(\mu_n)_{n \in \mathbf{N}}$ of measures on $\mathrm{T}$ with compact support, such that $\mu = \sum_n \mu_n$ (§1, No. 9, Cor. 5 of Prop. 14), whence $\nu = \sum_n \lambda \otimes \mu_n$ (No. 5, Cor. 3 of Prop. 10). The assertion *b*), being valid for each of the measures $\nu_n = \lambda \otimes \mu_n$, is also valid for $\nu = \sum_n \nu_n$.

Let us prove *a*); denote by $\mathrm{N}$ the set of $s \in \mathrm{S}$ such that $f_s$ is not $\mu$-measurable; for every compact subset $\mathrm{L}$ of $\mathrm{T}$, similarly denote by $\mathrm{N_L}$ the set of $s \in \mathrm{S}$ such that $f_s \varphi_{\mathrm{L}}$ is not $\mu$-measurable. If $\mathrm{K}$ and $\mathrm{L}$ are compact sets in $\mathrm{S}$ and $\mathrm{T}$ respectively, then $f_{\mathrm{K} \times \mathrm{L}}$ is measurable with respect to the measure $\nu_{\mathrm{K} \times \mathrm{L}} = \lambda_{\mathrm{K}} \otimes \mu_{\mathrm{L}}$, and Prop. 2 of Ch. V, §8, No. 2 shows that *the set* $\mathrm{N_L}$ *is locally negligible for* $\lambda_{\mathrm{K}}$; since $\mathrm{K}$ is arbitrary, $\mathrm{N_L}$ is locally $\lambda$-negligible.

Suppose that $f$ is zero outside a compact set of the form $\mathrm{K} \times \mathrm{L}$; then $\mathrm{N} = \mathrm{N_L}$, and $\mathrm{N}$ is contained in $\mathrm{K}$; it follows that $\mathrm{N}$ is $\lambda$-negligible. Similarly, if $f$ is $\nu$-negligible, Cor. 2 of Prop. 11 implies that $\mathrm{N}$ is $\lambda$-negligible. The case that $f$ is $\nu$-moderated can then be treated as above, on combining the preceding two cases.

Suppose that $\mu$ is carried by a compact subset L of T; then $N = N_L$ again, therefore N is locally $\lambda$-negligible. Since every moderated measure is the sum of a sequence of measures with compact support (§1, No. 9, Cor. 5 of Prop. 14), this result extends at once to the case that $\mu$ is moderated, on using Prop. 8 of §1, No. 7.

*Remark.* — Let $(K_\alpha)_{\alpha \in A}$ be a crushing of S for $\lambda$ and let $M = S - \bigcup_{\alpha \in A} K_\alpha$; define in an analogous way $(L_\beta)_{\beta \in B}$ and N for the measure $\mu$ on T. We denote by S′ the locally compact space that is the sum of the subspaces $K_\alpha$ of S and the *discrete* space M; the space T′ is defined analogously, and we set $X' = S' \times T'$. The locally compact space X′ is the sum of the family $(K_\alpha \times L_\beta)_{(\alpha,\beta) \in A \times B}$ of compact subspaces of X and the subspace $P = (M \times T) \cup (S \times N)$ that is a locally $\nu$-negligible subset of X (one observes that P is in general not a discrete space). We saw in the *Scholium* of §1, No. 8 that there exists a measure $\lambda'$ on S′ such that the measurable functions, the essential upper integral of positive functions, the essentially integrable functions and their integrals, are the same for $\lambda$ and $\lambda'$. We associate the measure $\mu'$ on T′ with $\mu$, and the measure $\nu'$ on X′ with $\nu$, in conformity with the cited Scholium; one sees immediately that $\nu'^{\bullet}(f \otimes g) = \lambda'^{\bullet}(f)\mu'^{\bullet}(g)$ for $f \in \mathscr{F}_+(S)$ and $g \in \mathscr{F}_+(T)$; therefore $\nu' = \lambda' \otimes \mu'$ by Prop. 10 of No. 5. Since the topology of X′ is finer than that of X, every $\nu$-moderated function is $\nu'$-moderated. This procedure permits extending without new proof the Lebesgue–Fubini theorem (Ch. V, §8, No. 4, Th. 1) to the present situation.

### 7. A result on the disintegration of measures

PROPOSITION 13. — *Let X be a topological space, $\nu$ a moderated measure on X, $p$ a $\nu$-proper mapping of X into a topological space T, and $\mu = p(\nu)$. Assume that every compact subspace of X is metrizable. Then there exists a mapping $t \mapsto \lambda_t$ of T into $\mathscr{M}_+(X)$ having the following properties:*

*a) for every $t \in T$, the measure $\lambda_t$ is carried by $\overset{-1}{p}(t)$;*

*b) for every universally measurable*[1] *positive function $f$ on X, the function $t \mapsto \lambda_t^{\bullet}(f)$ is universally measurable on T and*

$$\int_X^{\bullet} f(x)\, d\nu(x) = \int_T^{\bullet} d\mu(t) \int_X^{\bullet} f(x)\, d\lambda_t(x); \tag{12}$$

[1] A mapping of a topological space X into a topological space Y is said to be *universally measurable* if it is $\mu$-measurable for every measure $\mu$ on X (cf. Ch. V, §3, No. 4).

*c) the set of* $t \in \mathrm{T}$ *such that* $\lambda_t(1) \neq 1$ *is locally* $\mu$*-negligible.*

*Moreover, if* $t \mapsto \lambda'_t$ *is a mapping of* T *into* $\mathscr{M}_+(\mathrm{X})$ *satisfying the conditions a) and b), the set of* $t \in \mathrm{T}$ *such that* $\lambda_t \neq \lambda'_t$ *is locally* $\mu$*-negligible.*

We will need an auxiliary result:

*Lemma 4. — Let* X *be a topological space,* $\nu$ *a measure on* X, *and* $f$ *a* $\nu$*-measurable mapping of* X *into a topological space* F *(Hausdorff or not). There exists a universally measurable mapping* $f'$ *of* X *into* F, *equal to* $f$ *locally* $\nu$*-almost everywhere.*

The proof is identical to that of Prop. 7 of Ch. V, §3, No. 4, on taking into account Prop. 10 of §1, No. 8.

Let us pass to the proof of Prop. 13.

A) *Suppose that* X *is compact and metrizable and that* $p$ *is continuous and surjective*:

The space T is then compact and metrizable (GT, IX, §2, No. 10). By Th. 1 of Ch. VI, §3, No. 1, there exists a mapping $\mathrm{H} : t \mapsto \eta_t$ of T into $\mathscr{M}_+(\mathrm{X})$, vaguely $\mu$-measurable and scalarly essentially $\mu$-integrable, such that $\nu = \int_{\mathrm{T}} \eta_t \, d\mu(t)$ and such that $\eta_t$ has total mass 1 and is carried by $\overset{-1}{p}(t)$ for every $t \in \mathrm{T}$. Let $(\mathrm{S}_n)_{n \in \mathbf{N}}$ be a crushing of T for $\mu$, such that the restriction of H to each of the sets $\mathrm{S}_n$ is continuous (§1, No. 8, Props. 10 and 11); we denote by $\Lambda : t \mapsto \lambda_t$ the mapping of T into $\mathscr{M}_+(\mathrm{X})$ that is equal to H on $\mathrm{S} = \bigcup_{n \in \mathbf{N}} \mathrm{S}_n$ and to 0 on $\mathrm{T} - \mathrm{S}$. It is clear that $\nu = \int_{\mathrm{T}} \lambda_t \, d\mu(t)$ and that $\Lambda$ satisfies condition *a*) of the statement.

Let $\theta$ be a measure on T; the mapping $\Lambda$ is vaguely $\theta$-measurable and scalarly essentially $\theta$-integrable, hence also $\theta$-adequate (Ch. V, §3, No. 1, Prop. 2 *b*)). Let $f$ be a universally measurable positive function on X; by Prop. 5 of Ch. V, §3, No. 2 applied to $\int \lambda_t \, d\theta(t)$, the mapping $t \mapsto \lambda_t^\bullet(f)$ is $\theta$-measurable, hence universally measurable in view of the arbitrariness of $\theta$.

The formula (12) results from Prop. 5 of Ch. V, §3, No. 2.

B) *Suppose that there exists a compact subset* X′ *of* X *carrying the measure* $\nu$ *and such that* $p_{\mathrm{X}'}$ *is continuous*:

Set $\mathrm{T}' = p(\mathrm{X}')$, and $p' = p_{\mathrm{X}'}$; let us denote by $\nu'$ the measure $\nu_{\mathrm{X}'}$, and by $\mu'$ the image measure $p'(\nu')$ on T′. Since $p'$ is continuous and surjective and since X′ is compact and metrizable, by A) there exists a mapping $\Lambda' : t' \mapsto \lambda'_{t'}$ of T′ into $\mathscr{M}_+(\mathrm{X}')$ satisfying the following conditions:

*a*′) for every $t' \in \mathrm{T}'$, the measure $\lambda'_{t'}$ is carried by $\mathrm{X}' \cap \overset{-1}{p}(t')$;

*b*′) for every universally measurable positive function $f'$ on X′, the function $t' \mapsto \lambda'^\bullet_{t'}(f')$ is universally measurable on T′ and

$$\int_{\mathrm{X}'}^\bullet f'(x') \, d\nu'(x') = \int_{\mathrm{T}'}^\bullet d\mu'(t') \int_{\mathrm{X}'}^\bullet f'(x') \, d\lambda'_{t'}(x') .$$

Let $t \in \mathrm{T}$ ; if $t$ belongs to $\mathrm{T}'$, let us denote by $\lambda_t$ the image of $\lambda'_t$ under the canonical injection of $\mathrm{X}'$ into $\mathrm{X}$, and if $t$ belongs to $\mathrm{T} - \mathrm{T}'$ we set $\lambda_t = 0$. The reader will verify without difficulty that the mapping $t \mapsto \lambda_t$ satisfies the conditions $a)$ and $b)$ of the statement.

C) *Existence in the general case*:

The measure $\nu$ on $\mathrm{X}$ being moderated, we may choose a covering $(\mathrm{U}_m)_{m\in\mathbf{N}}$ of $\mathrm{X}$ consisting of $\nu$-integrable open sets. Let in addition $(\mathrm{X}_n)_{n\in\mathbf{N}}$ be a $\nu$-crushing of $\mathrm{X}$ such that the restriction of $p$ to each set $\mathrm{X}_n$ is continuous (§1, No. 8, Props. 10 and 11); denote by $\nu_n$ the measure $\varphi_{\mathrm{X}_n} \cdot \nu$ on $\mathrm{X}$ and by $\mu_n$ its image under $p$. By B) there exists, for every integer $n \in \mathbf{N}$, a mapping $t \mapsto \alpha_t^n$ of $\mathrm{T}$ into $\mathscr{M}_+(\mathrm{X})$ satisfying the following conditions:

$a'')$ The measure $\alpha_t^n$ is carried by $\overset{-1}{p}(t)$ for all $t \in \mathrm{T}$.

$b'')$ If $f$ is a universally measurable positive function on $\mathrm{X}$, the positive function $t \mapsto (\alpha_t^n)^\bullet(f)$ on $\mathrm{T}$ is universally measurable and

$$\int_{\mathrm{X}}^{\bullet} f(x)\, d\nu_n(x) = \int_{\mathrm{T}}^{\bullet} d\mu_n(t) \int_{\mathrm{X}}^{\bullet} f(x)\, d\alpha_t^n(x). \tag{13}$$

One has $\nu = \sum_{n\in\mathbf{N}} \nu_n$ and $\mu = \sum_{n\in\mathbf{N}} \mu_n$; it follows immediately from Prop. 3 of No. 2 and the above Lemma 4 that there exists a sequence $(g_n)_{n\in\mathbf{N}}$ of universally measurable positive functions on $\mathrm{T}$ such that $\mu_n = g_n \cdot \mu$ for all $n \in \mathbf{N}$ and such that $\sum_{n\in\mathbf{N}} g_n = 1$. For every $t \in \mathrm{T}$, let us denote by $\beta_t^n$ the measure $g_n(t) \cdot \alpha_t^n$ on $\mathrm{X}$ and by $q_t$ the encumbrance $\sum_{n\in\mathbf{N}} (\beta_t^n)^\bullet$ on $\mathrm{X}$. Let $f$ be a universally measurable positive function on $\mathrm{X}$; using Prop. 2 of No. 2 and summing over $n$ in (13), we obtain

$$\int_{\mathrm{X}}^{\bullet} f(x)\, d\nu(x) = \int_{\mathrm{T}}^{\bullet} q_t(f)\, d\mu(t)\,; \tag{14}$$

it is moreover clear that the function $t \mapsto q_t(f)$ on $\mathrm{T}$ is universally measurable.

For every $m \in \mathbf{N}$, let $\mathrm{E}_m$ be the set of $t \in \mathrm{T}$ such that $q_t(\mathrm{U}_m) = +\infty$; the set $\mathrm{E}_m$ is universally measurable because this is true of the mapping $t \mapsto q_t(\mathrm{U}_m)$, and $\mathrm{E}_m$ is locally $\mu$-negligible by the formula (14) applied to $f = \varphi_{\mathrm{U}_m}$, since $\nu^\bullet(\mathrm{U}_m)$ is finite. The set $\mathrm{E} = \bigcup_{m\in\mathbf{N}} \mathrm{E}_m$ is therefore universally measurable and locally $\mu$-negligible. We set $\lambda_t = 0$ for $t \in \mathrm{E}$. Moreover, let $t \in \mathrm{T} - \mathrm{E}$; the encumbrance $q_t$ is locally bounded since the open sets $\mathrm{U}_m$ cover $\mathrm{X}$ and since $q_t(\mathrm{U}_m)$ is finite; by Prop. 7 of §1, No. 7, there exists a measure $\lambda_t$ on $\mathrm{X}$ such that $q_t = \lambda_t^\bullet$ and $\lambda_t = \sum_{n\in\mathbf{N}} \beta_t^n$. It is immediate that the mapping $t \mapsto \lambda_t$ satisfies the conditions $a)$ and $b)$ of the statement.

D) *Proof of c)*:

Let $f$ be a universally measurable function on X that is positive and bounded; we are going to show that the universally measurable function $h_f : t \mapsto \lambda_t^\bullet(f)$ on T is a density for the measure $\mu_f = p(f \cdot \nu)$ with respect to $\mu = p(\nu)$. Let K be a compact subset of T and let $\mathrm{A} = \overset{-1}{p}(\mathrm{K})$. For every $t \in \mathrm{T}$, the measure $\lambda_t$ is carried by $\overset{-1}{p}(t)$; if $t$ belongs to K then $\overset{-1}{p}(t) \subset \mathrm{A}$, whence $\lambda_t^\bullet(f\varphi_\mathrm{A}) = \lambda_t^\bullet(f)$; on the other hand, if $t$ belongs to $\mathrm{T} - \mathrm{K}$ then $\overset{-1}{p}(t) \subset \mathrm{X} - \mathrm{A}$, whence $\lambda_t^\bullet(f\varphi_\mathrm{A}) = 0$. Applying the formula (12) to $f \cdot \varphi_\mathrm{A}$,[1] we obtain

$$\mu_f(\mathrm{K}) = \int_\mathrm{A}^\bullet f(x)\,d\nu(x) = \int_\mathrm{K}^\bullet d\mu(t) \int_\mathrm{X}^\bullet f(x)\,d\lambda_t(x) = \int_\mathrm{K}^\bullet h_f(t)\,d\mu(t)\,,$$

which establishes the relation $\mu_f = h_f \cdot \mu$.

Letting $f = 1$, one sees that the function $h_1 : t \mapsto \|\lambda_t\|$ is a density of the measure $\mu_1 = \mu$ with respect to $\mu$, hence is equal to 1 locally $\mu$-almost everywhere in T.

E) *Uniqueness*:

Let $t \mapsto \lambda_t^i$ (for $i = 1, 2$) be two mappings of T into $\mathscr{M}_+(\mathrm{X})$ satisfying the conditions $a)$ and $b)$ of the statement. As in C), choose a $\mu$-crushing $(\mathrm{X}_n)_{n\in\mathbf{N}}$ of X such that $p_{\mathrm{X}_n}$ is continuous for every $n \in \mathbf{N}$, and set $\mathrm{N} = \mathrm{X} - \bigcup_{n\in\mathbf{N}} \mathrm{X}_n$. For every integer $n \in \mathbf{N}$, choose a countable set $\mathrm{D}_n$ of positive functions on X, zero outside $\mathrm{X}_n$, whose restrictions to $\mathrm{X}_n$ form a dense set in the normed space $\mathscr{C}(\mathrm{X}_n)$ (apply Th. 1 of GT, X, §3, No. 3 to the metrizable compact space $\mathrm{X}_n$). We set $\mathrm{D} = \bigcup_{n\in\mathbf{N}} \mathrm{D}_n$.

Let $f \in \mathrm{D}$; by D), the functions $t \mapsto (\lambda_t^1)^\bullet(f)$ and $t \mapsto (\lambda_t^2)^\bullet(f)$ are densities of the measure $\mu_f$ with respect to $\mu$, and so there exists a locally $\mu$-negligible set $\mathrm{E}_f$ in T such that $(\lambda_t^1)^\bullet(f) = (\lambda_t^2)^\bullet(f)$ for $t \in \mathrm{T} - \mathrm{E}_f$. Moreover, by (12), the set $\mathrm{F}_i$ of $t \in \mathrm{T}$ such that $(\lambda_t^i)^\bullet(\mathrm{N}) \neq 0$ is locally $\mu$-negligible for $i = 1, 2$. Since D is countable, the set $\mathrm{G} = (\bigcup_{f\in\mathrm{D}} \mathrm{E}_f) \cup \mathrm{F}_1 \cup \mathrm{F}_2$ is locally $\mu$-negligible; for $t \in \mathrm{T} - \mathrm{G}$, we have $(\lambda_t^1)^\bullet(\mathrm{N}) = (\lambda_t^2)^\bullet(\mathrm{N}) = 0$ and $(\lambda_t^1)_{\mathrm{X}_n} = (\lambda_t^2)_{\mathrm{X}_n}$, whence $\lambda_t^1 = \lambda_t^2$ by Prop. 9 of §1, No. 8.

Q.E.D.

*Remarks.* — 1) If X is a Souslin space, then every compact subspace of X is a Souslin space, hence is metrizable (TG, IX, Appendix I, Cor. 2 of Prop. 3),[2]

---

[1] In case $f \cdot \varphi_\mathrm{A}$ is not universally measurable, make use of *Remark* 2) below.

[2] This result does not appear in GT (cf. the footnote to *Remark* 1 of §1, No. 9).

and every measure on X is moderated (§1, No. 9, *Remark* 1). By Prop. 13, every measure $\nu$ on X therefore admits a disintegration with respect to every $\nu$-proper mapping.

2) With the notations of Prop. 13, let $f$ be a positive $\nu$-measurable function. One can prove, as in Ch. V, §3, No. 2, Prop. 5, that the set of $t \in \mathrm{T}$ such that $f$ is not $\lambda_t$-measurable is locally $\mu$-negligible, that $t \mapsto \lambda_t^\bullet(f)$ is $\mu$-measurable, and that the relation (12) again holds.

## §3. MEASURES AND ADDITIVE SET FUNCTIONS

*In this section, we shall denote by* $\mathfrak{K}(\mathrm{T})$ *and* $\mathfrak{B}(\mathrm{T})$, *respectively, the set of compact subsets of a Hausdorff topological space* T *and the Borel tribe of* T.

### 1. Measures and additive set functions of compact sets

THEOREM 1. — *Let* T *be a topological space, and* I *a mapping of* $\mathfrak{K}(\mathrm{T})$ *into* $\mathbf{R}_+$. *In order that there exist a measure* $\mu$ *on* T *such that* $\mathrm{I}(\mathrm{K}) = \mu^\bullet(\mathrm{K})$ *for all* $\mathrm{K} \in \mathfrak{K}(\mathrm{T})$, *it is necessary and sufficient that* I *satisfy the following conditions*:

1) *If* K *and* L *are compact subsets of* T *such that* $\mathrm{K} \subset \mathrm{L}$, *then* $\mathrm{I}(\mathrm{K}) \leqslant \mathrm{I}(\mathrm{L})$ ('I *is increasing*').

2) *If* K *and* L *are compact subsets of* T, *then* $\mathrm{I}(\mathrm{K} \cup \mathrm{L}) \leqslant \mathrm{I}(\mathrm{K}) + \mathrm{I}(\mathrm{L})$.

3) *If* K *and* L *are disjoint compact subsets of* T, *then* $\mathrm{I}(\mathrm{K} \cup \mathrm{L}) = \mathrm{I}(\mathrm{K}) + \mathrm{I}(\mathrm{L})$ ('I *is additive*').

4) *For every decreasing directed family* $(\mathrm{K}_\alpha)_{\alpha \in \mathrm{A}}$ *of compact subsets of* T, *one has* $\mathrm{I}(\bigcap_{\alpha \in \mathrm{A}} \mathrm{K}_\alpha) = \inf_{\alpha \in \mathrm{A}} \mathrm{I}(\mathrm{K}_\alpha)$.

5) *For every* $x \in \mathrm{T}$, *there exists a neighborhood* V *of* $x$ *such that*

$$\sup_{\substack{\mathrm{K} \in \mathfrak{K}(\mathrm{T}) \\ \mathrm{K} \subset \mathrm{V}}} \mathrm{I}(\mathrm{K}) < +\infty$$

('I *is locally bounded*').

*The measure* $\mu$ *is then unique.*

The uniqueness of $\mu$ follows from the Cor. of Prop. 2 of §1, No. 2. The above conditions are necessary, the first three in obvious fashion, the last from the fact that $\mu^\bullet$ is a locally bounded encumbrance, and condition 4) by the Cor. of Prop. 5 of §1, No. 6.

To show that these conditions are sufficient, we begin by treating the case that T *is compact.*

*Lemma* 1. — *Assume that* T *is compact, and set* $l = \mathrm{I(T)}$. *For every* $\mathrm{A \subset T}$, *set*

$$\mathrm{J(A)} = \sup_{\substack{\mathrm{K} \in \mathfrak{K}(\mathrm{T}) \\ \mathrm{K} \subset \mathrm{A}}} \mathrm{I(K)} \tag{1}$$

*and let* $\Phi$ *be the set of* $\mathrm{A \subset T}$ *such that* $\mathrm{J(A) + J(\complement A)} = l$. *The set* $\Phi$ *is then a clan that contains* $\mathfrak{K}(\mathrm{T})$, *and the function* J *on* $\Phi$ *is increasing and additive.*

It is clear that J is an increasing set function, extending I, and that $\mathrm{J(A) + J(\complement A)} \leqslant l$ for every $\mathrm{A \subset T}$.

Let K and S be two compact sets in T; we are first going to show that

$$\mathrm{J(K \cap S) + J(\complement K \cap S) = J(S)}. \tag{2}$$

On considering the restrictions of I to $\mathfrak{K}(\mathrm{S})$ and of J to $\mathfrak{P}(\mathrm{S})$, one reduces immediately to the case that $\mathrm{S = T}$. Since T is normal, K is the intersection of the decreasing directed family of its compact neighborhoods, and condition 4) implies the existence, for every $\varepsilon > 0$, of a compact neighborhood H of K such that $\mathrm{I(H)} \leqslant \mathrm{I(K)} + \varepsilon$. Let L be the closure of $\mathrm{T - H}$; L is compact, $\mathrm{L \cap K} = \varnothing$ and $\mathrm{H \cup L = T}$, therefore $l = \mathrm{I(H \cup L)} \leqslant \mathrm{I(H) + I(L)} \leqslant \mathrm{I(K) + I(L)} + \varepsilon$ (condition 2)), whence the relation $\mathrm{J(K) + J(\complement K)} \geqslant \mathrm{I(K) + I(L)} \geqslant l - \varepsilon$. Since $\varepsilon$ is arbitrary, $\mathrm{J(K) + J(\complement K)} = l$. This proves the formula (2), as well as the inclusion $\mathfrak{K}(\mathrm{T}) \subset \Phi$.

Let us now prove that $\Phi$ is a clan. Since $\Phi$ is obviously stable under passage to the complement, it suffices to show that if $\mathrm{A_1}$ and $\mathrm{A_2}$ denote elements of $\Phi$, then $\mathrm{A_1 \cup A_2} \in \Phi$, or again that

$$\mathrm{J(A_1 \cup A_2) + J\big(\complement (A_1 \cup A_2)\big)} \geqslant l. \tag{3}$$

Denote by $\varepsilon$ a number $> 0$, and, for $i = 1, 2$, let $\mathrm{K}_i$ be a compact set contained in $\mathrm{A}_i$, and $\mathrm{L}_i$ a compact set contained in $\complement \mathrm{A}_i$, such that

$$\mathrm{I(K}_i) \geqslant \mathrm{J(A}_i) - \varepsilon, \quad \mathrm{I(L}_i) \geqslant \mathrm{J(\complement A}_i) - \varepsilon.$$

Set $\mathrm{M_1 = K_1 \cup L_1}$; the relations $l = \mathrm{J(M_1) + J(\complement M_1)}$ and

$$\mathrm{J(M_1) = I(K_1) + I(L_1)} \geqslant \mathrm{J(A_1) + J(\complement A_1)} - 2\varepsilon = l - 2\varepsilon$$

imply that $\mathrm{J(\complement M_1)} \leqslant 2\varepsilon$. Then if S is a compact subset of T, the relation (2) (applied to $\mathrm{K = M_1}$) implies $\mathrm{J(S)} \leqslant \mathrm{J(M_1 \cap S)} + 2\varepsilon$, whence

$$\mathrm{J(S)} \leqslant \mathrm{J(K_1 \cap S) + J(L_1 \cap S)} + 2\varepsilon.$$

Let us add the inequalities obtained by making $S = K_2$ and $S = L_2$, and take into account the inequality $J(K_2) + J(L_2) \geqslant l - 2\varepsilon$ and the fact that $K_1 \cap K_2$, $L_1 \cap K_2$ and $K_1 \cap L_2$ are three disjoint compact sets contained in $A_1 \cup A_2$. Denoting by $C$ the union of these three compact sets, it follows that

$$l - 2\varepsilon \leqslant J(K_2) + J(L_2) \leqslant J(C) + J(L_1 \cap L_2) + 4\varepsilon$$
$$\leqslant J(A_1 \cup A_2) + J(\complement(A_1 \cup A_2)) + 4\varepsilon,$$

whence immediately the desired formula (3), in view of the arbitrariness of $\varepsilon$. This having been established, the preceding inequalities imply that $J(C) \geqslant J(A_1 \cup A_2) - 6\varepsilon$; if $A_1$ and $A_2$ are disjoint, then $C$ is the union of $K_1 \cap L_2 \subset A_1$ and $K_2 \cap L_1 \subset A_2$, from which one deduces that $J(A_1 \cup A_2) \leqslant J(A_1) + J(A_2)$. The reverse inequality being obvious, $J$ is indeed additive on $\Phi$, and the lemma is established.

Let us complete the proof of the theorem for the case that $T$ is compact. Let $\mathscr{E}(\Phi)$ be the vector space of $\Phi$-step functions on $T$, equipped with the topology of uniform convergence (Ch. IV, §4, No. 9, Def. 4); we shall again denote by $J$ the positive linear form on $\mathscr{E}(\Phi)$ associated with the additive function $J$ (*loc. cit.*, Prop. 18). Since $J(T) = l$, $J$ is continuous and has norm $l$. Let then $\mathscr{H}$ be the closure of $\mathscr{E}(\Phi)$ for the topology of uniform convergence; one verifies at once that $J$ may be extended by continuity to a *positive* linear form on $\mathscr{H}$, again denoted $J$. Since $\mathscr{H}$ contains $\mathscr{C}(T)$ (*loc. cit.*, No. 10, Prop. 19) the restriction of $J$ to $\mathscr{C}(T)$ is a positive measure $\mu$. It remains to show that $\mu^\bullet(K) = I(K)$ for every compact subset $K$ of $T$. Now, we have $\mu^\bullet(K) = \inf_{f \in S_K} \mu^\bullet(f)$, where $S_K$ denotes the set of elements of $\mathscr{C}(T)$ that are $\geqslant \varphi_K$ (§1, No. 6, Prop. 5). Since $J(f) = \mu^\bullet(f)$ for $f \in \mathscr{C}(T)$, it clearly suffices to show that $J(K) \geqslant \inf_{f \in S_K} J(f)$. As in the proof of Lemma 1, let $H$ be a compact neighborhood of $K$ such that $J(H) \leqslant J(K) + \varepsilon$, and let $f$ be a continuous function on $T$, between 0 and 1, equal to 1 on $K$ and to 0 outside $H$ (GT, IX, §4, No. 1, Prop. 1). Then

$$J(f) \leqslant J(H) \leqslant J(K) + \varepsilon;$$

$\varepsilon$ being arbitrary, the desired inequality is proved, and the theorem is thus established when $T$ is compact.

Let us now pass to the general case. For every compact set $L$ in $T$, let $I_L$ be the restriction of $I$ to $\mathfrak{K}(L)$. By the special case just treated, there exists a measure $\mu_L$ on $L$, unique, such that $\mu_L(K) = I_L(K)$ for every compact set $K \subset L$. Let then $L'$ be a compact set contained in $L$; we have $(\mu_L)^\bullet_{L'}(K) = \mu^\bullet_L(K) = \mu^\bullet_{L'}(K)$ for every compact set $K \subset L'$, therefore $\mu_{L'} = (\mu_L)_{L'}$; the mapping $\mu : L \mapsto \mu_L$ is a premeasure. The condition 5) expresses that $\mu$ is a measure, and the relation $I(K) = \mu^\bullet(K)$ for all compact $K \subset T$ is obvious.

*Remarks.* — 1) The condition 4) may be replaced, in the statement of Theorem 1, by the following condition ('right-continuity'):

4′) *For every* $K \in \mathfrak{K}(T)$ *and every* $\varepsilon > 0$, *there exists an open set* U *containing* K, *such that* $I(H) \leqslant I(K) + \varepsilon$ *for every compact set* $H \subset U$.

For, if $\mu$ is a measure, the function $I : K \mapsto \mu^\bullet(K)$ satisfies 4′) (§1, No. 9, Prop. 13). Conversely, suppose that I satisfies 1) and 4′); let us show that I then satisfies 4). With notations as in the statement of Theorem 1, choose an $\varepsilon > 0$ and an open set U containing the compact set $K = \bigcap_{\alpha \in A} K_\alpha$ and such that 4′) is satisfied. There then exists an index $\beta \in A$ such that $K_\beta \subset U$, and this implies

$$\inf_{\alpha \in A} I(K_\alpha) \leqslant I(K_\beta) \leqslant I(K) + \varepsilon$$

and 4) is indeed verified.

2) The set of conditions 2) and 3) may be replaced, in the statement of Theorem 1, by the following condition:

*If* K *and* L *are compact subsets of* T, *then*

$$I(K \cup L) + I(K \cap L) = I(K) + I(L).$$

Indeed, this condition implies 2) and 3), and on the other hand

$$\mu^\bullet(K \cup L) = \mu^\bullet(K \cap L) = \mu^\bullet(K) + \mu^\bullet(L)$$

for every measure $\mu$, by virtue of the relation $\varphi_{K \cup L} + \varphi_{K \cap L} = \varphi_K + \varphi_L$ between the characteristic functions.

## 2. Inner regular set functions

DEFINITION 1. — *Let* T *be a topological space, and let* $\mathfrak{B}(T)$ *be the Borel tribe of* T; *let* I *be a mapping of* $\mathfrak{B}(T)$ *into* $\overline{\mathbf{R}}_+$.

*a*) I *is said to be countably additive if, for every sequence* $(A_n)$ *of pairwise disjoint elements of* $\mathfrak{B}(T)$,

$$I\Big(\bigcup_n A_n\Big) = \sum_n I(A_n). \tag{4}$$

*b*) I *is said to be inner regular if, for every set* $A \in \mathfrak{B}(T)$,

$$I(A) = \sup_K I(K), \tag{5}$$

*where* K *runs over the set of compact subsets of* A.

*c)* I *is said to be bounded* (resp. *locally bounded*) *if* $\mathrm{I(T)} < +\infty$ (resp. *if every point* $x \in \mathrm{T}$ *admits an open neighborhood* V *such that* $\mathrm{I(V)} < +\infty$).

*Remarks.* — 1) The condition *a)* clearly implies that I is an increasing mapping of $\mathfrak{B}(\mathrm{T})$ (ordered by inclusion) into $\overline{\mathbf{R}}_+$.

2) Suppose that I is countably additive; let $(\mathrm{A}_n)_{n\in\mathbf{N}}$ be an increasing sequence of Borel sets, and let $\mathrm{A} = \bigcup_{n\in\mathbf{N}} \mathrm{A}_n$. The sets $\mathrm{D}_0 = \mathrm{A}_0$, $\mathrm{D}_n = \mathrm{A}_n - \mathrm{A}_{n-1}$ being pairwise disjoint, and their union being A, we have $\mathrm{I(A)} = \sum_n \mathrm{I(D}_n) = \lim_{n\to\infty} \mathrm{I(A}_n)$. Similarly, if $(\mathrm{B}_n)$ is a decreasing sequence of Borel sets, and if $\mathrm{I(B_0)} < +\infty$, then $\mathrm{I}(\bigcap_n \mathrm{B}_n) = \lim_{n\to\infty} \mathrm{I(B}_n)$: it suffices to apply the preceding to the sets $\mathrm{A}_n = \mathrm{B}_0 - \mathrm{B}_n$.

3) Let $(\mathrm{A}_n)$ be any sequence of Borel sets of T. If I is countably additive, then $\mathrm{I}(\bigcup_n \mathrm{A}_n) \leqslant \sum_n \mathrm{I(A}_n)$. By the preceding remark, it suffices to establish this inequality for a finite sequence. One immediately reduces to the case of two sets $\mathrm{A}_1$ and $\mathrm{A}_2$; but the relation (4) implies that

$$\mathrm{I(A_1 \cup A_2) = I(A_1) + I(A_2 - (A_1 \cap A_2)) \leqslant I(A_1) + I(A_2)}.$$

The desired inequality then follows immediately.

4) If I is a countably additive and locally bounded function, the preceding remark implies at once that $\mathrm{I(K)} < +\infty$ for every compact set $\mathrm{K} \subset \mathrm{T}$.

5) One can show that if I is *additive*, that is, satisfies (4) for *finite* sequences, and if I is inner regular, then I is countably additive (Exer. 7). The reader can also ascertain that only additivity and inner regularity are used in the proof of Th. 2 below.

THEOREM 2. — *Let* T *be a topological space, and let* I *be a function defined on* $\mathfrak{B}(\mathrm{T})$, *with values in* $\overline{\mathbf{R}}_+$. *In order that there exist a measure* $\mu$ *on* T *such that* $\mu^\bullet(\mathrm{A}) = \mathrm{I(A)}$ *for every* $\mathrm{A} \in \mathfrak{B}(\mathrm{T})$, *it is necessary and sufficient that* I *be countably additive, locally bounded and inner regular. The measure* $\mu$ *is then unique.*

These three conditions are necessary: for, the mapping $\mathrm{A} \mapsto \mu^\bullet(\mathrm{A})$ on $\mathfrak{B}(\mathrm{T})$ is countably additive (§1, No. 5, Cor. of Prop. 4), locally bounded by the definition of measures (§1, No. 2, Def. 5), and inner regular by *Remark* 3 of §1, No. 2.

We pass to existence. It is clear that the restriction of I to $\mathfrak{K}(\mathrm{T})$ satisfies conditions 1), 2), 3) and 5) of the statement of Th. 1; let us show that 4) is satisfied as well. Let K be a compact subset of T, the intersection of a decreasing directed family $(\mathrm{K}_\alpha)_{\alpha\in\mathrm{A}}$ of compact sets, and let $\varepsilon$ be a number $> 0$; I being locally bounded, there exists an open (hence Borel) neighborhood V of K such that $\mathrm{I(V)} < +\infty$, and then there exists an index $\alpha$ such that $\mathrm{K}_\alpha \subset \mathrm{V}$; changing notation if necessary, we can suppose that $\mathrm{K}_\alpha \subset \mathrm{V}$ for all $\alpha \in \mathrm{A}$. By the inner regularity of I, there exists a compact set $\mathrm{L} \subset \mathrm{V} - \mathrm{K}$ such that $\mathrm{I(L)} \geqslant \mathrm{I(V - K)} - \varepsilon$; since L does not intersect K, there exists an index $\alpha$ such that $\mathrm{L} \cap \mathrm{K}_\alpha = \varnothing$, and one then has $I(V - K_\alpha) \geqslant I(L) \geqslant I(V - K) - \varepsilon$. Since $K_\alpha \subset V$, it follows that $I(K_\alpha) \leqslant I(K) + \varepsilon$ and the condition 4) is verified.

By Th. 1, there exists a measure $\mu$ such that $\mu^\bullet(K) = I(K)$ for all $K \in \mathfrak{K}(T)$. The inner regularity of the set functions $\mu^\bullet$ and I on $\mathfrak{B}(T)$ then implies that $\mu^\bullet(A) = I(A)$ for all $A \in \mathfrak{B}(T)$, and existence is proved. The uniqueness of $\mu$ follows from the uniqueness assertion of Th. 1.

## 3. Radon spaces

DEFINITION 2. — *Let* T *be a topological space. One says that* T *is a Radon* (resp. *strongly Radon*) *space if* T *is Hausdorff and if every function defined on the Borel tribe* $\mathfrak{B}(T)$ *of* T, *with values in* $\overline{\mathbf{R}}_+$, *that is countably additive and bounded* (resp. *locally bounded*) *is inner regular.*

For example, we shall see later on (Prop. 3) that every Polish space is strongly Radon. In particular, every locally compact space with a countable base is strongly Radon.

There exist Radon spaces that are not strongly Radon.

PROPOSITION 1. — *Every Lindelöf*[(1)] *Radon space is strongly Radon.*

Let T be a Lindelöf space that is Radon, and let I be a set function on the tribe $\mathfrak{B}(T)$ that is positive, countably additive and locally bounded. The open sets V such that $I(V) < +\infty$ form a covering of T, from which one can extract a countable covering $(V_n)_{n\in\mathbf{N}}$. Set $G_n = V_0 \cup V_1 \cup \cdots \cup V_n$ for every $n \in \mathbf{N}$; set $H_0 = G_0$ and $H_n = G_n - G_{n-1}$ for $n \geqslant 1$; finally, denote by $I_n$ the set function $A \mapsto I(A \cap H_n)$ on $\mathfrak{B}(T)$, which is obviously countably additive and bounded. Since the sets $H_n$ form a partition of T, we have $I = \sum_n I_n$. The space T being Radon, for every $n \in \mathbf{N}$ there exists a bounded measure $\mu_n$ on T such that $\mu_n^\bullet(A) = I_n(A)$ for all $A \in \mathfrak{B}(T)$; therefore also $\sum_n \mu_n^\bullet(A) = I(A)$. Since I is locally bounded, the family $(\mu_n)$ is summable (§1, No. 7, Prop. 7); if $\mu$ denotes $\sum_n \mu_n$, we have $\mu^\bullet(A) = I(A)$ for all $A \in \mathfrak{B}(T)$, and it follows that I is inner regular. In other words, T is strongly Radon.

Recall that a subset A of a topological space T is said to be *universally measurable* if A is $\mu$-measurable for every measure $\mu$ on T. This is equivalent to saying that A is $\mu$-measurable for every measure $\mu$ on T with *compact support* (§1, No. 8, Prop. 9).

---

(1) Recall (GT, I, §9, Exer. 14; TG, IX, Appendix I) that a *Lindelöf space* is a topological space T such that every open covering of T contains a countable covering.

PROPOSITION 2. — *Let* X *be a topological space and* T *a subspace of* X.

*a) Suppose that* T *is a Radon space. Then, for every function* I *defined on* $\mathfrak{B}(X)$ *that is positive, countably additive and bounded, one has*

$$\sup_{\substack{K \text{ compact} \\ K\subset T}} I(K) = \inf_{\substack{B\in\mathfrak{B}(X) \\ B\supset T}} I(B). \tag{6}$$

*Moreover,* T *is universally measurable in* X.

*b) Conversely, suppose that* X *is a Radon space and that* T *is universally measurable in* X; *then* T *is a Radon space.*

Let us prove *a*). We denote by $\alpha$ the right side of (6); for every $n \in \mathbf{N}$, there exists a set $C_n \in \mathfrak{B}(X)$ containing T, such that $I(C_n) \leqslant \alpha + 2^{-n}$. Setting $C = \bigcap_n C_n$, we then have $T \subset C$, $I(C) = \alpha$. If $A \in \mathfrak{B}(T)$, let us choose a Borel set B of X such that $A = B \cap T$ (GT, IX, §6, No. 3) and set $J(A) = I(B \cap C)$. This number does not depend on the choice of B, for if B′ is a second Borel set in X such that $A = B' \cap T$, then $B \cap C$ and $B' \cap C$ differ only by a Borel set M contained in $C - T$, and $I(M) = 0$ by the construction of C. Clearly $J(K) = I(K)$ for every compact set $K \subset T$. Let $(A_n)$ be a sequence of Borel sets of T, pairwise disjoint, and, for each $n$, let $B_n$ be a Borel set of X such that $B_n \cap T = A_n$. Replacing $B_n$ by $B_n - \left(\bigcup_{k<n} B_k\right)$ if necessary, we can suppose that the sets $B_n$ are pairwise disjoint. Set $A = \bigcup_n A_n$ and $B = \bigcup_n B_n$; then

$$J(A) = I(B \cap C) = \sum_n I(B_n \cap C) = \sum_n J(A_n);$$

J is thus a countably additive and bounded function on $\mathfrak{B}(T)$. Since T is by hypothesis a Radon space, there exists a bounded measure $\mu$ on T such that $J(A) = \mu^\bullet(A)$ for every $A \in \mathfrak{B}(T)$; consequently

$$\alpha = J(T) = \mu^\bullet(T) = \sup_K \mu^\bullet(K) = \sup_K J(K),$$

by the definition of $\mu^\bullet$. Formula (6) is thus established.

Let us now show that T is universally measurable. Let $\lambda$ be a bounded measure on X; the preceding argument may be applied to the set function $I : A \mapsto \lambda^\bullet(A)$ on $\mathfrak{B}(X)$, thus there exists a sequence $(K_n)$ of compact subsets of T such that (with the above notations)

$$\sup_n \lambda^\bullet(K_n) = J(T) = \lambda^\bullet(C).$$

Set $K' = \bigcup_{n\in N} K_n$; $K'$ is Borel in X, $K' \subset T \subset C$, $\lambda^\bullet(K') = \lambda^\bullet(C)$, therefore these three sets differ only by $\lambda$-negligible sets, and so T is $\lambda$-measurable. This completes the proof of *a*).

Let us pass to *b*). Suppose that X is a Radon space, and that T is universally measurable in X. Let I be a positive function on $\mathfrak{B}(T)$ that is countably additive and bounded; the function $A \mapsto I(A \cap T)$ on $\mathfrak{B}(X)$ is then positive, countably additive and bounded, therefore there exists a bounded measure $\nu$ on X such that $I(A \cap T) = \nu^\bullet(A)$ for all $A \in \mathfrak{B}(X)$. Now, T is $\nu$-measurable; the preceding relation shows that $\nu^\bullet(K) = 0$ for every compact subset K of X that is disjoint from T, therefore $\nu$ is concentrated on T. Consequently, for every Borel set A of X, we have $I(A \cap T) = \nu^\bullet(A \cap T) = \mu^\bullet(A \cap T)$, where $\mu$ is the measure induced by $\nu$ on T. Finally, it follows that $I(B) = \mu^\bullet(B)$ for every set $B \in \mathfrak{B}(T)$ (GT, IX, §6, No. 3, *Remark* 2), and I is indeed inner regular.

COROLLARY. — *If* X *is a Radon space, then every Borel subset* T *of* X *is Radon.*

For, T is universally measurable in X.

PROPOSITION 3. — *Every Souslin space (in particular, every Polish or Lusin space) is strongly Radon.*

Let T be a Souslin space; since T is a Lindelöf space (TG, IX, Appendix I, Cor. of Prop. 1),[2] it suffices to show that T is Radon (Prop. 1). Let I be a function defined on $\mathfrak{B}(T)$, positive, countably additive and bounded. We extend I to $\mathfrak{P}(T)$ by setting, for every subset A of T,

$$I(A) = \inf_{\substack{B\in\mathfrak{B}(T)\\ B\supset A}} I(B).$$

Let us show that this extension is a *capacity* on T (GT, IX, §6, No. 9). It is clear that the relation $A \subset A'$ implies $I(A) \leqslant I(A')$. Let $(A_n)$ be an increasing sequence of subsets of T, and let $A = \bigcup_n A_n$. The set of Borel sets that contain $A_n$ being stable for countable intersections, there exists for each $n$ a Borel set $B_n$ such that $A_n \subset B_n$ and $I(A_n) = I(B_n)$ (cf. the proof of Prop. 2). Set $C_n = \bigcap_{p\geqslant n} B_p$; $C_n$ is Borel, and $A_n \subset C_n \subset B_n$, therefore $I(A_n) = I(C_n)$. On the other hand, the sequence $(C_n)$ is increasing. Let $C = \bigcup_n C_n$: the relation $A \subset C$ implies that

$$I(A) \leqslant I(C) = \lim_n I(C_n) = \lim_n I(A_n),$$

[2] Every Souslin space has a countable base for open sets (GT, IX, §6, No. 2, Prop. 4), hence is Lindelöf (GT, I, §9, Exer. 14).

whence the equality $I(A) = \lim_n I(A_n)$ is immediate. Consequently, I is a capacity.

If $(H_n)$ is a decreasing sequence of closed sets in T, obviously $I(\bigcap_n H_n) = \inf_n I(H_n)$. It follows that every Souslin subset F of T is capacitable for I (TG, IX, §6, No. 10, Prop. 15). In particular, every Borel set A of T is capacitable (*loc. cit.*, §6, No. 3, Prop. 10).[3] In other words,

$$I(A) = \sup_K I(K),$$

where K runs over the set of compact sets contained in A; we have proved that I is inner regular.

*Remark.* — Let X be a Lusin space (in particular, any Polish space), and $f$ a bijective continuous mapping of X onto a (Lusin) regular space Y. One knows (TG, IX, §6, No. 7, Prop. 14) that the mapping $B \mapsto f^{-1}(B)$ is a bijection of the Borel tribe of Y onto the Borel tribe of X. The spaces X and Y are Lusin, hence strongly Radon (Prop. 3). It follows immediately that the mapping $\mu \mapsto f(\mu)$ is a bijection of the set of bounded measures on X onto the set of bounded measures on Y.[4]

## §4. INVERSE LIMITS OF MEASURES

*Throughout this section,* I *denotes a nonempty set, equipped with a preorder relation, denoted* $i \leqslant j$, *and directed for this relation. Recall* (GT, I, §4, No. 4) *that an inverse system of topological spaces indexed by* I *is a family* $(T_i, p_{ij})$ *where* $T_i$ *is a topological space and* $p_{ij}$ *is a continuous mapping of* $T_j$ *into* $T_i$ *for* $i \leqslant j$, *where* $p_{ii}$ *is the identity mapping of* $T_i$, *and where* $p_{ik} = p_{ij} \circ p_{jk}$ *for* $i \leqslant j \leqslant k$. *Let* T *be a topological space and* $(p_i)_{i \in I}$ *a family of continuous mappings* $p_i : T \to T_i$. *The family* $(p_i)_{i\in I}$ *is said to be coherent if* $p_i = p_{ij} \circ p_j$ *for* $i \leqslant j$, *and it is said to be separating if for distinct* $x, y$ *in* T *there exists an* $i \in I$ *such that* $p_i(x) \neq p_i(y)$. *When* $T = \varprojlim T_i$ *and* $p_i$ *is the canonical mapping of* T *into* $T_i$, *the family* $(p_i)_{i\in I}$ *is coherent and separating.*

---

(3) In a Souslin space, every Borel set is a Souslin set (GT, IX, §6, No. 3, Prop. 11).

(4) More generally, if $f : X \to Y$ is a continuous bijection between Souslin spaces, then $\mu \mapsto f(\mu)$ is a bijective mapping of the set of bounded measures on X onto the set of bounded measures on Y (§2, No. 4, Prop. 9).

## 1. Complements on compact spaces and inverse limits

PROPOSITION 1. — *Let* X *and* Y *be two topological spaces and* $f$ *a continuous mapping of* X *into* Y. *Let* $(\mathrm{K}_\alpha)_{\alpha\in\mathrm{A}}$ *be a decreasing directed family of compact subsets of* X, *with intersection* K. *Then* $f(\mathrm{K}) = \bigcap_{\alpha\in\mathrm{A}} f(\mathrm{K}_\alpha)$.

For, let $y$ be a point of $\bigcap_{\alpha\in\mathrm{A}} f(\mathrm{K}_\alpha)$; for every $\alpha \in \mathrm{A}$, the set $\mathrm{L}_\alpha = \mathrm{K}_\alpha \cap \overset{-1}{f}(y)$ is compact and nonempty. The family $(\mathrm{L}_\alpha)_{\alpha\in\mathrm{A}}$ is directed downward, therefore its intersection L is nonempty. Now, $\mathrm{L} = \mathrm{K} \cap \overset{-1}{f}(y)$, whence $y \in f(\mathrm{K})$. We have thus proved the inclusion $f(\mathrm{K}) \supset \bigcap_{\alpha\in\mathrm{A}} f(\mathrm{K}_\alpha)$, and the reverse inclusion is obvious.

PROPOSITION 2. — *Let there be given an inverse system* $(\mathrm{T}_i, p_{ij})$ *of topological spaces indexed by* I, *a topological space* T, *and a coherent and separating family of continuous mappings* $p_i : \mathrm{T} \to \mathrm{T}_i$. *Then:*

*a) For every compact subset* K *of* T, *one has* $\mathrm{K} = \bigcap_{i\in\mathrm{I}} \overset{-1}{p_i}(p_i(\mathrm{K}))$.

*b) Let* K *and* L *be two disjoint compact subsets of* T. *There exists an* $i \in \mathrm{I}$ *such that* $p_j(\mathrm{K})$ *and* $p_j(\mathrm{L})$ *are disjoint for* $j \geqslant i$.

*a)* Let $x$ be a point of $\bigcap_{i\in\mathrm{I}} \overset{-1}{p_i}(p_i(\mathrm{K}))$; for every $i \in \mathrm{I}$, the set $\mathrm{K}_i$ of points $y$ in K such that $p_i(y) = p_i(x)$ is a nonempty closed subset of K. For $i \leqslant j$ we have $\mathrm{K}_i \supset \mathrm{K}_j$, and, since K is compact, the set $\bigcap_{i\in\mathrm{I}} \mathrm{K}_i$ is therefore nonempty. Let $y$ be a point of $\bigcap_{i\in\mathrm{I}} \mathrm{K}_i$; we have $y \in \mathrm{K}$ and $p_i(y) = p_i(x)$ for all $i \in \mathrm{I}$, whence $y = x$; finally, $x \in \mathrm{K}$, which proves the inclusion $\mathrm{K} \supset \bigcap_{i\in\mathrm{I}} \overset{-1}{p_i}(p_i(\mathrm{K}))$; the reverse inclusion is obvious.

*b)* For every $i \in \mathrm{I}$, set $\mathrm{M}_i = \overset{-1}{p_i}(p_i(\mathrm{K})) \cap \mathrm{L}$; this is a closed subset of the compact space L, we have $\mathrm{M}_i \supset \mathrm{M}_j$ for $i \leqslant j$, and, by *a)*,

$$\bigcap_{i\in\mathrm{I}} \mathrm{M}_i = \mathrm{K} \cap \mathrm{L} = \varnothing .$$

Consequently, there exists an index $i$ such that $\mathrm{M}_i = \varnothing$. For $j \geqslant i$, we have $\overset{-1}{p_j}(p_j(\mathrm{K})) \cap \mathrm{L} = \mathrm{M}_j \subset \mathrm{M}_i = \varnothing$, whence $p_j(\mathrm{K}) \cap p_j(\mathrm{L}) = \varnothing$.

## 2. Inverse systems of measures

DEFINITION 1. — *Let* $\mathscr{T} = (\mathrm{T}_i, p_{ij})$ *be an inverse system of topological spaces indexed by* I. *One calls inverse system* (resp. *sub-inverse system*) *of measures on* $\mathscr{T}$ *a family* $(\mu_i)_{i\in I}$, *where* $\mu_i$ *is a bounded measure on* $T_i$ *for all* $i \in I$, *and where* $\mu_i = p_{ij}(\mu_j)$ (resp. $\mu_i \geqslant p_{ij}(\mu_j)$) *for* $i \leqslant j$.

PROPOSITION 3. — *Let there be given an inverse system of topological spaces* $\mathscr{T} = (T_i, p_{ij})$ *indexed by* I, *a topological space* T, *a coherent and separating family of continuous mappings* $p_i : T \to T_i$ (*for* $i \in I$) *and a sub-inverse system* $(\mu_i)_{i\in I}$ *of measures on* $\mathscr{T}$. *For every compact subset* K *of* T, *set*

$$J(K) = \inf_{i\in I} \mu_i^\bullet(p_i(K)). \tag{1}$$

*Then there exists a bounded measure* $\pi$ *on* T, *and only one, such that* $\pi^\bullet(K) = J(K)$ *for every compact subset* K *of* T. *One has* $\mu_i \geqslant p_i(\pi)$ *for all* $i \in I$, *and* $\pi$ *is the largest measure on* T *satisfying this condition.*

Let us first prove that J(K) is the limit of $\mu_i^\bullet((p_i(K))$ with respect to the section filter $\mathfrak{F}$ of the directed preordered set I: for this, it suffices (GT, IV, §5, No. 2, Th. 2) to show that $\mu_i^\bullet(p_i(K)) \geqslant \mu_j^\bullet(p_j(K))$ for $i \leqslant j$; now, setting $\mu'_{ij} = p_{ij}(\mu_j)$, we have $\mu'_{ij} \leqslant \mu_i$ and $p_j(K) \subset \overset{-1}{p}_{ij}(p_i(K))$, whence

$$\mu_j^\bullet(p_j(K)) \leqslant \mu_j^\bullet(\overset{-1}{p}_{ij}(p_i(K))) = (\mu'_{ij})^\bullet(p_i(K)) \leqslant \mu_i^\bullet(p_i(K)).$$

Let us now pass to the study of the properties of the function J:

1) It is clear that $J(K) \leqslant J(L)$ when $K \subset L$.

2) Let K and L be two compact subsets of T. For every $i \in I$, we have $p_i(K \cup L) = p_i(K) \cup p_i(L)$, whence

$$\mu_i^\bullet(p_i(K \cup L)) \leqslant \mu_i^\bullet(p_i(K)) + \mu_i^\bullet(p_i(L));$$

passing to the limit with respect to the filter $\mathfrak{F}$, we obtain $J(K \cup L) \leqslant J(K) + J(L)$.

3) Suppose that the compact sets K and L are disjoint. By Prop. 2 of No. 1, there exists an $i \in I$ such that $p_j(K) \cap p_j(L) = \varnothing$ for $j \geqslant i$. For $j \geqslant i$ we therefore have

$$\mu_j^\bullet(p_j(K \cup L)) = \mu_j^\bullet(p_j(K)) + \mu_j^\bullet(p_j(L)),$$

whence $J(K \cup L) = J(K) + J(L)$ on passing to the limit with respect to the filter $\mathfrak{F}$.

4) Let $(K_\alpha)_{\alpha\in A}$ be a decreasing directed family of compact subsets of T, with intersection K. By Prop. 1 of No. 1, $p_i(K) = \bigcap_{\alpha\in A} p_i(K_\alpha)$ and

therefore $\mu_i^\bullet(p_i(\mathrm{K})) = \inf_{\alpha\in\mathrm{A}} \mu_i^\bullet(p_i(\mathrm{K}_\alpha))$ for all $i \in \mathrm{I}$ (§1, No. 6, Cor. of Prop. 5). From this, one deduces

$$\begin{aligned}\mathrm{J}(\mathrm{K}) &= \inf_{i\in\mathrm{I}} \mu_i^\bullet(p_i(\mathrm{K})) = \inf_{i\in\mathrm{I}}\inf_{\alpha\in\mathrm{A}} \mu_i^\bullet(p_i(\mathrm{K}_\alpha)) \\ &= \inf_{\alpha\in\mathrm{A}}\inf_{i\in\mathrm{I}} \mu_i^\bullet(p_i(\mathrm{K}_\alpha)) = \inf_{\alpha\in\mathrm{A}} \mathrm{J}(\mathrm{K}_\alpha).\end{aligned}$$

5) Let us choose an $i \in \mathrm{I}$ and set $c = \mu_i^\bullet(\mathrm{T}_i)$. Then $c$ is finite and $\mathrm{J}(\mathrm{K}) \leqslant \mu_i^\bullet(p_i(\mathrm{K})) \leqslant \mu_i^\bullet(\mathrm{T}_i)$, thus $\mathrm{J}(\mathrm{K}) \leqslant c$ for every compact set $\mathrm{K}$ in $\mathrm{T}$.

The preceding properties permit applying Th. 1 of §3, No. 1; we conclude that there exists one and only one bounded measure $\pi$ on $\mathrm{T}$ such that $\pi^\bullet(\mathrm{K}) = \mathrm{J}(\mathrm{K})$ for every compact subset $\mathrm{K}$ of $\mathrm{T}$. For every $i \in \mathrm{I}$, let us denote by $\nu_i$ the measure on $\mathrm{T}_i$ that is the image of $\pi$ under $p_i$. Let $i \in \mathrm{I}$, $\mathrm{A}$ a compact subset of $\mathrm{T}_i$, and $\mathfrak{L}$ the set of compact subsets of $\overset{-1}{p_i}(\mathrm{A})$. By *Remark* 3 of §1, No. 2, we have $\pi^\bullet(\overset{-1}{p_i}(\mathrm{A})) = \sup_{\mathrm{K}\in\mathfrak{L}} \pi^\bullet(\mathrm{K})$; moreover, $\nu_i^\bullet(\mathrm{A}) = \pi^\bullet(\overset{-1}{p_i}(\mathrm{A}))$ and $\mathrm{J}(\mathrm{K}) = \pi^\bullet(\mathrm{K})$ for $\mathrm{K} \in \mathfrak{L}$, whence $\nu_i^\bullet(\mathrm{A}) = \sup_{\mathrm{K}\in\mathfrak{L}} \mathrm{J}(\mathrm{K})$. For $\mathrm{K} \in \mathfrak{L}$, we have $p_i(\mathrm{K}) \subset \mathrm{A}$, whence $\mathrm{J}(\mathrm{K}) \leqslant \mu_i^\bullet(p_i(\mathrm{K})) \leqslant \mu_i^\bullet(\mathrm{A})$ and finally $\nu_i^\bullet(\mathrm{A}) \leqslant \mu_i^\bullet(\mathrm{A})$. Since $\mathrm{A}$ is an arbitrary compact set in $\mathrm{T}_i$, we conclude that $\nu_i \leqslant \mu_i$. The last assertion of the proposition is obvious.

Q.E.D.

THEOREM 1 (Prokhorov). — *Let $\mathscr{T} = (\mathrm{T}_i, p_{ij})$ be an inverse system of topological spaces indexed by* $\mathrm{I}$, $\mathrm{T}$ *a topological space and* $(p_i)_{i\in\mathrm{I}}$ *a coherent and separating family of continuous mappings* $p_i : \mathrm{T} \to \mathrm{T}_i$. *Finally, let* $(\mu_i)_{i\in\mathrm{I}}$ *be an inverse system of measures on* $\mathscr{T}$.

*For there to exist a bounded measure $\mu$ on $\mathrm{T}$ such that $p_i(\mu) = \mu_i$ for all $i \in \mathrm{I}$, it is necessary and sufficient that the following condition be satisfied:*

(P) *for every $\varepsilon > 0$, there exists a compact subset $\mathrm{K}$ of $\mathrm{T}$ such that $\mu_i^\bullet(\mathrm{T}_i - p_i(\mathrm{K})) \leqslant \varepsilon$ for all $i \in \mathrm{I}$.*

*When this is so, the measure $\mu$ is uniquely determined and*

$$\mu^\bullet(\mathrm{K}) = \inf_i \mu_i^\bullet(p_i(\mathrm{K})) \tag{2}$$

*for every compact set* $\mathrm{K}$ *in* $\mathrm{T}$.

Let us first prove the uniqueness of $\mu$. Let $\mu$ be a bounded measure on $\mathrm{T}$ such that $p_i(\mu) = \mu_i$ for all $i \in \mathrm{I}$. Let $\mathrm{K}$ be a compact subset of $\mathrm{T}$; by Prop. 2 of No. 1, the set $\mathrm{K}$ is the intersection of the decreasing directed family $(\overset{-1}{p_i}(p_i(\mathrm{K})))_{i\in\mathrm{I}}$ of *closed* subsets of $\mathrm{T}$. By the Cor. of Prop. 5 of §1, No. 6, we therefore have

$$\mu^\bullet(\mathrm{K}) = \inf_{i\in\mathrm{I}} \mu^\bullet(\overset{-1}{p_i}(p_i(\mathrm{K}))) = \inf_{i\in\mathrm{I}} \mu_i^\bullet(p_i(\mathrm{K})),$$

which establishes the formula (2). Since two measures that coincide on the set of compact sets are equal (§1, No. 2, Cor. of Prop. 2), it follows that $\mu$ is unique.

By Prop. 3, there exists a bounded measure $\pi$ on T such that $\pi^\bullet(\mathrm{K}) = \inf_{i\in\mathrm{I}} \mu_i^\bullet(p_i(\mathrm{K}))$ for every compact subset K of T. By formula (2), the existence of a bounded measure $\mu$ on T such that $p_i(\mu) = \mu_i$ for all $i \in \mathrm{I}$ is therefore equivalent to the relation:

(P′) $p_i(\pi) = \mu_i$ *for all* $i \in \mathrm{I}$.

For $i \leqslant j$, we have $\mu_i = p_{ij}(\mu_j)$, whence $\mu_i^\bullet(\mathrm{T}_i) = \mu_j^\bullet(\mathrm{T}_j)$; since I is directed, there exists a finite number $c \geqslant 0$ such that $\mu_i^\bullet(\mathrm{T}_i) = c$ for all $i \in \mathrm{I}$. By Prop. 3, the measure $\mu_i - p_i(\pi)$ is positive, hence is zero if and only if its total mass is zero, that is, if $\mu_i(\mathrm{T}_i) = p_i(\pi)^\bullet(\mathrm{T}_i)$. Since $p_i(\pi)^\bullet(\mathrm{T}_i) = \pi^\bullet(\mathrm{T})$, the condition (P′) is thus equivalent to $\pi^\bullet(\mathrm{T}) = c$, that is (§1, No. 2, *Remark* 3) to the property:

(P″) $\sup_{\mathrm{K}\in\mathfrak{K}} \pi^\bullet(\mathrm{K}) = c$, *where* $\mathfrak{K}$ *is the set of compact subsets of* T.

Now, for $\mathrm{K} \in \mathfrak{K}$, we have

$$\pi^\bullet(\mathrm{K}) = \inf_{i\in\mathrm{I}} \mu_i^\bullet(p_i(\mathrm{K})) = c - \sup_{i\in\mathrm{I}} \mu_i^\bullet(\mathrm{T}_i - p_i(\mathrm{K}))$$

and this formula immediately implies the equivalence of (P) and (P″).

Q.E.D.

Let $(\mathrm{T}_i, p_{ij})$ be an inverse system of topological spaces. Set $\mathrm{T} = \varprojlim \mathrm{T}_i$ and denote by $p_i$ the canonical mapping of T into $\mathrm{T}_i$. Generalizing Def. 2 of Ch. III, §4, No. 5, we shall say that a bounded measure $\mu$ on T is the *inverse limit of an inverse system* $(\mu_i)_{i\in\mathrm{I}}$ *of measures* if $\mu_i = p_i(\mu)$ for all $i \in \mathrm{I}$. Th. 1 provides a criterion for the existence of inverse limits of measures. When the spaces $\mathrm{T}_i$ are *compact*, and the mappings $p_{ij}$ surjective, T is compact and $p_i(\mathrm{T}) = \mathrm{T}_i$ for every $i \in \mathrm{I}$; the condition (P) is therefore fulfilled, and in this case we recover Prop. 8, (iv) of Ch. III, §4, No. 5.

*Remark.* — Let $(\mu_i)_{i\in\mathrm{I}}$ be an inverse system of measures on the inverse system of spaces $\mathscr{T} = (\mathrm{T}_i, p_{ij})$. Assume given a topological space T′ and continuous mappings $p'_i : \mathrm{T}' \to \mathrm{T}_i$; assume that the family $(p'_i)_{i\in\mathrm{I}}$ is coherent, but not necessarily separating. *If Prokhorov's condition* (P) *is satisfied by the family* $(p'_i)_{i\in\mathrm{I}}$, *there exists a measure* $\mu'$ *(not necessarily unique) on* T′ *such that* $p'_i(\mu') = \mu_i$ *for all* $i \in \mathrm{I}$.

For, set $\mathrm{T} = \varprojlim \mathrm{T}_i$ and $p' = (p'_i)_{i\in\mathrm{I}}$, and denote by $p_i$ the canonical mapping of T into $\mathrm{T}_i$; Prokhorov's condition is satisfied by T and the $p_i$, because $p_i(p'(\mathrm{K}')) = p'_i(\mathrm{K}')$ and $p'(\mathrm{K}')$ is compact in T for every compact subset K′ of T′. By Th. 1, there exists a bounded measure $\mu$ on T such that $p_i(\mu) = \mu_i$ for all $i \in \mathrm{I}$. Let K′ be a compact set in T′; then $\mu^\bullet(p'(\mathrm{K}')) = \inf_{i\in\mathrm{I}} \mu_i^\bullet(p'_i(\mathrm{K}'))$, whence

$$\mu^\bullet(\mathrm{T} - p'(\mathrm{K}')) = \sup_{i\in \mathrm{I}} \mu_i^\bullet(\mathrm{T}_i - p_i'(\mathrm{K}')).$$

Let $\varepsilon > 0$; since Prokhorov's condition (P) is satisfied by the $p_i'$, one can therefore find a compact subset $\mathrm{K}'$ of $\mathrm{T}'$ such that $\mu^\bullet(\mathrm{T} - p'(\mathrm{K}')) \leqslant \varepsilon$. Prop. 8 of §2, No. 4 then establishes the existence of a bounded measure $\mu'$ on $\mathrm{T}'$ with $\mu = p'(\mu')$, whence $\mu_i = p_i(\mu) = p_i(p'(\mu')) = p_i'(\mu')$ for all $i \in \mathrm{I}$.

## 3. The case of countable inverse systems

THEOREM 2. — *Assume that the directed preordered set* I *has a* countable *cofinal subset. Let* $\mathscr{T} = (\mathrm{T}_i, p_{ij})$ *be an inverse system of topological spaces,* $\mathrm{T} = \varprojlim \mathrm{T}_i$ *and* $p_i$ *the canonical mapping of* T *into* $\mathrm{T}_i$. *Then every inverse system* $(\mu_i)_{i\in\mathrm{I}}$ *of measures on* $\mathscr{T}$ *admits an inverse limit.*

We shall first treat the case that $\mathrm{I} = \mathbf{N}$ and set $q_n = p_{n,n+1}$. Let $\varepsilon > 0$. Define recursively a sequence of compact sets $\mathrm{L}_n \subset \mathrm{T}_n$ as follows: $\mathrm{L}_0$ is a compact subset of $\mathrm{T}_0$ such that $\mu_0^\bullet(\mathrm{T}_0 - \mathrm{L}_0) \leqslant \varepsilon/2$, and for $n \geqslant 0$ the compact set $\mathrm{L}_{n+1}$ is contained in $\overset{-1}{q_n}(\mathrm{L}_n)$ and satisfies

$$\mu_{n+1}^\bullet\big(\overset{-1}{q_n}(\mathrm{L}_n) - \mathrm{L}_{n+1}\big) \leqslant \varepsilon/2^{n+2}.$$

This construction is possible by virtue of *Remark* 3 of §1, No. 2. We have

$$\begin{aligned}\mu_{n+1}^\bullet(\mathrm{T}_{n+1} - \mathrm{L}_{n+1}) &= \mu_{n+1}^\bullet\big(\mathrm{T}_{n+1} - \overset{-1}{q_n}(\mathrm{L}_n)\big) + \mu_{n+1}^\bullet\big(\overset{-1}{q_n}(\mathrm{L}_n) - \mathrm{L}_{n+1}\big)\\ &\leqslant \mu_{n+1}^\bullet\big(\mathrm{T}_{n+1} - \overset{-1}{q_n}(\mathrm{L}_n)\big) + \varepsilon/2^{n+2}\\ &= \mu_n^\bullet(\mathrm{T}_n - \mathrm{L}_n) + \varepsilon/2^{n+2}\end{aligned}$$

because $\mu_n = q_n(\mu_{n+1})$; by induction on $p$, one deduces that

$$\mu_p^\bullet(\mathrm{T}_p - \mathrm{L}_p) \leqslant \varepsilon(1 - 1/2^{p+1}) \leqslant \varepsilon.$$

Since T is a closed subspace of $\prod_{n\in\mathbf{N}} \mathrm{T}_n$ and the product space $\prod_{n\in\mathbf{N}} \mathrm{L}_n$ is compact, the subset $\mathrm{L} = \mathrm{T} \cap \prod_{n\in\mathbf{N}} \mathrm{L}_n = \bigcap_{n\in\mathbf{N}} \overset{-1}{p_n}(\mathrm{L}_n)$ of T is compact. Let $n \in \mathbf{N}$; we have $p_n(\mathrm{L}) = \bigcap_{m\geqslant n} p_{nm}(\mathrm{L}_m)$ (GT, I, §9, No. 6, Prop. 8) and $p_{nm}(\mathrm{L}_m) \supset p_{nm'}(\mathrm{L}_{m'})$ for $m' \geqslant m \geqslant n$, whence

$$\mu_n^\bullet\big(\mathrm{T}_n - p_n(\mathrm{L})\big) = \lim_{m\to\infty} \mu_n^\bullet\big(\mathrm{T}_n - p_{nm}(\mathrm{L}_m)\big).$$

But, for $m \geqslant n$, the measure $\mu_n$ is the image of $\mu_m$ under $p_{nm}$, whence

$$\mu_n^\bullet(\mathrm{T}_n - p_{nm}(\mathrm{L}_m)) = \mu_m^\bullet(\mathrm{T}_m - \overset{-1}{p}_{nm}(p_{nm}(\mathrm{L}_m))) \leqslant \mu_m^\bullet(\mathrm{T}_m - \mathrm{L}_m) \leqslant \varepsilon\,;$$

passing to the limit with respect to $m$, we obtain $\mu_n^\bullet(\mathrm{T}_n - p_n(\mathrm{L})) \leqslant \varepsilon$. In other words, Prokhorov's condition (P) is satisfied, and there exists a bounded measure $\mu$ on T such that $\mu_n = p_n(\mu)$ for all $n \in \mathbf{N}$ (No. 2, Th. 1).

Let us pass to the general case: there exists in I an increasing cofinal sequence $(i_n)_{n\in\mathbf{N}}$. The mapping $t \mapsto (p_{i_n}(t))_{n\in\mathbf{N}}$ is a homeomorphism of T onto the inverse limit of the inverse system $(\mathrm{T}_{i_n}, p_{i_n i_m})$ (GT, I, §4, No. 4). By the first part of the proof, there exists therefore a bounded measure $\mu$ on T such that $\mu_{i_n} = p_{i_n}(\mu)$ for all $n \in \mathbf{N}$. Let $i \in \mathrm{I}$; there exists an $n \in \mathbf{N}$ with $i \leqslant i_n$, whence

$$p_i(\mu) = p_{ii_n}(p_{i_n}(\mu)) = p_{ii_n}(\mu_{i_n}) = \mu_i\,.$$

Q.E.D.

Theorem 2 is often used in the following situation: let D be a countable set and $(\mathrm{X}_t)_{t\in\mathrm{D}}$ a family of topological spaces. Let $\mathfrak{F}$ be the set of finite subsets of D, ordered by inclusion. For J in $\mathfrak{F}$, set $\mathrm{X_J} = \prod_{t\in\mathrm{J}} \mathrm{X}_t$, and for $\mathrm{J} \subset \mathrm{J'}$ let $p_{\mathrm{JJ'}}$ be the canonical projection of $\mathrm{X_{J'}}$ onto the partial product $\mathrm{X_J}$. Also set $\mathrm{X} = \prod_{t\in\mathrm{D}} \mathrm{X}_t$ and denote by $p_\mathrm{J}$ the canonical projection of X onto the partial product $\mathrm{X_J}$. One shows easily (cf. S, III, §7, No. 2, *Remark* 3) that the family $(p_\mathrm{J})_{\mathrm{J}\in\mathfrak{F}}$ defines a homeomorphism of X onto $\varprojlim \mathrm{X_J}$. An inverse system of measures is then a family of bounded measures $\mu_\mathrm{J}$ on $\mathrm{X_J}$ such that $\mu_\mathrm{J} = p_{\mathrm{JJ'}}(\mu_{\mathrm{J'}})$ for $\mathrm{J} \subset \mathrm{J'}$. There exists one and only one bounded measure $\mu$ on X such that $\mu_\mathrm{J} = p_\mathrm{J}(\mu)$ for every finite subset J of D (*Kolmogoroff's theorem*). One sometimes says that $\mu$ is the measure on $\prod_{t\in\mathrm{D}} \mathrm{X}_t$ having *margins* $\mu_\mathrm{J}$.

In particular, suppose given, for every $t \in \mathrm{D}$, a measure $\nu_t$ on $\mathrm{X}_t$ of total mass 1. Set $\mu_\mathrm{J} = \bigotimes_{t\in\mathrm{J}} \nu_t$ for every finite subset J of D. Let $\mathrm{J} \subset \mathrm{J'}$ be two finite subsets of D and let $\mathrm{K} = \mathrm{J'} - \mathrm{J}$; identifying $\mathrm{X_{J'}}$ with $\mathrm{X_J} \times \mathrm{X_K}$, one has $\mu_{\mathrm{J'}} = \mu_\mathrm{J} \otimes \mu_\mathrm{K}$, and since the measure $\mu_\mathrm{K}$ has total mass 1, the projection of $\mu_\mathrm{J} \otimes \mu_\mathrm{K}$ on $\mathrm{X_J}$ is equal to $\mu_\mathrm{J}$. The measure on X admitting the margins $\mu_\mathrm{J}$ is denoted $\bigotimes_{t\in\mathrm{D}} \nu_t$ and is called the *product of the family* $(\nu_t)_{t\in\mathrm{D}}$. When the spaces $\mathrm{X}_t$ are compact, we recover the construction of Ch. III, §4, No. 6.

## §5. MEASURES ON COMPLETELY REGULAR SPACES

*If* T *is a topological space, and* F *is a Banach space, the notation* $\mathscr{C}^b(\mathrm{T};\mathrm{F})$ *indicates the space of bounded continuous functions on* T *with values in* F, *equipped with the norm of uniform convergence. If* $\mathrm{F}=\mathbf{R}$, *this notation is abbreviated to* $\mathscr{C}^b(\mathrm{T})$, *or to* $\mathscr{C}^b$ *if there is no ambiguity, and one denotes by* $\mathscr{C}^b_+(\mathrm{T})$ *or* $\mathscr{C}^b_+$ *the cone of positive functions in* $\mathscr{C}^b(\mathrm{T})$. *The space of bounded complex measures on* T *will be denoted* $\mathscr{M}^b(\mathrm{T};\mathbf{C})$, *the space of bounded real measures by* $\mathscr{M}^b(\mathrm{T})$ *or* $\mathscr{M}^b$, *and the cone of bounded positive measures by* $\mathscr{M}^b_+(\mathrm{T})$ *or* $\mathscr{M}^b_+$.

### 1. Measures and bounded continuous functions

Recall (GT, IX, §1, No. 5, Def. 4) that a topological space T is said to be *completely regular* if it is uniformizable and Hausdorff. This is equivalent to saying (*loc. cit.*, Prop. 3) that T is homeomorphic to a subspace of a compact space. If T is completely regular, then every positive lower semi-continuous function $f$ on T is the upper envelope of the increasing directed set of elements of $\mathscr{C}^b_+(\mathrm{T})$ that are $\leqslant f$, and every positive and bounded upper semi-continuous function $g$ is the lower envelope of the decreasing directed set of elements of $\mathscr{C}^b_+(\mathrm{T})$ that are $\geqslant g$ (*loc. cit.*, §1, No. 6, Prop. 5). We shall need the following lemma:

*Lemma. — Let* T *be a completely regular space,* K *a compact subset of* T, *and* U *an open subset of* T *containing* K.

*a) There exists an open subset* U′ *of* T *such that* $\mathrm{K}\subset\mathrm{U}'\subset\overline{\mathrm{U}'}\subset\mathrm{U}$.

*b) Let* $f$ *be a continuous function defined on* K *with values in an interval* I *of* **R** (resp. *in* **C**). *There exists a bounded continuous function* $f'$ *on* T, *with values in* I (resp. *in* **C**), *that extends* $f$ *and is zero on* $\mathrm{T}-\mathrm{U}$.

It suffices to treat the case that T is a subspace of a compact space X. Let V be an open subset of X such that $\mathrm{V}\cap\mathrm{T}=\mathrm{U}$; denote by V′ an open set in X containing K such that $\overline{\mathrm{V}'}\subset\mathrm{V}$, by $g$ a continuous function on X with values in I (resp. in **C**) extending $f$ and zero on $\mathrm{X}-\mathrm{V}$ (GT, IX, §4, No. 1, Prop. 1). The condition *a*) is satisfied by taking $\mathrm{U}'=\mathrm{V}'\cap\mathrm{T}$, and *b*) by taking $f'$ to be the restriction of $g$ to T.

PROPOSITION 1. — *Let* T *be a completely regular space.*

*a) Let* $\mu$ *be a positive measure on* T, *and* $f$ *a numerical function* $\geqslant 0$ *defined on* T *and lower semi-continuous* (resp. *upper semi-continu-*

*ous, finite, with compact support). Then*

$$\mu^\bullet(f) = \sup_{g\in \mathrm{I}_f} \mu^\bullet(g) \qquad (\text{resp.}\ \mu^\bullet(f) = \inf_{g\in \mathrm{S}_f} \mu(g))\,, \tag{1}$$

*where* $\mathrm{I}_f$ (resp. $\mathrm{S}_f$) *denotes the set of bounded continuous functions* $g$ *such that* $0 \leqslant g \leqslant f$ (resp. $g \geqslant f$).

*b) Let* $\theta$ *be a complex measure on* $\mathrm{T}$, *and* $f$ *a numerical function* $\geqslant 0$ *defined on* $\mathrm{T}$ *and lower semi-continuous. Then*

$$|\theta|^\bullet(f) = \sup_g |\theta(g)|\,, \tag{2}$$

*where* $g$ *runs over the set of bounded and* $|\theta|$*-integrable continuous complex functions such that* $|g| \leqslant f$.

The first of the formulas (1) is obvious, because $\mathrm{I}_f$ is an increasing directed set of continuous functions whose upper envelope is $f$, and one can apply Prop. 5 of §1, No. 6. The same proposition will imply the second formula, if we show that $\mathrm{S}_f$ contains a $\mu$-integrable bounded continuous function. Thus, let $\mathrm{K}$ be the support of $f$, and $\mathrm{M}$ the supremum of $f$; since $\mathrm{K}$ is compact, $\mathrm{M}$ is finite (GT, IV, §6, No. 2, Th. 3). Let $\mathrm{U}$ be an open set containing $\mathrm{K}$ and such that $\mu^\bullet(\mathrm{U}) < +\infty$; there exists (Lemma) a continuous function $g$ with values in $[0, \mathrm{M}]$, equal to $\mathrm{M}$ on $\mathrm{K}$ and zero outside $\mathrm{U}$; then $g \in \mathrm{S}_f$ and $\mu^\bullet(g) \leqslant \mathrm{M}\mu^\bullet(\mathrm{U}) < +\infty$.

Let us pass to *b*). It clearly suffices to show that $|\theta|^\bullet(f) \leqslant \sup_g |\theta(g)|$. Let $a$ and $b$ be two real numbers such that $a < b < |\theta|^\bullet(f)$. By (1), there exists a function $h \in \mathscr{C}^b_+(\mathrm{T})$ such that $h \leqslant f$ and $|\theta|^\bullet(h) > b$; denote by $\mathrm{M}$ the supremum of $h$. By the definition of $|\theta|^\bullet$ (§1, No. 2, Def. 4), there exists a compact subset $\mathrm{K}$ of $\mathrm{T}$ such that $|\theta|^\bullet_{\mathrm{K}}(h_{\mathrm{K}}) > b$. There then exists a continuous complex function $j$ on $\mathrm{K}$ such that $|j| \leqslant h_{\mathrm{K}}$ and $|\theta_{\mathrm{K}}(j)| > b$ (Ch. III, §1, No. 6). Let us choose an open set $\mathrm{U}$ containing $\mathrm{K}$ and such that $|\theta|^\bullet(\mathrm{U} - \mathrm{K}) \leqslant \dfrac{b-a}{\mathrm{M}}$ (§1, No. 9, Props. 13 and 14); extend $j$ to a continuous complex function $k$ on $\mathrm{T}$, zero outside $\mathrm{U}$ (Lemma); for every $t \in \mathrm{T}$, set

$$g(t) = \begin{cases} k(t) & \text{if } |k(t)| \leqslant h(t) \\ \dfrac{k(t)}{|k(t)|}h(t) & \text{if } |k(t)| > h(t)\,. \end{cases} \tag{3}$$

Clearly $|g| \leqslant h \leqslant f$, and $g = j$ on $\mathrm{K}$, therefore $\big||\theta_{\mathrm{K}}(j)| - |\theta(g)|\big| = \big||\theta(j^0)| - |\theta(g)|\big| \leqslant |\theta|^\bullet(|j^0 - g|) \leqslant \mathrm{M}\cdot|\theta|^\bullet(\mathrm{U} - \mathrm{K}) \leqslant b - a$, consequently $|\theta(g)| > a$. Let us show on the other hand that $g$ is a continuous function:

since $a$ is subject to the sole condition $a < |\theta|^\bullet(f)$, this will imply that the second member of (2) is $\geqslant$ the first, whence the proposition. Now, let F (resp. F′) be the set of $t \in \mathrm{T}$ such that $|k(t)| \leqslant h(t)$ (resp. $|k(t)| \geqslant h(t)$). These sets being closed, and their union being T, it will suffice to show that $g_{\mathrm{F}}$ and $g_{\mathrm{F}'}$ are continuous: now, this property is obvious for $g_{\mathrm{F}} = k_{\mathrm{F}}$, and it is so for $g_{\mathrm{F}'}$ at the points where $k(t) \neq 0$; on the other hand, if $t \in \mathrm{F}'$ is such that $k(t) = 0$, then also $h(t) = 0$, and the inequality $|g| \leqslant h$ implies that $g$ is continuous at the point $t$.

*Remarks.* — 1) Let $f$ be a positive lower semi-continuous function, and let $\mathrm{J}_f$ be the set of positive bounded continuous functions *zero outside a $\mu$-integrable open set* and bounded above by $f$. One can show that $f$ is the upper envelope of $\mathrm{J}_f$ and that $\mu^\bullet(f) = \sup_{g \in \mathrm{J}_f} \mu(g)$.

2) If the measure $\mu$ is bounded, the formula $\mu^\bullet(f) = \inf_{g \in \mathrm{S}_f} \mu(g)$ is obviously valid for every function $f$ that is upper semi-continuous, positive and bounded.

PROPOSITION 2. — *Let $\eta$ and $\eta'$ be two complex measures on a completely regular space* T, *such that $\eta(f) = \eta'(f)$ for every function $f \in \mathscr{C}^b(\mathrm{T})$ that is integrable for $|\eta|$ and $|\eta'|$. Then $\eta = \eta'$.*

Let us take up again the proof of the second part of Proposition 1, on setting $\theta = \eta - \eta'$. We can require the open set U to be integrable for $|\eta|$ and $|\eta'|$. The function $g$ is then integrable for these two measures, and the relation $\theta(g) = 0$ implies $a < 0$; therefore $|\theta|^\bullet(f) = 0$ for every positive lower semi-continuous function $f$, whence finally $|\theta| = 0$, on taking $f = +\infty$.

PROPOSITION 3. — *Let $\mu$ be a positive measure on a completely regular space* T, *and let $p \in [1, +\infty[$. The space $\mathscr{H}$ of functions $f \in \mathscr{C}^b(\mathrm{T})$, whose support is contained in a $\mu$-integrable open set, is dense in $\mathscr{L}^p(\mu)$.*

By Prop. 15 of §1, No. 10, it suffices to show that if K is compact in T, and if $g$ is the extension to T by 0 of a function in $\mathscr{C}_+(\mathrm{K})$ between 0 and 1, then there exists a function $f \in \mathscr{C}_+^b(\mathrm{T})$, with support contained in a $\mu$-integrable open set, such that $\|f - g\|_p$ is arbitrarily small. Now, let $\varepsilon$ be a number $> 0$, U an open neighborhood of K such that $\mu^\bullet(\mathrm{U} - \mathrm{K}) < \varepsilon$, V an open neighborhood of K such that $\overline{\mathrm{V}} \subset \mathrm{U}$, and $f$ a function with values in $[0, 1]$, continuous, equal to $g$ on K and to 0 outside V (Lemma). The function $|f - g|^p$ is then bounded above by $\varphi_{\mathrm{U}-\mathrm{K}}$; therefore $\|f - g\|_p \leqslant \varepsilon^{1/p}$, which establishes the proposition.

*Remark* 3). — There is an analogous statement for functions with values in a Banach space F: the subspace $\mathscr{H} \otimes \mathrm{F}$ of $\mathscr{C}^b(\mathrm{T}; \mathrm{F})$ is dense in $\mathscr{L}_{\mathrm{F}}^p(\mu)$.

PROPOSITION 4. — *In order that a bounded complex measure $\theta$ on a completely regular space* T *be positive, it is necessary and sufficient that $\theta(f) \geqslant 0$ for every function $f \in \mathscr{C}_+^b(\mathrm{T})$.*

Necessity is obvious. To establish sufficiency, let us take up again the proof of the preceding proposition, on taking $p = 1$ and $\mu = |\theta|$; the notations being the same, the relation $\mu^\bullet(|f - g|) \leqslant \varepsilon$ and the inequality $\theta(f) \geqslant 0$ imply $\theta_K(g_K) = \theta(g) \geqslant -\varepsilon$; since $g_K$ is an arbitrary element of $\mathscr{C}(K)$ between $0$ and $1$, the measure $\theta_K$ is positive; the compact set K being arbitrary, this means that $\theta$ is positive.

## 2. Bounded measures and linear forms on $\mathscr{C}^b(T)$

PROPOSITION 5. — *Let* T *be a completely regular space, and* I *a continuous complex linear form on the normed space* $\mathscr{C}^b(T; \mathbf{C})$. *In order that there exist a bounded complex measure* $\theta$ *on* T *such that* $\theta(f) = I(f)$ *for all* $f \in \mathscr{C}^b(T; \mathbf{C})$, *it is necessary and sufficient that the following condition be satisfied:*

(M) *For every number* $\varepsilon > 0$, *there exists a compact subset* K *of* T *such that the relations* $g \in \mathscr{C}^b(T; \mathbf{C})$, $|g| \leqslant 1$, $g_K = 0$ *imply* $|I(g)| \leqslant \varepsilon$.

*The measure* $\theta$ *is then unique.*

Uniqueness follows from Prop. 2 of No. 1. Let us show that the condition (M) is necessary. Let $\theta$ be a bounded complex measure; let K be a compact set such that $|\theta|^\bullet(T - K) \leqslant \varepsilon$ (§1, No. 2, *Remark* 3). The hypotheses $|g| \leqslant 1$, $g_K = 0$ imply $|g| \leqslant \varphi_{\complement K}$, therefore $|\theta(g)| \leqslant |\theta|^\bullet(\varphi_{\complement K}) \leqslant \varepsilon$.

Let us pass to the proof of sufficiency. Let X be the Stone–Čech compactification of T (GT, IX, §1, Exer. 7; or TG, IX, §1, No. 6). For every function $f \in \mathscr{C}(X; \mathbf{C})$, set $\nu(f) = I(f_T)$; we define in this way a continuous linear form on $\mathscr{C}(X; \mathbf{C})$, that is, a complex measure on the compact space X. Let $\varepsilon$ be a number $> 0$, K a compact set satisfying (M); the function $\varphi_{\complement K}$ being lower semi-continuous and positive on X, the formula (2) gives us the following relations, where $\mathscr{G}$ denotes the set of functions $g \in \mathscr{C}(X; \mathbf{C})$ such that $|g| \leqslant \varphi_{\complement K}$:

$$|\nu|^\bullet(X - K) = \sup_{g \in \mathscr{G}} |\nu(g)| = \sup_{g \in \mathscr{G}} |I(g_T)| \leqslant \varepsilon .$$

Let $(K_n)_{n \geqslant 1}$ be a sequence of compact subsets of T, such that each $K_n$ satisfies (M) for $\varepsilon = 1/n$, and let $S = \bigcup_n K_n$; S is a Borel set in X, contained in T, and $|\nu|^\bullet(X - T) \leqslant |\nu|^\bullet(X - S) \leqslant |\nu|^\bullet(X - K_n) \leqslant 1/n$ for all $n$, so that T is $\nu$-measurable and $\nu$ is concentrated on T. Let $f$ be a bounded continuous function on T; since X is the Stone–Čech compactification of T, $f$ may be extended by continuity to a function $g \in \mathscr{C}(X; \mathbf{C})$. Now let $\mu$

be the measure induced by $\nu$ on $\mathrm{T}$; one has $\mu(f) = \nu(f^0)$.[1] Since $\nu$ is concentrated on $\mathrm{T}$, the functions $f^0$ and $g$ are equal $\nu$-almost everywhere, therefore $\mu(f) = \nu(g) = \mathrm{I}(g_{\mathrm{T}}) = \mathrm{I}(f)$, which completes the proof.

COROLLARY. — *With notations as in Prop.* 5, *suppose that there exists a bounded positive measure* $\mu$ *on* $\mathrm{T}$ *such that* $|\mathrm{I}(f)| \leqslant \mu(|f|)$ *for all* $f \in \mathscr{C}^b(\mathrm{T};\mathbf{C})$; *then there exists a complex measure* $\theta$ *on* $\mathrm{T}$ *such that* $\theta(f) = \mathrm{I}(f)$ *for all* $f \in \mathscr{C}^b(\mathrm{T};\mathbf{C})$.

## 3. Tight convergence of bounded measures

Let $\mathrm{T}$ be a completely regular space; the bilinear form

$$(f,\mu) \mapsto \int f(t)\,d\mu(t)$$

on $\mathscr{C}^b(\mathrm{T}) \times \mathscr{M}^b(\mathrm{T})$ puts these two spaces in a separating duality. For, it is clear that the duality is separating in $\mathscr{C}^b(\mathrm{T})$ from the fact that the measures $\varepsilon_x$ $(x \in \mathrm{T})$ belong to $\mathscr{M}^b(\mathrm{T})$; it is separating in $\mathscr{M}^b(\mathrm{T})$ by Prop. 2 of No. 1.

DEFINITION 1. — *The weak topology on* $\mathscr{M}^b(\mathrm{T})$ *associated with the preceding duality between* $\mathscr{C}^b(\mathrm{T})$ *and* $\mathscr{M}^b(\mathrm{T})$ *is called the topology of tight convergence* (*or the tight topology*) *on* $\mathscr{M}^b(\mathrm{T})$.

The tight topology is Hausdorff, by the remarks preceding the definition. We shall often employ the adverb 'tightly' to mean 'in the sense of the tight topology'. Absent mention to the contrary, $\mathscr{M}^b(\mathrm{T})$ will be equipped with the tight topology throughout the rest of this section.

Every element of $\mathscr{C}^b(\mathrm{T})$ is a linear combination of elements of $\mathscr{C}^b_+(\mathrm{T})$. For a filter $\mathfrak{F}$ on $\mathscr{M}^b(\mathrm{T})$ to converge tightly to a bounded measure $\lambda$, it is necessary and sufficient that

$$\lim_{\mu} \mu(f) = \lambda(f) \text{ with respect to } \mathfrak{F} \text{ for every } f \in \mathscr{C}^b_+(\mathrm{T}). \tag{4}$$

*Remarks.* — 1) If $\mathrm{T}$ is locally compact, the tight topology is finer than the topology induced on $\mathscr{M}^b(\mathrm{T})$ by the vague topology, and these two topologies coincide only when $\mathrm{T}$ is compact. For, if $\mathrm{T}$ is not compact, the mapping $t \mapsto \varepsilon_t$ converges vaguely to $0$ with respect to the filter of complements of relatively compact subsets of $\mathrm{T}$, but does not converge tightly to $0$, because the function $1$ belongs to $\mathscr{C}^b(\mathrm{T})$ (for the relations between vague convergence and tight convergence, see Prop. 9).

---

[1] This relation was only established above (§2, No. 1, Prop. 1) in the case that $f$ and $\nu$ are positive. The extension to the present situation, where $f$ and $\nu$ are complex and bounded, is immediate by linearity.

2) It follows at once from Prop. 4 that $\mathscr{M}_+^b(\mathrm{T})$ is *closed* in $\mathscr{M}^b(\mathrm{T})$.

3) If T is completely regular, the mapping $t \mapsto \varepsilon_t$ of T into $\mathscr{M}^b(\mathrm{T})$ is a *homeomorphism* (GT, IX, §1, No. 5).

PROPOSITION 6. — *Let* T *be a completely regular space.*

*a) Let* $f$ *be a lower semi-continuous numerical function* $\geqslant 0$ *defined on* T; *then the function* $\mu \mapsto |\mu|^\bullet(f)$ *is lower semi-continuous on* $\mathscr{M}^b(\mathrm{T})$.

*b) Let* $f$ *be an upper semi-continuous bounded function defined on* T; *then the function* $\mu \mapsto \mu(f)$ *is upper semi-continuous on* $\mathscr{M}_+^b(\mathrm{T})$.

For, one sees by Prop. 1 *b*) of No. 1 that $\mu \mapsto |\mu|^\bullet(f)$ is the upper envelope of a family of functions of the form $\mu \mapsto |\mu(g)|$ with $g \in \mathscr{C}^b(\mathrm{T})$, hence continuous for the tight topology. This establishes *a*). To prove *b*), it suffices to choose a constant upper bound C for $f$, and to write $\mu(f) = \mu(\mathrm{C}) - \mu(\mathrm{C} - f)$; the function $\mu \mapsto \mu(\mathrm{C})$ is continuous, and the function $\mu \mapsto \mu(\mathrm{C} - f)$ is lower semi-continuous on $\mathscr{M}_+^b(\mathrm{T})$ by the foregoing.

PROPOSITION 7. — *Let* T *be a completely regular space. Let* $\mu$ *be a bounded positive measure on* T, *and let* $f$ *be a bounded positive function on* T, *such that the set of points of* T *where* $f$ *is not continuous is locally* $\mu$*-negligible. Then the mapping* $\lambda \mapsto \lambda^\bullet(f)$ *of* $\mathscr{M}_+^b(\mathrm{T})$ *into* **R** *is continuous at the point* $\mu$.

For every $t \in \mathrm{T}$, set $f'(t) = \liminf_{s\to t} f(s)$, $f''(t) = \limsup_{s \to t} f(s)$. Obviously $f' \leqslant f \leqslant f''$, with equality at every point of T where $f$ is continuous (hence $\mu$-almost everywhere). On the other hand, $f'$ is lower semi-continuous, $f''$ is upper semi-continuous and bounded (GT, IV, §6, No. 2, Prop. 4). We therefore have the following relations by Prop. 6,

$$\begin{aligned}\mu^\bullet(f') \leqslant \liminf_{\lambda\to\mu} \lambda^\bullet(f') \leqslant \liminf_{\lambda\to\mu} \lambda^\bullet(f) \leqslant \mu^\bullet(f) \leqslant \limsup_{\lambda\to\mu} \lambda^\bullet(f) \\ \leqslant \limsup_{\lambda\to\mu} \lambda^\bullet(f'') \leqslant \mu^\bullet(f'').\end{aligned}$$

One concludes by observing that $\mu^\bullet(f') = \mu^\bullet(f'')$, because $f'$ and $f''$ are equal locally $\mu$-almost everywhere.

PROPOSITION 8. — *Let* X *be a completely regular space,* T *a subspace of* X, *and* $i$ *the canonical injection of* T *into* X. *Denote by* W *the set of bounded positive measures on* X *that are concentrated on* T, *equipped with the topology induced by* $\mathscr{M}^b(\mathrm{X})$. *Then the mapping* $\mu \mapsto i(\mu)$ *of* $\mathscr{M}_+^b(\mathrm{T})$ *into* $\mathscr{M}^b(\mathrm{X})$ *is a homeomorphism of* $\mathscr{M}_+^b(\mathrm{T})$ *onto* W.

We denote again by $i$ the mapping $\mu \mapsto i(\mu)$ of $\mathscr{M}_+^b(\mathrm{T})$ into $\mathscr{M}_+^b(\mathrm{X})$; $i$ is injective (§2, No. 4, Prop. 8) and maps $\mathscr{M}_+^b(\mathrm{T})$ into W (§2, No. 3, Prop. 7). If $\lambda \in \mathrm{W}$, then $\lambda = i(\lambda_{\mathrm{T}})$ (§2, No. 3, Prop. 7 *b*)). Consequently, $i$ is a bijection of $\mathscr{M}_+^b(\mathrm{T})$ onto W, and the inverse bijection of $i$ is the mapping $r : \lambda \mapsto \lambda_{\mathrm{T}}$ on W. On the other hand, $i$ is continuous: for, if $f \in \mathscr{C}^b(\mathrm{X})$, then $\langle i(\mu), f\rangle = \langle \mu, f \circ i\rangle$, and $f \circ i$ belongs to $\mathscr{C}^b(\mathrm{T})$. Thus, everything comes down to showing that, for every measure $\mu \in \mathrm{W}$ and every function $f \in \mathscr{C}^b_+(\mathrm{T})$, one has

$$\lim_{\lambda \to \mu,\, \lambda \in \mathrm{W}} \lambda_{\mathrm{T}}(f) = \mu_{\mathrm{T}}(f),$$

or again

$$\lim_{\lambda \to \mu,\, \lambda \in \mathrm{W}} \lambda(f^0) = \mu(f^0).$$

Let $f^\infty$ be the function on X that coincides with $f$ on T and with $+\infty$ on $\mathrm{X} - \mathrm{T}$, and let $f'$ and $f''$ be, respectively, the upper semi-continuous regularization of $f^0$ and the lower semi-continuous regularization of $f^\infty$ (GT, IV, §6, No. 2). The relations

$$f'(x) = \limsup_{y \to x} f^0(y), \quad f''(x) = \liminf_{y \to x} f^\infty(y)$$

immediately imply that $f'$ and $f''$ both coincide with $f$ and $f^0$ on T. Prop. 6 then yields

$$\mu^\bullet(f') \geqslant \limsup_{\lambda \to \mu,\, \lambda \in \mathrm{W}} \lambda^\bullet(f'), \quad \mu^\bullet(f'') \leqslant \liminf_{\lambda \to \mu,\, \lambda \in \mathrm{W}} \lambda^\bullet(f'').$$

But one can replace $f'$ and $f''$ by $f^0$ in these two formulas, since the measures $\lambda$ and $\mu$ are carried by T; we have thus obtained the desired relation.

The statement of Prop. 8 is only valid for *positive* measures: the mapping $\mu \mapsto i(\mu)$ of $\mathscr{M}^b(\mathrm{T})$ into $\mathscr{M}^b(\mathrm{X})$ is injective and continuous, but is not in general a homeomorphism of $\mathscr{M}^b(\mathrm{T})$ onto its image. For example, take $\mathrm{X} = \mathbf{R}$, $\mathrm{T} = \mathbf{R} - \{0\}$; the measures $\lambda_t = \varepsilon_t - \varepsilon_{-t}$ $(t > 0)$ converge tightly to 0 in X as $t$ tends to 0, but do not converge tightly to 0 in T (the characteristic function of $]0, +\infty[$ belongs to $\mathscr{C}^b(\mathrm{T})$) (cf. however the Cor. of Th. 1 of No. 5).

PROPOSITION 9. — *Let* T *be a locally compact space, and let* $\mathfrak{F}$ *be a filter on* $\mathscr{M}^b_+(\mathrm{T})$ *that converges vaguely to a bounded measure* $\mu$. *For* $\mathfrak{F}$ *to converge tightly to* $\mu$, *it is necessary and sufficient that* $\lim_\lambda \lambda(1) = \mu(1)$ *with respect to* $\mathfrak{F}$.

The condition is obviously necessary. To show that it is sufficient, let us denote by X the Alexandroff compactification of T (GT, I, §9, No. 8) and by $i$ the canonical injection of T into X. By Prop. 8, everything comes down to showing that $\lambda \mapsto i(\lambda)$ converges tightly to $i(\mu)$ in $\mathscr{M}^b(\mathrm{X})$ with respect to $\mathfrak{F}$. Since $\mu(1) < +\infty$, there exists a set $\mathrm{A} \in \mathfrak{F}$ such that the total masses of the measures in A are bounded by a number M; it therefore suffices to verify that

$$\lim_{\lambda,\,\mathfrak{F}} \int_{X} g\, d(i(\lambda)) = \int_{X} g\, d(i(\mu)) \tag{5}$$

for functions $g \in \mathscr{C}^b(X)$ forming a total set in $\mathscr{C}^b(X)$. Now, this equality is satisfied when $g$ has compact support in T, because of the vague convergence of $\mathfrak{F}$ to $\mu$, and also when $g$ is a constant function on X, from the fact that $\lim\limits_{\lambda,\,\mathfrak{F}} \lambda(1) = \mu(1)$. Since the functions of the preceding two types form a total set in $\mathscr{C}^b(X)$ (Ch. III, §1, No. 2, Prop. 3), this completes the proof.

### 4. Application: topological properties of the space $\mathscr{M}_+^b(T)$

We first observe that if T is completely regular, then $\mathscr{M}^b(T)$ is a Hausdorff topological vector space, hence is completely regular. Consequently, $\mathscr{M}_+^b(T)$ is completely regular.

PROPOSITION 10. — *Let* T *be a Polish space; the space* $\mathscr{M}_+^b(T)$ *is then Polish for the tight topology.*

We begin by treating the case that T is Polish and *compact*. The set U of positive measures with mass $\leqslant 1$ is then compact (Ch. III, §1, No. 9, Cor. 2 of Prop. 15), and the topology induced on U by the tight topology (which here coincides with the vague topology) is also induced by the topology of pointwise convergence on a total subset of $\mathscr{C}(T)$ (*loc. cit.*, No. 10, Prop. 17). Now, there exists in $\mathscr{C}(T)$ a countable total set (GT, X, §3, No. 3, Th. 1); consequently, U is a metrizable compact space. The set V of positive measures of mass $< 1$ is open in U, hence is a Polish locally compact space. Now, the mapping $\mu \mapsto \dfrac{1}{1+\mu(1)}\mu$ of $\mathscr{M}_+^b(T)$ onto V is a homeomorphism, the mapping $\lambda \mapsto \dfrac{1}{1-\lambda(1)}\lambda$ being the inverse homeomorphism.

Let us pass to the case that T is Polish; we can suppose that T is the intersection of a decreasing sequence $(G_n)$ of open sets in a metrizable compact space X (GT, IX, §6, No. 1, Cor. 1 of Th. 1); the space $\mathscr{M}_+^b(T)$ is then homeomorphic to the subspace W of $\mathscr{M}_+^b(X)$ consisting of the measures concentrated on T (No. 3, Prop. 8), and it will suffice to show that W is the intersection of a sequence of open sets in the Polish space $\mathscr{M}_+^b(X)$ (GT, *loc. cit.*, Th. 1). Now, let $W_n$ be the set of measures $\mu \in \mathscr{M}_+^b(X)$ concentrated on $G_n$; the mapping $h_n : \mu \mapsto \mu^\bullet(X - G_n)$ on $\mathscr{M}_+^b(X)$ is upper semi-continuous (No. 3, Prop. 6), and the set $A_k^n$ of measures $\mu \in \mathscr{M}_+^b(X)$ such that $h_n(\mu) < 1/k$ is therefore open for every $k \geqslant 1$ and every $n \in \mathbf{N}$. The proof is completed by observing that $W = \bigcap_n W_n = \bigcap_{n,k} A_k^n$.

COROLLARY 1. — *If* T *is a metrizable space of countable type, then* $\mathscr{M}_+^b(T)$ *is metrizable of countable type for the tight topology.*

For, let $\widehat{T}$ be the completion of T for a metric defining the topology of T; the space $\widehat{T}$ is Polish, and $\mathscr{M}_+^b(T)$ is homeomorphic to the subspace of the Polish space $\mathscr{M}_+^b(\widehat{T})$ consisting of the measures concentrated on T (No. 3, Prop. 8). But every subspace of a Polish space is metrizable of countable type (GT, IX, §2, No. 8).

COROLLARY 2. — *If* T *is a completely regular Souslin* (resp. *Lusin*) *space, then the space* $\mathscr{M}_+^b(T)$ *is Souslin* (resp. *Lusin*).

For, consider a Polish space P and a continuous mapping $f$ of P onto T (GT, IX, §6, No. 2, Def. 2). Let $\widetilde{f}$ be the continuous mapping $\mu \mapsto f(\mu)$ of $\mathscr{M}_+^b(P)$ into $\mathscr{M}_+^b(T)$; the space $\mathscr{M}_+^b(P)$ is Polish by Prop. 10, and $\widetilde{f}$ is surjective (§2, No. 4, Prop. 9); the space $\mathscr{M}_+^b(T)$ is therefore Souslin. Similarly, if T is Lusin, then $f$ may be assumed to be injective (GT, *loc. cit.*, No. 4, Prop. 12); then $\widetilde{f}$ is injective (§2, No. 4, Prop. 8), and so $\mathscr{M}_+^b(T)$ is Lusin (GT, *loc. cit.*, No. 4, Prop. 12).

Let T be a completely regular Souslin space (recall that for this, it suffices that T be Souslin and *regular* (TG, App. 1, Cor. of Prop. 2)), and let H be a compact subset of $\mathscr{M}_+^b(T)$; then H is compact and Souslin, hence *metrizable*, for the tight topology (*loc. cit.*, App. 1, Cor. 2 of Prop. 3).

## 5. Compactness criterion for tight convergence

DEFINITION 2. — *Let* T *be a topological space, and let* H *be a subset of* $\mathscr{M}^b(T)$; *one says that* H *satisfies Prokhorov's condition if*

*a)* $\sup\limits_{\mu \in H} |\mu|(1) < +\infty$;

*b) for every number* $\varepsilon > 0$, *there exists a compact subset* $K_\varepsilon$ *of* T *such that*

$$|\mu|(T - K_\varepsilon) \leqslant \varepsilon \quad \textit{for every measure } \mu \in H. \tag{6}$$

It can be shown that if T is completely regular, the set of conditions *a)* and *b)* is equivalent to the following condition: there exists a real function $f \geqslant 1$ on T, such that the set of points $t$ of T satisfying $f(t) \leqslant c$ is compact for every $c \in \mathbf{R}_+$ (which in particular implies that $f$ is lower semi-continuous), and such that $\sup_{\mu \in \mathrm{H}} |\mu|(f) < +\infty$. Moreover, when T is locally compact, one obtains an equivalent statement by requiring $f$ to be continuous (cf. Exer. 10).

PROPOSITION 11. — *Let* T *be a completely regular space, and let* H *be a subset of* $\mathscr{M}^b(\mathrm{T})$ *that satisfies Prokhorov's condition; then its closure* $\overline{\mathrm{H}}$ *in* $\mathscr{M}^b(\mathrm{T})$ *satisfies Prokhorov's condition.*

For, the functions $\mu \mapsto |\mu|^{\bullet}(1)$, $\mu \mapsto |\mu|^{\bullet}(\mathrm{T} - \mathrm{K}_\varepsilon)$ are lower semi-continuous on $\mathscr{M}^b(\mathrm{T})$ by Prop. 6 of No. 3.

The interest of Prokhorov's condition comes from the following theorem, whose converse will be studied later on (Th. 2).

THEOREM 1 (Prokhorov). — *Let* T *be a completely regular space, and let* H *be a subset of* $\mathscr{M}^b(\mathrm{T})$ *that satisfies Prokhorov's condition; then* H *is relatively compact in* $\mathscr{M}^b(\mathrm{T})$ *for the tight topology.*

We can suppose that T is a subspace of a compact space X; let $i$ be the canonical injection of T into X. We can on the other hand suppose that H is *closed* in $\mathscr{M}^b(\mathrm{T})$, by Prop. 11. It will then suffice to show that every ultrafilter $\mathfrak{U}$ on H converges in $\mathscr{M}^b(\mathrm{T})$.

We shall begin with the case that $\mathrm{H} \subset \mathscr{M}_+^b(\mathrm{T})$. The total masses of the measures $\mu \in \mathrm{H}$ being bounded by hypothesis, $i(\mu)$ converges vaguely with respect to $\mathfrak{U}$, in $\mathscr{M}_+(\mathrm{X})$, to a measure $\nu \in \mathscr{M}_+(\mathrm{X})$ (Ch. III, §1, No. 9, Cor. 2 of Prop. 15); by Prop. 8 of No. 3, everything comes down to proving that $\nu$ is concentrated on T. Now, let $\varepsilon$ be a number $> 0$, and let $\mathrm{K}_\varepsilon$ be a compact subset of T satisfying the formula (6). Since $\mathrm{X} - \mathrm{K}_\varepsilon$ is open in X, we have, by Prop. 6 of No. 3 applied in X, the inequalities

$$\begin{aligned}\nu^{\bullet}(\mathrm{X} - \mathrm{T}) \leqslant \nu^{\bullet}(\mathrm{X} - \mathrm{K}_\varepsilon) &\leqslant \liminf_{\mu,\,\mathfrak{U}} i(\mu)^{\bullet}(\mathrm{X} - \mathrm{K}_\varepsilon) \\ &= \liminf_{\mu,\,\mathfrak{U}} \mu^{\bullet}(\mathrm{T} - \mathrm{K}_\varepsilon) \leqslant \varepsilon\,;\end{aligned}$$

since $\varepsilon > 0$ is arbitrary, the theorem is established in this special case.

Let us pass to the general case; for every measure $\mu$ on T, set

$$a_1(\mu) = \mathscr{R}(\mu)^+,\ a_2(\mu) = \mathscr{R}(\mu)^-,\ a_3(\mu) = \mathscr{I}(\mu)^+,\ a_4(\mu) = \mathscr{I}(\mu)^-\,;$$

since $\mu = a_1(\mu) - a_2(\mu) + ia_3(\mu) - ia_4(\mu)$, it will suffice to show that the mappings $a_j$ $(j = 1, 2, 3, 4)$ converge tightly with respect to $\mathfrak{U}$. But the set $\mathrm{H}_j$ of measures $a_j(\mu)$, where $\mu$ runs over H, satisfies Prokhorov's condition by virtue of the relation $|a_j(\mu)| \leqslant |\mu|$, and is contained in $\mathscr{M}_+^b(\mathrm{T})$; it is therefore relatively compact in $\mathscr{M}_+^b(\mathrm{T})$ by the special case, and the theorem then follows at once.

COROLLARY. — *Let* T *be a subspace of a completely regular space* X, *and let* H *be a subset of* $\mathscr{M}^b(\mathrm{T})$ *that satisfies Prokhorov's condition. If* $i$ *denotes the canonical injection of* T *into* X, *then the restriction to* H *of the mapping* $\mu \mapsto i(\mu)$ *of* $\mathscr{M}^b(\mathrm{T})$ *into* $\mathscr{M}^b(\mathrm{X})$ *is a homeomorphism of* H *onto its image.*

It suffices to treat the case that H is closed (Prop. 11), hence compact; the conclusion then follows from the fact that $\mu \mapsto i(\mu)$ is continuous and injective.

Recall that this result is also valid for an arbitrary subset of $\mathscr{M}_+^b(\mathrm{T})$ (No. 3, Prop. 8).

THEOREM 2. — *Let* T *be a locally compact space, or a Polish space, and let* H *be a relatively compact subset of* $\mathscr{M}_+^b(\mathrm{T})$; *then* H *satisfies Prokhorov's condition.*

We may restrict ourselves to the case that H is closed, hence compact. The total masses of the measures $\mu \in \mathrm{H}$ are obviously bounded, because the mapping $\mu \mapsto \mu(1)$ is continuous, and everything comes down to proving the assertion *b*) of Def. 2.

Suppose first that T is locally compact. Let $\varepsilon$ be a number $> 0$. Let us associate to every measure $\mu \in \mathrm{H}$ a compact set $\mathrm{K}_\mu$ in T such that $\mu^\bullet(\mathrm{T} - \mathrm{K}_\varepsilon) < \varepsilon$, then a relatively compact open neighborhood $\mathrm{U}_\mu$ of $\mathrm{K}_\mu$. The function $\lambda \mapsto \lambda^\bullet(\mathrm{T} - \mathrm{U}_\mu)$ being upper semi-continuous on $\mathscr{M}_+^b(\mathrm{T})$ (No. 3, Prop. 6), the set $\mathrm{V}^\mu$ of measures $\lambda \in \mathrm{H}$ such that $\lambda^\bullet(\mathrm{T} - \mathrm{U}_\mu) < \varepsilon$ is a neighborhood of $\mu$ in H. Therefore there exists a finite subset H′ of H such that the sets $\mathrm{V}^\mu$ $(\mu \in \mathrm{H}')$ cover H. Denoting by K the compact set $\bigcup_{\mu \in \mathrm{H}'} \overline{\mathrm{U}}_\mu$, we have $\lambda^\bullet(\mathrm{T} - \mathrm{K}) < \varepsilon$ for all $\lambda \in \mathrm{H}$.

Suppose next that T is Polish. We do not restrict the generality by assuming that T is the intersection of a decreasing sequence $(\mathrm{T}_p)_{p \geqslant 1}$ of open subsets of a compact space X (GT, IX, §6, No. 1, Cor. 1 of Th. 1). Let $i_p$ be the injection of T into $\mathrm{T}_p$, and let $\mathrm{H}_p$ be the set of measures of the form $i_p(\lambda)$ for $\lambda \in \mathrm{H}$; since $\mathrm{H}_p$ is compact in $\mathscr{M}_+^b(\mathrm{T}_p)$, it follows that there exists a compact set $\mathrm{K}_p \subset \mathrm{T}_p$ such that $\nu^\bullet(\mathrm{T}_p - \mathrm{K}_p) \leqslant \varepsilon 2^{-p}$ for every measure $\nu \in \mathrm{H}_p$, by the preceding result applied to the locally compact space $\mathrm{T}_p$. Therefore also $\nu^\bullet\big(\mathrm{T} - (\mathrm{T} \cap \mathrm{K}_p)\big) \leqslant \varepsilon 2^{-p}$, and finally $\lambda^\bullet\big(\mathrm{T} - (\mathrm{T} \cap \mathrm{K}_p)\big) \leqslant \varepsilon 2^{-p}$ for every measure $\lambda \in \mathrm{H}$. Now set $\mathrm{K} = \bigcap_p \mathrm{K}_p$; the set K is compact and is contained in T, and, for every measure $\lambda \in \mathrm{H}$, we have $\lambda^\bullet(\mathrm{T} - \mathrm{K}) \leqslant \sum_p \lambda^\bullet\big(\mathrm{T} - (\mathrm{T} \cap \mathrm{K}_p)\big) \leqslant \sum_p \varepsilon 2^{-p} = \varepsilon$. Prokhorov's condition is thus verified.

## 6. Tight convergence of measures and compact convergence of functions

PROPOSITION 12. — *Let* T *be a completely regular space, and let* B *be the unit ball of the normed space* $\mathscr{C}^b(\mathrm{T};\mathbf{C})$. *Let* I *be a linear form on* $\mathscr{C}^b(\mathrm{T};\mathbf{C})$. *In order that there exist a bounded complex measure* $\theta$ *on* T *such that* $\mathrm{I}(f) = \theta(f)$ *for all* $f \in \mathscr{C}^b(\mathrm{T};\mathbf{C})$, *it is necessary and sufficient that the restriction of* I *to* B *be continuous for the topology of compact convergence. The measure* $\theta$ *is then unique.*

Let us show that the condition in the statement is necessary. Let $\theta$ be a bounded complex measure on T, $\varepsilon$ a number $> 0$, and K a compact subset of T such that $|\theta|^\bullet(\mathrm{T}-\mathrm{K}) < \varepsilon$. Let $f \in \mathrm{B}$; we denote by U the neighborhood of $f$ in B for the topology of compact convergence, formed by the functions $g \in \mathrm{B}$ such that $\sup_{x\in\mathrm{K}} |g(x) - f(x)| \leqslant \varepsilon$. Then, for every $g \in \mathrm{U}$,

$$|\theta(g) - \theta(f)| \leqslant \int_{\mathrm{T}} |g - f|\, d|\theta| \leqslant \varepsilon|\theta|^\bullet(\mathrm{K}) + 2|\theta|^\bullet(\mathrm{T}-\mathrm{K}) \leqslant (\|\theta\| + 2)\varepsilon,$$

because $|g - f|$ is bounded above by $\varepsilon$ on K and by 2 on $\mathrm{T}-\mathrm{K}$.

Conversely, consider a linear form I on $\mathscr{C}^b(\mathrm{T};\mathbf{C})$ whose restriction to B is continuous for the topology of compact convergence. Then, for every number $\varepsilon > 0$, there exist a number $a > 0$ and a compact subset K of T such that the relations $f \in \mathrm{B}$, $\sup_{x\in\mathrm{K}} |f(x)| \leqslant a$ imply $|\mathrm{I}(f)| \leqslant \varepsilon$. Prop. 5 of No. 2 then implies the existence of a unique bounded complex measure $\theta$ such that $\mathrm{I}(f) = \theta(f)$ for all $f \in \mathscr{C}^b(\mathrm{T};\mathbf{C})$.

PROPOSITION 13. — *Let* T *be a locally compact space, and* H *a bounded subset of the normed space* $\mathscr{C}^b(\mathrm{T};\mathbf{C})$. *The mapping* $(\mu, f) \mapsto \mu(f)$ *of* $\mathscr{M}^b_+(\mathrm{T}) \times \mathrm{H}$ *into* $\mathbf{C}$ *is then continuous, when* $\mathscr{M}^b_+(\mathrm{T})$ *is equipped with the tight topology, and* H *with the topology of compact convergence.*

Let $\mu \in \mathscr{M}^b_+(\mathrm{T})$, $f \in \mathrm{H}$, and let M be a real number such that $\|\mu\| < \mathrm{M}$, and $|g| \leqslant \mathrm{M}$ for all $g \in \mathrm{H}$. Let $\varepsilon$ be a number $> 0$ and choose a compact subset K of T such that $\mu^\bullet(\mathrm{T}-\mathrm{K}) < \varepsilon$, then a relatively compact open neighborhood S of K. The set U of measures $\lambda \in \mathscr{M}^b_+(\mathrm{T})$ satisfying the inequalities

$$\lambda^\bullet(\mathrm{T}) < \mathrm{M}, \quad \lambda^\bullet(\mathrm{T}-\mathrm{S}) < \varepsilon, \quad |\lambda(f) - \mu(f)| < \varepsilon$$

is then a neighborhood of $\mu$ in $\mathscr{M}^b_+(\mathrm{T})$ (No. 3, Prop. 6). In addition, let V be the neighborhood of $f$ in H consisting of the functions $g \in \mathrm{H}$ such that

$$\sup_{x\in\mathrm{S}} |g(x) - f(x)| < \varepsilon.$$

Let $\lambda \in \mathrm{U}$ and $g \in \mathrm{V}$; since the function $|g - f|$ is bounded above by $\varepsilon$ in S, and by 2M in T − S, we have

$$|\lambda(g) - \lambda(f)| \leqslant \int_{\mathrm{T}} |g - f|\, d\lambda \leqslant \varepsilon\lambda^{\bullet}(\mathrm{S}) + 2\mathrm{M}\lambda^{\bullet}(\mathrm{T} - \mathrm{S}) \leqslant 3\mathrm{M}\varepsilon\,,$$

from which one deduces

$$|\lambda(g) - \mu(f)| \leqslant |\lambda(g) - \lambda(f)| + |\lambda(f) - \mu(f)| \leqslant (3\mathrm{M} + 1)\varepsilon\,.$$

This proves the continuity of the mapping $(\lambda, g) \mapsto \lambda(g)$ at the point $(\mu, f)$ of $\mathscr{M}_+^b(\mathrm{T}) \times \mathrm{H}$.

*Remark.* — Let T be a completely regular space, M a subset of $\mathscr{M}^b(\mathrm{T})$ that satisfies Prokhorov's condition, H a bounded subset of $\mathscr{C}^b(\mathrm{T})$. An argument very close to the one just made may be used to prove that the mapping $(\lambda, g) \mapsto \lambda(g)$ of $\mathrm{M} \times \mathrm{H}$ into $\mathbf{C}$ is continuous when M is equipped with the tight topology and H with the topology of compact convergence.

COROLLARY. — *Let* T *be a completely regular space,* X *a topological space, and* $f$ *a complex-valued function defined on* $\mathrm{T} \times \mathrm{X}$, *continuous and bounded. For every bounded measure* $\mu$ *on* T, *let* $\mathrm{F}_\mu$ *be the function on* X *defined by* $\mathrm{F}_\mu(x) = \int_{\mathrm{T}} f(t, x)\, d\mu(t)$ *for all* $x \in \mathrm{X}$.

*a) The function* $\mathrm{F}_\mu$ *is continuous and bounded for every bounded measure* $\mu$.

*b) Suppose that* T *is locally compact. The mapping* $\mu \mapsto \mathrm{F}_\mu$ *of* $\mathscr{M}_+^b(\mathrm{T})$ *into* $\mathscr{C}^b(\mathrm{X}; \mathbf{C})$ *is then continuous, if* $\mathscr{M}_+^b(\mathrm{T})$ *is equipped with the tight topology, and* $\mathscr{C}^b(\mathrm{X}; \mathbf{C})$ *with the topology of compact convergence.*

For every $x \in \mathrm{X}$, denote by $f_x$ the continuous and bounded function $t \mapsto f(t, x)$ on T; the mapping $x \mapsto f_x$ of X into $\mathscr{C}^b(\mathrm{T}; \mathbf{C})$ has bounded image, and it is continuous if $\mathscr{C}^b(\mathrm{T}; \mathbf{C})$ is equipped with the topology of compact convergence (GT, X, §3, No. 4, Th. 3). Since $\mathrm{F}_\mu(x) = \mu(f_x)$, the function $\mathrm{F}_\mu$ is continuous by Prop. 12. Suppose T is locally compact; Prop. 13 shows that the mapping $(\mu, x) \mapsto \mathrm{F}_\mu(x)$ of $\mathscr{M}_+^b(\mathrm{T}) \times \mathrm{X}$ into $\mathbf{C}$ is continuous; the assertion b) follows from this (*loc. cit.*).

## 7. Application: the Laplace transformation

In this No., we denote by M a commutative monoid, whose law of composition is written additively, equipped with the topology of a *locally compact* space, for which the mapping $(m, m') \mapsto m + m'$ of $\mathrm{M} \times \mathrm{M}$ into M is continuous. The neutral element of M is denoted by 0. One calls *character* of M every bounded continuous complex function $\chi$ on M satisfying the relations

$$\chi(m + m') = \chi(m) \cdot \chi(m')\,, \quad \chi(0) = 1\,, \quad |\chi(m)| \leqslant 1 \tag{7}$$

for $m, m'$ in M. If $\chi$ and $\chi'$ are characters, then so is $\chi\chi'$. The set of characters of M is a monoid, denoted X; equip it with the topology of compact convergence, for which the mapping $(\chi, \chi') \mapsto \chi\chi'$ of $X \times X$ into X is continuous. The neutral element of X is the constant function 1.

For every bounded complex measure $\mu$ on M, one calls *Laplace transform* of $\mu$ the function $\mathscr{L}\mu$ on X defined by

$$(\mathscr{L}\mu)(\chi) = \int_M \chi(m)\, d\mu(m). \tag{8}$$

By Th. 3 of GT, X, §3, No. 4, the mapping $(m, \chi) \mapsto \chi(m)$ of $M \times X$ into $\mathbf{C}$ is continuous and bounded. The corollary of Prop. 13 of No. 6 then implies the following result:

PROPOSITION 14. — *For every bounded complex measure $\mu$ on* M, *the function $\mathscr{L}\mu$ on* X *is continuous and bounded. If $\mathscr{M}_+^b(M)$ is equipped with the tight topology and $\mathscr{C}^b(X; \mathbf{C})$ with the topology of compact convergence, the mapping $\mu \mapsto \mathscr{L}\mu$ of $\mathscr{M}_+^b(M)$ into $\mathscr{C}^b(X; \mathbf{C})$ is continuous.*

The set of characters of M that tend to 0 at infinity will be denoted $X_0$; this set is stable under multiplication. We shall say that a submonoid[1] S of X is *full* if S is stable for the mapping $\chi \mapsto \overline{\chi}$, $S \cap X_0$ separates the points of M (GT, X, §4, No. 1, Def. 1) and if, given any $m \in M$, there exists an element $\chi$ of $S \cap X_0$ such that $\chi(m) \neq 0$.

Suppose in addition that M is a noncompact abelian *group*. Let $f$ be a function on M that tends to 0 at infinity; the same is then true of the function $x \mapsto f(x)f(-x)$ on M, whereas every character $\chi$ of M satisfies $\chi(x)\chi(-x) = \chi(0) = 1$. It follows that $X_0$ is empty, and that X does not contain any full submonoid. Thus, Theorem 3 below does not apply to locally compact groups that are not compact.

THEOREM 3. — *Let* S *be a full submonoid of* X.

*a) If $\mu$ and $\mu'$ are two bounded complex measures on* M, *such that $\mathscr{L}\mu$ and $\mathscr{L}\mu'$ have the same restriction to $S \cap X_0$, then $\mu = \mu'$.*

*b) Let $\mathfrak{F}$ be a filter on $\mathscr{M}_+^b(M)$, such that $\mathscr{L}\lambda(s)$ has a limit $\Phi(s) \in \mathbf{C}$ with respect to $\mathfrak{F}$ for every $s \in S$. Then the filter $\mathfrak{F}$ converges vaguely to a bounded positive measure $\mu$, and $\Phi(s) = \mathscr{L}\mu(s)$ for all $s \in S \cap X_0$.*

*c) Under the hypotheses of b), suppose in addition that the closure of $S \cap X_0$ contains* 1, *and that the function $\Phi$ on* S *is continuous at the point* 1. *Then $\mathfrak{F}$ converges tightly to $\mu$, and $\Phi(s) = \mathscr{L}\mu(s)$ for all $s \in S$.*

We shall denote by E the algebra of continuous complex functions tending to 0 at infinity on M, and by $\mathfrak{A}$ the linear subspace of E generated

[1] Recall that a submonoid of a monoid A contains by definition the neutral element of A (A, I, §2, No. 1).

by $S \cap X_0$; then $\mathfrak{A}$ is a subalgebra of $E$ stable under the mapping $f \mapsto \overline{f}$; since $S$ is a full submonoid of $X$, Cor. 2 of Prop. 7 of GT, X, §4, No. 4 implies that $\mathfrak{A}$ is dense in $E$.

Let us prove *a*): by hypothesis, $\mu(f) = \mu'(f)$ for every $f \in \mathfrak{A}$; since $\mu$ and $\mu'$ are continuous linear forms on $E$, this implies that $\mu(f) = \mu'(f)$ for $f \in E$, and in particular for every continuous function $f$ with compact support, whence $\mu = \mu'$.

Let us place ourselves under the hypotheses of *b*). The number $\Phi(1) = \lim_{\lambda, \mathfrak{F}} \lambda(1)$ is real and positive; let there be given a real number $a > \Phi(1)$; since $\|\lambda\| = \mathscr{L}\lambda(1)$ for $\lambda \in \mathscr{M}^b_+(M)$, the relation $\lim_{\lambda, \mathfrak{F}} \mathscr{L}\lambda(1) = \Phi(1)$ implies that the set $H$ of measures $\lambda \in \mathscr{M}^b_+(M)$ such that $\|\lambda\| \leqslant a$ belongs to $\mathfrak{F}$. Since $\mathscr{M}^b(M; \mathbf{C})$ may be identified with the dual of the normed space $E$ (Ch. III, §1, No. 8 & §1, No. 2, Prop. 3), the space $H$ is compact for the topology $\sigma(\mathscr{M}^b(M; \mathbf{C}), E)$. On the other hand (TVS, III, §3, No. 4, Prop. 5), this topology coincides on $H$ with the topology of pointwise convergence in any total subset of $E$. In particular, since $\mathfrak{A}$ is dense in $E$, and the same is true of the space of continuous functions with compact support (Ch. III, §1, No. 2, Prop. 3), the topology of pointwise convergence in $S \cap X_0$ coincides on $H$ with the vague topology, and $H$ is compact for this topology. It follows at once that $\mathfrak{F}$ converges vaguely to a measure $\mu \in H$, and that $\mathscr{L}\mu(s) = \lim_{\lambda, \mathfrak{F}} \mathscr{L}\lambda(s)$ for all $s \in S \cap X_0$.

Finally, let us pass to *c*). Since the functions $\Phi$ and $\mathscr{L}\mu$ are continuous at the point $1 \in S$, and equal on $S \cap X_0$, and since $1$ is in the closure of $S \cap X_0$, we have $\Phi(1) = \mathscr{L}\mu(1)$. In other words, $\lim_{\lambda, \mathfrak{F}} \lambda(1) = \mu(1)$. Prop. 9 of No. 3 then shows that $\mu$ is the tight limit of the filter $\mathfrak{F}$. Every element of $S$ being a bounded continuous function on $M$, this implies that $\Phi(s) = \lim_{\lambda, \mathfrak{F}} \lambda(s) = \mu(s) = \mathscr{L}\mu(s)$ for all $s \in S$.

COROLLARY. — *Let $S$ be a full submonoid of $X$, such that the closure of $S \cap X_0$ contains $1$. Let $L$ be the subset of $\mathscr{C}^b(S; \mathbf{C})$ consisting of the restrictions to $S$ of the Laplace transforms of the measures $\lambda \in \mathscr{M}^b_+(M)$.*

*a*) *The set $L$ is closed in the space $\mathscr{C}^b(S; \mathbf{C})$ equipped with the topology of pointwise convergence.*

*b*) *The mapping $\lambda \mapsto (\mathscr{L}\lambda)_S$ is a homeomorphism of $\mathscr{M}^b_+(M)$ onto $L$, if $\mathscr{M}^b_+(M)$ is equipped with the tight topology and $L$ with the topology of pointwise convergence.*

*c*) *The topology of pointwise convergence and the topology of compact convergence coincide on $L$.*

The assertions *a*) and *b*) are immediate consequences of Th. 3; the assertion *c*) follows from *b*) and Prop. 14, since the topology of compact convergence is finer than that of pointwise convergence.

One must be on guard that L is not closed in the set of all bounded complex functions on S, equipped with the topology of pointwise convergence. Assume for example the notations of *Example* 2 below ( $\mathrm{M} = \mathbf{R}_+$ , S identified with $\mathbf{R}_+$ ). The Laplace transforms of the measures $\varepsilon_n$ $(n \in \mathbf{N})$ are the functions $t \mapsto e^{-nt}$ on $\mathbf{R}_+$ ; as $n$ tends to $+\infty$, these functions converge pointwise to the function equal to 1 for $t = 0$ and to 0 for $t \neq 0$, which does not belong to L.

*Example* 1). — Take for M the set $\mathbf{N}$ of positive integers, equipped with the law of addition and with the discrete topology. Let D be the unit disc of $\mathbf{C}$ (the set of complex numbers of absolute value $\leqslant 1$ ) equipped with the topology induced by $\mathbf{C}$ and with the law induced by multiplication. For every $z \in \mathrm{D}$, let us denote by $f(z)$ the character $n \mapsto z^n$ of $\mathbf{N}$. For every character $\chi$ of $\mathbf{N}$, denote by $g(\chi)$ the complex number $\chi(1) \in \mathrm{D}$. One verifies immediately that $f$ and $g$ are mutually inverse homeomorphisms between D and X, which will permit us, from now on, to *identify* X and D. The set of characters tending to 0 at infinity may then be identified with the set $\mathrm{D}_0$ of complex numbers of absolute value $< 1$. Finally, the interval $]0,1]$ of $\mathbf{R}$ is a full submonoid of D, and 1 is in the closure of $]0,1] \cap \mathrm{D}_0 = ]0,1[$.

Every measure $\mu$ on $\mathbf{N}$ may be written in a unique way in the form $\mu = \sum_{n \in \mathbf{N}} u_n \cdot \varepsilon^n$, and $\mu$ is bounded if and only if $\sum_n |u_n| < +\infty$; one then has $\mathscr{L}\mu(z) = \sum_{n \in \mathbf{N}} u_n z^n$ for $z \in \mathrm{D}$. This function is continuous on D; it is customary to call it the *generating function* of the summable sequence $(u_n)_{n \in \mathbf{N}}$. Transcribed into this language, Th. 3 yields the following result (taking into account Prop. 9 of No. 3):

PROPOSITION 15. — *Let* A *be a set equipped with a filter* $\mathfrak{F}$. *For every* $\alpha \in \mathrm{A}$, *let* $(u_{\alpha,n})_{n \in \mathbf{N}}$ *be a summable sequence of positive numbers, and let* $\Phi_\alpha$ *be the function defined on the interval* $]0,1]$ *of* $\mathbf{R}$ *by* $\Phi_\alpha(x) = \sum_{n \in \mathbf{N}} u_{\alpha,n} x^n$. *In order that there exist a summable sequence* $(u_n)_{n \in \mathbf{N}}$ *of positive numbers such that*

$$\lim_{\alpha, \mathfrak{F}} u_{\alpha,n} = u_n \quad \textit{for all } n, \qquad \lim_{\alpha,\mathfrak{F}} \sum_{n \in \mathbf{N}} u_{\alpha,n} = \sum_{n \in \mathbf{N}} u_n,$$

*it is necessary and sufficient that* $\Phi_\alpha$ *converge pointwise on* $]0,1]$, *with respect to* $\mathfrak{F}$, *to a function* $\Phi$ *continuous at the point* 1. *In this case,* $\Phi(x) = \sum_{n \in \mathbf{N}} u_n x^n$ *for all* $x \in ]0,1]$.

Analogous results are obtained by taking M to be the monoid $\mathbf{N}^n$, where $n$ denotes an integer $> 1$; the space X may then be identified with $\mathrm{D}^n$, and one can choose $]0,1]^n$ as the full submonoid. We leave to the reader the task of transcribing Th. 3 in this case.

*Example* 2). — Let us take for M the set $\mathbf{R}_+$, equipped with the law of addition and with the usual topology. Let P be the set of complex numbers $z$ with positive real part, equipped with the topology induced by $\mathbf{C}$, and with the law induced by addition in $\mathbf{C}$. For every $p \in \mathrm{P}$, denote by $f(p)$ the character $x \mapsto e^{-px}$ of $\mathbf{R}_+$; it is easily verified that $f$ is an isomorphism of the topological monoid structure of P onto that of X; we shall identify X with P by means of $f$. It is clear that $\mathbf{R}_+$ is a full submonoid of P, and Th. 3 yields the following result:

PROPOSITION 16. — *Let* A *be a set equipped with a filter* $\mathfrak{F}$. *For every* $\alpha \in \mathrm{A}$, *let* $\mu_\alpha$ *be a bounded positive measure on* $\mathbf{R}_+$, *and let* $\Phi_\alpha$ *be the function defined on* $\mathbf{R}_+$ *by* $\Phi_\alpha(p) = \int_0^{+\infty} e^{-px}\, d\mu_\alpha(x)$. *In order that the mapping* $\alpha \mapsto \mu_\alpha$ *converge tightly with respect to* $\mathfrak{F}$ *to a bounded positive measure* $\mu$, *it is necessary and sufficient that* $\Phi_\alpha$ *converge pointwise on* $\mathbf{R}_+$, *with respect to* $\mathfrak{F}$, *to a function* $\Phi$ *continuous at the point* 0. *In this case,* $\Phi(p) = \int_0^{+\infty} e^{-px}\, d\mu(x)$ *for all* $p \in \mathbf{R}_+$.

There are analogous results for the additive monoids $\mathbf{R}_+^n$ ($n$ an integer $> 1$); we leave to the reader the transcription of Th. 3 in this case.

## §6. PROMEASURES AND MEASURES ON A LOCALLY CONVEX SPACE

*Throughout this section, only vector spaces over the field of real numbers are considered. By locally convex space, is meant a topological vector space over* $\mathbf{R}$ *that is Hausdorff and locally convex. The topological dual of a locally convex space* E *will be denoted* E′; *for* $x \in \mathrm{E}$ *and* $x' \in \mathrm{E}'$, *one writes* $\langle x, x' \rangle = x'(x)$.

### 1. Promeasures on a locally convex space

Let E be a locally convex space. We denote by $\mathscr{F}(\mathrm{E})$ the set of closed linear subspaces of E of finite codimension, ordered by the relation $\supset$. For every $\mathrm{V} \in \mathscr{F}(\mathrm{E})$, $p_\mathrm{V}$ denotes the canonical mapping of E onto E/V. Let V and W be two elements of $\mathscr{F}(\mathrm{E})$ such that $\mathrm{V} \supset \mathrm{W}$; we denote by $p_{\mathrm{VW}}$ the mapping of E/W into E/V deduced from the identity mapping of E by passage to the quotients. The family $\mathscr{Q}(\mathrm{E}) = (\mathrm{E/V}, p_{\mathrm{VW}})$ is an inverse system of locally convex spaces, indexed by $\mathscr{F}(\mathrm{E})$. It is called the *inverse system of finite-dimensional quotients of* E.

It can be shown that the inverse limit of the inverse system $\mathscr{Q}(\mathrm{E})$ is canonically isomorphic to the algebraic dual $\mathrm{E}'^*$ of $\mathrm{E}'$, equipped with the weak topology $\sigma(\mathrm{E}'^*, \mathrm{E}')$.

DEFINITION 1. — *Let* E *be a locally convex space. One calls promeasure on* E *every inverse system*[1] *of measures* (§4, No. 2, Def. 1) *on the inverse system of finite-dimensional quotients of* E.

In other words, a promeasure $\mu$ on E is a family $(\mu_{\mathrm{V}})_{\mathrm{V}\in\mathscr{F}(\mathrm{E})}$, where $\mu_{\mathrm{V}}$ is a bounded (positive) measure on the finite-dimensional space $\mathrm{E}/\mathrm{V}$, and where $\mu_{\mathrm{V}} = p_{\mathrm{VW}}(\mu_{\mathrm{W}})$ when $\mathrm{V} \supset \mathrm{W}$. All of the measures $\mu_{\mathrm{V}}$ have the same total mass, which is called the *total mass* of the promeasure $\mu$.

For a subspace V of E to belong to $\mathscr{F}(\mathrm{E})$, it is necessary and sufficient that there exist a finite number of elements $x'_1, \ldots, x'_n$ of $\mathrm{E}'$ such that V consists of the $x \in \mathrm{E}$ satisfying $\langle x, x'_i\rangle = 0$ for $1 \leqslant i \leqslant n$ (TVS, II, §6, No. 3, Cor. 2 of Th. 1 and No. 5, Cor. 2 of Prop. 7). Moreover, on a finite-dimensional vector space there exists one and only one Hausdorff topological vector space topology (TVS, I, §2, No. 3, Th. 2). Consequently, the concept of promeasure on E depends only on the dual $\mathrm{E}'$ of E.

Let $\lambda$ be a bounded measure on E. For every $\mathrm{V} \in \mathscr{F}(\mathrm{E})$, let us denote by $\widetilde{\lambda}_{\mathrm{V}}$ the image of $\lambda$ under the canonical mapping $p_{\mathrm{V}}$ of E onto $\mathrm{E}/\mathrm{V}$. One has $p_{\mathrm{V}} = p_{\mathrm{VW}} \circ p_{\mathrm{W}}$ for any two elements V and W of $\mathscr{F}(\mathrm{E})$ such that $\mathrm{V} \supset \mathrm{W}$; consequently, the family $\widetilde{\lambda} = (\widetilde{\lambda}_{\mathrm{V}})_{\mathrm{V}\in\mathscr{F}(\mathrm{E})}$ is a promeasure on E. We shall say that $\widetilde{\lambda}$ is the promeasure *associated* with the measure $\lambda$. One sees immediately that $\lambda$ and $\widetilde{\lambda}$ have the same total mass.

PROPOSITION 1. — *Let* E *be a locally convex space. The mapping* $\lambda \mapsto \widetilde{\lambda}$ *is a bijection of the set of bounded measures on* E *onto the set of promeasures* $(\mu_{\mathrm{V}})_{\mathrm{V}\in\mathscr{F}(\mathrm{E})}$ *on* E *satisfying the following condition:*

*For every* $\varepsilon > 0$, *there exists a compact subset* K *of* E *such that* $\mu_{\mathrm{V}}\big(\mathrm{E}/\mathrm{V} - p_{\mathrm{V}}(\mathrm{K})\big) \leqslant \varepsilon$ *for all* $\mathrm{V} \in \mathscr{F}(\mathrm{E})$.

One knows that the intersection of the kernels of the continuous linear forms on E is equal to 0 (TVS, II, §4, No. 1, Cor. 1 of Prop. 2); consequently $\bigcap_{\mathrm{V}\in\mathscr{F}(\mathrm{E})} \mathrm{V} = \{0\}$ and the family $(p_{\mathrm{V}})_{\mathrm{V}\in\mathscr{F}(\mathrm{E})}$ is coherent and separating. The proposition then follows from Th. 1 of §4, No. 2.

In particular, the mapping $\lambda \mapsto \widetilde{\lambda}$ is injective. If $\mu$ is a promeasure on E, and if there exists a bounded measure $\lambda$ on E such that $\mu = \widetilde{\lambda}$, we shall say, by an abuse of language, that $\mu$ is a measure. If E is finite-dimensional, every promeasure $\mu = (\mu_{\mathrm{V}})_{\mathrm{V}\in\mathscr{F}(\mathrm{E})}$ is a measure: for, $\{0\} \in \mathscr{F}(\mathrm{E})$, $\mathrm{E}/\{0\} = \mathrm{E}$ and $p_{\mathrm{V},\{0\}} = p_{\mathrm{V}}$, whence $\mu_{\mathrm{V}} = p_{\mathrm{V}}(\mu_{\{0\}})$ for all $\mathrm{V} \in \mathscr{F}(\mathrm{E})$; in other words, $\mu = \widetilde{\lambda}$ with $\lambda = \mu_{\{0\}}$.

---

[1] Also called a 'projective system'.

PROPOSITION 2. — *Let* T *be a countable set, and* E *the vector space of real functions on* T, *equipped with the topology of pointwise convergence. Every promeasure on* E *is a measure.*

For every $t \in \mathrm{T}$, let $\varepsilon_t$ be the linear form $f \mapsto f(t)$ on E. One knows (TVS, II, §6, No. 6, Cor. 2 of Prop. 8) that the family $(\varepsilon_t)_{t \in \mathrm{T}}$ is a basis of the vector space E′. Denote by $\Phi$ the set of finite subsets of T, and for every $\mathrm{J} \in \Phi$ let $\mathrm{E_J}$ be the set of functions on T that are zero at every point of J. Let $\mathrm{F} \in \mathscr{F}(\mathrm{E})$; since the orthogonal $\mathrm{F}^\circ$ of F is a finite-dimensional subspace of E′, there exists a $\mathrm{J} \in \Phi$ such that $\mathrm{F}^\circ$ is contained in the linear subspace G of E′ generated by the $\varepsilon_t$ for $t \in \mathrm{J}$. Since $\mathrm{F}^\circ \subset \mathrm{G}$, we have $\mathrm{E_J} = \mathrm{G}^\circ \subset \mathrm{F}^{\circ\circ} = \mathrm{F}$ and the *countable* family $(\mathrm{E_J})_{\mathrm{J} \in \Phi}$ is cofinal in $\mathscr{F}(\mathrm{E})$. The proposition then follows from Th. 2 of §4, No. 3.

## 2. Image of a promeasure

Let E and $\mathrm{E_1}$ be two locally convex spaces, and $u$ a continuous linear mapping of E into $\mathrm{E_1}$. For every $\mathrm{V_1} \in \mathscr{F}(\mathrm{E_1})$, the subspace $\mathrm{V} = \overset{-1}{u}(\mathrm{V_1})$ of E belongs to $\mathscr{F}(\mathrm{E})$, and $u$ defines, by passage to the quotients, a linear mapping $u_{\mathrm{V_1}}$ of E/V into $\mathrm{E_1/V_1}$. Let $\mathrm{V_1}$ and $\mathrm{W_1}$ in $\mathscr{F}(\mathrm{E_1})$ be such that $\mathrm{V_1} \supset \mathrm{W_1}$; set $\mathrm{V} = \overset{-1}{u}(\mathrm{V_1})$ and $\mathrm{W} = \overset{-1}{u}(\mathrm{W_1})$. We have $\mathrm{V} \supset \mathrm{W}$, and a commutative diagram

$$\begin{array}{ccccc}
\mathrm{E} & \xrightarrow{p_{\mathrm{W}}} & \mathrm{E/W} & \xrightarrow{p_{\mathrm{VW}}} & \mathrm{E/V} \\
\big\downarrow{\scriptstyle u} & & \big\downarrow{\scriptstyle u_{\mathrm{W_1}}} & & \big\downarrow{\scriptstyle u_{\mathrm{V_1}}} \\
\mathrm{E_1} & \xrightarrow{p_{\mathrm{W_1}}} & \mathrm{E_1/W_1} & \xrightarrow{p_{\mathrm{V_1W_1}}} & \mathrm{E_1/V_1}
\end{array}$$

Now let $\mu = (\mu_{\mathrm{V}})_{\mathrm{V} \in \mathscr{F}(\mathrm{E})}$ be a promeasure on E. For every $\mathrm{V_1} \in \mathscr{F}(\mathrm{E_1})$, set

$$\nu_{\mathrm{V_1}} = u_{\mathrm{V_1}}\big(\mu_{u^{-1}(\mathrm{V_1})}\big). \tag{1}$$

The commutativity of the preceding diagram shows that the family $\nu = (\nu_{\mathrm{V_1}})_{\mathrm{V_1} \in \mathscr{F}(\mathrm{E_1})}$ is a promeasure on $\mathrm{E_1}$. We say that $\nu$ *is the image of* $\mu$ *under* $u$, and denote it by $u(\mu)$.

Let $\lambda$ be a bounded measure on E, and $u(\lambda)$ the measure on $\mathrm{E_1}$ that is the image of $\lambda$ under $u$. If the promeasure $\mu$ is associated with $\lambda$, then the promeasure $u(\mu)$ is associated with $u(\lambda)$. This follows from the commutativity of the preceding diagram.

Let $V \in \mathscr{F}(E)$. It is immediate that the image promeasure on $E/V$ of the promeasure $\mu$ under the canonical mapping $p_V : E \to E/V$ is associated with the measure $\mu_V$.

Let $u_1$ be a continuous linear mapping of $E_1$ into a locally convex space $E_2$. One establishes without difficulty the relation

$$(u_1 \circ u)(\mu) = u_1\big(u(\mu)\big)$$

(*transitivity of the images of promeasures*).

### 3. Fourier transform of a promeasure

Let E be a locally convex space and $\mu = (\mu_V)_{V \in \mathscr{F}(E)}$ a promeasure on E. For every continuous linear form $x'$ on E, we denote by $\mu_{x'}$ the measure on $\mathbf{R}$ that is the image under $x'$ of the promeasure $\mu$ on E. The Fourier transform of $\mu$ is the function $\mathscr{F}\mu$ on $E'$ defined by

$$(\mathscr{F}\mu)(x') = \int_{\mathbf{R}} e^{it}\, d\mu_{x'}(t)\,. \tag{2}$$

Let $\lambda$ be a bounded measure on E. The Fourier transform of $\lambda$ is the function on $E'$ defined by

$$(\mathscr{F}\lambda)(x') = \int_{E} e^{i\langle x,x'\rangle}\, d\lambda(x)\,. \tag{3}$$

Let $\mu$ be the promeasure associated with $\lambda$. For every $x' \in E'$, the measure $\mu_{x'}$ on $\mathbf{R}$ is the image under $x' : E \to \mathbf{R}$ of the measure $\lambda$ on E; from the formulas (2) and (3), one immediately deduces $\mathscr{F}\mu = \mathscr{F}\lambda$.

Let $\mu$ be any promeasure on E, and $u$ a continuous linear mapping of E into a locally convex space $E_1$. Denote by ${}^t u$ the linear mapping of $E_1'$ into $E'$ that is the transpose of $u$, and by $\nu$ the promeasure $u(\mu)$ on $E_1$. For every $x_1' \in E_1'$, we have ${}^t u(x_1') = x_1' \circ u$, whence

$$\nu_{x_1'} = x_1'(\nu) = x_1'\big(u(\mu)\big) = (x_1' \circ u)(\mu) = \mu_{{}^t u(x_1')}\,.$$

Consequently,

$$\mathscr{F}\big(u(\mu)\big) = (\mathscr{F}\mu) \circ {}^t u\,. \tag{4}$$

In particular, let us take for $u$ the canonical mapping $p_V$ of E onto $E/V$ (for $V \in \mathscr{F}(E)$). The promeasure $p_V(\mu)$ on $E/V$ is associated with the measure $\mu_V$, and ${}^t p_V$ is an isomorphism of the dual of $E/V$ onto the subspace $V^\circ$ of $E'$ orthogonal to $V$. If $(E/V)'$ is identified with $V^\circ$ by means of ${}^t p_V$, then

$$(\mathscr{F}\mu)(x') = \int_{E/V} e^{i\langle x,x'\rangle}\, d\mu_V(x) \tag{5}$$

for all $x' \in V^\circ$. One has $E' = \bigcup_{V\in\mathscr{F}(E)} V^\circ$, so that the preceding formula characterizes the function $\mathscr{F}\mu$ on $E'$. Finally, if one sets $x' = 0$ in (5), one sees that the total mass of $\mu$ is equal to $(\mathscr{F}\mu)(0)$.

PROPOSITION 3. — *Let* E *be a locally convex space. The mapping* $\mu \mapsto \mathscr{F}\mu$ *of the set of promeasures on* E *into the set of functions on* E′ *is injective.*

The formula (5) permits reducing to the case that E is finite-dimensional; since every finite-dimensional space is isomorphic to a space $\mathbf{R}^n$, we can even suppose that there exists an integer $n \geqslant 0$ such that $E = \mathbf{R}^n$. We therefore have to prove that if $\mu$ is a bounded measure (not necessarily positive) on $\mathbf{R}^n$ and if

$$\int_{\mathbf{R}^n} e^{i\langle x,y\rangle}\, d\mu(x) = 0$$

for every linear form $y$ on $\mathbf{R}^n$, then $\mu = 0$.

For every integer $m \geqslant 0$, let $G_m$ be the subgroup $m \cdot \mathbf{Z}^n$ of $\mathbf{R}^n$. Denote by $\mathscr{C}_m$ the vector space of continuous functions $f$ on $\mathbf{R}^n$ such that $f(x+a) = f(x)$ for $x \in \mathbf{R}^n$ and $a \in G_m$. By Prop. 8 of GT, X, §4, No. 4, every function in $\mathscr{C}_m$ is the uniform limit of finite linear combinations of functions of the type $x \mapsto e^{2\pi i\langle x,q\rangle}$ with $q \in m^{-1}\cdot\mathbf{Z}^n$. Therefore $\mu(f) = 0$ for every function $f \in \mathscr{C}_m$.

Let $f$ be a continuous function on $\mathbf{R}^n$ with compact support. For every integer $m \geqslant 0$, set $f_m(x) = \sum_{q\in G_m} f(x+q)$. It is immediate that for every $x \in \mathbf{R}^n$, the preceding series has only finitely many terms, and that $f_m$ belongs to $\mathscr{C}_m$. Moreover, it is easy to see that the sequence $(f_m)$ tends to $f$ uniformly on every compact set, and that there exists a constant $C \geqslant 0$ such that $|f_m| \leqslant C$ for all $m$. Consequently, $\mu(f) = \lim_{m\to\infty} \mu(f_m)$ by Prop. 12 of §5, No. 6. Since $f_m \in \mathscr{C}_m$, we have $\mu(f_m) = 0$, whence finally $\mu(f) = 0$. Thus $\mu = 0$.

**Remark.* — When E is finite-dimensional, every character of E is of the form $x \mapsto e^{i\langle x,x'\rangle}$ with $x' \in E'$ (*Théor. spect.*, Ch. II, §1, No. 9, Cor. 3 of Prop. 12). In this case, Prop. 3 follows from the uniqueness theorem for the Fourier transformation (*loc. cit.*, §1, No. 6, Cor. of Prop. 6).$_*$

## 4. Calculation of Gaussian integrals

*Lemma* 1. — *For every integer* $n \geqslant 0$,

$$\int_{\mathbf{R}} |x|^n e^{-x^2/2}\, dx = 2^{\frac{n+1}{2}}\, \Gamma\Big(\frac{n+1}{2}\Big) \tag{6}$$

$$\int_{\mathbf{R}} x^{2n} e^{-x^2/2}\, dx = (2\pi)^{1/2}\frac{(2n)!}{2^n n!} \tag{7}$$

$$\int_{\mathbf{R}} x^{2n+1} e^{-x^2/2}\, dx = 0\,. \tag{8}$$

Recall the formula

$$\Gamma(s) = \int_0^\infty u^{s-1}e^{-u}\, du \tag{9}$$

valid for every real number $s > 0$ (FRV, VII, §1, No. 3, Prop. 3). On making the change of variable $x = (2u)^{1/2}$, it follows from (9) that

$$\int_0^\infty x^n e^{-x^2/2}\, dx = \int_0^\infty (2u)^{n/2}e^{-u}\, \tfrac{1}{2}\, 2^{1/2}u^{-1/2}\, du = 2^{\frac{n-1}{2}}\, \Gamma\Big(\frac{n+1}{2}\Big),$$

whence the formula (6) since

$$\int_{\mathbf{R}} |x|^n e^{-x^2/2}\, dx = 2\int_0^\infty x^n e^{-x^2/2}\, dx\,.$$

The formula (7) follows from (6) and the relation

$$\Gamma\Big(n + \frac{1}{2}\Big) = \pi^{1/2}\,\frac{(2n)!}{2^{2n}n!}\,. \tag{10}$$

For $n = 0$, this relation reduces to $\Gamma(\frac{1}{2}) = \pi^{1/2}$, that is, to the formula (21) of FRV, VII, §1, No. 3. The general case then follows by induction on $n$, on taking into account the relation $\Gamma(x+1) = x \cdot \Gamma(x)$ (*loc. cit.*, §1, No. 1).

Finally, the formula (8) follows from the fact that the function $x \mapsto x^{2n+1}e^{-x^2/2}$ is odd.

*Lemma* 2. — *For every complex number* $y$,

$$(2\pi)^{-1/2}\int_{\mathbf{R}} e^{-x^2/2}e^{ixy}\, dx = e^{-y^2/2}\,. \tag{11}$$

*In particular,*

$$(2\pi)^{-1/2}\int_{\mathbf{R}} e^{-x^2/2}\, dx = 1\,.$$

The change of variable $x \mapsto -x$ yields

$$(2\pi)^{-1/2} \int_{\mathbf{R}} e^{-x^2/2} e^{ixy}\, dx = (2\pi)^{-1/2} \int_{\mathbf{R}} e^{-x^2/2} e^{-ixy}\, dx\,;$$

since $\cos u = \dfrac{e^{iu} + e^{-iu}}{2}$ for every complex number $u$, it follows that

$$(12) \qquad (2\pi)^{-1/2} \int_{\mathbf{R}} e^{-x^2/2} e^{ixy}\, dx = (2\pi)^{-1/2} \int_{\mathbf{R}} e^{-x^2/2} \cos xy\, dx\,.$$

For every integer $n \geqslant 0$, set

$$g_n(x) = (-1)^n (2\pi)^{-1/2} \frac{(xy)^{2n}}{(2n)!} e^{-x^2/2}\,.$$

By (7),

$$(13) \qquad \int_{\mathbf{R}} |g_n(x)|\, dx = \frac{1}{n!} \Big( \frac{|y|^2}{2} \Big)^n$$

$$(14) \qquad \int_{\mathbf{R}} g_n(x)\, dx = \frac{1}{n!} \Big( -\frac{y^2}{2} \Big)^n,$$

whence

$$\sum_{n=0}^{\infty} \int_{\mathbf{R}} |g_n(x)|\, dx = e^{|y|^2/2} < +\infty\,.$$

Since, moreover,

$$(2\pi)^{-1/2} e^{-x^2/2} \cos xy = \sum_{n=0}^{\infty} g_n(x)\,,$$

this equality can be integrated term by term, whence

$$(2\pi)^{-1/2} \int_{\mathbf{R}} e^{-x^2/2} \cos xy\, dx = \sum_{n=0}^{\infty} \int_{\mathbf{R}} g_n(x)\, dx = e^{-y^2/2}$$

by (14). The formula (11) then follows from (12).

## 5. Gaussian promeasures and measures

PROPOSITION 4. — *Let* E *be a locally convex space. For every positive quadratic form* Q *on* $\mathrm{E}'$, *there exists one and only one promeasure* $\Gamma_{\mathrm{Q}}$ *on* E *such that* $\mathscr{F}\Gamma_{\mathrm{Q}} = e^{-\mathrm{Q}/2}$. *The total mass of* $\Gamma_{\mathrm{Q}}$ *is equal to* $1$.

The *uniqueness* of $\Gamma_Q$ follows from Prop. 3 of No. 3. The total mass of $\Gamma_Q$ is equal to $(\mathscr{F}\Gamma_Q)(0) = e^{-Q(0)/2} = 1$. We will prove *existence* in stages.

A) E *of finite dimension* $n$, *and* Q *nondegenerate.*

By Lemma 2 of No. 4, the measure $\gamma_1$ on $\mathbf{R}$ having density $t \mapsto (2\pi)^{-1/2}e^{-t^2/2}$ is bounded, of total mass 1. Set $\gamma = \gamma_1 \otimes \cdots \otimes \gamma_1$ ($n$ factors). From Lemma 2 of No. 4, one deduces

$$\begin{aligned}\int_{\mathbf{R}^n} e^{i(a_1t_1+\cdots+a_nt_n)}\,d\gamma(t_1,\ldots,t_n) &= \prod_{j=1}^n \int_{\mathbf{R}} e^{ia_jt}\,d\gamma_1(t)\\ &= \prod_{j=1}^n (2\pi)^{-1/2}\int_{\mathbf{R}} e^{ia_jt}e^{-t^2/2}\,dt\\ &= \prod_{j=1}^n e^{-a_j^2/2}\\ &= \exp\big(-\tfrac{1}{2}(a_1^2+\cdots+a_n^2)\big).\end{aligned}$$

Since Q is positive and nondegenerate, there exists a basis $(e'_1,\ldots,e'_n)$ of E′ orthonormal for Q (*Alg.*, Ch. IX, §7, No. 1). Let us denote by $f$ the isomorphism $x \mapsto (e'_1(x),\ldots,e'_n(x))$ of E onto $\mathbf{R}^n$, and by $\Gamma_Q$ the measure $f^{-1}(\gamma)$ on E. Let $x' = a_1e'_1 + \cdots + a_ne'_n$ be in E′; then $x'(f^{-1}(t_1,\ldots,t_n)) = \sum_{j=1}^n t_ja_j$ for $t_1,\ldots,t_n$ real, whence

$$\begin{aligned}\int_{\mathrm{E}} e^{i\langle x,x'\rangle}\,d\Gamma_Q(x) &= \int_{\mathbf{R}^n} e^{i(a_1t_1+\cdots+a_nt_n)}\,d\gamma(t_1,\ldots,t_n)\\ &= \exp\big(-\tfrac{1}{2}(a_1^2+\cdots+a_n^2)\big) = \exp\big(-\tfrac{1}{2}Q(x')\big).\end{aligned}$$

Consequently, $\mathscr{F}\Gamma_Q = e^{-Q/2}$.

B) E *finite-dimensional.*

Let N be the linear subspace of E′ formed by the $x'$ such that $Q(x') = 0$. Denote by M the orthogonal of N in E, and by $j$ the canonical injection of M into E. The linear mapping ${}^tj : \mathrm{E}' \to \mathrm{M}'$ is surjective, with kernel N, therefore there exists on M′ a nondegenerate positive quadratic form $q$ such that $Q = q \circ {}^tj$. By the foregoing, there exists a bounded measure Γ on M such that $\mathscr{F}\Gamma = e^{-q/2}$. Setting $\Gamma_Q = j(\Gamma)$, we have

$$\mathscr{F}\Gamma_Q = (\mathscr{F}\Gamma)\circ {}^tj = \exp(-q\circ {}^tj/2) = e^{-Q/2}$$

by formula (4) of No. 3.

C) *The general case.*

Let $V \in \mathscr{F}(E)$. Denote by $p_V$ the canonical mapping of E onto E/V, and by $Q_V$ the positive quadratic form $Q \circ {}^t p_V$ on $(E/V)'$; finally, let $\mu_V$ be the measure on E/V with Fourier transform $e^{-Q_V/2}$ (cf. B)). If $W \in \mathscr{F}(E)$ is contained in V, then $p_V = p_{VW} \circ p_W$, whence $Q_V = Q_W \circ {}^t p_{VW}$; by formula (4) of No. 3, the measure $p_{VW}(\mu_W)$ has as Fourier transform the function $(e^{-Q_W/2}) \circ {}^t p_{VW} = e^{-Q_V/2}$, hence is equal to $\mu_V$. The family $(\mu_V)_{V \in \mathscr{F}(E)}$ is therefore a promeasure $\mu$ on E. Formula (5) of No. 3 shows that $\mathscr{F}\mu$ is equal to $e^{-Q/2}$.

DEFINITION 2. — *Let* E *be a locally convex space. For every positive quadratic form* Q *on* E′, *the promeasure on* E *whose Fourier transform is equal to* $e^{-Q/2}$ *is called the Gaussian promeasure on* E *with variance* Q, *and is denoted* $\Gamma_Q$. *A promeasure* $\mu$ *on* E *is said to be Gaussian if there exists a positive quadratic form* Q *on* E′ *such that* $\mu = \Gamma_Q$.

By an abuse of language, a bounded measure $\mu$ on E will be said to be Gaussian with variance Q if the associated promeasure $\widetilde{\mu}$ is equal to $\Gamma_Q$.

*Remarks.* — 1) Let E be a finite-dimensional vector space, and let $\mu$ be a positive measure on E of mass 1, such that every linear form on E belongs to $\mathscr{L}^2(E, \mu)$. One defines an element $m$ of E and a positive quadratic form V on E′ by the formulas

$$\langle m, x' \rangle = \int_E \langle x, x' \rangle \, d\mu(x), \qquad V(x') = \int_E \langle x - m, x' \rangle^2 \, d\mu(x).$$

In the traditional terminology of Probability Theory, $m$ is called the *mean* and V the *variance* of $\mu$; $\mu$ is said to be *centered* if $m = 0$.

Now let $a$ be an element of E and Q a positive quadratic form on E′. Let us denote by $\Gamma_{a,Q}$ the image of the measure $\Gamma_Q$ under the translation $x \mapsto x + a$. It is easily seen that $\Gamma_{a,Q}$ is a positive measure on E of mass 1, with Fourier transform $x' \mapsto e^{i\langle a, x' \rangle - \frac{1}{2}Q(x')}$ and mean $a$. Moreover, Prop. 6 below implies that Q is the variance of $\Gamma_{a,Q}$. One traditionally says that $\Gamma_{a,Q}$ is the Gaussian measure with mean $a$ and variance Q, and that $\Gamma_Q = \Gamma_{0,Q}$ is the *centered* Gaussian measure with variance Q. Since we shall only be considering *centered* Gaussian measures, we shall omit this qualifier.

2) Let Q be a quadratic form on the dual E′ of a locally convex space E. If there exists a promeasure on E with Fourier transform $e^{-Q/2}$, the quadratic form Q is necessarily positive: for, the function $e^{-Q/2}$ is bounded on E′; therefore, for every $x' \in E'$, the function $t \mapsto e^{-t^2 Q(x')/2} = e^{-Q(tx')/2}$ on $\mathbf{R}$ is bounded, whence $Q(x') \geqslant 0$.

3) The dual of $\mathbf{R}$ is canonically isomorphic to $\mathbf{R}$ and the positive quadratic forms on $\mathbf{R}$ are the functions of the form $t \mapsto at^2$ with $a \geqslant 0$. Therefore, there exists for every $a \geqslant 0$ one and only bounded measure $\gamma_a$ on $\mathbf{R}$ whose Fourier transform is equal to the function $t \mapsto e^{-at^2/2}$; by an abuse of language, $\gamma_a$ is said to be *the Gaussian measure on* $\mathbf{R}$ *with variance* $a$.

The Fourier transform of $\gamma_0$ is the constant 1, whence $\gamma_0 = \varepsilon_0$ (unit mass at the origin of $\mathbf{R}$). Suppose $a > 0$ and denote by $u_a$ the linear mapping $x \mapsto a^{1/2}x$; then $\mathscr{F}\gamma_a = \mathscr{F}\gamma_1 \circ {}^t u_a$, whence $\gamma_a = u_a(\gamma_1)$. Lemma 2 shows that $\gamma_1$ is the measure with density $x \mapsto (2\pi)^{-1/2}e^{-x^2/2}$ with respect to Lebesgue measure: from this, one easily deduces

$$d\gamma_a(x) = (2\pi a)^{-1/2}e^{-x^2/2a}\,dx\,. \tag{15}$$

The image of a Gaussian promeasure under a continuous linear mapping is a Gaussian promeasure. More precisely, we have the following result:

PROPOSITION 5. — *Let* E *and* $E_1$ *be two locally convex spaces, and* $u$ *a continuous linear mapping of* E *into* $E_1$. *Let* Q *be a positive quadratic form on* $E'$, *and* $Q_1$ *the positive quadratic form* $Q \circ {}^t u$ *on* $E'_1$. *Then* $u(\Gamma_Q) = \Gamma_{Q_1}$.

Set $\mu = u(\Gamma_Q)$. By formula (4) of No. 3,

$$\mathscr{F}\mu = (\mathscr{F}\Gamma_Q) \circ {}^t u = e^{-Q/2} \circ {}^t u = e^{-Q_1/2} = \mathscr{F}\Gamma_{Q_1}\,,$$

whence $\mu = \Gamma_{Q_1}$ by Prop. 3 of No. 3.

COROLLARY. — *Let* E *be a locally convex space and* Q *a positive quadratic form on* $E'$. *For every* $x' \in E'$, *the image of* $\Gamma_Q$ *under* $x'$ *is the Gaussian measure on* **R** *with variance* $Q(x')$.

PROPOSITION 6. — *Let* E *be a locally convex space, and* $\mu$ *a Gaussian measure on* E, *with variance* Q. *For every integer* $n \geqslant 0$ *and every* $x' \in E'$, *one has the relations*

$$\int_E |\langle x, x'\rangle|^n\,d\mu(x) = \pi^{-1/2}\,2^{n/2}\,\Gamma\Big(\frac{n+1}{2}\Big)Q(x')^{n/2} \tag{16}$$

$$\int_E \langle x, x'\rangle^{2n}\,d\mu(x) = \frac{(2n)!}{2^n n!}\,Q(x')^n \tag{17}$$

$$\int_E \langle x, x'\rangle^{2n+1}\,d\mu(x) = 0\,. \tag{18}$$

*In particular,*

$$\int_E \langle x, x'\rangle^2\,d\mu(x) = Q(x') \qquad (x' \in E')\,. \tag{19}$$

If these formulas are true for an element $x'$ of $E'$, then they are true for all of its multiples $t \cdot x'$ (with $t$ real). We may therefore content ourselves with establishing them when $Q(x')$ is equal to $0$ or $1$.

*a*) Suppose $Q(x') = 0$. The measure $x'(\mu)$ is equal to $\gamma_0 = \varepsilon_0$, therefore $x'$ is zero $\mu$-almost everywhere; the formulas (16) to (19) are then obvious.

*b*) Suppose $\mathrm{Q}(x')=1$, whence $x'(\mu)=\gamma_1$. Then

$$\int_{\mathrm{E}} |\langle x,x'\rangle|^n \, d\mu(x) = \int_{\mathbf{R}} |t|^n \, d\gamma_1(t) = (2\pi)^{-1/2} \int_{\mathbf{R}} |t|^n e^{-t^2/2} \, dt$$

and (16) follows immediately from (6) (No 4, Lemma 1). Similarly, formulas (17) and (18) follow from (7) and (8). Finally, (19) is obtained by setting $n=1$ in (17).

We can now prove a converse of the Cor. of Prop. 5.

PROPOSITION 7. — *Let* E *be a locally convex space and* $\mu$ *a promeasure on* E. *Suppose that* $x'(\mu)$ *is a Gaussian measure on* $\mathbf{R}$ *for every* $x'\in \mathrm{E}'$. *Then* $\mu$ *is a Gaussian promeasure on* E.

For every $x'\in\mathrm{E}'$, let $\mathrm{Q}(x')$ be the variance of the Gaussian measure $x'(\mu)$ on $\mathbf{R}$. One has $x'(\mu)=\gamma_{\mathrm{Q}(x')}$, whence

$$(\mathscr{F}\mu)(x') = \int_{\mathbf{R}} e^{it\cdot 1} \, d\gamma_{\mathrm{Q}(x')}(t) = e^{-\mathrm{Q}(x')\cdot 1^2/2}$$

by the definition of $\mathscr{F}\mu$ (No. 3, formula (2)). In other words, $\mathscr{F}\mu = e^{-\mathrm{Q}/2}$, and it remains to prove that Q is a positive quadratic form on E′.

For every closed linear subspace V of E with finite codimension, denote by $p_{\mathrm{V}}$ the canonical mapping of E onto E/V, by $\mu_{\mathrm{V}}$ the measure $p_{\mathrm{V}}(\mu)$ on E/V, and set $\mathrm{Q}_{\mathrm{V}}=\mathrm{Q}\circ {}^t p_{\mathrm{V}}$. Since $\mathrm{E}' = \bigcup_{\mathrm{V}\in\mathscr{F}(\mathrm{E})} \mathrm{Im}({}^t p_{\mathrm{V}})$ and ${}^t p_{\mathrm{V}}$ is injective, it suffices to prove that $\mathrm{Q}_{\mathrm{V}}$ is a positive quadratic form on $(\mathrm{E}/\mathrm{V})'$. Let $u\in(\mathrm{E}/\mathrm{V})'$ and $x'={}^t p_{\mathrm{V}}(u)$. We have

$$u(\mu_{\mathrm{V}}) = u\big(p_{\mathrm{V}}(\mu)\big) = x'(\mu) = \gamma_{\mathrm{Q}(x')} ;$$

Prop. 6 then implies

$$\mathrm{Q}_{\mathrm{V}}(u) = \mathrm{Q}(x') = \int_{\mathbf{R}} t^2 \, d\gamma_{\mathrm{Q}(x')}(t) = \int_{\mathrm{E}/\mathrm{V}} u(x)^2 \, d\mu_{\mathrm{V}}(x) ,$$

thus $\mathrm{Q}_{\mathrm{V}}$ is a positive quadratic form on $(\mathrm{E}/\mathrm{V})'$.

## 6. Examples of Gaussian promeasures

1) Let E be a real Hilbert space. The mapping $x'\mapsto \|x'\|^2$ is a positive quadratic form on E′. The corresponding Gaussian promeasure is called the *canonical Gaussian promeasure* on E. It can be shown that this promeasure is not a measure if E is infinite-dimensional.

Let $A$ be a continuous linear operator on E. The mapping $x' \mapsto \|{}^t A \cdot x'\|^2$ is a positive quadratic form on E'. The corresponding promeasure $\mu_A$ on E is a measure if and only if $A$ is a Hilbert–Schmidt operator (cf. No. 11, Cor. 2 of Th. 3).

2) *Kernels of positive type.* Let T be a set and $E = \mathbf{R}^T$ the vector space of real functions on T, equipped with the topology of pointwise convergence. For every $t \in T$, one denotes by $\varepsilon_t$ the linear form $f \mapsto f(t)$ on E. The family $(\varepsilon_t)_{t\in T}$ is a basis of E' (TVS, II, §6, No. 6, Cor. 2 of Prop. 8).

One calls (real) kernel of positive type on T every real-valued function K on $T \times T$ satisfying the relations

$$\mathrm{K}(t,t') = \mathrm{K}(t',t) \quad \text{for } t, t' \text{ in } \mathrm{T}, \tag{20}$$

$$\sum_{i,j=1}^{p} c_i c_j \mathrm{K}(t_i, t_j) \geqslant 0 \tag{21}$$

for any positive integer $p$, elements $t_1, \dots, t_p$ of T, and real numbers $c_1, \dots, c_p$. If this is so, the formula

$$q\Big(\sum_{t\in \mathrm{T}} c_t \varepsilon_t\Big) = \sum_{t,t'\in \mathrm{T}} c_t c_{t'} \mathrm{K}(t,t') \tag{22}$$

defines a positive quadratic form on E'. Conversely, if $q$ is a positive quadratic form on E', then the formula

$$\mathrm{K}(t,t') = \tfrac{1}{2}[q(\varepsilon_t + \varepsilon_{t'}) - q(\varepsilon_t) - q(\varepsilon_{t'})] \tag{23}$$

defines a kernel K of positive type on T. One thus obtains two mutually inverse bijections between the set of kernels of positive type on T, and that of the positive quadratic forms on E'.

Let K be a kernel of positive type on T, and $q$ the associated quadratic form on E'. The Gaussian promeasure on E with variance $q$ is also called the *Gaussian promeasure on* E *with covariance* K. If T is countable, Prop. 2 of No. 1 implies that this promeasure is a measure.

3) Let T be a countable set. A kernel $\delta$ on T of positive type is defined by setting

$$\delta(t,t') = \begin{cases} 1 & \text{if } t = t' \\ 0 & \text{if } t \neq t'. \end{cases} \tag{24}$$

The corresponding quadratic form is given by $q\Big(\sum_{t\in \mathrm{T}} c_t\varepsilon_t\Big) = \sum_{t\in\mathrm{T}} c_t^2$. For every $t \in T$, let us denote by $\mu_t$ the Gaussian measure on $\mathbf{R}$ with variance 1;

one shows easily that the Gaussian measure on $\mathbf{R}^{\mathrm{T}}$ with covariance $\delta$ is equal to $\bigotimes_{t\in\mathrm{T}} \mu_t$.

4) Let $n \geqslant 1$ be an integer. A square matrix $C = (c_{ij})$ of order $n$ is said to be *positive symmetric* if it is symmetric and $\sum_{i,j=1}^{n} c_{ij}x_ix_j \geqslant 0$ for any real $x_1, \ldots, x_n$; it comes to the same to say that the mapping $(i,j) \mapsto c_{ij}$ is a kernel of positive type on the set $\{1, 2, \ldots, n\}$. We may therefore speak of the Gaussian measure $\gamma_C$ on $\mathbf{R}^n$, with covariance $C$; it is characterized by the formula

$$(25) \qquad \int_{\mathbf{R}^n} e^{i(x_1t_1+\cdots+x_nt_n)}\, d\gamma_C(t_1, \ldots, t_n) = \exp\Big(-\frac{1}{2}\sum_{j,k=1}^{n} c_{jk}x_jx_k\Big),$$

for $x_1, \ldots, x_n$ real. From Prop. 6 of No. 5 (formula (19)), one deduces

$$(26) \qquad \int_{\mathbf{R}^n} t_jt_k\, d\gamma_C(t_1, \ldots, t_n) = c_{jk} \qquad (1 \leqslant j, k \leqslant n).$$

From Prop. 5 of No. 5, one deduces the formula

$$(27) \qquad u(\gamma_C) = \gamma_{UC\,{}^tU},$$

where $u$ is a linear mapping of $\mathbf{R}^n$ into $\mathbf{R}^m$ with matrix $U$. Moreover, one sees easily (cf. the first part of the proof of Prop. 4 of No. 5) that if $I_n$ is the identity matrix of order $n$, then the measure $\gamma_{I_n}$ admits the density

$$(2\pi)^{-n/2}\exp\big(-\tfrac{1}{2}(t_1^2 + \cdots + t_n^2)\big)$$

with respect to the Lebesgue measure $\lambda_n$ on $\mathbf{R}^n$.

We are going to show that if the matrix $C$ is invertible, with inverse $D = (d_{jk})$, then

$$(28) \qquad d\gamma_C(t_1, \ldots, t_n) = (2\pi)^{-n/2}(\det D)^{1/2}\Big(\exp\Big(-\frac{1}{2}\sum_{j,k=1}^{n} d_{jk}t_jt_k\Big)\Big)dt_1 \cdots dt_n.$$

For, if $C$ is invertible, the quadratic form $q$ on $\mathbf{R}^n$ defined by

$$q(x_1, \ldots, x_n) = \sum_{j,k=1}^{n} c_{jk}x_jx_k$$

is nondegenerate. Using the existence of a basis of $\mathbf{R}^n$ orthonormal for $q$, one proves the existence of a square matrix $U$ of order $n$ such that $C = U \cdot {}^tU$, whence $\gamma_C = u(\gamma_{I_n})$ by (27) (where $u$ denotes the automorphism of $\mathbf{R}^n$ with matrix $U$). Let Q be the quadratic form on $\mathbf{R}^n$ defined by

$$Q(t_1, \ldots, t_n) = t_1^2 + \cdots + t_n^2 ;$$

then

$$\gamma_{I_n} = (2\pi)^{-n/2} e^{-Q/2} \cdot \lambda_n ,$$

whence

$$u(\gamma_{I_n}) = (2\pi)^{-n/2} e^{-(Q\circ u^{-1})/2} \cdot u(\lambda_n) . \tag{29}$$

It is immediate that the quadratic form $Q \circ u^{-1}$ on $\mathbf{R}^n$ takes the value $\sum_{j,k=1}^{n} d_{jk}t_jt_k$ at the point $(t_1, \ldots, t_n)$, and Prop. 15 of Ch. VII, §1, No. 10 shows that

$$u(\lambda_n) = (\det U)^{-1} \cdot \lambda_n = (\det D)^{1/2} \cdot \lambda_n . \tag{30}$$

Formula (28) then follows from this.

## 7. Wiener measure

In this No., we denote by T the interval $]0,1]$ of $\mathbf{R}$ and by $\mathscr{H}$ the Hilbert space of real functions on T square-integrable with respect to Lebesgue measure, where the scalar product is denoted $(f|g)$. We also denote by $\mathscr{C}$ the space of continuous real functions on T tending to 0 at the point 0; we equip $\mathscr{C}$ with the norm $\|f\| = \sup_{t\in \mathrm{T}} |f(t)|$. The compact interval $[0,1] = \mathrm{T} \cup \{0\}$ is the Alexandroff compactification of the locally compact but noncompact interval T; consequently, the set of continuous functions on T with compact support is dense in $\mathscr{C}$, and the dual of $\mathscr{C}$ may be identified with the space $\mathscr{M}^1$ of bounded measures (not necessarily positive) on T (Ch. III, §1, No. 8, Def. 3).

For every function $f \in \mathscr{H}$, one defines a function $Pf$ on T by

$$(Pf)(t) = \int_0^t f(x)\,dx = (f|\mathrm{I}_t) , \tag{31}$$

where $\mathrm{I}_t$ is the characteristic function of the interval $]0,t]$. The Cauchy–Schwarz inequality implies the inequalities

$$|(Pf)(t)| \leqslant \|f\|_2 \cdot t^{1/2} \tag{32}$$

$$|(Pf)(t) - (Pf)(t')| \leqslant \|f\|_2 \cdot |t - t'|^{1/2} ; \tag{33}$$

consequently, $Pf$ belongs to $\mathscr{C}$, and the linear mapping $P$ of $\mathscr{H}$ into $\mathscr{C}$ is continuous with norm $\leqslant 1$.

Let us identify the Hilbert space $\mathscr{H}$ with its dual (TVS, V, §1, No. 7, Th. 3), and denote by $\Pi : \mathscr{M}^1 \to \mathscr{H}$ the transpose of $P : \mathscr{H} \to \mathscr{C}$. For every measure $\mu \in \mathscr{M}^1$ and every function $f \in \mathscr{H}$, we have

$$\begin{aligned}(\Pi\mu|f) = \mu(Pf) &= \int_{\mathrm{T}} d\mu(t) \int_{\mathrm{T}} \mathrm{I}_t(x) f(x)\, dx \\ &= \int_{\mathrm{T}} f(x)\, dx \int_{\mathrm{T}} \mathrm{I}_t(x)\, d\mu(t)\end{aligned}$$

by the Lebesgue–Fubini theorem. Now,

$$\mathrm{I}_t(x) = \begin{cases} 1 & \text{if } 0 < x \leqslant t \leqslant 1 \\ 0 & \text{otherwise}, \end{cases}$$

whence finally

$$(\Pi\mu)(x) = \mu([x,1]) \qquad \text{for } x \in \mathrm{T}. \tag{34}$$

Let $\mu, \nu$ be in $\mathscr{M}^1$. Then

$$\begin{aligned}(\Pi\mu|\Pi\nu) &= \int_{\mathrm{T}} \Pi\mu(x)\, \Pi\nu(x)\, dx = \int_{\mathrm{T}} dx \int_{\mathrm{T}} \mathrm{I}_t(x)\, d\mu(t) \int_{\mathrm{T}} \mathrm{I}_{t'}(x)\, d\nu(t') \\ &= \int_{\mathrm{T}}\int_{\mathrm{T}} d\mu(t)\, d\nu(t') \int_{\mathrm{T}} \mathrm{I}_t(x)\, \mathrm{I}_{t'}(x)\, dx.\end{aligned}$$

Now, $\mathrm{I}_t \cdot \mathrm{I}_{t'}$ is the characteristic function of the interval $]0,t] \cap ]0,t']$, whence immediately

$$\int_{\mathrm{T}} \mathrm{I}_t(x) \mathrm{I}_{t'}(x)\, dx = \inf(t,t'). \tag{35}$$

It follows that

$$(\Pi\mu|\Pi\nu) = \int_{\mathrm{T}}\int_{\mathrm{T}} \inf(t,t')\, d\mu(t)\, d\nu(t'). \tag{36}$$

By the preceding result, one defines a positive quadratic form W on $\mathscr{M}^1$ by the formula

$$\mathrm{W}(\mu) = \int_{\mathrm{T}}\int_{\mathrm{T}} \inf(t,t')\, d\mu(t)\, d\mu(t') = \|\Pi\mu\|_2^2. \tag{37}$$

In particular, if $t_1, \ldots, t_n$ are elements of T, and $c_1, \ldots, c_r$ are real numbers, then

$$\mathrm{W}\Big(\sum_{j=1}^{n} c_j \varepsilon_{t_j}\Big) = \sum_{j,k=1}^{n} c_j c_k \inf(t_j, t_k)$$

and since W is positive, the function $(t, t') \mapsto \inf(t, t')$ is a kernel of positive type on T.

THEOREM 1 (Wiener). — *Let $w$ be the image under $P : \mathscr{H} \to \mathscr{C}$ of the canonical Gaussian promeasure on the Hilbert space $\mathscr{H}$. Then $w$ is a Gaussian* measure *on $\mathscr{C}$ with variance* W.

By construction, $\mathrm{W}(\mu) = \|{}^tP(\mu)\|_2^2$; Prop. 5 of No. 5 shows that $w$ is a Gaussian promeasure with variance W. It remains to prove that $w$ is a measure on $\mathscr{C}$.

A) *Construction of an auxiliary measured space*(2) $(\Omega, m)$:

For every integer $n \geqslant 0$, denote by $\mathrm{D}_n$ the set of numbers of the form $k/2^n$ with $k = 1, 2, 3, \ldots, 2^n$. Set $\mathrm{D} = \bigcup_{n \geqslant 0} \mathrm{D}_n$ (the set of dyadic numbers contained in T) and $\Omega = \mathbf{R}^{\mathrm{D}}$. For every $t \in \mathrm{D}$, denote by $\mathrm{X}(t)$ the linear form $f \mapsto f(t)$ on $\Omega$.

For $t, t'$ in D, set $\mathrm{M}(t, t') = \inf(t, t')$; we have seen that M is a kernel of positive type on D. Since the set D is countable, one can define the Gaussian *measure* $m$ on $\Omega$ with covariance M (No. 6, *Example* 2).

*Lemma* 3. — *For any $t, t'$ in* D,

$$\int_\Omega \Big|\mathrm{X}\Big(\frac{t+t'}{2}\Big) - \frac{\mathrm{X}(t) + \mathrm{X}(t')}{2}\Big|^3 dm = \frac{1}{(8\pi)^{1/2}} |t - t'|^{3/2}. \tag{38}$$

Note that $\dfrac{t+t'}{2}$ belongs to D. One knows (No. 6, *Example* 2) that the family $(\mathrm{X}(t))_{t \in \mathrm{D}}$ is a basis of the topological dual $\Omega'$ of $\Omega$; therefore there exists a symmetric bilinear form $\widehat{\mathrm{M}}$ on $\Omega' \times \Omega'$ characterized by $\widehat{\mathrm{M}}(\mathrm{X}(t), \mathrm{X}(t')) = \inf(t, t')$. By construction, the variance of the Gaussian measure $m$ on $\Omega$ is the quadratic form $\xi \mapsto \widehat{\mathrm{M}}(\xi, \xi)$ on $\Omega'$. Set, in particular,

$$\xi = \mathrm{X}\Big(\frac{t+t'}{2}\Big) - \frac{\mathrm{X}(t) + \mathrm{X}(t')}{2}; \tag{39}$$

an easy calculation yields

$$\widehat{\mathrm{M}}(\xi, \xi) = \frac{|t - t'|}{4}. \tag{40}$$

(2) *Espace mesuré*: a locally compact space equipped with a measure (Ch. III, 1st edn., §2, No. 2, p. 52).

By Prop. 6 of No. 5 (formula (16)),

$$(41)\qquad \int_\Omega |\xi|^3\, dm = \pi^{-1/2}\, 2^{3/2}\, \Gamma(2)\, \widehat{\mathrm{M}}(\xi,\xi)^{3/2}\,;$$

the lemma follows immediately from formulas (40) and (41).

B) *Construction of a mapping* $u$ *of* $\Omega$ *into* $\mathscr{C}$:

For every integer $n \geqslant 0$, denote by $\mathrm{E}_n$ the subspace of $\mathscr{C}$ formed by the functions that are affine on each of the intervals $\left[\frac{k-1}{2^n}, \frac{k}{2^n}\right]$ for $1 \leqslant k \leqslant 2^n$. An affine function on a compact interval I of $\mathbf{R}$ attains its bounds at the endpoints of I; consequently,

$$(42)\qquad \|f\| = \sup_{1\leqslant k\leqslant 2^n} \left|f\Big(\frac{k}{2^n}\Big)\right|$$

for $f \in \mathrm{E}_n$.

For every function $g \in \Omega$ and every integer $n \geqslant 0$, there exists one and only one function $u_n(g)$ that belongs to $\mathrm{E}_n$ and coincides with $g$ at every point of $\mathrm{D}_n$; we shall write $\mathrm{T}_n g = u_{n+1}(g) - u_n(g)$. Since $\mathrm{D}_n$ is finite, the mapping $\mathrm{T}_n$ of $\Omega$ into $\mathscr{C}$ is continuous, hence $m$-measurable.

*Lemma 4. — For every integer* $n \geqslant 0$,

$$(43)\qquad \int_\Omega \|\mathrm{T}_n g\|^3\, dm(g) \leqslant \frac{1}{(8\pi)^{1/2}}\, 2^{-n/2}.$$

Let $g \in \Omega$ and $n \in \mathbf{N}$. One has $\mathrm{E}_n \subset \mathrm{E}_{n+1}$; consequently, the function $\mathrm{T}_n g$ belongs to $\mathrm{E}_{n+1}$ and is zero at every point of $\mathrm{D}_n$; therefore, by (42),

$$(44)\qquad \|\mathrm{T}_n g\|^3 = \sup_{1\leqslant k\leqslant 2^n} \left|\mathrm{T}_n g\Big(\frac{2k-1}{2^{n+1}}\Big)\right|^3 \leqslant \sum_{k=1}^{2^n} \left|\mathrm{T}_n g\Big(\frac{2k-1}{2^{n+1}}\Big)\right|^3.$$

Let us make the convention $g(0) = 0$. The construction of $u_n(g)$ by linear interpolation of $g$ implies the relations

$$(45)\qquad \mathrm{T}_n g\Big(\frac{2k-1}{2^{n+1}}\Big) = g\Big(\frac{2k-1}{2^{n+1}}\Big) - \frac{1}{2}\Big(g\Big(\frac{k-1}{2^n}\Big) + g\Big(\frac{k}{2^n}\Big)\Big)$$

for $1 \leqslant k \leqslant 2^n$. From this, one deduces, by integration,

$$\int_\Omega \left|\mathrm{T}_n g\Big(\frac{2k-1}{2^{n+1}}\Big)\right|^3 dm(g) = \int_\Omega \left|\mathrm{X}\Big(\frac{2k-1}{2^{n+1}}\Big) - \frac{1}{2}\Big(\mathrm{X}\Big(\frac{k-1}{2^n}\Big) + \mathrm{X}\Big(\frac{k}{2^n}\Big)\Big)\right|^3 dm\,;$$

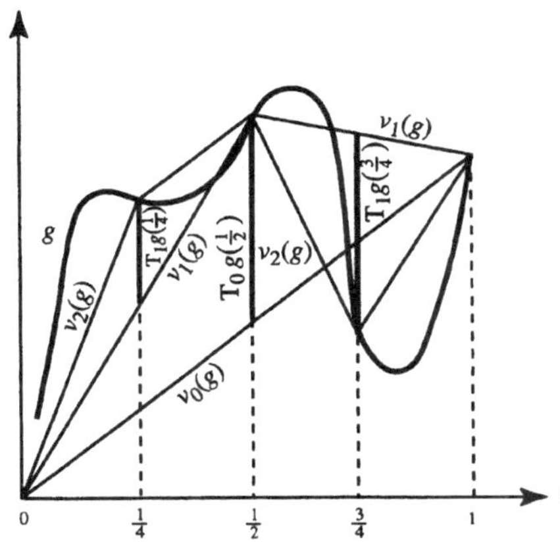


FIGURE 1

one can then apply Lemma 3 with $t = \dfrac{k-1}{2^n}$, $t' = \dfrac{k}{2^n}$, whence

$$\int_\Omega \left|\mathrm{T}_n g\left(\frac{2k-1}{2^{n+1}}\right)\right|^3 dm(g) = \frac{1}{(8\pi)^{1/2}}\, 2^{-\frac{3n}{2}}. \tag{46}$$

By (44), we then have

$$\int_\Omega \|\mathrm{T}_n g\|^3\, dm(g) \leqslant \sum_{k=1}^{2^n} \int_\Omega \left|\mathrm{T}_n g\left(\frac{2k-1}{2^{n+1}}\right)\right|^3 dm(g) = \frac{1}{(8\pi)^{1/2}}\, 2^n \cdot 2^{-\frac{3n}{2}},$$

whence the lemma.

By Lemma 4, the mapping $\mathrm{T}_n$ of $\Omega$ into the Banach space $\mathscr{C}$ belongs to $\mathrm{L}^3_{\mathscr{C}}(\Omega, m)$ and $\mathrm{N}_3(\mathrm{T}_n) \leqslant \dfrac{1}{(8\pi)^{1/6}}\,(2^{-1/6})^n$, whence $\sum\limits_{n=0}^{\infty} \mathrm{N}_3(\mathrm{T}_n) < +\infty$. By Prop. 6 of Ch. IV, §3, No. 3, there exists a set $\Omega_0 \subset \Omega$ such that $\Omega - \Omega_0$ is $m$-negligible and such that the series $\sum\limits_{n=0}^{\infty} \mathrm{T}_n(g)$ converges absolutely in $\mathscr{C}$ for every $g \in \Omega_0$. One then defines an $m$-measurable mapping $u$ of $\Omega$ into $\mathscr{C}$ by

$$u(g) = \begin{cases} \displaystyle\sum_{n=0}^{\infty} \mathrm{T}_n g = \lim_{n\to\infty} u_n(g) & \text{for } g \in \Omega_0 \\ 0 & \text{for } g \in \Omega - \Omega_0 . \end{cases} \tag{47}$$

Since $u_n(g)$ and $g$ coincide on $\mathrm{D}_m \subset \mathrm{D}_n$ for $0 \leqslant m \leqslant n$, it is immediate that the restriction of $u(g)$ to D is equal to $g$ for every $g \in \Omega_0$.

C) *Construction of a Gaussian measure on* $\mathscr{C}$:

Let $w'$ be the bounded measure on $\mathscr{C}$ that is the image of $m$ under the $m$-measurable mapping $u : \Omega \to \mathscr{C}$. We are going to show that $w'$ is a Gaussian measure on $\mathscr{C}$, with variance W, whence $w = w'$. Denote by $\mathscr{D}$ the linear subspace of $\mathscr{M}^1$ generated by the measures $\varepsilon_t$ for $t$ running over D.

*Lemma* 5. — *For every measure* $\mu \in \mathscr{D}$,

$$\int_{\mathscr{C}} e^{i\langle f,\mu\rangle}\, dw'(f) = e^{-\mathrm{W}(\mu)/2}. \tag{48}$$

Set $\mu = c_1\varepsilon_{t_1} + c_2\varepsilon_{t_2} + \cdots + c_n\varepsilon_{t_n}$ with $t_1, \ldots, t_n$ in D and $c_1, \ldots, c_n$ in $\mathbf{R}$. For every $g \in \Omega_0$, the function $u(g)$ coincides with $g$ on D; therefore

$$\langle u(g), \mu\rangle = \sum_{j=1}^{n} c_j g(t_j) \qquad (g \in \Omega_0). \tag{49}$$

Also,

$$\mathrm{W}(\mu) = \sum_{j,k=1}^{n} c_j c_k \inf(t_j, t_k), \tag{50}$$

and, since $m$ is the Gaussian measure on $\Omega$ with covariance M, and $\Omega - \Omega_0$ is $m$-negligible, we have

$$\int_{\Omega_0} e^{i\sum_{j=1}^{n} c_j g(t_j)}\, dm(g) = \exp\Big(-\frac{1}{2}\sum_{j,k=1}^{n} c_j c_k \inf(t_j, t_k)\Big). \tag{51}$$

Now, $\Omega - \Omega_0$ is $m$-negligible and $w' = u(m)$; it follows that

$$\int_{\mathscr{C}} e^{i\langle f,\mu\rangle}\, dw'(f) = \int_{\Omega_0} e^{i\langle u(g),\mu\rangle}\, dm(g). \tag{52}$$

The formula (48) follows immediately from the formulas (49) to (52).

*Lemma* 6. — *Let* $\mu \in \mathscr{M}^1$. *There exists a sequence of measures* $\mu_n \in \mathscr{D}$ *such that* $\mu(f) = \lim_{n\to\infty} \mu_n(f)$ *for all* $f \in \mathscr{C}$ *and* $\mathrm{W}(\mu) = \lim_{n\to\infty} \mathrm{W}(\mu_n)$.

Let $\mathrm{I} = [0,1]$. The space $\mathscr{M}^1$ of bounded measures on $\mathrm{T} = ]0,1]$ will be identified with the subspace of $\mathscr{M}(\mathrm{I})$ formed by the measures that place no weight at 0.(3) We equip $\mathscr{M}(\mathrm{I})$ with the vague topology. The mapping

(3) That is, the measures on I that are concentrated on $\mathrm{T} = \mathrm{I} - \{0\}$.

$t \mapsto \varepsilon_t$ of I into $\mathscr{M}(\mathrm{I})$ is continuous (Ch. III, §1, No. 9, Prop. 13); since D is dense in I, the closure $\overline{\mathscr{D}}$ of $\mathscr{D}$ contains all of the point measures. Let A be the set of measures $\nu \in \mathscr{D}$ such that $\|\nu\| \leqslant \|\mu\|$; the measure $\mu$ is in the closure of A (Ch. III, §2, No. 4, Cor. 1 of Th. 1). The set A is relatively compact in $\mathscr{M}(\mathrm{I})$ (Ch. III, §1, No. 9, Prop. 15) and the compact subsets of $\mathscr{M}(\mathrm{I})$ are metrizable (TVS, III, §3, No. 4, Cor. 2 of Prop. 6,[4] and GT, X, §3, No. 3, Th. 1). Therefore there exists a sequence of measures $\mu_n \in \mathrm{A}$ converging to $\mu$ in $\mathscr{M}(\mathrm{I})$. Since $\mathscr{C}$ is identified with the subspace of continuous functions on I zero at the origin, we have $\mu(f) = \lim_{n\to\infty} \mu_n(f)$ for all $f \in \mathscr{C}$. Moreover, since $\mathscr{C}(\mathrm{I})\otimes\mathscr{C}(\mathrm{I})$ is dense in the normed space $\mathscr{C}(\mathrm{I}\times\mathrm{I})$ (Ch. III, §4, No. 1, Lemma 1), the relations $\lim_{n\to\infty} \mu_n = \mu$ and $\|\mu_n\| \leqslant \|\mu\|$ imply that $\lim_{n\to\infty} (\mu_n \otimes \mu_n) = \mu \otimes \mu$ (Ch. III, §1, No. 10, Prop. 17); since the measures $\mu_n$ and $\mu$ place no weight at 0, we have

$$\mathrm{W}(\mu_n) = \int_{\mathrm{I}}\int_{\mathrm{I}} \inf(t,t')\, d\mu_n(t)\, d\mu_n(t'),$$

$$\mathrm{W}(\mu) = \int_{\mathrm{I}}\int_{\mathrm{I}} \inf(t,t')\, d\mu(t)\, d\mu(t'),$$

whence $\lim_{n\to\infty} \mathrm{W}(\mu_n) = \mathrm{W}(\mu)$.

It remains to prove that the Fourier transform of $w'$ is equal to $e^{-\mathrm{W}/2}$. Let $\mu \in \mathscr{M}^1$; choose measures $\mu_n \in \mathscr{D}$ as in Lemma 6. The measure $w'$ is bounded, and $|e^{i\langle f,\mu_n\rangle}| = 1$ for all $n$; Lemma 5 and Lebesgue's convergence theorem (Ch. IV, §4, No. 3, Th. 2) then imply

$$\begin{aligned}\int_{\mathscr{C}} e^{i\langle f,\mu\rangle}\, dw'(f) &= \lim_{n\to\infty} \int_{\mathscr{C}} e^{i\langle f,\mu_n\rangle}\, dw'(f) \\ &= \lim_{n\to\infty} e^{-\mathrm{W}(\mu_n)/2} = e^{-\mathrm{W}(\mu)/2}.\end{aligned}$$

Q.E.D.

The measure $w$ on $\mathscr{C}$ whose Fourier transform is equal to $e^{-\mathrm{W}/2}$ is called the *Wiener measure on* $\mathscr{C}$.

*Remark.* — For every semi-open interval $\mathrm{J} = ]a,b]$ contained in T, let us set $l(\mathrm{J}) = b - a$ (the length of J) and denote by $\mathrm{A_J}$ the linear form $f \mapsto f(b) - f(a)$ on $\mathscr{C}$. It can be shown that the Wiener measure is characterized by the following property:

*Let* $\mathrm{J}_1, \ldots, \mathrm{J}_n$ *be semi-open intervals contained in* T *and pairwise disjoint. The image of the measure* $w$ *under the linear mapping* $f \mapsto (\mathrm{A}_{\mathrm{J}_1}(f), \ldots, \mathrm{A}_{\mathrm{J}_n}(f))$ *of* $\mathscr{C}$ *into* $\mathbf{R}^n$ *is equal to* $\gamma_{a_1} \otimes \cdots \otimes \gamma_{a_n}$ *with* $a_i = l(\mathrm{J}_i)^{1/2}$ *for* $1 \leqslant i \leqslant n$.

(4) In the cited Cor. 2, read 'second' (axiom of countability) instead of 'first'.

## 8. Continuity of the Fourier transform

PROPOSITION 8. — *Let* E *be a locally convex space,* $\mu$ *a promeasure on* E, *and* $\Phi$ *the Fourier transform of* $\mu$. *One has the inequalities*

$$|\Phi(x')| \leqslant \Phi(0) \tag{53}$$

$$|\Phi(x') - \Phi(y')|^2 \leqslant 2\Phi(0)\big(\Phi(0) - \mathscr{R}\Phi(x' - y')\big) \tag{54}$$

*for* $x', y'$ *in* E'.

Formula (5) of No. 3 permits reducing to the case that E is finite-dimensional and $\mu$ is a measure. Then

$$|\Phi(x')| = \Big| \int_{\mathrm{E}} e^{i\langle x,x'\rangle}\, d\mu(x)\Big| \leqslant \int_{\mathrm{E}} |e^{i\langle x,x'\rangle}|\, d\mu(x) = \int_{\mathrm{E}} d\mu(x) = \Phi(0)\,,$$

whence (53). Moreover, if $a$ and $b$ are real numbers, then

$$|e^{ia} - e^{ib}|^2 = |e^{ib}|^2|e^{i(a-b)} - 1|^2 = (e^{i(a-b)} - 1)(e^{-i(a-b)} - 1) = 2 - 2\cos(a-b)\,;$$

by the Cauchy–Schwarz inequality, we then have

$$\begin{aligned}|\Phi(x') - \Phi(y')|^2 &= \Big|\int_{\mathrm{E}} (e^{i\langle x,x'\rangle} - e^{i\langle x,y'\rangle})\, d\mu(x)\Big|^2 \\ &\leqslant \int_{\mathrm{E}} |e^{i\langle x,x'\rangle} - e^{i\langle x,y'\rangle}|^2\, d\mu(x) \int_{\mathrm{E}} 1^2\, d\mu(x) \\ &= \int_{\mathrm{E}} (2 - 2\cos\langle x, x' - y'\rangle)\, d\mu(x) \cdot \Phi(0) \\ &= 2\Phi(0)\big(\Phi(0) - \mathscr{R}\Phi(x' - y')\big),\end{aligned}$$

whence (54).

COROLLARY. — *Equip* E' *with a topology compatible with its vector space structure. For* $\Phi$ *to be continuous, it is necessary and sufficient that its real part* $\mathscr{R}\Phi$ *be continuous at the origin, in which case* $\Phi$ *is uniformly continuous.*

This follows from the inequality (54).

Let F be a locally convex space. We equip the dual F' of F with a topology compatible with the duality between F and F', and we identify F with the dual of F'. Consequently, the Fourier transform of a bounded measure $\mu$ on F' is the function $\mathscr{F}\mu$ on F defined by

$$(\mathscr{F}\mu)(x) = \int_{\mathrm{F}'} e^{i\langle x,x'\rangle}\, d\mu(x')\,.$$

PROPOSITION 9. — *If* F *is barreled, then the Fourier transform of every bounded measure on* F′ *is a uniformly continuous function on* F.

Let $\mu$ be a bounded measure on F′ and $\Phi$ its Fourier transform. Let $\varepsilon > 0$. There exists a compact subset K of F′ such that $\mu(\mathrm{F}' - \mathrm{K}) \leqslant \varepsilon$. Now, K is compact for the weak topology $\sigma(\mathrm{F}', \mathrm{F})$, hence is equicontinuous because F is barreled (TVS, III, §4, No. 2, Th. 1). Therefore there exists a symmetric neighborhood U of 0 in F whose polar U° contains K. Let $x$ be in $\varepsilon$U; then

$$\Phi(0) - \mathscr{R}\Phi(x) = \int_{\mathrm{F}'} (1 - \cos\langle x, x'\rangle)\, d\mu(x').$$

Now, $0 \leqslant 1 - \cos\langle x, x'\rangle \leqslant 2$ for every $x' \in \mathrm{F}' - \mathrm{K}$, and

$$1 - \cos\langle x, x'\rangle \leqslant \tfrac{1}{2}\langle x, x'\rangle^2 \leqslant \frac{\varepsilon^2}{2}$$

for $x' \in \mathrm{K} \subset \mathrm{U}^\circ$; it follows that

$$0 \leqslant \Phi(0) - \mathscr{R}\Phi(x) \leqslant 2\,\mu(\mathrm{F}' - \mathrm{K}) + \frac{\varepsilon^2}{2}\,\mu(\mathrm{K}) \leqslant 2\varepsilon + \frac{\varepsilon^2}{2}\,\mu(\mathrm{F}').$$

The second member of this inequality tends to 0 with $\varepsilon$; thus $\mathscr{R}\Phi$ is continuous at 0 and the proposition follows from the Cor. of Prop. 8.

## 9. Minlos's lemma

Let T be a finite-dimensional vector space and $\mu$ a bounded measure on T′; we shall identify T with the dual of T′, so that the Fourier transform $\Phi$ of $\mu$ is a function on T. We assume given two positive quadratic forms $h$ and $q$ on T and a number $\varepsilon > 0$. For every real number $r > 0$, we denote by $\mathrm{C}_r$ the set of $x' \in \mathrm{T}'$ such that $\langle x, x'\rangle^2 \leqslant r^2 h(x)$ for all $x \in \mathrm{T}$.

PROPOSITION 10. — *Under the hypothesis* $\Phi(0) - \mathscr{R}\Phi \leqslant \varepsilon + q$, *we have*

$$\mu(\mathrm{T}' - \mathrm{C}_r) \leqslant 3\big(\varepsilon + r^{-2}\,\mathrm{Tr}\,(q/h)\big) \tag{55}$$

*for every* $r > 0$.

One writes $\mathrm{Tr}\,(q/h)$ for the trace of $q$ with respect to $h$ (cf. Annex, No. 1). The formula (55) is trivial when $\mathrm{Tr}\,(q/h)$ is infinite. We assume henceforth that $\mathrm{Tr}\,(q/h)$ is finite, hence that $h(x) = 0$ implies $q(x) = 0$ for $x \in \mathrm{T}$.

Let $a_1, \ldots, a_n$ be elements of T, and D the set of $x' \in \mathrm{T}'$ such that $\sum_{j=1}^{n} \langle a_j, x'\rangle^2 > 1$. For every real $t \geqslant 0$ we have $3(1 - e^{-t/2}) \geqslant 0$, and we even have

$$3(1 - e^{-t/2}) \geqslant 3(1 - e^{-1/2}) \geqslant 3\big(1 - (\tfrac{9}{4})^{-1/2}\big) = 1$$

for $t > 1$, because $e > \frac{9}{4}$. Applying these inequalities to $t = \sum_{j=1}^{n} \langle a_j, x'\rangle^2$, we obtain

$$\text{(56)} \qquad \mu(\mathrm{D}) \leqslant 3 \int_{\mathrm{T}'} \Big(1 - \exp\Big(-\frac{1}{2}\sum_{j=1}^{n} \langle a_j, x'\rangle^2\Big)\Big)\, d\mu(x').$$

Let $\gamma$ be the measure on $\mathbf{R}$ having density $t \mapsto (2\pi)^{-1/2}\, e^{-t^2/2}$ with respect to Lebesgue measure. By Lemma 2 of No. 4,

$$\int_{\mathbf{R}} e^{iut}\, d\gamma(t) = e^{-u^2/2}$$

for all real $u$. Consequently,

$$\text{(57)} \qquad 1 - \exp\Big(-\frac{1}{2}\sum_{j=1}^{n} \langle a_j, x'\rangle^2\Big) = \int \cdots \int \Big(1 - e^{i\sum_{j=1}^{n} \langle a_j, x'\rangle t_j}\Big)\, d\gamma(t_1) \ldots d\gamma(t_n)$$

for all $x' \in \mathrm{T}'$. The function of $x', t_1, \ldots, t_n$ to be integrated in the second member is continuous and is bounded above in absolute value by 2, and the measures $\mu$ and $\gamma$ are bounded; one can therefore integrate the two members of (57) with respect to $d\mu(x')$ and interchange the integrations with respect to $\mu$ and $\gamma$; one obtains

$$\text{(58)} \qquad \int_{\mathrm{T}'} \Big(1 - \exp\Big(-\frac{1}{2}\sum_{j=1}^{n} \langle a_j, x'\rangle^2\Big)\Big)\, d\mu(x') = \int \cdots \int \Big(\Phi(0) - \Phi\Big(\sum_{j=1}^{n} t_j a_j\Big)\Big)\, d\gamma(t_1) \ldots d\gamma(t_n).$$

Since $q$ is a quadratic form on T, there exist real numbers $q_{jk}$ such that

$$q\Big(\sum_{j=1}^{n} t_j a_j\Big) = \sum_{j,k} q_{jk} t_j t_k$$

for $t_1, \ldots, t_n$ real; in particular, $q_{jj} = q(a_j)$ for $1 \leqslant j \leqslant n$. Moreover, the integral $\int_{\mathbf{R}} t^n \, d\gamma(t)$ has the values $1, 0, 1$ for $n = 0, 1, 2$, respectively (No. 4, Lemma 1). From this, one deduces immediately

$$\int \cdots \int \Big(\varepsilon + q\Big(\sum_{j=1}^{n} t_j a_j\Big)\Big) d\gamma(t_1) \ldots d\gamma(t_n) = \varepsilon + \sum_{j=1}^{n} q(a_j). \tag{59}$$

Now, the first member of (58) and $\Phi(0)$ are real numbers; one can therefore replace $\Phi$ by $\mathscr{R}\Phi$ in the second member of (58). The inequality $\Phi(0) - \mathscr{R}\Phi \leqslant \varepsilon + q$ and the formulas (56), (58) and (59) then imply

$$\mu(\mathrm{D}) \leqslant 3\Big(\varepsilon + \sum_{j=1}^{n} q(a_j)\Big). \tag{60}$$

Let us fix the number $r > 0$. Since the quadratic form $h$ is positive, there exist a basis $(e_1, \ldots, e_n)$ of T and an integer $m$ between 0 and $n$ such that

$$h\Big(\sum_{j=1}^{n} t_j e_j\Big) = \sum_{j=1}^{m} t_j^2$$

for $t_1, \ldots, t_n$ real (Annex, No. 1, Prop. 2). It is then immediate that $\mathrm{C}_r$ consists of the $x' \in \mathrm{T}'$ such that

$$\sum_{j=1}^{m} \langle e_j, x' \rangle^2 \leqslant r^2, \qquad \sum_{j=m+1}^{n} \langle e_j, x' \rangle^2 = 0.$$

For every integer $l \geqslant 1$, let $\mathrm{D}_l$ be the set of $x' \in \mathrm{T}'$ satisfying the inequality

$$\sum_{j=1}^{m} \langle r^{-1} e_j, x' \rangle^2 + \sum_{j=m+1}^{n} \langle l e_j, x' \rangle^2 > 1.$$

One sees easily that the sequence $(\mathrm{D}_l)_{l \geqslant 1}$ is increasing with union $\mathrm{T}' - \mathrm{C}_r$, whence

$$\mu(\mathrm{T}' - \mathrm{C}_r) = \lim_{l \to \infty} \mu(\mathrm{D}_l). \tag{61}$$

But by (60),

$$\mu(D_l) \leqslant 3\Big(\varepsilon + \sum_{j=1}^{m} r^{-2} q(e_j) + \sum_{j=m+1}^{n} l^2 q(e_j)\Big); \tag{62}$$

for $j = m+1, \ldots, n$ we have $h(e_j) = 0$, therefore $q(e_j) = 0$. Moreover, $\operatorname{Tr}(q/h) = \sum_{j=1}^{m} q(e_j)$ (Annex, No. 1, Prop. 2). The relation (55) then follows from (61) and (62).

Q.E.D.

## 10. Measures on the dual of a nuclear space

Let F be a locally convex space. Let $\mathscr{T}_s$ be the weak topology $\sigma(F', F)$ on $F'$, and $\mathscr{T}_c$ the topology of uniform convergence on the compact convex subsets of F. By Mackey's theorem (TVS, IV, §1, No. 1, Th. 1) the topologies $\mathscr{T}_s$ and $\mathscr{T}_c$ on $F'$ are compatible with the duality between F and $F'$; the same is therefore true of every locally convex topology $\mathscr{T}$ on $F'$ intermediate to $\mathscr{T}_s$ and $\mathscr{T}_c$. If $\mathscr{T}$ is such a topology, and $F'_{\mathscr{T}}$ denotes the space $F'$ equipped with $\mathscr{T}$, we shall identify F with the dual of $F'_{\mathscr{T}}$. The promeasures on $F'$ are therefore the same for all topologies $\mathscr{T}$ of the preceding type, and if $\mu$ is such a promeasure then its Fourier transform is a function on F.

One calls *Sazonov's topology* on F the locally convex topology $\mathscr{S}$ defined by the continuous seminorms N satisfying the following condition: $N^2$ *is a positive quadratic form on* F *and there exists a continuous positive quadratic form* H *on* F *such that* $\operatorname{Tr}(N^2/H) < +\infty$. The topology $\mathscr{S}$ is coarser than the given topology on F; if these topologies are identical, F is said to be *nuclear*. This class of spaces will be studied in detail later on.

THEOREM 2 (Minlos). — *Let* F *be a locally convex space,* $\mathscr{T}$ *a locally convex topology on* $F'$ *intermediate to* $\mathscr{T}_s$ *and* $\mathscr{T}_c$, *and* $\mu$ *a promeasure on* $F'_{\mathscr{T}}$. *Assume that the Fourier transform* $\Phi$ *of* $\mu$ *is continuous on* F *for the Sazonov topology. Then* $\mu$ *is a measure on* $F'_{\mathscr{T}}$.

Let $\varepsilon > 0$. Since $\Phi$ is continuous for the Sazonov topology on F, there exist two continuous positive quadratic forms Q and H on F such that $\operatorname{Tr}(Q/H) < +\infty$ and

$$\Phi(0) - \mathscr{R}\Phi(x) \leqslant \varepsilon/6$$

for every $x \in F$ such that $Q(x) \leqslant 1$. By Prop. 8 of No. 8, $|\mathscr{R}\Phi(x)| \leqslant \Phi(0)$ for all $x \in F$, whence

$$\Phi(0) - \mathscr{R}\Phi(x) \leqslant \varepsilon/6 + 2\,\Phi(0)Q(x) \tag{63}$$

for all $x \in F$.

Set $r = \big(12\,\Phi(0)\,\mathrm{Tr}(\mathrm{Q}/\mathrm{H})\,\varepsilon^{-1}\big)^{1/2}$ and denote by K the set of $x' \in \mathrm{F}'_{\mathscr{T}}$ such that $\langle x, x'\rangle^2 \leqslant r^2\mathrm{H}(x)$ for all $x \in \mathrm{F}$. Since $\mathrm{H}^{1/2}$ is a continuous semi-norm on F, the set K is equicontinuous and closed in $\mathrm{F}'_{\mathscr{T}}$; it is therefore compact in $\mathrm{F}'_{\mathscr{T}}$ by Ascoli's theorem (GT, X, §2, No. 5, Cor. 1 of Th. 2).

Let V be a closed linear subspace of $\mathrm{F}'_{\mathscr{T}}$ with finite codimension; then, V is the orthogonal of a finite-dimensional linear subspace T of F. Let $\mu_{\mathrm{V}}$ be the measure on T′ that is the image of the promeasure $\mu$ on $\mathrm{F}'_{\mathscr{T}}$ under the mapping $p_{\mathrm{V}}$ that is the transpose of the canonical injection of T into F; its Fourier transform is the restriction of $\Phi$ to T. Finally, by the Hahn–Banach theorem (TVS, II, §3, No. 2, Cor. 1 of Th. 1), $p_{\mathrm{V}}(\mathrm{K})$ is equal to the set $\mathrm{C}_r$ of $x' \in \mathrm{T}'$ such that $\langle x, x'\rangle^2 \leqslant r^2\mathrm{H}(x)$ for all $x \in \mathrm{T}$. By the inequality (63), one can apply Prop. 10 of No. 9 to the measure $\mu_{\mathrm{V}}$ on T′, on taking for $q$ the restriction of $2\Phi(0)\mathrm{Q}$ to T and for $h$ that of H. Then $\mathrm{Tr}(q/h) \leqslant 2\,\Phi(0)\,\mathrm{Tr}\mathrm{Q}/\mathrm{H})$, whence

$$\mu_{\mathrm{V}}(\mathrm{T}' - \mathrm{C}_r) \leqslant 3\Big(\frac{\varepsilon}{6} + 2\,\Phi(0)\,\mathrm{Tr}(\mathrm{Q}/\mathrm{H})\,r^{-2}\Big) = \varepsilon\,.$$

Since $p_{\mathrm{V}}$ defines, by passage to the quotient, an isomorphism of $\mathrm{F}'_{\mathscr{T}}/\mathrm{V}$ onto T′, Prop. 1 of No. 1 then shows that $\mu$ is a measure on $\mathrm{F}'_{\mathscr{T}}$.

Q.E.D.

COROLLARY. — *Let* F *be a barreled nuclear space,* $\mathscr{T}$ *a locally convex topology on* F′ *intermediate to* $\mathscr{T}_s$ *and* $\mathscr{T}_c$, $\mu$ *a promeasure on* $\mathrm{F}'_{\mathscr{T}}$, *and* $\Phi$ *the Fourier transform of* $\mu$. *For* $\mu$ *to be a measure, it is necessary and sufficient that* $\Phi$ *be continuous on* F.

Necessity follows from Prop. 9 of No. 8 and sufficiency from Th. 2.

*Remark.* — Let F be a barreled space and $\mathscr{T}$ a locally convex topology on F′ intermediate to $\mathscr{T}_s$ and $\mathscr{T}_c$. Every subset of F′ compact for $\mathscr{T}$ is compact for the coarser topology $\mathscr{T}_s$. Conversely, let K be a subset of F′ compact for $\mathscr{T}_s$. Since F is barreled, K is equicontinuous (TVS, III, §4, No. 2, Th. 1); but by Ascoli's theorem, every equicontinuous subset of F′ is relatively compact for $\mathscr{T}_c$ and *a fortiori* for $\mathscr{T}$, therefore K is contained in a subset of F′ compact for $\mathscr{T}$. It is not difficult to infer from this that the identity mapping of $\mathrm{F}'_{\mathscr{T}}$ onto $\mathrm{F}'_{\mathscr{T}_s}$ defines a bijection between the sets of measures on these two spaces.

## 11. Measures on a Hilbert space

Let E be a real Hilbert space, in which the scalar product is denoted $(x|y)$. There exists an isomorphism $j$ of E onto its dual E′, characterized by the formula $\langle x, j(y)\rangle = (x|y)$ for $x, y$ in E (TVS, V, §1, No. 7, Th. 3). We will identify E and E′ by means of $j$. The Fourier transform of a promeasure $\mu$ on E is therefore a function $\mathscr{F}\mu$ on E; when $\mu$ is a measure, we have

$$(\mathscr{F}\mu)(x) = \int_{\mathrm{E}} e^{i(x|y)}\, d\mu(y) \qquad (x \in \mathrm{E})\,. \tag{64}$$

THEOREM 3 (Prokhorov–Sazonov) — *Let* E *be a Hilbert space and* $\mathrm{E}_s$ *the space* E *equipped with the weakened topology. Let* $\mu$ *be a promeasure on* E, *and* $\Phi$ *its Fourier transform. The following conditions are equivalent:*

*a) The function* $\Phi$ *is continuous on* E *for the Sazonov topology.*

*b) For every* $\varepsilon > 0$, *there exists a nuclear positive quadratic form* Q *on* E *such that* $\Phi(0) - \mathscr{R}\Phi \leqslant \varepsilon + \mathrm{Q}$.

*c) The promeasure* $\mu$ *is a measure on* $\mathrm{E}_s$.

*b)* $\Rightarrow$ *a)*: This follows from Prop. 8 of No. 8 (cf. the inequality (54)).

*a)* $\Rightarrow$ *c)*: This follows from Theorem 2 of No. 10.

*c)* $\Rightarrow$ *b)*: Suppose that $\mu$ is a measure on $\mathrm{E}_s$. Let $\varepsilon > 0$. For every integer $n \geqslant 1$, the set $\mathrm{B}_n$ of $x \in \mathrm{E}$ with norm $\leqslant n$ is a closed subset of $\mathrm{E}_s$, and $\mathrm{E} = \bigcup_{n\geqslant 1} \mathrm{B}_n$. Therefore there exists an integer $n \geqslant 1$ such that $\mu(\mathrm{E} - \mathrm{B}_n) < \frac{\varepsilon}{2}$. The formula

$$\mathrm{Q}(x) = \frac{1}{2}\int_{\mathrm{B}_n} (x|y)^2\, d\mu(y) \tag{65}$$

defines a positive quadratic form Q on E. Set $\mathrm{C} = \frac{n^2}{2}\mu(\mathrm{B}_n)$. If $(e_1, \dots, e_p)$ is a finite orthonormal sequence in E, then

$$\sum_{j=1}^{p} (e_j|y)^2 \leqslant \|y\|^2 \leqslant n^2$$

for every $y \in \mathrm{B}_n$ by Bessel's inequality. It follows by integration that

$$\sum_{j=1}^{p} \mathrm{Q}(e_j) = \frac{1}{2}\int_{\mathrm{B}_n} \sum_{j=1}^{p} (e_j|y)^2\, d\mu(y) \leqslant \frac{n^2}{2}\,\mu(\mathrm{B}_n) = \mathrm{C},$$

therefore Q is nuclear.

Moreover, $1 - \cos t \leqslant \inf\left(2, \frac{t^2}{2}\right)$ for every real number $t$, whence

$$\begin{aligned}\Phi(0) - \mathscr{R}\Phi(x) &= \int_{\mathrm{E}} \big(1 - \cos(x|y)\big)\, d\mu(y)\\ &\leqslant \int_{\mathrm{B}_n} \tfrac{1}{2}(x|y)^2\, d\mu(y) + \int_{\mathrm{E}-\mathrm{B}_n} 2\cdot d\mu(y)\\ &< \mathrm{Q}(x) + \varepsilon\end{aligned}$$

for all $x \in \mathrm{E}$. Thus $b$) is verified.

Q.E.D.

COROLLARY 1. — *Let* $\mathrm{E}_1$ *and* $\mathrm{E}_2$ *be two Hilbert spaces,* $u$ *a Hilbert–Schmidt mapping of* $\mathrm{E}_1$ *into* $\mathrm{E}_2$, *and* $\mu$ *a promeasure on* $\mathrm{E}_1$. *Assume that the Fourier transform* $\Phi$ *of* $\mu$ *is continuous on* $\mathrm{E}_1$. *Then the promeasure* $\nu = u(\mu)$ *is a measure on* $\mathrm{E}_2$ *equipped with the weak topology.*

With the identifications of $\mathrm{E}_1$ and $\mathrm{E}_2$ with their duals introduced in this No., the Fourier transform of $\nu$ is equal to $\Phi \circ u^*$, where $u^*$ is the adjoint of $u$. Now, $u^*$ is a Hilbert–Schmidt mapping of $\mathrm{E}_2$ into $\mathrm{E}_1$ (Annex, No. 2), and the quadratic form $y \mapsto \|u^*(y)\|^2$ on $\mathrm{E}_2$ is therefore nuclear. If $(\mathrm{E}_2)_{\mathscr{S}}$ denotes $\mathrm{E}_2$ equipped with the Sazonov topology, $u^*$ is therefore a continuous linear mapping of $(\mathrm{E}_2)_{\mathscr{S}}$ into $\mathrm{E}_1$, and $\mathscr{F}_{\mathrm{V}} = \Phi \circ u^*$ is continuous on $(\mathrm{E}_2)_{\mathscr{S}}$; Theorem 3 then shows that $\nu$ is a measure on the space $\mathrm{E}_2$ equipped with the weak topology.

COROLLARY 2. — *Let* $\mathrm{Q}$ *be a nuclear positive quadratic form on the Hilbert space* $\mathrm{E}$. *The Gaussian promeasure* $\Gamma_{\mathrm{Q}}$ *on* $\mathrm{E}$ *with variance* $\mathrm{Q}$ *is a measure on* $\mathrm{E}_s$.

The Fourier transform $\Phi$ of $\Gamma_{\mathrm{Q}}$ is equal to $e^{-\mathrm{Q}/2}$. Now, $e^t \geqslant 1 + t$ for every real number $t$, whence $\Phi(0) - \mathscr{R}\Phi \leqslant \mathrm{Q}/2$. The condition $b$) of Theorem 3 is therefore verified and $\Gamma_{\mathrm{Q}}$ is a measure on $\mathrm{E}_s$.

*Remarks.* — 1) Let $\mathrm{E}$ be a Hilbert space, $\mathrm{E}_s$ the space $\mathrm{E}$ equipped with the weak topology, and $j$ the identity mapping of $\mathrm{E}$ into $\mathrm{E}_s$. One knows that $j$ defines a bijection of the set of promeasures on $\mathrm{E}$ onto the corresponding set for $\mathrm{E}_s$. Moreover, if $\mathrm{E}$ is separable, it is a Polish space and $j$ defines a bijection of the set of bounded measures on $\mathrm{E}$ onto the set of bounded measures on $\mathrm{E}_s$ (§3, No. 3, *Remark*); it can be shown (the theorem of Phillips) that this theorem still holds if $\mathrm{E}$ is not separable. Consequently, Theorem 3 furnishes crieteria for a promeasure on $\mathrm{E}$ to be a measure.

2) One can show (Annex, Exer. 7) that the Sazonov topology on a Hilbert space $\mathrm{E}$ is defined by the semi-norms of the type $\mathrm{Q}^{1/2}$ where $\mathrm{Q}$ is a *nuclear* positive quadratic form on $\mathrm{E}$.

## *12. Relations with functions of positive type

DEFINITION 3. — *Let* $\mathrm{G}$ *be a group. A complex-valued function* $\Phi$ *on* $\mathrm{G}$ *is said to be of positive type if the inequality*

$$\sum_{j,k=1}^{p} c_j \bar{c}_k \Phi(x_j x_k^{-1}) \geqslant 0 \tag{66}$$

*holds for any* $x_1, \ldots, x_p$ *in* $\mathrm{G}$ *and any complex numbers* $c_1, \ldots, c_p$.

This concept will be studied in detail later on.

PROPOSITION 11. — *Let* E *be a finite-dimensional vector space,* $\mu$ *a bounded (positive) measure on* E, *and* $\Phi$ *the Fourier transform of* $\mu$. *The function* $\Phi$ *is continuous and of positive type on* E'.

The continuity of $\Phi$ follows from Prop. 9 of No. 8.

Let us show that $\Phi$ is of positive type. Let $x'_1, \ldots, x'_p$ be in E' and $c_1, \ldots, c_p$ complex numbers. Then

$$\sum_{j,k} c_j \bar{c}_k \Phi(x'_j - x'_k) = \int_{\mathrm{E}} \sum_{j,k} c_j \bar{c}_k e^{i\langle x, x'_j - x'_k\rangle} \, d\mu(x)$$
$$= \int_{\mathrm{E}} \left| \sum_{j=1}^{p} c_j e^{i\langle x, x'_j\rangle} \right|^2 d\mu(x) \geqslant 0 .$$

Q.E.D.

One can prove a converse known as *Bochner's theorem*: every continuous function on E' of positive type is the Fourier transform of a bounded (positive) measure.(*) We shall assume this result for the rest of No. 12.

THEOREM 4. — *Let* E *be a locally convex space. The Fourier transformation is a bijection of the set of promeasures on* E *onto the set of functions of positive type on* E' *whose restriction to every finite-dimensional subspace is continuous.*

We know (No. 3, Prop. 3) that the Fourier transformation is injective. Let $\mu = (\mu_{\mathrm{V}})_{\mathrm{V} \in \mathscr{F}(\mathrm{E})}$ be a promeasure on E and $\Phi$ its Fourier transform. Let T be a finite-dimensional subspace of E' and let V be the orthogonal of T in E. One can identify T with the dual of E/V; the restriction $\Phi_{\mathrm{T}}$ of $\Phi$ to T is the Fourier transform of the bounded measure $\mu_{\mathrm{V}}$ on E/V. By Prop. 11, $\Phi_{\mathrm{T}}$ is continuous and of positive type on T. Since T is arbitrary, it is clear that $\Phi$ is of positive type on E'.

Conversely, let $\Phi$ be a function of positive type on E' whose restriction to every finite-dimensional subspace of E' is continuous. For every $\mathrm{V} \in \mathscr{F}(\mathrm{E})$, we identify the dual of E/V with the orthogonal $\mathrm{V}^\circ$ of V in E'; the restriction $\Phi_{\mathrm{V}}$ of $\Phi$ to $\mathrm{V}^\circ$ is continuous and of positive type and so, by Bochner's theorem, there exists a bounded (positive) measure $\mu_{\mathrm{V}}$ on E/V whose Fourier transform is $\Phi_{\mathrm{V}}$. Let V and W in $\mathscr{F}(\mathrm{E})$ be such that $\mathrm{W} \subset \mathrm{V}$, and let $p_{\mathrm{VW}}$ be the canonical mapping of E/W onto E/V; with the identifications made, ${}^t p_{\mathrm{VW}}$ is the injection of $\mathrm{V}^\circ$ into $\mathrm{W}^\circ$. By formula (4) of No. 3, we then have

$$\mathscr{F}(p_{\mathrm{VW}}(\mu_{\mathrm{W}})) = (\mathscr{F}\mu_{\mathrm{W}}) \circ {}^t p_{\mathrm{VW}} = \Phi_{\mathrm{W}} \circ {}^t p_{\mathrm{VW}} = \Phi_{\mathrm{V}} = \mathscr{F}\mu_{\mathrm{V}} ,$$

(*) This question will be studied in a forthcoming chapter of the Book *Théories spectrales*. The reader may consult for this subject the book of L.H. LOOMIS, *Abstract harmonic analysis*, Van Nostrand, New York, 1953.

whence $p_{\mathrm{VW}}(\mu_{\mathrm{W}}) = \mu_{\mathrm{V}}$ by Prop. 3 of No. 3. Consequently, the family $\mu = (\mu_{\mathrm{V}})_{\mathrm{V} \in \mathscr{F}(\mathrm{E})}$ is a promeasure on E; it is clear that $\Phi$ is the Fourier transform of $\mu$.

COROLLARY. — *Let* F *be a barreled nuclear space; equip* $\mathrm{F}'$ *with a locally convex topology* $\mathscr{T}$ *intermediate to the weak topology* $\sigma(\mathrm{F}', \mathrm{F})$ *and the topology of uniform convergence on the compact convex subsets of* F. *The Fourier transformation is a bijection of the set of bounded (positive) measures on* $\mathrm{F}'$ *onto the set of continuous functions of positive type on* F.

This follows immediately from Theorem 4 and the Cor. of Th. 2 of No. 10.$_*$

ANNEX

# Complements on Hilbert spaces

## 1. Trace of a quadratic form with respect to another[1]

In this No., we denote by E a real vector space and by Q, H two positive quadratic forms on E. There exist two symmetric bilinear forms $(x,y) \mapsto (x|y)_{\mathrm{Q}}$ and $(x,y) \mapsto (x|y)_{\mathrm{H}}$ characterized by

$$\mathrm{Q}(x) = (x|x)_{\mathrm{Q}}\,, \qquad \mathrm{H}(x) = (x|x)_{\mathrm{H}}$$

for all $x \in \mathrm{E}$.

One calls *trace of* Q *with respect to* H and one denotes by $\mathrm{Tr}\,(\mathrm{Q}/\mathrm{H})$ the positive real number, finite or not, defined as follows:

*a*) If there exists an $x \in \mathrm{E}$ such that $\mathrm{H}(x) = 0$ and $\mathrm{Q}(x) \neq 0$, one sets $\mathrm{Tr}\,(\mathrm{Q}/\mathrm{H}) = +\infty$.

*b*) In the contrary case, $\mathrm{Tr}\,(\mathrm{Q}/\mathrm{H})$ is the supremum of the set of numbers of the form $\sum_{i=1}^{p} \mathrm{Q}(e_i)$, where $(e_1, \dots, e_p)$ runs over the set of finite sequences of elements of E orthonormal for H.

Let E be a real Hilbert space and Q a positive quadratic form on E. Set $\mathrm{H}(x) = \|x\|^2$ for all $x \in \mathrm{E}$; then H is a positive quadratic form on E. One says that Q is *nuclear* if $\mathrm{Tr}\,(\mathrm{Q}/\mathrm{H})$ is finite. For every $x \in \mathrm{E}$ of norm 1, one has $\mathrm{Q}(x) \leqslant \mathrm{Tr}\,(\mathrm{Q}/\mathrm{H})$, whence $\mathrm{Q} \leqslant \mathrm{Tr}\,(\mathrm{Q}/\mathrm{H}) \cdot \mathrm{H}$; in particular, every nuclear form Q is continuous.

*Remarks.* — 1) For every subspace F of E, denote by $\mathrm{Q_F}$ the restriction of Q to F, and by $\mathrm{H_F}$ that of H. Then $\mathrm{Tr}\,(\mathrm{Q_F}/\mathrm{H_F}) \leqslant \mathrm{Tr}\,(\mathrm{Q}/\mathrm{H})$ and $\mathrm{Tr}\,(\mathrm{Q}/\mathrm{H})$ is the supremum of the numbers $\mathrm{Tr}\,(\mathrm{Q_F}/\mathrm{H_F})$ for $\mathrm{F} \subset \mathrm{E}$ finite-dimensional.

2) Let $\mathrm{E}_1$ be a real vector space, $\mathrm{Q}_1$ and $\mathrm{H}_1$ two positive quadratic forms on $\mathrm{E}_1$, and $\pi : \mathrm{E} \to \mathrm{E}_1$ a surjective linear mapping. If $\mathrm{Q} = \mathrm{Q}_1 \circ \pi$ and $\mathrm{H} = \mathrm{H}_1 \circ \pi$, then $\mathrm{Tr}\,(\mathrm{Q}/\mathrm{H}) = \mathrm{Tr}(\mathrm{Q}_1/\mathrm{H}_1)$.

---

[1] Cf. TVS, V, §4, No. 9.

PROPOSITION 1. — *Assume that* E *is finite-dimensional and* H *is nondegenerate.*

*a) There exists an endomorphism* $u$ *of* E *characterized by* $(x|y)_{\mathrm{Q}} = (u(x)|y)_{\mathrm{H}}$ *for* $x, y$ *in* E.

*b)* $\mathrm{Tr}\,(\mathrm{Q}/\mathrm{H}) = \mathrm{Tr}\,(u)$.

*c)* $\mathrm{Tr}\,(\mathrm{Q}/\mathrm{H}) = \sum_{i=1}^{m} \mathrm{Q}(e_i)$ *for every basis* $(e_1, \ldots, e_m)$ *of* E *orthonormal for* H.

*a)* follows from the fact that the bilinear form $(x, y) \mapsto (x|y)_{\mathrm{H}}$ is nondegenerate. Every sequence in E orthonormal for H may be completed to a basis of E orthonormal for H. Consequently, $\mathrm{Tr}\,(\mathrm{Q}/\mathrm{H})$ is the supremum of the set of numbers of the form $\sum_{i=1}^{m} \mathrm{Q}(e_i)$ over all bases $(e_1, \ldots, e_m)$ of E orthonormal for H. To prove *b)* and *c)*, it suffices to show that $\sum_{i=1}^{m} \mathrm{Q}(e_i) = \mathrm{Tr}\,(u)$ for every basis of this kind. Set $a_{ij} = (u(e_i)|e_j)_{\mathrm{H}} = (e_i|e_j)_{\mathrm{Q}}$ for $1 \leqslant i, j \leqslant m$; then $u(e_i) = \sum_{j=1}^{m} a_{ij} e_j$ for $1 \leqslant i \leqslant m$, whence

$$\mathrm{Tr}\,(u) = \sum_{i=1}^{m} a_{ii} = \sum_{i=1}^{m} (e_i|e_i)_{\mathrm{Q}} = \sum_{i=1}^{m} \mathrm{Q}(e_i).$$

Q.E.D.

PROPOSITION 2. — *Assume that* E *is finite-dimensional. There exist a basis* $(e_1, \ldots, e_n)$ *of* E *and an integer* $m$ *with* $0 \leqslant m \leqslant n$ *such that*

$$\mathrm{H}\Big(\sum_{i=1}^{n} t_i e_i\Big) = \sum_{i=1}^{m} t_i^2 \tag{1}$$

*for* $t_1, \ldots, t_n$ *real. If, moreover, the relation* $\mathrm{H}(x) = 0$ *implies* $\mathrm{Q}(x) = 0$ *for* $x \in \mathrm{E}$, *then* $\mathrm{Tr}\,(\mathrm{Q}/\mathrm{H}) = \sum_{i=1}^{m} \mathrm{Q}(e_i)$.

There exists a basis $(e'_1, \ldots, e'_n)$ of E orthogonal for H. One can assume the basis to be indexed in such a way that $\mathrm{H}(e'_i) > 0$ for $1 \leqslant i \leqslant m$ and $\mathrm{H}(e'_i) = 0$ for $m < i \leqslant n$. Then set $e_i = e'_i/\mathrm{H}(e'_i)^{1/2}$ for $1 \leqslant i \leqslant m$ and $e_i = e'_i$ for $m < i \leqslant n$; the relation (1) is satisfied.

Let F be the subspace of E generated by $e'_{m+1}, \ldots, e'_n$; it is the set of $x \in \mathrm{E}$ such that $\mathrm{H}(x) = 0$. Denote by $\pi$ the canonical mapping of E onto $\mathrm{E}_1 = \mathrm{E}/\mathrm{F}$. Since Q and H are zero on F, there exist two positive quadratic forms $\mathrm{Q}_1$ and $\mathrm{H}_1$ on $\mathrm{E}_1$ such that $\mathrm{Q} = \mathrm{Q}_1 \circ \pi$ and $\mathrm{H} = \mathrm{H}_1 \circ \pi$. Moreover, $(\pi(e_1), \ldots, \pi(e_m))$ is a basis of $\mathrm{E}_1$ orthonormal for $\mathrm{H}_1$, and therefore $\mathrm{H}_1$ is nondegenerate.

By Prop. 1 and *Remark* 2,

$$\operatorname{Tr}(Q/H) = \operatorname{Tr}(Q_1/H_1) = \sum_{i=1}^{m} Q_1\big(\pi(e_i)\big) = \sum_{i=1}^{m} Q(e_i).$$

Q.E.D.

*Remark* 3). — Assume E finite-dimensional and H nondegenerate. Let $(e_1, \dots, e_n)$ be a basis of E. Set $q = \big((e_i|e_j)_Q\big)_{1\leqslant i,j\leqslant n}$ and $h = \big((e_i|e_j)_H\big)_{1\leqslant i,j\leqslant n}$. With notations as in Prop. 1, the matrix of $u$ for the basis $(e_1, \dots, e_n)$ of E is equal to $h^{-1}q$, whence

$$\text{(2)} \qquad \operatorname{Tr}(Q/H) = \operatorname{Tr}(h^{-1}q) = \operatorname{Tr}(qh^{-1}).$$

## 2. Hilbert–Schmidt mappings[2]

Let E be a real Hilbert space, in which the scalar product is denoted by $(x|y)$. There exists an isometry $j_E$ of E onto its dual, characterized by the formula

$$\text{(3)} \qquad (x|y) = \langle x, j_E(y)\rangle \qquad \text{for } x, y \text{ in E}$$

(TVS, V, §1, No. 7, Th. 3).

Let $E_1$ and $E_2$ be two real Hilbert spaces and $u$ a continuous linear mapping of $E_1$ into $E_2$. One calls *adjoint of* $u$ the continuous linear mapping $u^* = j_{E_1}^{-1} \circ {}^t u \circ j_{E_2}$ of $E_2$ into $E_1$. The mapping $u^*$ is characterized by the relation

$$\text{(4)} \qquad (u(x_1)|x_2) = (x_1|u^*(x_2)) \qquad \text{for } x_1 \in E_1,\ x_2 \in E_2.$$

If $v$ is a continuous linear mapping of $E_2$ into a Hilbert space $E_3$, one has $(v \circ u)^* = u^* \circ v^*$.

Let $E_1$ and $E_2$ be two real Hilbert spaces and $u$ a linear mapping of $E_1$ into $E_2$. One defines two positive quadratic forms H and Q on $E_1$ by the formulas

$$H(x) = \|x\|^2, \quad Q(x) = \|u(x)\|^2 \qquad (x \in E_1).$$

PROPOSITION 3. — *Suppose* $u$ *is continuous. Let* $(e_i)_{i\in I}$ *be an orthonormal basis of* $E_1$ *and* $(f_j)_{j\in J}$ *an orthonormal basis of* $E_2$. *Then*

$$\operatorname{Tr}(Q|H) = \sum_{i\in I} \|u(e_i)\|^2 = \sum_{j\in J} \|u^*(f_j)\|^2 = \sum_{i\in I}\sum_{j\in J} (u(e_i)|f_j)^2.$$

[2] Cf. TVS, V, §4, No. 7.

For every $x \in E_1$, we have $\|x\|^2 = \sum\limits_{i \in I}(x|e_i)^2$, and similarly $\|y\|^2 = \sum\limits_{j \in J}(y|f_j)^2$ for every $y \in E_2$. Consequently,

$$\begin{aligned}\sum_{i\in I}\|u(e_i)\|^2 &= \sum_{i\in I}\sum_{j\in J}(u(e_i)|f_j)^2 \\ &= \sum_{j\in J}\sum_{i\in I}(e_i|u^*(f_j))^2 \\ &= \sum_{j\in J}\|u^*(f_j)\|^2 .\end{aligned}$$

In particular, the number $\sum\limits_{i\in I}\|u(e_i)\|^2$ is independent of the orthonormal basis $(e_i)_{i\in I}$ of $E_1$.

Set $t = \mathrm{Tr}\,(Q|H)$. For every finite subset $I'$ of $I$, by definition

$$\sum_{i\in I'}\|u(e_i)\|^2 = \sum_{i\in I'} Q(e_i) \leqslant t ,$$

whence $\sum\limits_{i\in I}\|u(e_i)\|^2 \leqslant t$. Let $(e'_1, \ldots, e'_p)$ be a finite orthonormal sequence in E. This sequence can be completed to an orthonormal basis $(e'_\alpha)_{\alpha\in A}$ of $E_1$. Then

$$\sum_{\alpha=1}^{p}\|u(e'_\alpha)\|^2 \leqslant \sum_{\alpha\in A}\|u(e'_\alpha)\|^2 = \sum_{i\in I}\|u(e_i)\|^2$$

and, passing to the supremum over all $(e'_1, \ldots, e'_p)$, one obtains the inequality $t \leqslant \sum\limits_{i\in I}\|u(e_i)\|^2$. We have thus established the equality $t = \sum\limits_{i\in I}\|u(e_i)\|^2$.

Q.E.D.

One says that $u$ is a *Hilbert–Schmidt mapping* of $E_1$ into $E_2$ if the positive quadratic form $Q : x \mapsto \|u(x)\|^2$ on $E_1$ is nuclear. When this is so, one has $Q \leqslant \mathrm{Tr}\,(Q/H)\cdot H$, therefore $u$ is continuous and

$$\|u\| \leqslant \mathrm{Tr}\,(Q/H)^{1/2} .$$

Let $u : E_1 \to E_2$ be a continuous linear mapping. By Prop. 3, $u$ is a Hilbert–Schmidt mapping if and only if there exists an orthonormal basis $(e_i)_{i\in I}$ of $E_1$ such that $\sum\limits_{i\in I}\|u(e_i)\|^2 < +\infty$. When this is so, every orthonormal basis of $E_1$ has the same property. Moreover, if $u$ is a Hilbert–Schmidt mapping, then so is its adjoint $u^*$ by virtue of the formula $\sum\limits_{i\in I}\|u(e_i)\|^2 = \sum\limits_{j\in J}\|u^*(f_j)\|^2$ (Prop. 3).

# Exercises

## §1

1) Let T be a set and $p$ an encumbrance on T. For every subset A of T, set $p(\mathrm{A}) = p(\varphi_{\mathrm{A}})$.

*a*) $p(\mathrm{A}) \leqslant p(\mathrm{B})$ when A is contained in B.

*b*) For every sequence $(\mathrm{A}_n)$, finite or not, of subsets of T, $p(\bigcup\limits_n \mathrm{A}_n) \leqslant \sum\limits_n p(\mathrm{A}_n)$.

*c*) For every increasing sequence $(\mathrm{A}_n)_{n\geqslant 1}$ of subsets of T, $p(\bigcup\limits_n \mathrm{A}_n) = \lim\limits_{n\to\infty} p(\mathrm{A}_n)$.

2) Let T be a set and $p$ an encumbrance on T. A function $f \in \mathscr{F}_+(\mathrm{T})$ (resp. a subset A of T) is said to be $p$-negligible if $p(f) = 0$ (resp. $p(\mathrm{A}) = 0$).

*a*) Let $f$ and $g$ be in $\mathscr{F}_+(\mathrm{T})$ and let $t$ be a positive real number such that $f \leqslant t\cdot g$. If $g$ is $p$-negligible, then so is $f$. The sum and upper envelope of a sequence of $p$-negligible functions are $p$-negligible.

*b*) Every subset of a $p$-negligible set is $p$-negligible, as is the union of a sequence (finite or infinite) of $p$-negligible sets.

*c*) Let $f \in \mathscr{F}_+(\mathrm{T})$. For every finite real number $a > 0$, denote by $\mathrm{T}_a$ the set of $t \in \mathrm{T}$ such that $f(t) \geqslant a$. Show that $p(\mathrm{T}_a) \leqslant a^{-1} \cdot p(f)$. From this, deduce that $f$ is $p$-negligible if and only if the set of $t \in \mathrm{T}$ such that $f(t) \neq 0$ is $p$-negligible.

*d*) Let $f \in \mathscr{F}_+(\mathrm{T})$. If $p(f)$ is finite, the set of $t \in \mathrm{T}$ such that $f(t)$ is infinite is $p$-negligible.

*e*) Let $f$ and $g$ be in $\mathscr{F}_+(\mathrm{T})$. If the set of $t \in \mathrm{T}$ such that $f(t) \neq g(t)$ is $p$-negligible, then $p(f) = p(g)$.

3) Let T be a set and $p$ an encumbrance on T. For every sequence $(f_n)_{n\geqslant 1}$ of elements of $\mathscr{F}_+(\mathrm{T})$, $p(\liminf\limits_{n\to\infty} f_n) \leqslant \liminf\limits_{n\to\infty} p(f_n)$.

4) Let T be a set, M an encumbrance on T, E a real or complex Banach space, and $p$ a finite real number such that $p \geqslant 1$. Denote by $\mathscr{F}$ the set of mappings of T into E, and, for $f$ in $\mathscr{F}$ or in $\mathscr{F}_+(\mathrm{T})$, set $\mathrm{N}_p(f) = \mathrm{M}(|f|^p)^{1/p}$.

a) For every sequence $(f_n)_{n\geqslant 1}$ of elements of $\mathscr{F}_+(\mathrm{T})$, $\mathrm{N}_p\Big(\sum_{n\geqslant 1} f_n\Big) \leqslant \sum_{n\geqslant 1} \mathrm{N}_p(f_n)$.

b) Let $\mathscr{F}^p$ be the set of $f \in \mathscr{F}$ such that $\mathrm{N}_p(f)$ is finite. Show that $\mathscr{F}^p$ is a linear subspace of $\mathscr{F}$, and $\mathrm{N}_p$ is a semi-norm on $\mathscr{F}^p$. Show that the set $\mathscr{N}$ of $f \in \mathscr{F}^p$ such that $\mathrm{N}_p(f) = 0$ is equal to the set of $f \in \mathscr{F}$ that are zero outside an M-negligible set. Show that the normed space $\mathscr{F}^p/\mathscr{N}$ is complete (cf. Ch. IV, §3, No. 3).

5) Let T be a set. Denote by $\mathscr{B}_+(\mathrm{T})$ the set of bounded positive real functions on T, and by $p_0$ a mapping of $\mathscr{B}_+(\mathrm{T})$ into the interval $[0,+\infty]$ of $\overline{\mathbf{R}}$ satisfying the conditions a) to d) of Def. 1 of No. 1. For every function $f \in \mathscr{F}_+(\mathrm{T})$, denote by $p(f)$ the supremum of the set of numbers $p_0(f_0)$, where $f_0$ runs over the set of bounded functions $\leqslant f$. Show that $p$ is an encumbrance on T. Analogous question, with $\mathscr{B}_+(\mathrm{T})$ replaced by the set of (finite) positive real functions on T.

6) Let T be a set, $\mathscr{I}_+$ a subset of $\mathscr{F}_+(\mathrm{T})$, and M a mapping of $\mathscr{I}_-$ into $\overline{\mathbf{R}}_+$. We make the following hypotheses:

a) For every $f \in \mathscr{I}_+$ and every real number $t \geqslant 0$, the function $t\cdot f$ belongs to $\mathscr{I}_+$ and $\mathrm{M}(t\cdot f) = t\cdot \mathrm{M}(f)$.

b) If $f$ and $g$ belong to $\mathscr{I}_+$, then so do $f+g$, $\sup(f,g)$ and $\inf f,g)$, and one has $\mathrm{M}(f+g) = \mathrm{M}(f) + \mathrm{M}(g)$.

c) For every increasing sequence $(f_n)_{n\geqslant 1}$ of elements of $\mathscr{I}_+$, the function $f = \sup_n f_n$ belongs to $\mathscr{I}_+$, and $\mathrm{M}(f) = \lim_{n\to\infty} \mathrm{M}(f_n)$.

For every function $f \in \mathscr{F}_+(\mathrm{T})$, we set $p(f) = \inf_{g\in\mathscr{I}_+,\, g\geqslant f} \mathrm{M}(g)$. Show that $p$ is an encumbrance on T (cf. Ch. IV, §1, No. 3). If $f$ and $g$ belong to $\mathscr{F}_+(\mathrm{T})$ and if A and B are subsets of T, then $p(\sup(f,g))+p(\inf(f,g)) \leqslant p(f)+p(g)$ and $p(\mathrm{A}\cup\mathrm{B})+p(\mathrm{A}\cap\mathrm{B}) \leqslant p(\mathrm{A}) + p(\mathrm{B})$.

7) On a Lindelöf space every measure, every function and every subset is moderated.

¶ 8) Let us take up again the notations of Exer. 5 of Ch. IV, §1. Recall that $\mu = \sum_{k=-\infty}^{\infty} \sum_{n=1}^{\infty} n^{-3}\varepsilon_{(1/n,k/n^2)}$ and $\mu^*(\mathrm{D}) = +\infty$. Show that $\mu^\bullet(\mathrm{D}) = 0$; deduce from this that $\mu$ is not moderated, even though it is the sum of a series of measures with pairwise disjoint compact supports. Show that there exists a partition $(\mathrm{X}_n)_{n\in\mathbf{N}}$ of X into Borel subsets $\mathrm{X}_n$ such that $\mu^\bullet(\mathrm{X}_n)$ is finite for every $n \in \mathbf{N}$.

9) Let T be a topological space. If $\lambda$ and $\mu$ are two positive measures on T such that $\lambda^\bullet(\mathrm{U}) = \mu^\bullet(\mathrm{U})$ for every open set U in T, then $\lambda = \mu$ (first prove that $\lambda^\bullet(f) = \mu^\bullet(f)$ for every lower semi-continuous function $f \geqslant 0$ on T, then that $\lambda^\bullet = \mu^\bullet$).

10) Let T be a topological space and $\mu$ a positive measure on T; assume that T is the union of a sequence of $\mu$-integrable sets $\mathrm{T}_n$ $(n \geqslant 0)$. Denote by $\mathfrak{C}$ the clan of subsets of T generated by the compact subsets.

a) Let X be a Lusin space. For every $\mu$-measurable mapping $f$ of T into X, there exists a sequence of mappings $f_m : \mathrm{T} \to \mathrm{X}$ having the following properties: $\alpha$) for $\mu$-almost every $t \in \mathrm{T}$, $f(t) = \lim_{m\to\infty} f_m(t)$; $\beta$) for every integer $m$, there exists a finite partition of T into sets $\mathrm{A}_j \in \mathfrak{C}$ such that $f_m$ is constant on each of the sets $\mathrm{A}_j$ (reduce to the case that the sets $\mathrm{T}_n$ form a partition of T, where $\mathrm{T}_0$ is $\mu$-negligible and $\mathrm{T}_n$ is compact for $n \geqslant 1$, and where X is a Polish space).

b) Let $\mathrm{X}_1, \ldots, \mathrm{X}_p$ be Lusin spaces, Y a topological space, $g$ a mapping of $\mathrm{X}_1 \times\cdots\times \mathrm{X}_p$ into Y, and $f_i$ (for $1 \leqslant i \leqslant p$) a $\mu$-measurable mapping of T into $\mathrm{X}_i$. Assume that, for $1 \leqslant i \leqslant p$, the partial mapping $x_i \mapsto g(x_1, \ldots, x_p)$ of $\mathrm{X}_i$ into Y is continuous for any $x_1, \ldots, x_{i-1}, x_{i+1}, \ldots, x_p$. Show that the mapping $t \mapsto g(f_1(t), \ldots, f_p(t))$ of T into Y is $\mu$-measurable.

## §2

1) Let $T_1$ and $T_2$ be two topological spaces, and $f$ an *injective* continuous mapping of $T_1$ into $T_2$. Show that the mapping $\mu \mapsto f(\mu)$ is a bijection of $\mathscr{M}^b(T_1)$ onto the subspace of $\mathscr{M}^b(T_2)$ consisting of the measures carried by $f(T_1)$.

2) Give an example of two topological spaces $T_1$ and $T_2$ and a bijective continuous mapping $f$ of $T_1$ onto $T_2$, such that the mapping $\mu \mapsto f(\mu)$ of $\mathscr{M}_+^b(T_1)$ into $\mathscr{M}_+^b(T_2)$ is not surjective.

3) Let X be a set, and $\mathscr{T}$, $\mathscr{T}'$ two Hausdorff topologies on X, with $\mathscr{T}'$ coarser than $\mathscr{T}$. Denote by T (resp. T′) the set X equipped with the topology $\mathscr{T}$ (resp. $\mathscr{T}'$), and by $j$ the identity mapping of T into T′ (which is continuous). Assume that every bounded positive measure on T′ is carried by the union of a sequence of compact subsets of T. For every topology $\mathscr{S}$ intermediate to $\mathscr{T}$ and $\mathscr{T}'$, denote by $T_{\mathscr{S}}$ the set X equipped with the topology $\mathscr{S}$. Show that the set of bounded measures on $T_{\mathscr{S}}$ is independent of $\mathscr{S}$.

## §3

1) Let E be a set, $\Phi$ a clan of subsets of E, and $m$ an additive set function defined on $\Phi$. Denote by $\mathscr{E}(\Phi)$ the vector space of real $\Phi$-step functions on E, and by J the linear form on $\mathscr{E}(\Phi)$ such that $J(\varphi_A) = m(A)$ for all $A \in \Phi$.

*a*) For every $A \in \Phi$, one sets

$$m_+(A) = \sup_{B\in\Phi,\, B\subset A} m(B) \quad \text{and} \quad m_-(A) = \sup_{B\in\Phi,\, B\subset A} (-m(B)).$$

Show that $|m|(A) = m_+(A) + m_-(A)$ is the supremum of the set of numbers of the form $\sum_{i=1}^{p} |m(A_i)|$ for all the finite partitions $(A_i)_{1\leqslant i\leqslant p}$ of A into sets belonging to $\Phi$.

*b*) One says that $m$ is *relatively bounded* if the set of numbers of the form $m(B)$ with $B \in \Phi$, $B \subset A$ is bounded for any set $A \in \Phi$. Show that the following conditions are equivalent: $\alpha$) $m$ is relatively bounded; $\beta$) the linear form J on the Riesz space $\mathscr{E}(\Phi)$ is relatively bounded; $\gamma$) $m$ is the difference of two positive additive set functions on $\Phi$; $\delta$) $|m|(A)$ is finite for all $A \in \Phi$ (argue in a circle $\alpha) \Rightarrow \beta) \Rightarrow \gamma) \Rightarrow \delta) \Rightarrow \alpha)$).

*c*) Suppose that $m$ is relatively bounded. Show that the set functions $|m|$, $m_+$ and $m_-$ are additive and that $m = m_+ - m_-$ (make use of Th. 1 of Ch. II, §2, No. 2). Moreover, if $m'$ and $m''$ are additive and positive set functions on $\Phi$ such that $m = m' - m''$, then $m' \geqslant m_+$ and $m'' \geqslant m_-$.

*d*) One says that $m$ is *countably additive* if $m(A) = \sum_{n\geqslant 1} m(A_n)$ for every countable partition $(A_n)_{n\geqslant 1}$ of a set $A \in \Phi$ into sets belonging to $\Phi$. Show that this condition signifies that $\lim_{n\to\infty} m(A_n) = 0$ for every decreasing sequence $(A_n)_{n\geqslant 1}$ of elements of $\Phi$ such that $\bigcap_{n\geqslant 1} A_n = \varnothing$. If $m$ is countably additive, show that $|m|$ is countably additive.

2) Let E be a set, $\Phi$ a clan of subsets of E, and V a fully lattice-ordered Riesz space. A mapping $m$ of $\Phi$ into V is said to be positive if $m(A) \geqslant 0$ for all $A \in \Phi$, that it is additive if $m(A \cup B) = m(A) + m(B)$ when the sets $A \in \Phi$ and $B \in \Phi$ are disjoint, and that it is relatively bounded if the set of elements $m(B)$ for $B \in \Phi$, $B \subset A$ is bounded above in V for any $A \in \Phi$.

*a*) Let $m$ be a relatively bounded mapping of $\Phi$ into V. One defines for every A the elements $m_+(A)$, $m_-(A)$ and $|m|(A)$ of V as in Exer. 1 *a*). Show that $|m|(A)$

is the supremum of the set of elements of V of the form $\sum_{i=1}^{p} |m(A_i)|$ for all the finite partitions $(A_i)_{1\leqslant i\leqslant p}$ of A into sets belonging to $\Phi$. From this, deduce that the mapping $|m|$ of $\Phi$ into V is additive, then that $m$ is the difference of the additive positive mappings $m_+$ and $m_-$ of $\Phi$ into V.

*b*) Let $m'$ and $m''$ be two additive positive mappings of $\Phi$ into V and let $m = m' - m''$. Show that $m$ is relatively bounded, and that $m' \geqslant m_+$ and $m'' \geqslant m_-$.

*c*) Generalize Exer. 1 *d*).

3) Let E be a set and $\mathfrak{T}$ a tribe of subsets of E. Denote by $\mathscr{M}$ the set of mappings $f$ of E into $\overline{\mathbf{R}}_+$ such that the set of $x \in E$ for which $f(x) \geqslant c$ belongs to $\mathfrak{T}$ for every $c \in \overline{\mathbf{R}}_+$.

*a*) Show that the limit superior and limit inferior of every sequence of elements of $\mathscr{M}$ belong to $\mathscr{M}$ (first study sequences that are increasing or decreasing).

*b*) Show that $\mathfrak{T}$ is the set of subsets of E whose characteristic function belongs to $\mathscr{M}$.

*c*) Let $f \in \mathscr{M}$. For every Borel subset B of $\overline{\mathbf{R}}_+$, the set $\overset{-1}{f}(B)$ belongs to $\mathfrak{T}$ (the set of subsets B of $\overline{\mathbf{R}}_+$ such that $\overset{-1}{f}(B) \in \mathfrak{T}$ is a tribe, containing the intervals $[c, +\infty]$).

*d*) Let $f_1, \ldots, f_n$ be in $\mathscr{M}$ and let $\varphi$ be a Borel mapping of $(\overline{\mathbf{R}}_+)^n$ into $\overline{\mathbf{R}}_+$; show that the mapping $x \mapsto \varphi(f_1(x), \ldots, f_n(x))$ of E into $\overline{\mathbf{R}}_+$ belongs to $\mathscr{M}$. Deduce from this that $\mathscr{M}$ contains the sum of every series of elements of $\mathscr{M}$.

*e*) Show that $\mathscr{M}$ is the set of limits of increasing sequences of finite positive $\mathfrak{T}$-step functions.

4) Notations and hypotheses are those of the preceding exercise. One calls *abstract measure* on $(E, \mathfrak{T})$ every mapping $m$ of $\mathfrak{T}$ into $\overline{\mathbf{R}}_+$ having the following property: for every sequence $(A_n)_{n\in\mathbf{N}}$ of sets belonging to $\mathfrak{T}$, pairwise disjoint, one has $m(\bigcup_{n\in\mathbf{N}} A_n) = \sum_{n\in\mathbf{N}} m(A_n)$. One calls *integral* on $(E, \mathfrak{T})$ every mapping J of $\mathscr{M}$ into $\overline{\mathbf{R}}_+$ satisfying the following conditions: $\alpha$) $J(\lambda \cdot f) = \lambda \cdot J(f)$ for $\lambda \in \overline{\mathbf{R}}_+$ and $f \in \mathscr{M}$ (with the usual convention $0 \cdot (+\infty) = 0$); $\beta$) $J(\sum_{n\in\mathbf{N}} f_n) = \sum_{n\in\mathbf{N}} J(f_n)$ for every sequence $(f_n)_{n\in\mathbf{N}}$ of elements of $\mathscr{M}$.

*a*) Let J be an integral on $(E, \mathfrak{T})$; for every subset $A \in \mathfrak{T}$, set $m_J(A) = J(\varphi_A)$; show that $m_J$ is an abstract measure and that the mapping $J \mapsto m_J$ is a bijection of the set of integrals on $(E, \mathfrak{T})$ onto the set of abstract measures on $(E, \mathfrak{T})$ (if $m$ is an abstract measure, first define $J(f)$ by linearity for the finite positive $\mathfrak{T}$-step functions and use Exer. 3 *e*)). If $m$ is an abstract measure, we will denote by $m^*$ the corresponding integral.

*b*) Let $m$ be an abstract measure on $(E, \mathfrak{T})$. For every positive function $f$ on E, finite or not, set $m^*(f) = \inf_{g\in\mathscr{M},\, g\geqslant f} m^*(g)$. Show that $m^*$ is an encumbrance on E (imitate the proof of Th. 3 of Ch. IV, §1, No. 3). One sets $m^*(A) = m^*(\varphi_A)$ for every subset A of E.

5) Let E be a set, $\mathfrak{T}$ a tribe of subsets of E, $m$ an abstract measure on $(E, \mathfrak{T})$, and $p \geqslant 1$ a finite real number. For every numerical function $f$ on E, one sets $N_p(f) = m^*(|f|^p)^{1/p}$ and one denotes by $\mathscr{F}^p$ the set of functions $f$ such that $N_p(f)$ is finite (cf. §1, Exer. 4). One equips the vector space $\mathscr{F}^p$ with the semi-norm $N_p$, for which it is complete.

*a*) Let $\mathscr{N}$ be the set of functions $f$ such that $m^*(|f|) = 0$, $\mathscr{V}$ the vector space generated by the finite functions belonging to $\mathscr{M}$, and $\mathscr{L}^p$ the smallest closed linear subspace of $\mathscr{F}^p$ containing the characteristic functions of the sets $A \in \mathfrak{T}$ such that $m^*(A)$ is finite. Show that $\mathscr{L}^p = \mathscr{N} + (\mathscr{V} \cap \mathscr{F}^p)$.

*b*) Extend Lebesgue's theorem (Ch. IV, §3, No. 7, Th. 6) to the present case, and examine in particular the case $p = 1$ (one will define the integral of a function $f \in \mathscr{L}^1$).

*c*) A subset A of E is said to be $m$-integrable if $\varphi_A \in \mathscr{L}^1$. Show that there then exists a set $A' \in \mathfrak{T}$ such that $m^*(A \cap \complement A') = m^*(A' \cap \complement A) = 0$. Converse, in the case that $m(E)$ is finite.

¶ 6) Let E be a set, $\mathfrak{T}$ a tribe of subsets of E, and $m$ an abstract measure on $(E, \mathfrak{T})$. Suppose that $m(E)$ is finite and that, for every $B \in \mathfrak{T}$ such that $m(B) > 0$, there exists a set $A \in \mathfrak{T}$ such that $A \subset B$ and $0 < m(A) < m(B)$. For every $B \in \mathfrak{T}$ and every number $t$ such that $0 \leqslant t \leqslant m(B)$, there exists a set $A \in \mathfrak{T}$ such that $m(A) = t$ and $A \subset B$.

¶ 7) Let T be a topological space, $\Phi$ a clan of subsets of T, and $m$ a relatively bounded additive mapping (cf. Exer. 1) of $\Phi$ into $\mathbf{R}$. Assume that there exists an open covering $\mathfrak{U}$ of T such that $\Phi$ consists of the Borel subsets of T contained in the union of a finite number of elements of $\mathfrak{U}$. Assume, moreover, that for every $A \in \Phi$ and every $\varepsilon > 0$, there exist a compact subset K of T and an open subset U of T such that $K \subset A \subset U$, $U \in \Phi$ and $|m|(U - K) < \varepsilon$. Show that there exists a measure $\mu$ on T (not necessarily positive) such that $m(A) = \mu^\bullet(A)$ for all $A \in \Phi$, and that such a measure $\mu$ is unique (on introducing the decomposition $m = m_+ - m_-$ of Exer. 1 *c*), reduce to the case that $m$ takes positive values; first treat the case that $T \in \Phi$ and observe that $m$ is then inner regular; finally, treat the general case by pasting together measures).

8) Let T be a topological space and $m$ a bounded, countably additive (cf. Exer. 1*d*)) mapping of $\mathfrak{B}(T)$ into $\mathbf{R}$. Denote by $\mathfrak{T}$ the set of Borel subsets A of T having the following property: for every $\varepsilon > 0$, there exist a closed set F and an open set U in T such that $F \subset A \subset U$ and $|m(B)| < \varepsilon$ for every Borel subset B of $U - F$ (in other words, $|m|(U - F) < \varepsilon$). Show that $\mathfrak{T}$ is a tribe of subsets of T (first observe that $T \in \mathfrak{T}$ and that $\mathfrak{T}$ is a clan; it then suffices to prove that if $(A_n)_{n \in \mathbf{N}}$ is a sequence of pairwise disjoint elements of $\mathfrak{T}$, the set $A = \bigcup_{n \in \mathbf{N}} A_n$ belongs to $\mathfrak{T}$; to this end, choose closed sets $F_n$ and open sets $U_n$ such that $F_n \subset A_n \subset U_n$ and $|m|(U_n - F_n) < \varepsilon/2^n$, then set $F = F_0 \cup \cdots \cup F_p$ for $p$ sufficiently large and $U = \bigcup_{n \in \mathbf{N}} U_n$).

9) Let X be a set; one calls *gauge* on X every countably additive mapping of $\mathfrak{P}(X)$ into $\mathbf{R}_+$; a gauge $m$ on X is said to be *diffuse* if $m(\{x\}) = 0$ for all $x \in X$, and to be *atomic* if there exists a positive function $f$ on X such that $m(A) = \sum_{x \in A} f(x)$ for every subset A of X.

*a*) Show that every gauge on X may be decomposed in a unique way as the sum of a diffuse gauge and an atomic gauge.

*b*) Let X′ be a subset of X, and $m'$ a gauge on X′; for every subset A of X, one sets $m(A) = m'(A \cap X')$. Show that $m$ is a gauge on X and that it is diffuse (resp. atomic) if and only if $m'$ has the same property.

*c*) Let $m$ be a gauge on X and let $(X_i)_{i \in I}$ be a family of pairwise disjoint subsets of X. If every gauge on I is atomic, then $m(\bigcup_{i \in I} X_i) = \sum_{i \in I} m(X_i)$.

10) Let X be an uncountable infinite set, equipped with a well-ordering relation denoted $x \leqslant y$. For every $x \in X$, one denotes by $I(x)$ the set of $y \in X$ such that $y < x$. We make the following hypotheses: $\alpha$) there exists a largest element $a$ in X; $\beta$) the set of cardinals strictly smaller than Card(X) has a largest element $\mathfrak{c}$; $\gamma$) for every $x$ in $I(a)$, one has $\mathrm{Card}(I(x)) \leqslant \mathfrak{c}$, and the set Y of $x \in X$ such that $\mathrm{Card}(I(x)) = \mathfrak{c}$ is nonempty;

$\delta$) every gauge on a set with cardinal $\leqslant \mathfrak{c}$ is atomic. Show that every gauge on X is atomic. One can argue by contradiction in the following way: let $m$ be a nonzero diffuse gauge on X. Denote by $b$ the smallest element of Y, and for every $x \in \mathrm{I}(a)$ let $f_x$ be an injection of $\mathrm{I}(x)$ into $\mathrm{I}(b)$. For every pair $(x, y) \in \mathrm{I}(a) \times \mathrm{I}(b)$, denote by $\mathrm{A}_{x,y}$ the set of elements $z \in \mathrm{X}$ such that $x < z < a$ and $f_z(x) = y$; for every $x \in \mathrm{I}(a)$, denote by $\mathrm{M}_x$ the set of $y \in \mathrm{I}(b)$ such that $m(\mathrm{A}_{x,y}) > 0$, and for every $y \in \mathrm{I}(b)$, denote by $\mathrm{N}_y$ the set of $x \in \mathrm{I}(a)$ such that $m(\mathrm{A}_{x,y}) > 0$. Using Exer. 9 $c$), show that $m(\mathrm{X}) = \sum_{y \in \mathrm{I}(b)} m(\mathrm{A}_{x,y})$; from this, deduce that the set $\mathrm{M}_x$ is countable and nonempty for every $x \in \mathrm{I}(a)$; then show that each of the sets $\mathrm{N}_y$ is countable and from this deduce the contradiction $\mathrm{Card}(\mathrm{X}) = \mathfrak{c}$.

11) A cardinal $\mathfrak{c}$ is said to be an *Ulam cardinal* (or to be *ulamian*) if every gauge on a set with cardinal $\mathfrak{c}$ is atomic. Show that every countable cardinal is ulamian, that if $\mathfrak{c}$ is an Ulam cardinal then so is every cardinal $\mathfrak{c}' \leqslant \mathfrak{c}$, and that if $(\mathfrak{c}_i)_{i \in \mathrm{I}}$ is a family of Ulam cardinals such that $\mathrm{Card}(\mathrm{I})$ is ulamian, then the cardinal $\sum_{i \in \mathrm{I}} \mathfrak{c}_i$ is ulamian. Finally, the smallest uncountable cardinal is ulamian (cf. Exer. 10).

12) Let X be a set, $\mathfrak{U}$ a subset of $\mathfrak{P}(\mathrm{X})$, and $m$ the characteristic function of $\mathfrak{U}$. For $m$ to be a gauge, it is necessary and sufficient that $\mathfrak{U}$ be an ultrafilter and that the intersection of every countable subset of $\mathfrak{U}$ belong to $\mathfrak{U}$. Under these conditions, $\mathfrak{U}$ is said to be an *Ulam ultrafilter* on X. Suppose that this is the case, and that $(\mathrm{H}_i)_{i \in \mathrm{I}}$ is a family of elements of $\mathfrak{U}$ such that $\mathrm{Card}(\mathrm{I})$ is an Ulam cardinal; show that $\bigcap_{i \in \mathrm{I}} \mathrm{H}_i$ belongs to $\mathfrak{U}$.

13) In this exercise, we admit the continuum hypothesis, that is, we adjoin to the axioms of set theory the axiom « $\mathrm{Card}(\mathbf{R})$ is the smallest uncountable cardinal ».(1)

*a*) Every gauge on $\mathbf{R}$ is atomic (cf. Exer. 11).

*b*) Let X be a set and $m$ a gauge on X satisfying the following property: for every subset A of X such that $m(\mathrm{A}) > 0$, there exists a subset B of A such that $0 < m(\mathrm{B}) < m(\mathrm{A})$; then $m$ is zero (for every integer $n \geqslant 1$, there exists a finite partition $(\mathrm{X}_{n,i})_{i \in \mathrm{I}_n}$ of X such that $m(\mathrm{X}_{n,i}) \leqslant 1/n$ for all $i \in \mathrm{I}_n$; set $\mathrm{I} = \prod_{n \geqslant 1} \mathrm{I}_n$ and $\mathrm{X}_\alpha = \bigcap_{n \geqslant 1} \mathrm{X}_{n,\alpha_n}$ for every $\alpha = (\alpha_n)_{n \geqslant 1}$ in I; the family $(\mathrm{X}_i)_{i \in \mathrm{I}}$ is a partition of X, $m(\mathrm{X}_i) = 0$ for every $i \in \mathrm{I}$, and every gauge on I is atomic by *a*); conclude by means of Exer. 9 *c*)).

*c*) Let X be a set on which there does not exist any nontrivial Ulam ultrafilter; then every gauge on X is atomic (otherwise, by *b*), there would exist a diffuse gauge $m$ on X and a subset A of X such that $m(\mathrm{A}) > 0$ and such that $m(\mathrm{B}) = 0$ or $m(\mathrm{A} - \mathrm{B}) = 0$ for every subset B of A; the set $\mathfrak{U}$ of subsets B of X such that $m(\mathrm{A}) = m(\mathrm{A} \cap \mathrm{B})$ would be a nontrivial Ulam ultrafilter on X).

*d*) Show that $2^{\mathfrak{c}}$ is an Ulam cardinal for every Ulam cardinal $\mathfrak{c}$ (by *c*), it suffices to show that if C is a set with cardinal $\mathfrak{c}$, then every Ulam ultrafilter $\mathfrak{U}$ on $\mathrm{X} = \{0, 1\}^{\mathrm{C}}$ is trivial; equip C with a well-ordering structure and define by transfinite induction a family $s = (s_\alpha)_{\alpha \in \mathrm{C}}$ of elements of $\{0, 1\}$ such that set of $x = (x_\alpha)_{\alpha \in \mathrm{C}}$ in X for which

(1) It comes to the same to say that $\mathbf{R}$ can be equipped with a well-ordering structure $\prec$ for which the set of $x \in \mathbf{R}$ such that $x \prec a$ is countable for every $a \in \mathbf{R}$. It has been shown (cf. P. Cohen, *Proc. Nat. Acad. Sci. U.S.A.* **50** (1963), 1143–1148 and **51** (1964), 105–110) that this axiom does not introduce any new contradiction into the theory of sets, and that it is moreover independent of the other axioms of set theory.

$x_\alpha = s_\alpha$ for all $\alpha \leqslant \beta$ belongs to $\mathfrak{U}$ for every $\beta \in C$ (cf. Exer. 12); then $s$ belongs to every element of $\mathfrak{U}$).[2]

14) Let $(T_i)_{i\in I}$ be a family of Radon topological spaces, and T the sum space of the family. If Card(I) is an Ulam cardinal, then the topological space T is a Radon space (show that for every positive and bounded countably additive function $m$ on the Borel tribe of T, there exists a countable subset J of I such that $m(\bigcup_{i\in I-J} T_i) = 0$.

15) Let T be a discrete space whose cardinal is the smallest uncountable cardinal. Show that T is a Radon locally compact space whose topology does not have a countable base (cf. Exer. 11). Deduce from this that there exist compact Radon spaces that are not metrizable.

16) Let I be a countable set, and $(T_i)_{i\in I}$ a family of Radon spaces. Assume that every compact subset of any of the spaces $T_i$ is metrizable. Show that the product space $T = \prod_{i\in I} T_i$ is Radon. Generalize to the case of inverse limits.

17) Let T be a topological space.

*a*) Let $m$ be a countably additive mapping of $\mathfrak{B}(T)$ into $\mathbf{R}_+$. Assume that there exists a sequence $(T_n)_{n\in\mathbf{N}}$ of Borel subsets of T such that $T = \bigcup_{n\in\mathbf{N}} T_n$, $m(T_0) = 0$ and the subspace $T_n$ of T is Radon for every $n \geqslant 1$. Show that there exists a bounded positive measure $\mu$ on T such that $m(A) = \mu^\bullet(A)$ for every Borel subset A of T (one can reduce to the case that the $T_n$ are pairwise disjoint on applying the Cor. of Prop. 2 of No. 3).

*b*) Extend the preceding result to the case that the sets $T_n$ for $n \geqslant 1$ are no longer assumed Borel but merely universally measurable in T (argue as in Prop. 2 of No. 3).

*c*) Every topological space that is the union of a sequence of metrizable compact subspaces is Radon.

18) Let $T_1$ and $T_2$ be two topological spaces and $f$ a bijective continuous mapping of $T_1$ onto $T_2$. Assume that, for every bounded countably additive function $m_1$ on the Borel tribe $\mathfrak{B}(T_1)$, there exists a sequence of compact subsets $K_n$ of $T_1$ such that $m_1(T_1 - \bigcup_n K_n) = 0$. Let $m_2$ be a bounded countably additive function on $\mathfrak{B}(T_2)$. In order that there exist a bounded countably additive function $m_1$ on $\mathfrak{B}(T_1)$ such that $m_2(A) = m_1(f^{-1}(A))$ for all $A \in \mathfrak{B}(T_2)$, it is necessary and sufficient that the following condition be fulfilled: there exists a sequence of compact subsets $K_n$ of $T_1$ such that $m_2(T_2 - \bigcup_n f(K_n)) = 0$.

## §4

1) Let $\mathscr{T} = (K_i, p_{ij})$ be an inverse system of compact spaces indexed by the set I, let K be a compact space, and let $(p_i)_{i\in I}$ be a coherent and separating family of continuous mappings $p_i : K \to K_i$.

---

(2) A cardinal $\mathfrak{c}$ is said to be *strongly inaccessible* if it is uncountable, $2^{\mathfrak{b}} < \mathfrak{c}$ for every cardinal $\mathfrak{b} < \mathfrak{c}$, and $\sum_{i\in I} \mathfrak{c}_i < \mathfrak{c}$ for every family of cardinals $\mathfrak{c}_i < \mathfrak{c}$ such that $\mathrm{Card}(I) < \mathfrak{c}$. Exercises 11 and 13 show that, assuming the continuum hypothesis, the smallest cardinal that is not an Ulam cardinal is strongly inaccessible. It is not known whether the axiom of the existence of strongly inaccessible cardinals is contradictory to the other axioms of set theory.

*a*) Let A be the set of continuous functions on K of the form $f_i \circ p_i$, where $i$ runs over I and $f_i$ runs over the set of continuous real functions on $K_i$. Show that A is a dense linear subspace of $\mathscr{C}(K)$.

*b*) Let $(\mu_i)_{i \in I}$ be a sub-inverse system of positive measures on $\mathscr{T}$. Assume that the $p_i$ are surjective. For every $f \in A$, let $I_f$ be the set of $i \in I$ such that $f$ is of the form $f_i \circ p_i$ with $f_i \in \mathscr{C}(K_i)$ (necessarily unique). Show that there exists one and only one measure $\pi$ on K such that $\pi(f) = \inf_{i \in I_f} \mu_i(f_i)$ for all $f \in A$. Show that $\pi$ is the greatest of the positive measures $\mu$ on K such that $p_i(\mu) \leqslant \mu_i$ for all $i \in I$.

¶ 2) Hypotheses and notations are those of Th. 1 of No. 2, for which we propose to indicate a new proof.

*a*) Let K be a compact subset of T; for every $i \in I$, set $K_i = p_i(T)$, and denote by $q_i$ the mapping of K onto $K_i$ that coincides on K with $p_i$; for $i \leqslant j$, denote by $q_{ij}$ the mapping of $K_j$ into $K_i$ that coincides on $K_j$ with $p_{ij}$. Deduce from the preceding exercise the existence of a greatest positive measure $\pi^K$ on K such that $q_i(\pi^K) = (\mu_i)_{K_i}$ for all $i \in I$.

*b*) If K and L are compact subsets of T such that $K \subset L$, show that $(\pi^L)_K \geqslant \pi^K$; deduce from this that the set of measures of the form $i^K(\pi^K)$ (where K runs over the set of compact subsets of T, and $i^K$ is the canonical injection of K into T) admits a supremum $\mu$ in $\mathscr{M}_+(T)$.

*c*) Show that $p_i(\mu) = \mu_i$ for every $i \in I$ (it suffices to observe that $p_i(\mu) \leqslant \mu_i$ and that the measures $p_i(\mu)$ and $\mu_i$ have the same total mass by the hypothesis (P)).

## §5

1) Let T be a topological space and $\mu$ a positive measure on T. Denote by $\mathfrak{S}_\mu$ the set of subsets of T whose boundary is $\mu$-negligible.

*a*) Show that $\mathfrak{S}_\mu$ is a clan.

*b*) If T is completely regular, every point of T has a fundamental system of neighborhoods contained in $\mathfrak{S}_\mu$ (observe that if $f$ is a continuous function on T, zero outside a $\mu$-integrable set, then the set of real numbers $r$ such that $\overset{-1}{f}(r)$ is not $\mu$-negligible is countable).

¶ 2) Let T be a completely regular space and $\mathfrak{F}$ a filter on $\mathscr{M}_+^b(T)$ that converges tightly to a bounded measure $\mu$. A Borel subset A of T is said to be a convergence set (for $\mathfrak{F}$) if $\lim_{\lambda, \mathfrak{F}} \lambda_A = \mu_A$.

*a*) If the disjoint sets $A_1$ and $A_2$ are convergence sets, then so is $A_1 \cup A_2$.

*b*) Let A be an open or closed set. For A to be a convergence set, it is necessary and sufficient that $\lim_{\lambda, \mathfrak{F}} \lambda^\bullet(A) = \mu^\bullet(A)$. If A is a convergence set, then so is $T - A$.

*c*) Let A be an open or closed convergence set in T, and B a convergence set such that $T - B$ is also a convergence set. Then $A \cup B$ and $A \cap B$ are convergence sets, as are their complements.

*d*) The clan generated by the open convergence sets consists of convergence sets.

*e*) Let A be a subset of T whose boundary is $\mu$-negligible. Then A is a convergence set.

*f*) Suppose that T is locally compact or Polish. Show that every compact subset K of T is contained in a compact convergence set (in case T is locally compact, use Exercise 1 *b*)). If T is Polish, use the same exercise to construct a sequence of finite coverings $\mathfrak{U}_p$ of K by open subsets of T, with $\mu$-negligible boundary and of diameter $< 2^{-p}$; show that $L = \bigcap_p \bigcup_{U \in \mathfrak{U}_p} \overline{U}$ is compact and conclude that it is a convergence set by *b*)).

*g*) Extend the result of *f*) to the case of a convergent *sequence* of measures on a complete metric space.

3) Let T be a Polish space and $(\mu_n)$ a sequence of bounded positive measures on T converging tightly to a measure $\mu$. Denote by $\mathfrak{C}$ the set of subsets A of T having the following property: for every $\varepsilon > 0$, there exists a compact subset K of A such that $\sup_n \mu_n(A - K) < \varepsilon$. Let $\mathfrak{D}$ be the set of subsets A of T that, together with their complement, belong to $\mathfrak{C}$.

*a*) If the sets A and A′ belong to $\mathfrak{C}$, then so do $A \cup A'$ and $A \cap A'$; deduce from this that $\mathfrak{D}$ is a clan of subsets of T.

*b*) Every subset A of T whose boundary is $\mu$-negligible belongs to $\mathfrak{D}$ (apply Prokhorov's theorem to the interior of A, and use Exer. 2 *e*)).

*c*) Every $A \in \mathfrak{D}$ is a convergence set for the sequence $(\mu_n)$ (cf. Exer. 2); conversely, every open or closed convergence set belongs to $\mathfrak{D}$.

*d*) Let $A \in \mathfrak{D}$. Show that there exist a sequence of disjoint compact sets $K_p$ and a subset N of T having the following properties: $\alpha$) each $K_p$ is a convergence set for the sequence $(\mu_n)$; $\beta$) $A \subset N \cup \bigcup_p K_p$ and $\bigcup_p K_p \subset A \cup N$; $\gamma$) for every $\varepsilon > 0$, there exists an open neighborhood U of N such that $\sup_n \mu_n(U) < \varepsilon$ (apply Exer. 2 *f*) and Prokhorov's theorem to a suitable subspace of T that is the intersection of a sequence of open sets).

4) Let T be a completely regular space, *t* a point of T, and $\mathfrak{U}$ a set of Borel subsets of T that generates the filter of neighborhoods of *t*. Let I be a set equipped with a filter $\mathfrak{F}$, and let $(\mu_i)_{i \in I}$ be a family of bounded positive measures of total mass 1 on T. In order that $\lim_{i,\mathfrak{F}} \mu_i = \varepsilon_t$, it is necessary and sufficient that $\lim_{i,\mathfrak{F}} \mu_i(U) = 1$ for every $U \in \mathfrak{U}$.

5) Let E be a real Hilbert space admitting an orthonormal basis $(x_n)_{n \in \mathbf{N}}$. Let T be the space E equipped with the weakened topology, and let $(a_n)_{n \in \mathbf{N}}$ be a sequence of real numbers such that $0 < a_n < 1$ for all *n* and $\lim_{n\to\infty} n^2 \log a_n = 0$. For every $n \in \mathbf{N}$, set $\mu_n = \sum_{p \in \mathbf{N}} a_n^p (1 - a_n) \cdot \varepsilon_{n \cdot x_p}$. Show that $\mu_n$ is a bounded positive measure on T of total mass 1 for every $n \in \mathbf{N}$, that $\lim_{n\to\infty} \mu_n = \varepsilon_0$ (tight convergence), and that $\lim_{n\to\infty} \mu_n(K) = 0$ for every compact subset K of T (apply the criterion of the preceding exercise to the set $\mathfrak{U}$ of neighborhoods of 0 of the form $\{x \mid (a|x) \leqslant 1\}$, where *a* runs over E). In particular, the set of elements of the sequence $(\mu_n)_{n \in \mathbf{N}}$ is a relatively compact subset of $\mathscr{M}_+^b(T)$ that does not satisfy Prokhorov's condition.

6) Let T be a topological space and H a set of positive measures on T; assume that the following condition is fulfilled:

(V) Every point of T admits an open neighborhood W such that $\sup_{\mu \in H} \mu^\bullet(W)$ is finite.

Finally, let $\mathfrak{U}$ be an ultrafilter on H.

*a*) Let K be a compact subset of T. Show that the induced measures $\mu_K$ $(\mu \in H)$ converge vaguely with respect to $\mathfrak{U}$ to a measure $\pi^K$ on K.

*b*) Let K and L be two compact subsets of T such that $K \subset L$; show that $(\pi^L)_K \geqslant \pi^K$.

*c*) For every compact subset K of T, let $i^K$ be the canonical injection of K into T. Show that the family of measures $i^K(\pi^K)$ admits a supremum $\pi$ in $\mathscr{M}_+(T)$.

*d*) Let *f* be a lower semi-continuous positive function on T, and *g* an upper semi-continuous positive function on T with compact support. Show that $\pi^\bullet(f) \leqslant \lim_{\mu,\mathfrak{U}} \mu^\bullet(f)$ and $\pi^\bullet(g) \geqslant \lim_{\mu,\mathfrak{U}} \mu^\bullet(g)$.

¶ 7) Let H be a set of bounded positive measures on a topological space T. Assume that $\sup_{\mu\in H} \mu^\bullet(T)$ is finite, and that for every $\varepsilon > 0$, there exists a compact subset K of T such that $\sup_{\mu\in H} \mu^\bullet(T - K) < \varepsilon$. Show that the condition (V) of the preceding exercise is verified. Let $\mathfrak{U}$ be an ultrafilter on H. The measure $\pi$ is defined as in the preceding exercise.

*a*) Show that $\pi^\bullet(g) \geqslant \lim_{\mu,\mathfrak{U}} \mu^\bullet(g)$ for every upper semi-continuous bounded positive function $g$ on T. Deduce from this that $\pi^\bullet(f) = \lim_{\mu,\mathfrak{U}} \mu^\bullet(f)$ for every bounded function $f$ on T whose set of points of discontinuity is $\pi$-negligible.

*b*) When T is completely regular, deduce from *a*) that H is relatively compact for the tight topology (this furnishes a new proof of Th. 1 of No. 5 for the case of positive measures).

¶ 8) Let T be a complete separable metric space, and $d$ its metric. For every closed subset F of T and every real number $a > 0$, denote by $F^a$ the set of $x \in T$ such that $d(x, F) < a$. Given two bounded positive measures $\lambda$ and $\mu$ on T, denote by $D(\lambda, \mu)$ the infimum of the set of real numbers $a > 0$ satisfying the inequalities $\lambda^\bullet(F) \leqslant \mu^\bullet(F^a) + a$, $\mu^\bullet(F) \leqslant \lambda^\bullet(F^a) + a$ for every closed subset F of T.

*a*) Show that D is a metric on $\mathscr{M}_+^b$ and that $D(\varepsilon_x, \varepsilon_y) = d(x, y)$ for two points $x$ and $y$ of T such that $d(x, y) < 1$.

*b*) Let $f$ be a Borel mapping of T into T (cf. TG, IX, §6, No. 3 or GT, IX, §6, Exer. 16) and $\lambda$ a bounded positive measure on T. Show that there exists a bounded positive measure $\mu$ on T such that $\mu^\bullet(A) = \lambda^\bullet(\overset{-1}{f}(A))$ for every Borel subset A of T. Let $a > 0$, and let H be the set of $x \in T$ such that $d(x, f(x)) \geqslant a$; show that H is Borel in T and that $D(\lambda, \mu) \leqslant \sup(a, \lambda(H))$ (for every closed subset F of T, one has $F \cap (T - H) \subset \overset{-1}{f}(F^a)$ and $\lambda^\bullet(F) \leqslant \lambda^\bullet(F \cap (T - H)) + \lambda^\bullet(H)$).

*c*) Let $\mu$ and $\mu_n$ (for $n \geqslant 1$) be bounded positive measures on T such that $\lim_{n\to\infty} D(\mu_n, \mu) = 0$. Show that the sequence $(\mu_n)$ converges tightly to $\mu$ (show that $\lim_{n\to\infty} \mu_n^\bullet(T) = \mu^\bullet(T)$ and $\mu^\bullet(F) \geqslant \limsup_{n\to\infty} \mu_n^\bullet(F)$ for every closed subset F of T; deduce from this that $\mu^\bullet(f) \leqslant \liminf_{n\to\infty} \mu_n^\bullet(f)$ for every lower semi-continuous function $f \geqslant 0$ by the method of Lemma 3 of §2, No. 6).

*d*) Conversely, show that for every sequence of bounded positive measures $\mu_n$ tending tightly to $\mu$ in $\mathscr{M}_+^b$, one has $\lim D(\mu_n, \mu) = 0$. One can proceed as follows: let $\varepsilon > 0$, and let K be a compact subset of T such that $\mu^\bullet(T - K) < \varepsilon$ and $\sup_n \mu_n^\bullet(T - K) < \varepsilon$; construct a finite family $(B_i)_{1\leqslant i\leqslant p}$ of sets with $\mu$-negligible boundary, of diameter $\leqslant \varepsilon/2$, pairwise disjoint, whose union contains K (cf. Exer. 1). Let $f$ be a mapping of T into T, constant on each of the sets $B_1, \ldots, B_p$ and $\complement(B_1 \cup \cdots \cup B_p)$ and such that $f(B_i) \subset B_i$ for $1 \leqslant i \leqslant p$. Define the measures $\pi_n$ and $\pi$ by $\pi_n^\bullet(A) = \mu_n^\bullet(\overset{-1}{f}(A))$ and $\pi^\bullet(A) = \mu^\bullet(\overset{-1}{f}(A))$ for every Borel subset A of T; deduce from *b*) the relations $D(\pi_n, \mu_n) \leqslant \varepsilon$, $D(\pi, \mu) \leqslant \varepsilon$ and show that $\lim_{n\to\infty} D(\pi_n, \pi) = 0$.

*e*) The metric D on $\mathscr{M}_+^b$ is compatible with the tight topology (observe that $\mathscr{M}_+^b$ is metrizable for the tight topology).

¶ 9) Notations and hypotheses are those of the preceding exercise. Let $(\mu_n)$ be a Cauchy sequence in $\mathscr{M}_+^b$ for the metric D; show that the sequence $(\mu_n)$ is convergent (this gives a new proof of the fact that the space $\mathscr{M}_+^b$ is Polish for the tight topology). One can proceed thus:

*a*) Let $\varepsilon > 0$ and $a > 0$ be two real numbers; there exists a finite subset F of T such that $\sup\limits_n \mu_n^\bullet(\mathrm{T} - \mathrm{F}^a) \leqslant \varepsilon$ (choose an integer $\mathrm{N} \geqslant 1$ and a compact subset K of T such that $\sup\limits_{n\geqslant \mathrm{N}} \mathrm{D}(\mu_n, \mu_\mathrm{N}) \leqslant \varepsilon/2$ and $\mu_\mathrm{N}^\bullet(\mathrm{T} - \mathrm{K}) \leqslant \varepsilon/2$; deduce from this that $\sup\limits_{n\geqslant \mathrm{N}} |\mu_n^\bullet(\mathrm{T}) - \mu^\bullet(\mathrm{T})| \leqslant \varepsilon$ and $\sup\limits_{n\geqslant \mathrm{N}} \mu_n^\bullet(\mathrm{T} - \mathrm{K}^{a/2}) \leqslant \varepsilon$; finally, choose a finite subset F of T such that $\mathrm{K}^{a/2} \subset \mathrm{F}^a$ and $\sup\limits_{n<\mathrm{N}} \mu_n^\bullet(\mathrm{T} - \mathrm{F}^a) \leqslant \varepsilon$).

*b*) Let $\varepsilon > 0$; there exists a compact subset K of T such that $\sup \mu_n^\bullet(\mathrm{T} - \mathrm{K}) \leqslant \varepsilon$ (for every integer $p \geqslant 1$ choose a finite subset $\mathrm{F}_p$ of T such that $\mu_n^\bullet(\mathrm{T} - (\mathrm{F}_p)^{2^{-p}}) \leqslant \varepsilon/2^p$, and set $\mathrm{K} = \bigcap\limits_{p\geqslant 1} \overline{(\mathrm{F}_p)^{2^{-p}}}$).

*c*) Deduce from *b*) that the Cauchy sequence $(\mu_n)$ has a convergent subsequence (for the metric D).

¶ 10) let T be a completely regular space; denote by $\mathscr{S}$ the set of real functions $f \geqslant 1$ on T, such that the set of points $t$ of T for which $f(t) \leqslant c$ is compact for every real number $c$. For every $f \in \mathscr{S}$, denote by $\mathscr{M}_f$ the set of bounded measures $\mu$ on T such that $|\mu|^\bullet(f) \leqslant 1$.

*a*) For a subset H of $\mathscr{M}^b(\mathrm{T})$ to satisfy Prokhorov's condition, it is necessary and sufficient that there exist an $f \in \mathscr{S}$ such that $\mathrm{H} \subset \mathscr{M}_f$ (for necessity, take $f$ to be of the form $c(1 + \sum\limits_{n\geqslant 1} n \cdot f_n)$ where $f_n$ is the characteristic function of a set $\mathrm{U}_n$ such that $\mathrm{T} - \mathrm{U}_n$ is compact and $\sup\limits_{\mu\in\mathrm{H}} |\mu|^\bullet(\mathrm{U}_n) \leqslant 2^{-n}$).

*b*) Suppose T is locally compact. For a subset H of $\mathscr{M}^b(\mathrm{T})$ to be relatively compact (for the tight topology), it is necessary and sufficient that there exist a continuous function $f \in \mathscr{S}$ such that $\mathrm{H} \subset \mathscr{M}_f$.

*c*) Let C be a closed convex cone in $\mathscr{M}_+^b(\mathrm{T})$. Show that $\mathrm{C} \cap \mathscr{M}_f$ is a cap of C (TVS, II, §7, No. 2, Def. 3) for every $f \in \mathscr{S}$, and deduce from this that C is the union of its caps. Denote by E the union of the extremal generators of C. Suppose T is Souslin. Show that for every $\pi \in \mathrm{C}$, there exist a Borel subset B of $\mathscr{M}_+^b(\mathrm{T})$ contained in E and a positive measure P on B of total mass 1, such that $\pi = \int_\mathrm{B} \mu\, d\mathrm{P}(\mu)$ (apply Choquet's integral representation theorem).

11) Let T be a completely regular space. Assume given, for every compact space K and every continuous mapping $f$ of T into K, a positive measure $\mu_{f,\mathrm{K}}$ on K; assume that $g(\mu_{f,\mathrm{K}}) = \mu_{g\circ f,\mathrm{L}}$ for any compact spaces K and L and any continuous mappings $f : \mathrm{T} \to \mathrm{K}$ and $g : \mathrm{K} \to \mathrm{L}$; assume moreover that the measure $\mu_{f,\mathrm{K}}$ is concentrated on $f(\mathrm{T})$ for any $f$ and K. Show that there exists one and only one bounded positive measure $\pi$ on T such that $\mu_{f,\mathrm{K}} = f(\pi)$ for any $f$ and K (make use of the universal property of the Stone–Čech compactification of T).

12) Let T be a completely regular space, a subspace of a locally compact space L. For every measure $\mu$ (positive or not) on T, there exists a locally compact subspace T′ of L containing T, and a measure $\mu'$ on T′ concentrated on T that induces $\mu$ on T.

¶ 13) *Let G be a locally compact abelian group. One denotes by $\widehat{\mathrm{G}}$ the dual group of G and by $\alpha$ a Haar measure on $\widehat{\mathrm{G}}$. For every bounded measure $\mu$ on G, one denotes by $\mathscr{F}\mu$ the Fourier transform of $\mu$, which is a function on the group $\widehat{\mathrm{G}}$ (cf. *Théor. spec.*, Ch. II, §1, No. 2).

*a*) The Fourier transformation $\mathscr{F}$ is a continuous mapping of $\mathscr{M}_+^b(\mathrm{G})$ equipped with the tight topology into $\mathscr{C}^b(\widehat{\mathrm{G}})$ equipped with the topology of compact convergence. (Make use of the Cor. of Prop. 13 of No. 6.)

*b*) Let $\mathfrak{F}$ be a filter on $\mathscr{M}_+^b(\mathrm{G})$ and $\Phi$ a bounded continuous function on $\widehat{\mathrm{G}}$; assume that $\lim_{\lambda, \mathfrak{F}}(\mathscr{F}\lambda) \cdot \alpha = \Phi \cdot \alpha$ (vague convergence) and $\lim_{\lambda, \mathfrak{F}}(\mathscr{F}\lambda)(0) = \Phi(0)$. Show that the filter $\mathfrak{F}$ converges tightly to a measure $\mu$ such that $\mathscr{F}\mu = \Phi$ (show by application of the Stone–Weierstrass theorem that the set of Fourier transforms of continuous functions on G with compact support is a dense linear subspace of $\mathscr{C}^0(\widehat{\mathrm{G}})$ for uniform convergence; next observe that $\lim_{\lambda, \mathfrak{F}} \int (\mathscr{F}\lambda) \cdot u\, d\alpha = \int \Phi u\, d\alpha$ for every $u \in \mathrm{E}$ and that the filter $\mathfrak{F}$ contains a vaguely compact set of bounded measures; deduce from this that the filter $\mathfrak{F}$ has at most one cluster point, then that it converges tightly).

*c*) Let $\mathfrak{F}$ be a filter on $\mathscr{M}_+^b(\mathrm{G})$ *having a countable base*. In order that $\mathfrak{F}$ converge tightly to a bounded positive measure $\mu$, it is necessary and sufficient that there exist a continuous function $\Phi$ on $\widehat{\mathrm{G}}$ such that $\mathscr{F}\lambda$ converges pointwise to $\Phi$ with respect to $\mathfrak{F}$, in which case $\mathscr{F}\mu = \Phi$ (*P. Lévy's theorem*).$_*$

## §6

¶ 1) For every compact interval K of $\mathbf{R}$, let $\mathscr{C}(\mathrm{K})$ be the real vector space of continuous functions on K, equipped with the topology of uniform convergence; denote by $\mathscr{D}(\mathrm{K})$ the vector space of infinitely differentiable real functions on $\mathbf{R}$ that are zero outside K. Equip $\mathscr{D}(\mathrm{K})$ with the coarsest topology for which the mappings $f \mapsto \mathrm{D}^p f|\mathrm{K}$ are continuous for every positive integer $p$ (D is the differentiation operator). Set $\mathscr{D} = \bigcup_{\mathrm{K}} \mathscr{D}(\mathrm{K})$, a space that one equips with the locally convex topology that is the direct limit of the topologies of the subspaces $\mathscr{D}(\mathrm{K})$.

*a*) For every integer $p \geqslant 0$ and every function $f \in \mathscr{D}(\mathrm{K})$, set

$$\mathrm{Q}_p(f) = \int_{\mathrm{K}} (\mathrm{D}^p f(x))^2\, dx .$$

Show that $\mathrm{Q}_p$ is a positive quadratic form on $\mathscr{D}(\mathrm{K})$, that the sequence of norms $\mathrm{Q}_p^{1/2}$ defines the topology of $\mathscr{D}(\mathrm{K})$, and that $\mathrm{Tr}\,(\mathrm{Q}_{p+1}/\mathrm{Q}_p)$ is finite for every $p \geqslant 0$. (For every real $t$, let $\mathrm{I}_t$ be the characteristic function of the interval $[t, +\infty[$; using the formula $\mathrm{D}^p f(t) = \int_{\mathrm{K}} \mathrm{D}^{p+1} f \cdot \mathrm{I}_t\, dx$ and Bessel's inequality, prove that $\sum_{i=1}^{n} \mathrm{Q}_p(f_i) \leqslant l^2$ for every finite sequence $f_1, \ldots, f_n$ of functions belonging to $\mathscr{D}(\mathrm{K})$, orthonormal for $\mathrm{Q}_{p+1}$; the number $l$ is the length of the interval K.) Deduce from this that $\mathscr{D}(\mathrm{K})$ is a nuclear space.

*b*) Prove that the space $\mathscr{D}$ is nuclear. (First establish the existence of a nonzero function in $\mathscr{D}$, and deduce from this the existence of a function $h \geqslant 0$ in $\mathscr{D}$ such that $\sum_{n \in \mathbf{Z}} h(x - n) = 1$ for every $x \in \mathbf{R}$. Let K be a compact interval of $\mathbf{R}$ such that $h$ is zero outside of K; for every integer $n$, set $h_n(x) = h(x - n)$ and $\mathrm{K}_n = \mathrm{K} + n$. Let V be a convex neighborhood of 0 in $\mathscr{D}$; there exist positive integers $p_n$ such that every function $f \in \mathscr{D}(\mathrm{K}_n)$ with $\mathrm{Q}_{p_n}(f) \leqslant 1$ belongs to V. Define the continuous quadratic forms Q and R on $\mathscr{D}$ by $\mathrm{Q}(f) = \sum_{n \in \mathbf{Z}} 2^{2n} \mathrm{Q}_{p_n}(f \cdot h_n)$ and $\mathrm{R}(f) = \sum_{n \in \mathbf{Z}} 2^{3n} \mathrm{Q}_{p_n+1}(f \cdot h_n)$; show that V contains the set of $f \in \mathscr{D}$ such that $\mathrm{Q}(f) \leqslant 1$ and that $\mathrm{Tr}\,(\mathrm{R}/\mathrm{Q})$ is finite.)

$^*$*c*) Generalize the foregoing to infinitely differentiable functions on $\mathbf{R}^n$ with compact support.$_*$

## ANNEX

1) Let E be a finite-dimensional vector space over a commutative field of characteristic $\neq 2$. Denote by H a nondegenerate quadratic form on E, and by S the symmetric bilinear form on $E \times E$ such that $H(x) = S(x, x)$ for all $x$ in E. Let Q be a quadratic form on E.

*a*) There exists an endomorphism $u$ of E characterized by the relations $Q(x) = S(u(x), x)$ and $S(u(x), y) = S(x, u(y))$ for all $x$ and $y$ in E. One sets $\mathrm{Tr}(Q/H) = \mathrm{Tr}(u)$.

*b*) Generalize *Remark* 3 of No. 1.

*c*) If $(e_i)_{1 \leqslant i \leqslant m}$ is a basis of E orthonormal for H, then $\mathrm{Tr}(Q/H) = \sum_{i=1}^{m} Q(e_i)$.

2) Let E be a real Hilbert space, Q a continuous positive quadratic form on E, and H the quadratic form $x \mapsto \|x\|^2$ on E.

*a*) Show that $\mathrm{Tr}(Q/H) = \sum_{i \in I} Q(e_i)$ for every orthonormal basis $(e_i)_{i \in I}$ of E. (Let $(a_1, \ldots, a_p)$ be a finite orthonormal family in E; for every $\varepsilon > 0$, there exist a finite subset J of I and elements $a'_1, \ldots, a'_p$ that are linear combinations of the $e_i$ for $i \in J$ and are such that $\|a_j - a'_j\| < \varepsilon$ for $1 \leqslant j \leqslant p$; then $\sum_{j=1}^{p} Q(a'_j) \leqslant \sum_{i \in J} Q(e_j) \leqslant \sum_{i \in I} Q(e_i)$. Deduce from this that $\sum_{j=1}^{p} Q(a_j) \leqslant \sum_{i \in I} Q(e_i)$.)

*b*) Deduce from *a*) a new proof of Prop. 3.

*c*) Let $E_0$ be a dense linear subspace of E, $Q_0$ (resp. $H_0$) the restriction of Q (resp. H) to $E_0$. Show that $\mathrm{Tr}(Q/H) = \mathrm{Tr}(Q_0/H_0)$. (Let $(e_1, \ldots, e_n)$ be a linearly independent sequence in E, generating a subspace F. Denote by $Q_F$ (resp. $H_F$) the restriction of Q (resp. H) to F. Using *Remark* 3 of No. 1, show that $\mathrm{Tr}(Q_F/H_F)$ is a continuous function of $(e_1, \ldots, e_n)$.)

3) Let E and F be two real vector spaces and $u$ a linear mapping of E into F; if Q and H are positive quadratic forms on F such that $H(x) = 0$ implies $Q(x) = 0$ for $x \in F$, then $\mathrm{Tr}(Q \circ u / H \circ u) \leqslant \mathrm{Tr}(Q/H)$.

4) Let I be a set and $l^2(I)$ the vector space of families $\mathbf{x} = (x_i)_{i \in I}$ of real numbers such that $\sum_{i \in I} x_i^2$ is finite. Let $(\lambda_i)_{i \in I}$ be a summable family of positive numbers. For every $\mathbf{x}$ in $l^2(I)$, set $Q(\mathbf{x}) = \sum_{i \in I} \lambda_i x_i^2$ and $H(\mathbf{x}) = \sum_{i \in I} x_i^2$. Show that $\mathrm{Tr}(Q/H) = \sum_{i \in I} \lambda_i$. (Make use of Exercise 2 *a*).)

5) Let E be a real vector space, $(\lambda_i)_{i \in I}$ a summable family of positive numbers, and $(y_i)_{i \in I}$ a family of linear forms on E. For every $x \in E$, set $H(x) = \sum_{i \in I} \langle x, y_i \rangle^2$ and $Q(x) = \sum_{i \in I} \lambda_i \langle x, y_i \rangle^2$; assume that $H(x)$ is finite for every $x \in E$. Show that $Q(x)$ is finite for every $x \in E$, that Q and H are positive quadratic forms on E, and that $\mathrm{Tr}(Q/H) \leqslant \sum_{i \in I} \lambda_i$. (Set $u(x) = (\langle x, y_i \rangle)_{i \in I}$ and apply Exer. 3 to the linear mapping $u$ of E into $l_2(I)$.)

6) Let E be a real vector space, Q, H and $H_0$ positive quadratic forms on E. Assume that $H \leqslant a \cdot H_0$, where $a$ is a positive real number. Prove that $\mathrm{Tr}(Q/H_0) \leqslant a \cdot \mathrm{Tr}(Q/H)$. (Reduce to the case that E is finite-dimensional by *Remark* 1 of No. 1, then conclude by Prop. 1, using the existence of a basis of E that is orthogonal for both H and $H_0$.)

7) Let E be a real Hilbert space. Show that the Sazonov topology on E is defined by the set of semi-norms $Q^{1/2}$, where Q runs over the set of nuclear positive quadratic forms on E. (Use Exer. 6.)

8) Let E and F be two real Hilbert spaces. There exists on $E \otimes F$ a Hausdorff pre-Hilbert space structure such that $(x_1 \otimes y_1 | x_2 \otimes y_2) = (x_1|x_2) \cdot (y_1|y_2)$ for $x_1, x_2$ in E and $y_1, y_2$ in F. One denotes by $E \otimes_2 F$ the Hilbert space completion of $E \otimes F$.

*a*) If E′ (resp. F′) is a closed linear subspace of E (resp. F), show that $E' \otimes_2 F'$ may be identified with the closed linear subspace of $E \otimes_2 F$ generated by the elements $x \otimes y$ such that $x \in E'$ and $y \in F'$.

*b*) Suppose that E (resp. F) is the hilbertian sum of a family $(E_\alpha)_{\alpha \in A}$ (resp. $(F_\beta)_{\beta \in B}$) of closed linear subspaces (TVS, V, §2, No. 2, Def. 2). Show that $E \otimes_2 F$ is the hilbertian sum of the family $(E_\alpha \otimes_2 F_\beta)_{(\alpha,\beta) \in A \times B}$ of closed linear subspaces.

*c*) If $(e_i)_{i \in I}$ (resp. $(f_j)_{j \in J}$) is an orthonormal basis of E (resp. F), then the family $(e_i \otimes f_j)_{(i,j) \in I \times J}$ is an orthonormal basis of $E \otimes_2 F$.

*d*) Let G be a real Hilbert space. Define canonical isomorphisms of $E \otimes_2 F$ onto $F \otimes_2 E$ and of $(E \otimes_2 F) \otimes_2 G$ onto $E \otimes_2 (F \otimes_2 G)$.

*e*) Let $E_1$ and $F_1$ be two real Hilbert spaces, and $u : E \to E_1$, $v : F \to F_1$ continuous linear mappings. Show that $u \otimes v$ may be extended by continuity to a continuous linear mapping $u \otimes_2 v$ of $E \otimes_2 F$ into $E_1 \otimes_2 F_1$, and that $\|u \otimes_2 v\| = \|u\| \cdot \|v\|$.

9) Let E and F be two real Hilbert spaces.

*a*) Show that there exists a continuous linear mapping $\varphi$ of $E \otimes_2 F$ into $\mathscr{L}(E; F)$ characterized by $\varphi(x \otimes y)(x') = (x|x') \cdot y$ for $x, x'$ in E and $y$ in F. Show that $\varphi$ has norm 1, and that $\varphi(E \otimes F)$ is the set of continuous linear mappings of finite rank of E into F.

*b*) Show that $\varphi$ is a bijection of $E \otimes_2 F$ onto the set HS (E, F) of Hilbert–Schmidt linear mappings of E into F (make use of Exer. 8 *c*) and Prop. 3). Equip HS (E, F) with the Hilbert space structure deduced from that of $E \otimes_2 F$ by means of the bijection $\varphi$; the corresponding norm is denoted $\|u\|_2$. Let $u \in HS(E, F)$; one defines positive quadratic forms $Q_u$ and H on E by $Q_u(x) = \|u(x)\|^2$ and $H(x) = \|x\|^2$. Show that $\|u\|_2^2 = \mathrm{Tr}(Q_u/H)$.

*c*) Let $E_1$ and $F_1$ be two real Hilbert spaces, and $u : E_1 \to E$, $v : F \to F_1$ continuous linear mappings. Let $\varphi_1$ be the isomorphism of $E_1 \otimes_2 F_1$ onto $HS(E_1, F_1)$ defined in a manner analogous to $\varphi$. Show that $v \circ \varphi(t) \circ u = \varphi_1((u^* \otimes v)(t))$ for all $t \in E \otimes_2 F$. From this, deduce that for every Hilbert–Schmidt mapping $w$ of E into F, the linear mapping $v \circ w \circ u$ of $E_1$ into $F_1$ is Hilbert–Schmidt, and that $\|v \circ w \circ u\|_2 \leqslant \|v\| \cdot \|w\|_2 \cdot \|u\|$.

# HISTORICAL NOTE

(N.B. — The Roman numerals refer to the bibliography at the end of this note.)

While the study of the connections between topology and measure theory goes back to the beginnings of the modern theory of functions of real variables, it is only very recently that integration in Hausdorff topological spaces has been thoroughly worked out in general fashion. Before setting out the history of the work that preceded the present synthesis, we review several stages in the evolution of the ideas regarding the relations between topology and measure.

For Lebesgue, it is only a question of integrating functions of one or several real variables. In 1913, Radon defined general measures on $\mathbf{R}^n$ and the corresponding integrals; this theory is exposed in detail in the book (I) of Ch. de la Vallée Poussin and rests, in constant fashion, on the topological properties of Euclidean spaces. A little later, in 1915, Fréchet defined in (II, *a*)) 'abstract' measures on a set equipped with a tribe, and the integrals with respect to these measures; he noted that one can establish in this way the principal results of the Lebesgue theory without the use of topological methods. He justified his undertaking with the following words, taken from the introduction of (II, *b*), first part): « *Que par exemple dans l'espace à une infinité de coordonnées où diverses applications de l'Analyse avaient conduit à diverses définitions non équivalentes d'une suite convergente, on remplace une de ces définitions par une autre, rien ne sera changé dans les propriétés des familles et fonctions additives d'ensembles dans ces espaces* ».(*) Fréchet's investigations were completed by Carathéodory, to whom is due an important theorem on the extension of a set function to a measure. The beginning of the book of Saks (III) gives a condensed exposition of this point of view.

The discovery of Haar measure on locally compact groups (cf. the His-

(*) "While, for example, in the space with infinitely many coordinates where various applications of Analysis had led to various inequivalent definitions of convergent sequence, if one replaces one of these definitions by another, nothing will be changed in the properties of the families and additive functions of sets in these spaces."

torical Note for Chs. VII and VIII) and the numerous applications that it immediately received, then the works of Weil and Gelfand on Harmonic Analysis, led around 1940 to a profound modification of this point of view: in this kind of question, it is most convenient to regard a measure as a linear form on a space of continuous functions. This method obliges one to restrict oneself to compact or locally compact spaces, but this is not an inconvenience for nearly all of the applications: better yet, the introduction of Harmonic Analysis on $p$-adic groups and adèle groups by J. Tate and A. Weil has permitted a spectacular renewal of Analytic Number Theory.

It is from an entirely different direction that the need arises for enlarging this point of view by considering measures on non locally compact topological spaces: gradually, Probability Theory leads to the study of such spaces and furnishes numerous nontrivial examples. Perhaps the reason for the tardy influence of these developments on the theory of measure is to be found in the relative isolation of Probability Theory, which remained until recently on the fringes of the traditional mathematical disciplines.

## Measures on spaces of sequences

One of the most highly developed branches of classical Probability Theory is that of limit theorems (law of large numbers, tendency towards the Gauss–Laplace law, ...); this concerns a deepening of the concept of statistical regularity manifested by phenomena that bring into play very large populations. The correct mathematical formulation of these problems requires the introduction of measures on sequence spaces; these spaces, which constitute the most obvious generalization of finite-dimensional spaces, were the subject of choice of investigations into 'General Analysis' undertaken around 1920 by Fréchet, Lévy, Lusin, .... Nor is it fortuitous that Khintchine and Kolmogoroff, the creators of the new methods of Probability Theory, were both disciples of Lusin, and that Lévy very quickly oriented himself towards probabilistic problems: these constituted the touchstone of the new methods.

The first implicit intervention of a measure on a sequence space appeared in the work devoted by E. Borel in 1909 to countable probabilities (IV). A highly original idea of Borel consists in the application of the probabilistic results he had just obtained to the proof of properties possessed by the decimal expansion of almost every real number between $0$ and $1$. This application rests on the following fundamental remark: let us define every real number between $0$ and $1$ by the sequence of figures (or 'digits') in its expansion in a given base $q$ $(q \geqslant 2)$; if one successively draws at random the various figures of a number $x$, independently of each other and with equal probability $1/q$ for $0, 1, \ldots, q-1$, the probability that $x$ lies in an interval of $[0, 1[$ is equal to the length of that interval.

In 1923, Steinhaus (V) established these results rigorously and described the precise mathematical model for the infinite sequence of random drawings considered by Borel: for simplicity let us take $q = 2$ and denote by I the set with two elements $\{0, 1\}$; one equips I with the measure $\mu$ defined by $\mu(0) = \mu(1) = \frac{1}{2}$; the elements of the product space $\mathrm{I}^{\mathbf{N}}$ are the sequences $\varepsilon = (\varepsilon(n))_{n \in \mathbf{N}}$ of numbers equal to 0 or 1, and the mapping $\varphi : \varepsilon \mapsto \sum_{n \geqslant 0} \varepsilon(n) \cdot 2^{-n-1}$ is, up to a countable set, a bijection of $\mathrm{I}^{\mathbf{N}}$ onto the interval $[0, 1]$; moreover, $\varphi^{-1}$ transforms the Lebesgue measure on $[0, 1]$ into the measure P on $\mathrm{I}^{\mathbf{N}}$ that is the product of the measures $\mu$ on each of the factors. Actually, Steinhaus did not have at his disposal a construction for product measures; he used the existence of the quasi-bijection $\varphi$ to *construct* the measure P on $\mathrm{I}^{\mathbf{N}}$ starting from Lebesgue measure on $[0, 1]$, and then gave an axiomatic characterization of P. The isomorphism so obtained made it possible to translate the language of probability into that of measure and to apply the known theorems on the Lebesgue integral.

In the same work, Steinhaus considered the random series $\sum_{n \geqslant 0} \sigma_n \cdot a_n$, where the signs $\sigma_n = \pm 1$ are chosen at random independently of each other and with equal probability $\frac{1}{2}$; between 1928 and 1935, he studied numerous other random series. From their side, Paley, Wiener and Zygmund considered random Fourier series[(1)] of the form $\sum_{n=-\infty}^{\infty} a_n \exp\big(2\pi i(nt + \Phi_n)\big)$; the 'amplitudes' $a_n$ are fixed, and the 'phases' $\Phi_n$ are independent random variables uniformly distributed on $[0, 1]$. While the analytic difficulties vary enormously from one of these problems to another, the translation in terms of measure theory is the same in all cases and represents an extension of the case treated by Borel and Steinhaus; it is a matter of constructing a measure on $\mathbf{R}^{\mathbf{N}}$ that is the product of a family of measures all identical to a same positive measure $\mu$ on $\mathbf{R}$ of mass 1; for example, the preceding random Fourier series correspond to the case that $\mu$ is the Lebesgue measure on $[0, 1]$.

To construct such product measures, two methods can be used. The first is a direct method, precisely exposed for the first time by Daniell (VI, *a*)) in 1918; it was rediscovered in 1934 by Jessen (VII), who made a detailed study of the case that $\mu$ is Lebesgue measure on $[0, 1]$. The second method is to seek artifices analogous to that of Steinhaus to reduce to Lebesgue measure on $[0, 1]$; this way of proceeding had the advantage of convenience

(1) For an account of random Fourier series, see the exposition of J. P. KAHANE in the *Séminaire Bourbaki* (No. 200, 12th year, 1959/60, Benjamin, New York).

so long as one did not have available a complete exposition of general measure theory, for it permitted using Lebesgue's theorems without the need for new proofs.(2)

## The theory of Brownian motion

This theory occupies a special position in contemporary scientific development, by the constant and fertile exchange between physical problems and 'pure' mathematics to which it bears witness. The study of Brownian motion, discovered in 1829 by the botanist Brown, had been conducted intensively in the 19th century by numerous physicists,(3) but the first satisfactory mathematical model was only invented by Einstein in 1905. In the simple case of a particle moving along a straight line, Einstein's fundamental hypotheses may be formulated as follows: if $x(t)$ is the abscissa of the particle at the instant $t$, and if $t_0 < t_1 < \cdots < t_{n-1} < t_n$, the successive displacements $x(t_i) - x(t_{i-1})$ (for $1 \leqslant i \leqslant n$) are independent Gaussian random variables. This is not the place to evoke in detail the important experimental work of J. Perrin that motivated Einstein's theory; for our purposes, we need only retain a remark of Perrin, according to which the observation of trajectories of Brownian motion irresistibly suggested to him the "mathematicians' functions without derivative". This remark was to be the initial spark for Wiener.

Quite another current of ideas has its origins in the kinetic theory of gases, developed between 1870 and 1900 by Boltzmann and Gibbs. Consider a gas formed by N molecules of mass $m$ at (absolute) temperature T, and denote by $\mathbf{v}_1, \ldots, \mathbf{v}_N$ the velocities of the N molecules of the gas; the kinetic energy of the system is equal to

$$\frac{m}{2}(\mathbf{v}_1^2 + \cdots + \mathbf{v}_N^2) = 3\mathrm{N}k\mathrm{T}, \tag{1}$$

where $k$ is Boltzmann's constant. According to the ideas of Gibbs, the multitude of shocks between molecules does not allow the precise determination of the velocities of the molecules, and it is convenient to introduce a probability law P on the sphere S of the space of dimension 3N defined by the equation (1). The 'micro-canonical' hypothesis consists in assuming

(2) Wiener also took care on numerous occasions (cf. for example (XI), Ch. IX) to show that the measure of Brownian motion is isomorphic to Lebesgue measure on $[0,1]$. The possibility of such artifices finds its explanation in a general theorem of von Neumann that gives an axiomatic characterization of the measures isomorphic to Lebesgue measure on $[0,1]$.

(3) A very lively account of this history may be found in the recent book of E. Nelson, Dynamical theories of Brownian motion, *Mathematical Notes*, Princeton, 1967.

that P is the measure on the sphere S with mass 1 invariant under rotation. Moreover, Maxwell's law of velocities states that the law of probability of a component of the velocity of a molecule is a Gaussian measure with variance $2k\mathrm{T}/m$ (§6, No. 5, *Remark* 3). Borel seems to have been the first to observe in 1914 that Maxwell's law is a consequence of Gibbs' hypotheses and the properties of the sphere when the number of molecules is very large. He considered a sphere S in a Euclidean space of large dimension and the measure P of mass 1 on S invariant under rotation; using the classical approximation methods based on Stirling's formula, he showed that the projection of P on a coordinate axis is approximately Gaussian. These results were sharpened a little later by Gâteaux and Lévy (IX, *a*)). Given an integer $m \geqslant 1$ and a number $r > 0$, denote by $\mathrm{S}_{m,r}$ the set of sequences of the form $(x_1, \ldots, x_m, 0, 0, \ldots)$ with $x_1^2 + \cdots + x_m^2 = r^2$; also, denote by $\sigma_{m,r}$ the measure with mass 1 on $\mathrm{S}_{m,r}$ invariant under rotation. Stated in modern language, the result of Gâteaux and Lévy is as follows: the sequence of measures $\sigma_{m,1}$ tends tightly to a unit mass at the origin $(0, 0, \ldots)$, and the sequence of measures $\sigma_{m,\sqrt{m}}$ tends tightly to a measure $\Gamma$ of the form

$$d\Gamma(x_1, x_2, \ldots) = \prod_{n=1}^{\infty} d\gamma(x_n)$$

($\gamma$ is the Gaussian measure on $\mathbf{R}$ with variance 1).

The preceding measure $\Gamma$ plays the role of a Gaussian measure in infinite dimensions. It seems indeed that Lévy had confusedly hoped to define in an intrinsic manner a Gaussian measure on every infinite-dimensional Hilbert space. In fact, as shown by Lévy and Wiener, the measure $\Gamma$ is invariant in a certain sense(4) under the automorphisms of $l^2$; unfortunately, the set $l^2$ of square-summable sequences $(x_1, x_2, \ldots, x_n, \ldots)$ has measure zero for $\Gamma$. It is now known that one must be content to have a Gaussian *promeasure* on an infinite-dimensional Hilbert space.(5)

We owe to Wiener the essential progress: if one does not have a reasonable Gaussian measure on an infinite-dimensional Hilbert space, one can

(4) More precisely, one has the following result. Let $U$ be an automorphism of the Hilbert space $l^2$, and $(u_{mn})$ the matrix of $U$. Let E be the vector space of all real sequences $(x_n)_{n\geqslant 1}$ and F the subspace of E formed by the sequences $(x_n)_{n\geqslant 1}$ for which the series $\sum_{n\geqslant 1} u_{mn}x_n$ converges for every $m \geqslant 1$. The formula $(\widetilde{U}x)_m = \sum_{n\geqslant 1} u_{mn}x_n$ defines a linear mapping $\widetilde{U}$ of F into E, the measure $\Gamma$ is concentrated on F, and $\widetilde{U}(\Gamma) = \Gamma$.

(5) This concept was introduced under the name "weak canonical distribution" by I. E. SEGAL (*Trans. Amer. Math. Soc.* **88** (1958), 12–42). One owes to this author a detailed study of Gaussian promeasures, and their application to certain problems in the quantum theory of fields.

construct by the operation of primitive a measure $w$ on a space of continuous functions starting from a Gaussian promeasure (cf. §6, No. 7, Th. 1 for the details). We shall explain succinctly Wiener's original construction of $w$ (X); it is directly influenced by the relation $\Gamma = \lim_{m\to\infty} \sigma_{m,\sqrt{m}}$ of Gâteaux and Lévy. For every integer $m \geqslant 1$, denote by $\mathrm{H}_m$ the set of functions on $\mathrm{T} = ]0,1]$ that are constant on each of the intervals $\left]\frac{k-1}{m}, \frac{k}{m}\right]$ (for $k = 1, 2, \ldots, m$), and by $\pi_m$ the measure of mass $1$, invariant under rotation, on the Euclidean sphere of radius 1 in $\mathbf{R}^m$. Let $f_m$ be the isomorphism of $\mathrm{H}_m$ onto $\mathbf{R}^m$ that associates, to each function taking the value $a_k$ on the interval $\left]\frac{k-1}{m}, \frac{k}{m}\right]$, the vector $(a_1, a_2 - a_1, \ldots, a_m - a_{m-1})$ (whence the term 'differential space' dear to Wiener); denote by $w_m$ the measure on $\mathrm{H}_m$ that is the image of $\pi_m$ under $f_m^{-1}$. Wiener defined the desired measure $w$ as the limit of the measures $w_m$. To be precise, let us denote by H the set of regulated functions on T, with the topology of uniform convergence (we have $\mathrm{H}_m \subset \mathrm{H}$ for every integer $m \geqslant 1$); for every bounded uniformly continuous function F on H, the limit $\mathrm{A}\{\mathrm{F}\} = \lim_{m\to\infty} \int_{\mathrm{H}_m} \mathrm{F}(x)\, dw_m(x)$ exists; next, Wiener obtained certain upper bounds by a subtle analysis of the fluctuations of the game of heads-or-tails, and taking up again the compactness arguments highlighted by Daniell, he showed that one is under the conditions for applying Daniell's extension theorem. One concludes the existence of a measure $w$ carried by $\mathscr{C}(\mathrm{T})$ and such that $\mathrm{A}\{\mathrm{F}\} = \int_{\mathscr{C}(\mathrm{T})} \mathrm{F}(x)\, dw(x)$. Wiener was then able to show that the measure $w$ corresponds to Einstein's hypotheses,[6] and his estimates allowed him to give a precise meaning to Perrin's remark on functions without derivatives: the set of functions satisfying a Lipschitz condition of order $\frac{1}{2}$ is negligible for $w$ (however, for every $a$ with $0 < a < \frac{1}{2}$, almost every function satisfies a Lipschitz condition of order $a$).

Today, numerous constructions of the Wiener measure are known. Thus, Paley and Wiener use random Fourier series (XI, Ch. IX): for every real

---

[6] This may be translated by the formula

$$\int_{\mathscr{C}(\mathrm{T})} f(x(t_1), \ldots, x(t_n))\, dw(x) =$$

$$(2\pi)^{-n/2} \prod_{i=1}^{n} (t_i - t_{i-1})^{-1/2} \int \cdots \int f(x_1, \ldots, x_n) \exp\left(-\frac{1}{2} \sum_{i=1}^{n} \frac{(x_i - x_{i-1})^2}{t_i - t_{i-1}}\right) dx_1 \cdots dx_n,$$

where $f$ is an arbitrary bounded continuous function on $\mathbf{R}^n$ and where $0 = t_0 < t_1 < \cdots < t_n \leqslant 1$ (one makes the convention $x_0 = 0$). Wiener, trained in analytical rigor by Hardy, and justly mistrustful of the foundations of Probability Theory at that time, was careful not to use probabilistic terminology or results. It follows that his memoirs are full of formidable formulas of which the foregoing is a sample; this circumstance is one of the factors that delayed the diffusion of Wiener's ideas.

sequence $\mathbf{a} = (a_n)_{n\geqslant 1}$ and every integer $m \geqslant 0$, let us define the function $f_{m,\mathbf{a}}$ on $]0,1]$ by

$$f_{m,\mathbf{a}}(t) = a_1 t + 2 \sum_{k=2}^{2^{m+1}} \frac{1}{\pi k} a_{k-1} \sin \pi k t\,;$$

one can show that for $\Gamma$-almost every sequence $\mathbf{a}$, the sequence of functions $f_{m,\mathbf{a}}$ tends to a continuous function $f_{\mathbf{a}}$, and that $w$ is the image of $\Gamma$ under the mapping (defined almost everywhere) $\mathbf{a} \mapsto f_{\mathbf{a}}$. Later on, Lévy gave in (IX, *b*), *c*)) a construction very near to that which we have exposed in §6, No. 7. Finally, Kac, Donsker and Erdös showed around 1950 how to replace the spherical measures $\pi_m$ on $\mathbf{R}^m$ in Wiener's original construction by more general measures. Their results establish a solid link between Wiener measure and the limit theorems of Probability Theory; they were to be completed and systematized by Prokhorov (XIII) in a work to which we shall return later on.

This is not the place to analyze the numerous and important probabilistic works brought about by Wiener's discovery; nowadays, Brownian motion appears only as one of the most important examples of a Markoff process. We shall only mention Kac's application of Wiener measure to the solution of certain parabolic partial differential equations; this is a question of adapting ideas of Feynmann in quantum field theory—yet another example of the reciprocal influence of mathematics and problems of physics.

## Inverse limits of measures

This is a theory that has developed mainly in response to the needs of Probability Theory. Problems concerning a finite sequence of random variables $X_1, \ldots, X_n$ are in principle solved when one knows the law $P_X$ of the sequence: this is a positive measure on $\mathbf{R}^n$ of mass 1 such that the probability of obtaining simultaneously the inequalities $a_1 \leqslant X_1 \leqslant b_1$, $\ldots$, $a_n \leqslant X_n \leqslant b_n$ is equal to $P_X(C)$, where C is the closed box $[a_1, b_1] \times \cdots \times [a_n, b_n]$ in $\mathbf{R}^n$. In practice, the measure $P_X$ either has discrete support or admits a density with respect to Lebesgue measure. When dealing with an infinite sequence $(X_n)_{n\geqslant 1}$ of random variables, one generally knows the law $P_n$ of the partial sequence $(X_1, \ldots, X_n)$ for every integer $n \geqslant 1$; these data satisfy a compatibility condition that expresses that the sequence $(P_n)_{n\geqslant 1}$ is an inverse system (or 'projective system') of measures. Until about 1920, the probabilities of events relating to the infinite sequence were defined in more or less implicit fashion by 'natural' passages to the limit from the probabilities of the finite case; one thus assumes that the probability that a game terminates is the limit, as $n$ tends to infinity, of the probability that it terminates in at most $n$ steps. Naturally, such a theory is not very coherent, and nothing excludes the presence of 'paradoxes', the same probability receiving two distinct estimates according as one evaluates by means of one or the other of two procedures, each as 'natural' as the other.

Steinhaus (V) appears to have been the first to have felt the need for considering (for the game of heads-or-tails) not only the inverse system $(\mathrm{P}_n)_{n\geqslant 1}$ but its limit as well. A little earlier, in 1919, Daniell (VI, *b*)) had proved in general the existence of such inverse limits,(7) but this result seems to have remained unknown in Europe. It was rediscovered in 1933 by Kolmogoroff in the work (XII) where the author formulates the axiomatic conception of Probability Theory. The proofs of Daniell and Kolmogoroff make use of a compactness argument, which is substantially the same as the one we have employed in Th. 2 of §4, No. 3 and which rests on Dini's theorem.

The Daniell–Kolmogoroff theorem left nothing to be desired for the case of random sequences $(\mathrm{X}_n)_{n\geqslant 1}$, but the study of random functions undertaken from 1935 on by Kolmogoroff, Feller and Doob harbors difficulties of quite another order. Consider for example an interval T of $\mathbf{R}$, representing the set of instants of observation of a 'stochastic process'; the set of possible trajectories is the product space $\mathbf{R}^{\mathrm{T}}$, regarded as the inverse limit of the partial products $\mathbf{R}^{\mathrm{H}}$, where H runs over the set of finite subsets of T; one generally assumes given an inverse system of measures $(\mu_{\mathrm{H}})$ (cf. §4, No. 2). Kolmogoroff's theorem does indeed yield a measure on $\mathbf{R}^{\mathrm{T}}$, but it is defined on a tribe notably smaller than the Borel tribe.(8) A variant of Kolmogoroff's construction, which yields a measure on a topological space, is due to Kakutani (*Proc. Imp. Acad. Tokyo* **19** (1943), 184–188), and has been rediscovered several times since: one regards $\mu_{\mathrm{H}}$ as a measure on $\overline{\mathbf{R}}^{\mathrm{H}}$ carried by $\mathbf{R}^{\mathrm{H}}$;(9) the compact space $\mathrm{E} = \overline{\mathbf{R}}^{\mathrm{T}}$ is the inverse limit of the finite products $\overline{\mathbf{R}}^{\mathrm{H}}$ and one can define a measure $\mu$ on E as the inverse limit of the $\mu_{\mathrm{H}}$ (cf. Ch. III, §4, No. 5). However, this procedure has a serious inconvenience; the elements of $\overline{\mathbf{R}}^{\mathrm{T}}$ possess no regularity property that permits advancing the probabilistic study of the process—or even simply

(7) Daniell treated the case of measures on a product $\prod_{n\geqslant 1} \mathrm{I}_n$ of compact intervals of $\mathbf{R}$, but his method extends immediately to the case of an arbitrary product of compact spaces; this is essentially the method we have used in Ch. III, §4, No. 5.

(8) Kolmogoroff's measure is only defined for the Borel sets in $\mathbf{R}^{\mathrm{T}}$ of the form $\mathrm{A} \times \mathbf{R}^{\mathrm{T}-\mathrm{D}}$, where D is a countable subset of T, and A is a Borel subset of $\mathbf{R}^{\mathrm{D}}$; because of this, Kolmogoroff's theorem for an arbitrary product $\mathbf{R}^{\mathrm{T}}$ is an immediate consequence of the case of countable products.

(9) One could replace $\overline{\mathbf{R}}$ by any compact space containing $\mathbf{R}$ as a dense subspace.

eliminating the parasitic values $\pm\infty$ introduced by the compactification $\overline{\mathbf{R}}$ of $\mathbf{R}$. This can be remedied by inducing the measure $\mu$ of $\overline{\mathbf{R}}^{\mathrm{T}}$ on a particular subspace (for example $\mathscr{C}(\mathrm{T})$ in the case of Brownian motion); the fundamental difficulty arises from the fact that a function space, even one of the usual type, is not necessarily $\mu$-measurable in $\overline{\mathbf{R}}^{\mathrm{T}}$, and even the choice of function space may be questionable.[10]

A decisive step was taken in 1956 by Prokhorov in a work (XIII) that has had a determining influence on the theory of stochastic processes. By putting into a suitable axiomatic form the methods used by Wiener in the article analyzed above, he established a general existence theorem for inverse limits of measures on function spaces that is a special case of Th. 1 of §4, No. 2 corresponding to Polish spaces.

A more restricted class of inverse systems was introduced by Bochner (XIV) in 1947; these are inverse systems formed by finite-dimensional real vector spaces and surjective linear mappings. The inverse limit of such a system may be identified in a natural way with the algebraic dual $\mathrm{E}^*$ of a real vector space $\mathrm{E}$, equipped with the weak topology $\sigma(\mathrm{E}^*, \mathrm{E})$; a corresponding inverse system of measures has a limit that is a measure $\mu$ defined on a tribe notably smaller than the Borel tribe of $\mathrm{E}^*$. Bochner completely characterized such 'promeasures' by their Fourier transform, which is a function on $\mathrm{E}$. But this result is scarcely usable in the absence of a topology on $\mathrm{E}$, in which case one has to examine the possibility of regarding $\mu$ as a measure on the topological dual $\mathrm{E}'$ of $\mathrm{E}$. In an independent way, R. Fortet and E. Mourier, while seeking to generalize to random variables with values in a Banach space certain classical results of Probability Theory (law of large numbers, central limit theorem) also brought into evidence the fundamental role played by the Fourier transformation in such questions. But substantial progress was not made until 1956 when Gelfand (XV, *b*)) suggested that the natural setting for the Fourier transformation is not that of Banach spaces or Hilbert spaces, but that of nuclear Fréchet spaces. He conjectured that every continuous function of positive type on such a space is the Fourier transform of a measure on its dual, a result established soon afterwards by Minlos (XVI). Its importance stems above all from the fact that it is applicable to spaces of distributions, and that the quasi-totality of function spaces are Borel subsets of the space of distributions (which thus constitutes a much better receptacle than $\mathbf{R}^{\mathrm{T}}$).[11] The theory of random distributions is an area in full expansion, and we shall content ourselves by referring the

---

[10] For a detailed discussion of the problem of constructing measures on function spaces, and the methods used prior to Prokhorov, see J. L. Doob, *Bull. Amer. Math. Soc.* **53** (1947), 15–30.

[11] One may consult the exposition of X. Fernique, *Ann. Inst. Fourier* **17** (1967), 1–92, which also contains numerous results on tight convergence.

reader to the book of Gelfand and Vilenkin (XVII).

The results on inverse limits just mentioned make use of the existence of topologies on the base spaces. One may ask whether there exists an analogous theory in the case of 'abstract' measures. Von Neumann proved in 1935 the existence of product measures in all cases, but the discovery of a counter-example by Jessen and Andersen (XVIII) dashed the hope that every inverse system of measures admits a limit. Two palliatives have been discovered: in 1949, C. Ionescu-Tulcea established the existence of countable inverse limits, by means of the existence of suitable disintegrations,[12] a highly interesting result for the study of Markoff processes; moreover, it had been observed that the topology of the spaces only intervenes through the intermediary of the set of compact subsets. It was therefore natural to try to axiomatize this situation within the abstract theory, by means of the concept of compact class of subsets of a set. This work was done in 1953 by Marczewski (who established an abstract inverse limit theorem by such means) and Ryll-Nardzewski (who treated the disintegration of measures).[13]

## Measures on general topological spaces and tight convergence

The study of the connections between topology and measure theory has above all been conceived of as the study of regularity properties of measures, and in particular that of 'outer' regularity and of 'inner' regularity;[14] inner regularity is equivalent to outer regularity on a locally compact space countable at infinity. The construction that Lebesgue gives to the measure of subsets of the line highlights these two kinds of regularity, and the outer regularity property of measures on a Polish space seems to have had public notoriety around 1935. But it was not until 1940, in an article whose diffusion was retarded by the war, that A. D. Alexandroff (XIX) gave prominence to the role of inner regularity and showed that it is possessed by the measures on a Polish space; this result was rediscovered later by Prokhorov (XIII) and is often erroneously attributed to this author. It was not perceived until very recently that this property extends to Souslin spaces; because of this, the

(12) It seems that it is the absence of a satisfactory theory of disintegration that marks the limit of the theory of 'abstract' measures. This difficulty reappears in insistent fashion in Probability Theory in connection with conditional probabilities.

(13) For an exposition of this theory, one may consult J. PFANZAGL and W. PIERLO, *Lecture Notes in Mathematics* (Springer–Verlag), Vol. 16 (1966).

(14) An 'abstract' measure $\mu$ on the Borel tribe of a Hausdorff topological space is said to be outer regular if the measure of every Borel set is the infimum of the measures of the open sets that contain it; the measure $\mu$ is said to be inner regular if the measure of every Borel set is the supremum of the measures of its compact subsets.

importance of these spaces has increased greatly, even more so as it was realized that their theory could be worked out without any metrizability hypothesis, and that the quasi-totality of the function spaces were Souslin (most often, even Lusin).(15) These are the reasons that moved us to place the accent on inner regular measures in this chapter.

The definition of a mode of convergence (vague or tight) for measures is most conveniently done by putting the space of measures into duality with a space of continuous functions. Generalizing an old result of F. Riesz, A. A. Markoff established in 1938 a one-to-one correspondence between the positive functionals on $\mathscr{C}(X)$ and the regular measures on a compact space X. In the work (XIX) already cited, A. D. Alexandroff extends these results to the case of a completely regular space: he introduces a hierarchy in the set of positive linear forms on the space $\mathscr{C}^b(X)$ of bounded continuous functions on a completely regular space X,(16) he defines tight convergence of bounded measures and proves among others the following two theorems:

*a*) if X is Polish, the set of linear forms on $\mathscr{C}^b(X)$ corresponding to the measures is closed for the weak convergence of sequences;

*b*) if a sequence of bounded measures has a tight limit, 'no mass escapes at infinity' (this is a weak form of the converse of Prokhorov's theorem on tight convergence).

From this abundance of concepts and theorems, Prokhorov was able to extract the results important for the theory of stochastic processes, and to present them in a simple and striking form. In his great work of 1956 already cited (XIII), a large part is devoted to bounded positive measures on a Polish space; generalizing a construction of Lévy, he defines a metric on the set of positive measures of mass 1 that makes it a Polish space, and then establishes an important compactness criterion for tight convergence (cf. §5, No. 5, Th. 1). Independently of Prokhorov, Le Cam (XX) obtained a number of compactness results for the tight convergence of measures; he makes no metrizability hypothesis on the spaces he considers, and his results reduce to earlier theorems of Dieudonné in the locally compact case.

---

(15) To attempt to resolve certain probabilistic difficulties (specifically the relations between various notions of stochastic independence or dependence), a number of authors introduced restricted classes of 'abstract' measures: the 'perfect' spaces of Kolmogoroff–Gnedenko, the 'Lusin' spaces of Blackwell, the 'Lebesgue' spaces of Rokhlin. In fact (at least assuming a rather weak countability hypothesis), all of these definitions give characterizations of 'abstract' measures isomorphic to a bounded positive measure on a Souslin space. On this subject, one may consult the work cited in footnote (13).

(16) He distinguishes by decreasing order of generality between the $\sigma$-measures ('abstract' measures on the Borel tribe of X), the $\tau$-measures (outer regular measures) and the taut measures (inner regular measures). When X is Polish, these three notions coincide. The terminology itself is due to McShane and Le Cam (XX). One can find an account of the works to which this classification has given rise in V. S. VARADARAJAN (*Amer. Math. Soc. Transl.* (2), **48**, 161–228).

# BIBLIOGRAPHY

(I) Ch. DE LA VALLÉE POUSSIN, *Intégrales de Lebesgue, Fonctions d'ensembles, Classes de Baire*, Paris (Gauthier–Villars), 1st. edn., 1916; 2nd. edn., 1936.

(II) M. FRÉCHET: *a*) Sur l'intégrale d'une fonctionelle étendue à un ensemble abstrait, *Bull. Soc. Math. France,* **43** (1915), pp. 248–265; *b*) Les familles et fonctions additives d'ensembles abstraits, *Fund. Math.*, **4** (1923), pp. 329–365; *ibid.* **5** (1924), pp. 206-251.

(III) S. SAKS, *Theory of the Integral*, 2nd. edn., New York (Stechert), 1937.

(IV) E. BOREL, Les probabilités dénombrables et leurs applications arithmétiques, *Rend. Circ. Math. Palermo*, **27** (1909), pp. 247–271.

(V) H. STEINHAUS, Les probabilités dénombrables et leur rapport à la théorie de la mesure, *Fund. Math.,* **4** (1923), pp. 286–310.

(VI) P. J. DANIELL: *a*) Integrals in an infinite number of dimensions, *Ann. of Math.,* **20** (1918–19), pp. 281–288; *b*) Functions of limited variation in an infinite number of dimensions, *ibid.* **21** (1919–20), pp. 30–38.

(VII) B. JESSEN, The theory of integration in a space of an infinite number of dimensions, *Acta Math.*, **63** (1934), pp. 249–323.

(VIII) A. EINSTEIN, *Investigations on the Theory of the Brownian Movement*, New York (Dover), 1956.

(IX) P. LÉVY: *a*) *Leçons d'Analyse Fonctionnelle*, Paris (Gauthier–Villars), 1922 (the second edition appeared in 1951 with the same publisher, under the title *Problèmes concrets d'Analyse Fonctionnelle*); *b*) *Processus stochastiques et mouvement brownien*, Paris (Gauthier–Villars), 1948; *c*) Le mouvement brownien, *Mémor. Sci. Math.*, **126** (1954).

(X) N. WIENER, Differential space, *J. Math. Phys. MIT*, **2** (1923), pp. 131–174 (= *Selecta*, pp. 55–98, Cambridge (MIT Press), 1964).

(XI) R. E. A. C. PALEY and N. WIENER, *Fourier transforms in the complex domain*, Amer. Math. Soc. Colloq. Publ. No. 19, New York, 1934.

(XII) A. N. KOLMOGOROFF, *Grundbegriffe der Wahrscheinlichkeitsrechnung*, Berlin (Springer), 1933.

(XIII) JU. V. PROKHOROV, Convergence of random processes and limit theorems in probability theory, *Theor. Probability Appl.*, **1** (1956), pp. 156–214.

(XIV) S. BOCHNER, *Harmonic Analysis and the Theory of Probability*, Berkeley (University of California Press), 1960.

(XV) I. M. GELFAND: *a*) Generalized stochastic processes (in Russian), *Dokl. Akad. Nauk SSSR*, **100** (1955), pp. 853–856; *b*) Some problems of functional analysis (in Russian), *Uspehi Mat. Nauk*, **11** (1956), pp. 3–12 (= *Amer. Math. Soc. Transl. (2)*, **16** (1960), pp. 315–324).

(XVI) R. A. MINLOS, Generalized random processes and their extension to a measure (in Russian), *Trudy Moskov. Math. Obšč.*, **8** (1959), pp. 497–518 (= *Selected translations in mathematical statistics and probability*, III (19), pp. 291–313).

(XVII) I. M. GELFAND and N. Ya. VILENKIN, *Generalized Functions*, Vol. IV, New York (Academic Press), 1964 (English translation).

(XVIII) E. SPARRE-ANDERSSEN and B. JESSEN, On the introduction of measures in infinite product sets, *Dansk. Vid. Selbskab. Mat. Fys. Medd.*, **25** (1948), No. 4, pp. 1-7.

(XIX) A. D. ALEXANDROFF, Additive set functions in abstract spaces. I (Ch. 1), *Mat. Sbornik*, **8** (1940), pp. 307–348; II (Chs. 2 and 3), *ibid.*, **9** (1941), pp. 563–628; III (Chs. 4 to 6), *ibid.*, **13** (1943), pp. 169–238.

(XX) L. LE CAM, Convergence in distribution of stochastic processes, *Univ. Cal. Publ. Statistics, No. 11* (1957), pp. 207–236.

# Index of notations

Reference numbers indicate, in order, the chapter, section and subsection.

*Chapter* IX :

# Index of terminology

Reference numbers indicate, in order, the chapter, section and subsection (or, exceptionally, exercise).

## Principal formulas of Chapter VII

*Formulas concerning the* $\boldsymbol{\gamma}(s)$ *and the* $\boldsymbol{\delta}(s)$

Let G be a topological group operating continuously on the left in a locally compact space X by $(s, x) \mapsto sx$.

| | |
|---|---|
| $\boldsymbol{\gamma}(s)x = sx$ | $(s \in \mathrm{G},\ x \in \mathrm{X})$ |
| $\boldsymbol{\gamma}(st) = \boldsymbol{\gamma}(s)\boldsymbol{\gamma}(t)$ | $(s,\ t$ in G$)$ |
| $(\boldsymbol{\gamma}(s)f)(x) = f(s^{-1}x)$ | ($f$ a function on X) |
| $\langle f, \boldsymbol{\gamma}(s)\mu\rangle = \langle \boldsymbol{\gamma}(s^{-1})f, \mu\rangle$ | ($\mu$ a measure on X) |
| $d(\boldsymbol{\gamma}(s)\mu)(x) = d\mu(s^{-1}x)$ | |
| $(\boldsymbol{\gamma}(s)\mu)(\mathrm{A}) = \mu(s^{-1}\mathrm{A})$ | (A a $\boldsymbol{\gamma}(s)\mu$-integrable set) |

If $\mu$ is relatively invariant with multiplier $\chi$,

$$\boldsymbol{\gamma}(s)\mu = \chi(s)^{-1}\mu$$
$$d\mu(sx) = \chi(s)\,d\mu(x).$$

Let G be a topological group operating continuously on the right in a locally compact space X by $(s, x) \mapsto xs$.

$$\boldsymbol{\delta}(s)x = xs^{-1}$$
$$\boldsymbol{\delta}(st) = \boldsymbol{\delta}(s)\boldsymbol{\delta}(t)$$
$$(\boldsymbol{\delta}(s)f)(x) = f(xs)$$
$$\langle f, \boldsymbol{\delta}(s)\mu\rangle = \langle \boldsymbol{\delta}(s^{-1})f, \mu\rangle$$
$$d(\boldsymbol{\delta}(s)\mu)(x) = d\mu(xs)$$
$$(\boldsymbol{\delta}(s)\mu)(\mathrm{A}) = \mu(\mathrm{A}s).$$

If $\mu$ is relatively invariant with multiplier $\chi'$,

$$\boldsymbol{\delta}(s)\mu = \chi'(s)\mu$$
$$d\mu(xs) = \chi'(s)\,d\mu(x).$$

*Formulas concerning Haar measures*

Let G be a locally compact group, $\Delta$ its modulus, $\mu$ a left Haar measure, $\nu$ a right Haar measure.

1) One has

$$\boldsymbol{\gamma}(s)\mu = \mu \qquad \boldsymbol{\delta}(s)\mu = \Delta(s)\mu \qquad \check{\mu} = \Delta^{-1}\cdot\mu$$
$$d\mu(sx) = d\mu(x) \qquad d\mu(xs) = \Delta(s)\,d\mu(x) \qquad d\mu(x^{-1}) = \Delta(x)^{-1}\,d\mu(x).$$

If $f$ is $\mu$-integrable,

$$\int f(sx)\,d\mu(x) = \int f(x)\,d\mu(x) \qquad \int f(xs)\,d\mu(x) = \Delta(s)^{-1}\int f(x)\,d\mu(x)$$
$$\int f(x^{-1})\Delta(x)^{-1}\,d\mu(x) = \int f(x)\,d\mu(x)\,.$$

If $\mathrm{A}\subset\mathrm{G}$ is $\mu$-integrable,

$$\mu(s\mathrm{A}) = \mu(\mathrm{A}) \qquad \mu(\mathrm{A}s) = \Delta(s)\mu(\mathrm{A})\,.$$

2) One has

$$\boldsymbol{\delta}(s)\nu = \nu \qquad \boldsymbol{\gamma}(s)\nu = \Delta(s)\nu \qquad \check{\nu} = \Delta\cdot\nu$$
$$d\nu(xs) = d\nu(x) \qquad d\nu(s^{-1}x) = \Delta(s)\,d\nu(x) \qquad d\nu(x^{-1}) = \Delta(x)\,d\nu(x).$$

If $f$ is $\nu$-integrable,

$$\int f(xs)\,d\nu(x) = \int f(x)\,d\mu(x) \qquad \int f(sx)\,d\nu(x) = \Delta(s)\int f(x)\,d\nu(x)$$
$$\int f(x^{-1})\Delta(x)\,d\nu(x) = \int f(x)\,d\nu(x)\,.$$

If $\mathrm{A}\subset\mathrm{G}$ is $\nu$-integrable,

$$\nu(\mathrm{A}s) = \nu(\mathrm{A}) \qquad \nu(s\mathrm{A}) = \Delta(s^{-1})\nu(\mathrm{A})\,.$$

3) $\nu$ is proportional to $\Delta^{-1}\cdot\mu$, $\mu$ is proportional to $\Delta\cdot\nu$.

## CONDITIONS SUFFICIENT FOR THE EXISTENCE OF THE CONVOLUTION PRODUCT

I. — The case that the convolution product $\mu * \nu$ of two measures exists:

($a$) $*$ is defined by a continuous mapping $\varphi : X \times Y \to Z$:
$\mu$, $\nu$ bounded (then $\mu * \nu$ is bounded and $\|\mu * \nu\| \leqslant \|\mu\| \cdot \|\nu\|$).
$\mu$, $\nu$ have compact support (then $\mu * \nu$ has compact support and $\operatorname{Supp}(\mu * \nu) \subset \varphi(\operatorname{Supp}\mu \times \operatorname{Supp}\nu)$).

($b$) $*$ is defined by a group operating continuously on the left in a space: $\mu$ with compact support, $\nu$ arbitrary.

($c$) $*$ is defined by the multiplication in a group G:
one of the two measures has compact support.
$\mu$, $\nu$ in $\mathscr{M}^{\rho}(G)$ (then $\mu * \nu \in \mathscr{M}^{\rho}(G)$, and $\|\mu * \nu\|_{\rho} \leqslant \|\mu\|_{\rho}\|\nu\|_{\rho}$).

II. — The case that the convolution product $\mu * f$ of a measure and a function exists:

($a$) $*$ is defined by a group G operating continuously on the left in a space X equipped with a measure $\beta \geqslant 0$ such that $\gamma(s)\beta = \chi(s^{-1}, \cdot)\beta$, $\chi$ being continuous:
$\mu$ with compact support, $f$ locally $\beta$-integrable (if $f$ is continuous, $\mu * f$ is continuous; if $f$ is continuous with compact support, $\mu * f$ is continuous with compact support).
G operates properly in X, $f \in \mathscr{K}(X)$ ($\mu * f$ is continuous).

($b$) the $\chi(s, \cdot)$ are bounded; let $\rho(s) = \sup_{x \in X} \chi(s^{-1}, x)$:
$\mu \in \mathscr{M}^{\rho}(G)$, $f \in \overline{L^{\infty}(X, \beta)}$ (then $\overline{\mu * f} \in L^{\infty}(X, \beta)$; if $f \in \mathscr{C}^{\infty}(X)$, $\mu * f \in \mathscr{C}^{\infty}(X)$; if $f \in \overline{\mathscr{K}(X)}$, $\mu * f \in \overline{\mathscr{K}(X)}$).
$\mu \in \mathscr{M}^{\rho^{1/q}}(G)$, $f \in L^p(X, \beta)$ where $1/p + 1/q = 1$ (then $\mu * f \in L^p(X, \beta)$ and $\|\mu * f\|_p \leqslant \|\mu\|_{\rho^{1/q}}\|f\|_p$).

III. — The case that the convolution product $f * g$ of two locally $\beta$-integrable functions exists ($\beta$ a relatively invariant measure $\geqslant 0$ on a group G, with left and right multipliers $\chi$ and $\chi'$):

$f$ or $g$ continuous, $f$ or $g$ with compact support (then $f * g$ is continuous; if $f$, $g$ are in $\mathscr{K}(G)$, then $f * g \in \mathscr{K}(G)$).

$f\chi^{-1/q} \in L^1(G,\beta)$ and $g \in L^p(G,\beta)$, where $1/p + 1/q = 1$ (then $f * g \in L^p(G,\beta)$ and $\|f * g\|_p \leqslant \|f\chi^{-1/q}\|_1 \|g\|_p$).

$f \in L^p(G,\beta)$ and $g\chi'^{-1/q} \in L^1(G,\beta)$ (then $f * g \in L^p(G,\beta)$ and $\|f * g\|_p \leqslant \|f\|_p \|g\chi'^{-1/q}\|_1$).

$f\chi^{-1} \in L^1(G,\beta)$ and $g \in \mathscr{C}^\infty(G)$ (resp. $\overline{\mathscr{K}(G)}$) (then $f * g \in \mathscr{C}^\infty(G)$ (resp. $\overline{\mathscr{K}(G)}$)).

$f \in \mathscr{C}^\infty(G,\beta)$ (resp. $\overline{\mathscr{K}(G)}$) and $g\chi'^{-1} \in L^1(G,\beta)$ (then $f * g \in \mathscr{C}^\infty(G)$ (resp. $\overline{\mathscr{K}(G)}$)).

$f \in L^p(G,\beta)$, $g \in L^q(G,\check{\beta})$ with $1/p + 1/q = 1$, $1 < p < +\infty$, $\beta$ left-invariant (then $f * g \in \overline{\mathscr{K}(G)}$ and $\|f * g\|_\infty \leqslant \|f\|_p \|\check{g}\|_q$).

# Contents

Made in United States
Troutdale, OR
05/27/2025

31708863R00189